The Early Universe

The Early Universe

Edward W. Kolb
Fermi National Accelerator Laboratory
and
The University of Chicago

Michael S. Turner
Fermi National Accelerator Laboratory
and
The University of Chicago

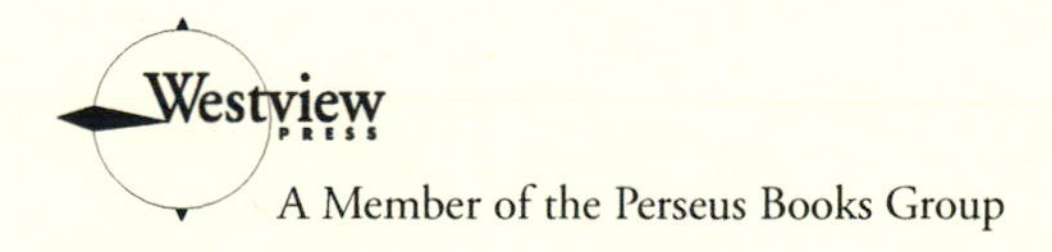

A Member of the Perseus Books Group

Library of Congress Cataloging-in-Publication Data

Kolb, Edward W.
The early universe / Edward W. Kolb, Michael Turner.
p. cm.—(Frontiers in physics ; v. 69)
Includes index.
1. Cosmology. I. Turner, Michael Stanley. II. Title.
III. Series.
QB981.K687 1990 523.1—dc19
ISBN 0-201-11603-0 (H), ISBN 0-201-62674-8 (P) 90-139

This book was prepared by the authors, using TEX typesetting language on a VAX computer.

14 13 12 11

FRONTIERS IN PHYSICS

David Pines/Editor

Volumes of the Series published from 1961 to 1973 are not officially numbered. The parenthetical numbers shown are designed to aid librarians and bibliographers to check the completeness of their holdings.

Titles published in this series prior to 1987 appear under either the W. A. Benjamin or the Benjamin/Cummings imprint; titles published since 1986 appear under the Addison-Wesley imprint.

(1)	N. Bloembergen	Nuclear Magnetic Relaxation: A Reprint Volume, 1961
(2)	G. F. Chew	S-Matrix Theory of Strong Interactions: A Lecture Note and Reprint Volume, 1961
(3)	R. P. Feynman	Quantum Electrodynamics: A Lecture Note and Reprint Volume, 1961
(4)	R. P. Feynman	The Theory of Fundamental Processes: A Lecture Note Volume, 1961
(5)	L. Van Hove, N. M. Hugenholtz, and L. P. Howland	Problems in Quantum Theory of Many-Particle Systems: A Lecture Note and Reprint Volume, 1961
(6)	D. Pines	The Many-Body Problem: A Lecture Note and Reprint Volume, 1961
(7)	H. Frauenfelder	The Mössbauer Effect: A Review—with a Collection of Reprints, 1962
(8)	L. P. Kadanoff G. Baym	Quantum Statistical Mechanics: Green's Function Methods in Equilibrium and Nonequilibrium Problems, 1962
(9)	G. E. Pake	Paramagnetic Resonance: An Introductory Monograph, 1962 [cf. (42)—2nd edition]
(10)	P. W. Anderson	Concepts in Solids: Lectures on the Theory of Solids, 1963
(11)	S. C. Frautschi	Regge Poles and S-Matrix Theory, 1963
(12)	R. Hofstadter	Electron Scattering and Nuclear and Nucleon Structure: A Collection of Reprints with an Introduction, 1963
(13)	A. M. Lane	Nuclear Theory: Pairing Force Correlations to Collective Motion, 1964
(14)	R. Omnès M. Froissart	Mandelstam Theory and Regge Poles: An Introduction for Experimentalists, 1963
(15)	E. J. Squires	Complex Angular Momenta and Particle Physics: A Lecture Note and Reprint Volume, 1963
(16)	H. L. Frisch J. L. Lebowitz	The Equilibrium Theory of Classical Fluids: A Lecture Note and Reprint Volume, 1964
(17)	M. Gell-Mann Y. Ne'eman	The Eightfold Way: (A Review—with a Collection of Reprints), 1964
(18)	M. Jacob G. F. Chew	Strong-Interaction Physics: A Lecture Note Volume, 1964
(19)	P. Nozières	Theory of Interacting Fermi Systems, 1964
(20)	J. R. Schrieffer	Theory of Superconductivity, 1964 (revised 3rd printing, 1983)
(21)	N. Bloembergen	Nonlinear Optics: A Lecture Note and Reprint Volume, 1965

(22)	R. Brout	Phase Transitions, 1965
(23)	I. M. Khalatnikov	An Introduction to the Theory of Superfluidity, 1965
(24)	P. G. deGennes	Superconductivity of Metals and Alloys, 1966
(25)	W. A. Harrison	Pseudopotentials in the Theory of Metals, 1966
(26)	V. Barger D. Cline	Phenomenological Theories of High Energy Scattering: An Experimental Evaluation, 1967
(27)	P. Choquàrd	The Anharmonic Crystal, 1967
(28)	T. Loucks	Augmented Plane Wave Method: A Guide to Performing Electronic Structure Calculations—A Lecture Note and Reprint Volume, 1967
(29)	Y. Ne'eman	Algebraic Theory of Particle Physics: Hadron Dynamics in Terms of Unitary Spin Currents, 1967
(30)	S. L. Adler R. F. Dashen	Current Algebras and Applications to Particle Physics, 1968
(31)	A. B. Migdal	Nuclear Theory: The Quasiparticle Method, 1968
(32)	J. J. J. Kokkedee	The Quark Model, 1969
(33)	A. B. Migdal V. Krainov	Approximation Methods in Quantum Mechanics, 1969
(34)	R. Z Sagdeev and A. A. Galeev	Nonlinear Plasma Theory, 1969
(35)	J. Schwinger	Quantum Kinematics and Dynamics, 1970
(36)	R. P. Feynman	Statistical Mechanics: A Set of Lectures, 1972
(37)	R. P. Feynman	Photo-Hadron Interactions, 1972
(38)	E. R. Caianiello	Combinatorics and Renormalization in Quantum Field Theory, 1973
(39)	G. B. Field, H. Arp, and J. N. Bahcall	The Redshift Controversy, 1973
(40)	D. Horn F. Zachariasen	Hadron Physics at Very High Energies, 1973
(41)	S. Ichimaru	Basic Principles of Plasma Physics: A Statistical Approach, 1973 (2nd printing, with revisions, 1980)
(42)	G. E. Pake T. L. Estle	The Physical Principles of Electron Paramagnetic Resonance, 2nd Edition, completely revised, enlarged, and reset, 1973 [cf. (9)—1st edition]

Volumes published from 1974 onward are being numbered as an integral part of the bibliography.

43	R. C. Davidson	Theory of Nonneutral Plasmas, 1974
44	S. Doniach E. H. Sondheimer	Green's Functions for Solid State Physicists, 1974
45	P. H. Frampton	Dual Resonance Models, 1974
46	S. K. Ma	Modern Theory of Critical Phenomena, 1976
47	D. Forster	Hydrodynamic Fluctuations, Broken Symmetry, and Correlation Functions, 1975
48	A. B. Migdal	Qualitative Methods in Quantum Theory, 1977
49	S. W. Lovesey	Condensed Matter Physics: Dynamic Correlations, 1980
50	L. D. Faddev A. A. Slavnov	Gauge Fields: Introduction to Quantum Theory, 1980

51	P. Ramond	Field Theory: A Modern Primer, 1981 [cf. 74—2nd edition]
52	R. A. Broglia A. Winther	Heavy Ion Reactions: Lecture Notes Vol. I: Elastic and Inelastic Reactions, 1981
53	R. A. Broglia A. Winther	Heavy Ion Reactions: Lecture Notes Vol. II, *in preparation*
54	H. Georgi	Lie Algebras in Particle Physics: From Isospin to Unified Theories, 1982
55	P. W. Anderson	Basic Notions of Condensed Matter Physics, 1983
56	C. Quigg	Gauge Theories of the Strong, Weak, and Electromagnetic Interactions, 1983
57	S. I. Pekar	Crystal Optics and Additional Light Waves, 1983
58	S. J. Gates, M. T. Grisaru, M. Roček, and W. Siegel	Superspace *or* One Thousand and One Lessons in Supersymmetry, 1983
59	R. N. Cahn	Semi-Simple Lie Algebras and Their Representations, 1984
60	G. G. Ross	Grand Unified Theories, 1984
61	S. W. Lovesey	Condensed Matter Physics: Dynamic Correlations, 2nd Edition, 1986
62	P. H. Frampton	Gauge Field Theories, 1986
63	J. I. Katz	High Energy Astrophysics, 1987
64	T. J. Ferbel	Experimental Techniques in High Energy Physics, 1987
65	T. Appelquist, A. Chodos, and P. G. O. Freund	Modern Kaluza-Klein Theories, 1987
66	G. Parisi	Statistical Field Theory, 1988
67	R. C. Richardson E. N. Smith	Techniques in Low-Temperature Condensed Matter Physics, 1988
68	J. W. Negele H. Orland	Quantum Many-Particle Systems, 1987
69	E.W.Kolb M.S. Turner	The Early Universe, 1990
70	E. W. Kolb M. S. Turner	The Early Universe: Reprints, 1988
71	V. Barger R. J. N. Phillips	Collider Physics, 1987
72	T. Tajima	Computational Plasma Physics, 1989
73	W. Kruer	The Physics of Laser Plasma Interactions, 1988
74	P. Ramond	Field Theory: A Modern Primer 2nd edition, 1989 [cf. 51—1st edition]
75	B. F. Hatfield	Quantum Field Theory of Point Particles and Strings, 1989
76	P. Sokolsky	Introduction to Ultrahigh Energy Cosmic Ray Physics, 1989

To

Adrienne and Barbara

CONTENTS

SERIES LISTING vii

EDITOR'S FOREWORD xvii

PREFACE TO THE PAPERBACK EDITION xix

PREFACE xxxix

1. THE UNIVERSE OBSERVED

1.1 Introduction **1**

1.2 The Expansion **2**

1.3 Large-Scale Isotropy and Homogeneity **8**

1.4 Age of the Universe **12**

1.5 Cosmic Microwave Background Radiation **14**

1.6 Light-Element Abundances **15**

1.7 The Matter Density: Dark Matter in the Universe **16**

1.8 The Large-Scale Structure of the Universe **21**

1.9 References **26**

2. ROBERTSON-WALKER METRIC

2.1 Open, Closed, and Flat Spatial Models 29

2.2 Particle Kinematics 36

2.3 Kinematics of the RW Metric 39

2.4 References 45

3. STANDARD COSMOLOGY

3.1 The Friedmann Equation 47

3.2 The Expansion Age of the Universe 52

3.3 Equilibrium Thermodynamics 60

3.4 Entropy 65

3.5 Brief Thermal History of the Universe 70

3.6 Horizons 82

3.7 References 86

4. BIG-BANG NUCLEOSYNTHESIS

4.1 Nuclear Statistical Equilibrium 87

4.2 Initial Conditions ($T \gg 1$ MeV, $t \ll 1$ sec) 89

4.3 Production of the Light Elements: 1–2–3 93

4.4 Primordial Abundances: Predictions 96

4.5 Primordial Abundances: Observations 100

4.6 Primordial Nucleosynthesis as a Probe 107

4.7 Concluding Remarks 109

4.8 References 111

5. THERMODYNAMICS IN THE EXPANDING UNIVERSE

5.1 The Boltzmann Equation 115

5.2 Freeze Out: Origin of Species 119

5.3 Out-of-Equilibrium Decay 130

5.4 Recombination Revisited 136

5.5 Neutrino Cosmology 139

5.6 Concluding Remarks 151

5.7 References 152

6. BARYOGENESIS

6.1 Overview 157

6.2 Evidence for a Baryon Asymmetry 158

6.3 The Basic Picture 160

6.4 Simple Boltzmann Equations 168

6.5 Damping of Pre-Existing Asymmetries 176

6.6 Lepton Numbers of the Universe 180

6.7 Way-Out-of-Equilibrium Decay 181

6.8 Sphalerons 184

6.9 Spontaneous Baryogenesis 190

6.10 Epilogue 191

6.11 References 193

7. PHASE TRANSITIONS

7.1 High-Temperature Symmetry Restoration 195

7.2 Domain Walls 213

7.3 Cosmic Strings 220

7.4 Magnetic Monopoles 233

7.5 The Kibble Mechanism 237

7.6 Monopoles, Cosmology, and Astrophysics 239

7.7 References 255

8. INFLATION

8.1 Shortcomings of the Standard Cosmology 261

8.2 Inflation—The Basic Picture 270

8.3 Inflation as Scalar Field Dynamics 275

8.4 Density Perturbations and Relic Gravitons 283

8.5 Specific Inflationary Models 291

8.6 Cosmic No-Hair Theorems 303

8.7 Testing the Inflationary Paradigm 309

8.8 Summary: A Paradigm in Search of as Model 313

8.9 References 317

9. STRUCTURE FORMATION

9.1 Overview 321

9.2 Notation, Definitions, and Preliminaries 324

9.3 The Evolution of Density Inhomogeneities: The Standard Lore 341

9.4 The Spectrum of Density Perturbations 364

9.5 Two Stories: Hot and Cold Dark Matter 369

9.6 Probing the Primeval Spectrum 378

9.7 The Ω Problem 390

9.8 Epilogue 395

9.9 References 397

10. AXIONS

10.1 The Axion and the Strong-CP Problem 401

10.2 Axions and Stars 408

10.3 Axions and Cosmology 422

10.4 Isocurvature Axion Fluctuations 436

10.5 Detection of Relic Axions 439

10.6 References 443

11. TOWARD THE PLANCK EPOCH

11.1 Overview 447

11.2 The Wheeler-De Witt Equation 451

11.3 The Wave Function of the Universe 458

11.4 Cosmology and Extra Dimensions 464

11.5 Limiting Temperature in Superstring Models 485

11.6 References 487

FINALE 491

APPENDIX A

A.1 Units 499

A.2 Physical Parameters 502

APPENDIX B

B.1 Quarks, Leptons, and Gauge Bosons 507

B.2 The Standard Model 509

B.3 The Higgs Sector 515

B.4 Beyond the Standard Model 521

B.5 References 532

INDEX 535

EDITOR'S FOREWORD

The problem of communicating in a coherent fashion recent developments in the most exciting and active fields of physics continues to be with us. The enormous growth in the number of physicists has tended to make the familiar channels of communication considerably less effective. It has become increasingly difficult for experts in a given field to keep up with the current literature; the novice can only be confused. What is needed is both a consistent account of a field and the presentation of a definite "point of view" concerning it. Formal monographs cannot meet such a need in a rapidly developing field, while the review article seems to have fallen into disfavor. Indeed, it would seem that the people most actively engaged in developing a given field are the people least likely to write at length about it.

FRONTIERS IN PHYSICS was conceived in 1961 in an effort to improve the situation in several ways. Leading physicists frequently give a series of lectures, a graduate seminar, or a graduate course in their special fields of interest. Such lectures serve to summarize the present status of a rapidly developing field and may well constitute the only coherent account available at the time. Often, notes on lectures exist (prepared by the lecturer himself, by graduate students, or by postdoctoral fellows) and are distributed on a limited basis. One of the principal purposes of the FRONTIERS IN PHYSICS Series is to make such notes available to a wider audience of physicists. A second principal purpose which has emerged is the concept of an *informal monograph*, in which authors would feel free to describe the present status of a rapidly developing field of research, in full knowledge, shared with the reader, that further developments might change aspects of that field in unexpected ways.

The Early Universe provides a fine example of what an informal monograph can accomplish in a frontier field of science. The authors, Edward W. Kolb and Michael S. Turner, are theoretical astrophysicists of great distinction who have made seminal contributions to our understanding of the early Universe. In the present volume they begin by treating those aspects of cosmology for which the fundamental physics is well established, that is events that occurred after the first 0.01 sec of the "big bang." In the second part of their book, they examine events that occurred before 0.01 sec for which the fundamental physics lies beyond the standard model of particle physics, and is therefore tied to speculation about the physics that lies beyond that model. In a joyously written finale, they give their personal views on the future of the particle physics–cosmology interface, while in a companion reprint volume, *Early Universe: Reprints*, they provide the reader with a collection of reprints, accompanied by lucid and lively commentaries. The present volume has been long awaited by the particle physics–astrophysics community, and I am confident that their hope—that together the two volumes will provide both the beginning graduate student and the interested outsider with a sound introduction to modern cosmology—will be realized.

David Pines
Urbana, Illinois
September, 1989

PREFACE TO THE PAPERBACK EDITION

In September 1989 we concluded *The Early Universe* with a very optimistic Finale, predicting that we would see not only the continued success of the basic framework of the hot big bang model, but also the confirmation of some of the speculations of early-Universe cosmology. Despite being four years older, we are no less optimistic; we are, in fact, even more optimistic!

There is very good reason for our continuing optimism: Important developments, both in observation and theory, have taken place in the last four years. In this update to the paperback edition we highlight some of these developments, and finish with another gaze into the crystal ball.

Before beginning our brief update we regret to report that, as we had feared, the flawless camera-ready manuscript we submitted was corrupted by the publishers, resulting in typos in the published version. We have not corrected these errors in this paperback edition (they weren't our fault anyway); early in 1994 we will make available a typo list. We thank alert readers for reporting problems.[1]

Observation

CMBR spectrum: After fifteen years in the planning, NASA's Cosmic Background Explorer (COBE) satellite was successfully launched in November 1989. A mere nine minutes of data from the Far InfraRed

[1] In the words of Umberto Eco, "I libri non sono fatti per crederci, ma per essere sottoposti a indagini."

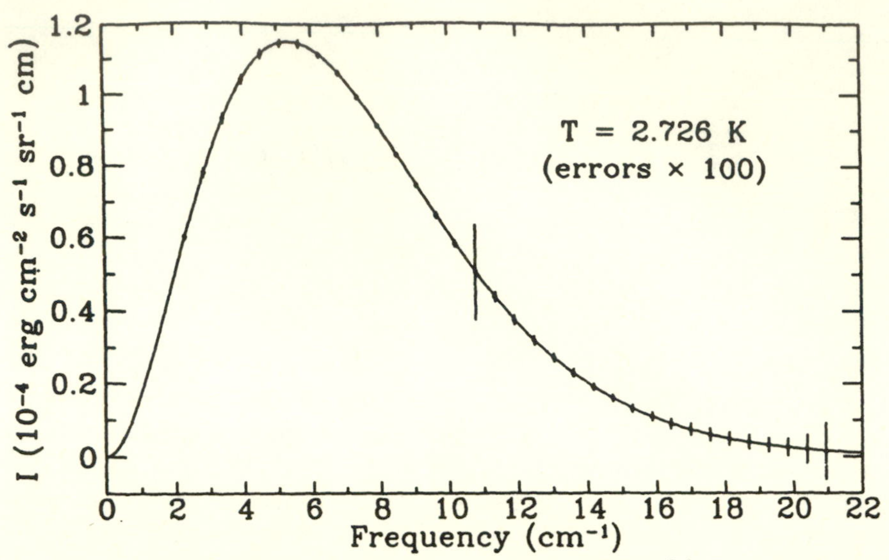

Fig. 1. The spectrum of the CMBR measured by FIRAS [2]; the solid curve is a $T = 2.726$ K blackbody.

Absolute Spectrophotometer (FIRAS) established that in the wavelength interval 0.05 cm to 0.5 cm the Cosmic Microwave Background Radiation (CMBR) has a black-body spectrum to an accuracy of 1% [1]. This put to rest claims of spectral distortions in the submillimeter range amounting to about 10% of the total CMBR energy density (see Fig. 1.8).

In Spring 1993 the FIRAS team published their final spectral data [2], illustrated in Fig. 1. Their results can be summarized as follows: (1) The best-fit black-body temperature is $T_0 = 2.726 \pm 0.01$ K;[2] (2) Deviations from a black-body spectrum over the wavelength interval 0.05 cm to 0.5 cm are less than 0.03% of the peak intensity; and (3) Limits to the distortion parameters y and μ are $y \leq 2.5 \times 10^{-5}$ and $|\mu/kT| \leq 3.3 \times 10^{-4}$ (95% CL).

These results severely constrain any process that distorts the CMBR, including radiative decays of relic particles (discussed in Section 5.5), energy release by a very early generation of stars, or the presence of hot gas between here and the last-scattering surface (e.g., as predicted in explosive scenarios of structure formation).

[2]In January 1990, a rocket-borne instrument probed a similar wavelength range and determined a black-body temperature of $T_0 = 2.736\,\mathrm{K} \pm 0.017\,\mathrm{K}$ [3].

CMBR anisotropy: Variation in the CMBR temperature in different directions is expected due to several effects: the motion of our local reference frame with respect to the cosmic rest frame (i.e., the FRW frame), rotation of the Universe, anisotropic expansion, and the presence of the density inhomogeneities presumed to have triggered the formation of structure. The Differential Microwave Radiometer (DMR) on COBE very accurately determined the amplitude of the dipole anisotropy, $\Delta T_{\rm dipole} = 3.365 \pm 0.027\,{\rm mK}$ (consistent with previous measurements), corresponding to a velocity of the local group of galaxies of $627 \pm 22\,{\rm km\,s^{-1}}$ in the general direction of Hydra-Centaurus (more precisely, $RA = 166° \pm 3°$ and $\delta = -27.1° \pm 3°$) [4].[3]

In April 1992 the DMR team announced the discovery of anisotropy in the CMBR temperature on angular scales from about 10° to 90° at the level of about 1 part in 10^5. Their strongest result is the measurement of the *rms* temperature variation on the sky averaged over a beam of FWHM 10°: $30 \pm 5\mu$K; additionally, they measured a quadrupole anisotropy of amplitude $11 \pm 3\mu$K [5]. The multipole amplitudes extracted from their data are shown in Fig. 2. (See Section 9.6.2 for a discussion of CMBR temperature anisotropies.)

This anisotropy is presumed to arise from the density inhomogeneities that triggered structure formation, providing the first evidence for their existence.[4] The DMR discovery is the culmination of the quest to find spatial anisotropy that began with Penzias and Wilson's anisotropy limit of around 10%. We hope that it represents the beginning of the mapping of anisotropy on angular scales from arcseconds upward, helping to reveal the primeval spectrum of density fluctuations as well as clarifying the post-recombination history of the Universe.

While measuring a temperature difference of order tens of microKelvins is in itself a technical challenge, even more daunting is shielding against sunshine, earthshine, and moonshine, and discriminating against foreground sources including synchrotron, bremsstrahlung and thermal dust emission from the Milky Way, as well as discrete sources between here and the last-scattering surface. With sampling at three frequencies (31.5 GHz, 53 GHz and 90 GHz), two sets of receivers at each frequency, and full-sky coverage, the DMR was able to discriminate against foreground

---

[3]The DMR also measured a yearly modulation of the dipole anisotropy due to Earth's motion around the sun: $\Delta T = 0.27\,{\rm mK}$, corresponding to a mean orbital velocity of $30\,{\rm km\,s^{-1}}$ (Galileo is vindicated!).

[4]It is possible that some portion of the anisotropy is due to long-wavelength gravitational waves produced during inflation [6]; see Section 8.4.

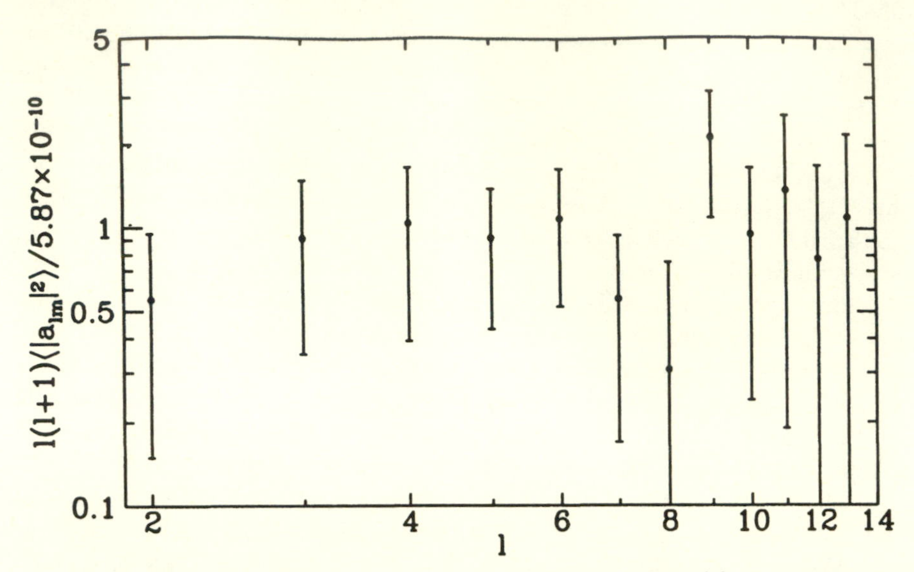

Fig. 2. The multipole amplitudes determined by the DMR [5]. See Section 9.6.2 for discussion and definitions.

sources and to present a very convincing case for CMBR anisotropy. Just recently, the DMR result received confirmation from a re-analysis of an earlier balloon-borne experiment on angular scales from about 4° to 100° [7]; moreover, this measurement was at a frequency of 170 GHz, providing further evidence that the anisotropy is thermal and associated with the CMBR.

Currently there are a dozen or so ongoing CMBR anisotropy experiments probing angular scales from about 10′ to 10° with sensitivities at the level of 10^{-5}. Five of these experiments have detected statistically significant anisotropy not associated with any *known* foreground source [8]. In the near future the anisotropy of the CMBR may well be measured on angular scales from 10′ to 90°, thereby probing the density field of the Universe on scales from about $10h^{-1}$ Mpc to 10^4h^{-1} Mpc (see Section 9.6.2).

Determination of Ω_0: A definitive measurement of the mean mass density of the Universe, one of the fundamental parameters of cosmology, still eludes us (see Section 1.7). While the bulk of the measurements are consistent with a value between 10% and 30% of critical density, recent work

[9] makes a very good case for $\Omega_0 = 1$. If this is true, it strongly suggests the existence of nonbaryonic dark matter since primordial nucleosynthesis constrains the baryonic fraction to be less than 10% of the critical density (see Section 4.6).

The physics underlying these recent measurements is simple: the peculiar velocities of galaxies are driven by the gravitational effect of the inhomogeneous distribution of matter, and thereby depend upon the level of inhomogeneity ($\delta\rho/\rho$) and the average density (Ω_0) [see Section 9.6.1, especially Eq. (9.135)]. Measurements of peculiar velocities (e.g., that of our own galaxy or those of thousands of nearby galaxies) and of the distribution of galaxies can thus be used to determine Ω_0. A crucial new ingredient are the red-shift surveys based upon the IRAS Catalogue of infrared selected galaxies. The IRAS Catalogue is especially useful since it is largely unaffected by galactic absorption, and thus provides a fair sample of galaxies in all directions which can be used to determine $\delta\rho/\rho$.[5]

Not only does this route to Ω_0 sample a very large volume, a cube of order $100h^{-1}$ Mpc on a side, but it also gives the "right" answer: $\Omega_0 = 1$ with an estimated standard error of about ± 0.2. While the Ω_0 issue is still far from being settled, this set of measurements provides a much needed shot in the arm for the FSU (Flat Universe Society).

Mapping the Universe: Red-shift surveys are the most widely used technique for mapping structure in the Universe (see Section 1.8). When the hardback edition went to press the total number of red shifts measured was around 30,000, and the largest survey, the CfA slices of the Universe, contained about 8,000 red shifts and probed the Universe out to $z \simeq 0.03$. The total number of red shifts measured is now about double that and growing rapidly. Several new red-shifts surveys exist [10]: the previously mentioned survey of IRAS galaxies; sparse surveys where the red shifts of only a small fraction of the galaxies in a catalogue are measured, e.g., that based upon the APM catalogue [11]; and pencil-beam surveys where a small patch of the sky (a square degree or so) is probed to great depth [12].

The most ambitious project in the works is the Sloan Digital Sky Survey [13]. A catalogue of over 100 million galaxies on the northern sky will be

[5] More precisely, the IRAS galaxies provide a means of determining $\delta n_{\rm GAL}/n_{\rm GAL}$. Assuming a simple relationship between the distribution of mass and light, $\delta n_{\rm GAL}/n_{\rm GAL} = b(\delta\rho/\rho)$ where b is the biasing factor (see Section 9.7), the peculiar-velocity field and the distribution of light can be used to infer $\Omega_0^{0.6}/b$.

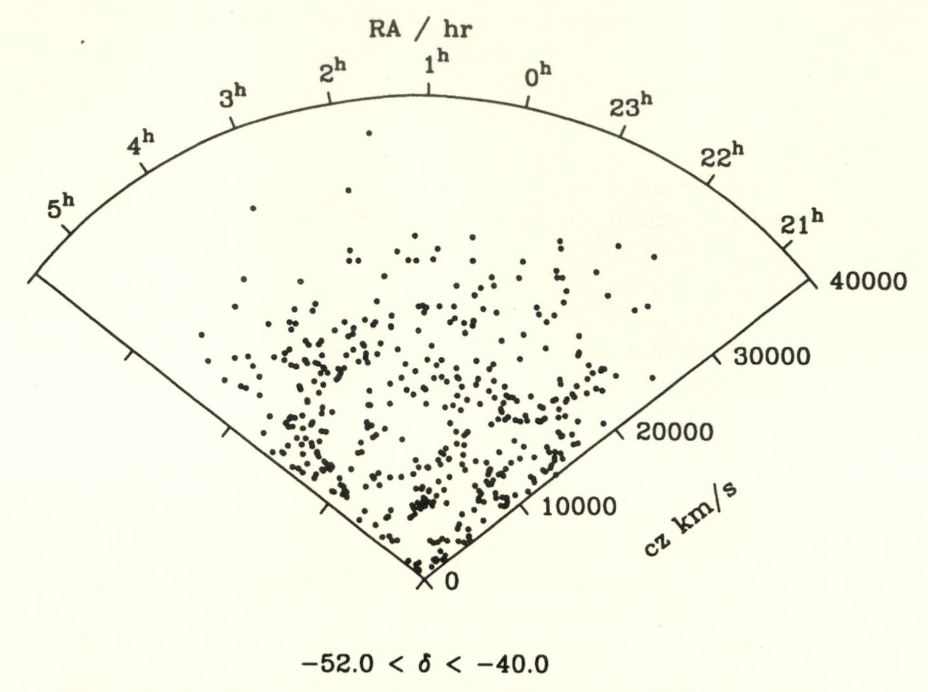

Fig. 3: A slice of a sparse sample of the APM catalogue [11]. The APM slices are about three times as deep as the CfA slices; cf. Fig. 1.12.

constructed using four-color CCD photometry, and red shifts of about a million galaxies and of about a hundred thousand QSOs will be measured. The red-shift survey will cover about π-sr on the sky and extend out to $z \sim 0.2$. This "π-in-the-sky" project is scheduled to be finished by the end of the millennium.

The picture of the Universe emerging from the deeper sparse surveys confirms the most prominent feature of the CfA slices—an abundance of voids of size $30h^{-1}$ Mpc—and reassures cosmologists that the Universe does indeed become smoother on large scales reflecting the smoothness implied by the isotropy of CMBR; see Fig. 3. This should end speculation that the Universe contains structures of ever increasing size or that the large-scale structure is a fractal. The pencil-beam surveys [12], which probe a small portion of the sky to great depth, have confirmed another feature in the CfA slices: the existence of sheet-like structures similar to the CfA "great wall."

Completing the standard model: Millions of Z^0 bosons have been produced and detected at CERN in the LEP collider, and trillions of protons and antiprotons have been collided at Fermilab in the Tevatron. These experiments have tested the electroweak model to very high precision, better than 1% in many instances. For example, precise measurements of the Z^0 resonance have determined the mass of the Z^0 to be $91.187\,\mathrm{GeV} \pm 7\,\mathrm{MeV}$ and the number of light neutrino species to be $N_\nu = 2.98 \pm 0.03$. The latter precludes the possibility of a fourth generation neutrino of mass less than about 45 GeV and all but eliminates a heavy neutrino as a dark matter candidate.

The search for the top quark goes on at this very moment at the Tevatron, with a current lower mass limit of 113 GeV. The search for the Higgs continues at LEP with a lower mass limit of about 64 GeV. Likewise, supersymmetry remains elusive, as all searches for the various supersymmetric partners continue only to set mass limits.

The good news is that the electroweak model is in great shape with only two key parameters to be determined: the top-quark mass and the Higgs mass. The good news is also the bad news: deviations from standard-model predictions are potential signals for physics beyond the standard model—and thus far there aren't any. While there continue to be hints of physics beyond the standard model in the neutrino sector, e.g., in solar neutrino experiments and atmospheric neutrino fluxes, there is no firm evidence yet.[6]

Theory

Inflation: Most of the discussion of inflation in Chapter 8 was devoted to "slow-rollover" inflation, the only viable model of inflation at that time. In slow-rollover inflation a very weakly coupled scalar field is initially displaced from the minimum of its potential, and the associated potential energy causes the Universe to inflate while the scalar field slowly evolves to the bottom of the potential (see Chapter 8). Though slow-rollover provides a viable implementation of inflation, the decoupled and disconnected nature of the scalar field responsible for inflation, and the very name given

[6] In the interim between the hardback and paperback editions, the specter of a 17 keV neutrino has come and gone. While experiment was the final arbitrator, consideration of the cosmological and astrophysical effects of a massive, unstable neutrino (as discussed in Chapter 5) proved important.

it, "inflaton," makes slow-rollover less than compelling.

In 1990 Steinhardt and La [14] proposed a new type of inflation, called extended inflation, where again inflation is associated with a first-order symmetry-breaking phase transition. The key difference between extended inflation and Guth's original model, which was also based on a first-order phase transition, is the underlying theory of gravity: Jordan–Brans–Dicke (JBD) rather than general relativity.

In JBD the gravitational "constant" is set by the value of a scalar field. During inflation this scalar field evolves and gravity becomes weaker; as a result the cosmic scale factor grows as a large power of time rather than exponentially. This means that in extended inflation the physical volume of space remaining in the false vacuum grows only as power of time and not exponentially, and unlike Guth's original model, bubble nucleation can convert all of space to the true vacuum.

On the face of it, extended inflation seems to combine the best features of old inflation (close connection with particle-physics models through a first-order phase transition) and new inflation (it works!).[7] Because reheating and the transition to the radiation-dominated era occur through bubble collisions some aspects of inflation are different, even leading to potentially observable signatures, e.g., gravitational waves from bubble collisions, creation of large voids, and the production of topological defects [15].

Extended inflation is not without problems, most notably "the big-bubble problem" [16]. Inflation ends through bubble nucleation and percolation; the faster the Universe expands, determined by the Brans–Dicke parameter ω, the broader the distribution of bubble sizes, and the higher the level of inhomogeneity. Achieving sufficient homogeneity requires ω to be less than about 20; unfortunately, solar-system tests of JBD theory constrain ω to be greater than about 500.

Clearly the La–Steinhardt model is not viable; it does provide a simple toy model that illustrates the general features of "first-order" inflation. Because attractive theoretical ideas such as superstrings, supergravity, and extra dimensions lead to a JBD-like description of gravity at high energies, there is both the motivation for pursuing extended inflation and the hope that a workable, realistic, and compelling model may be found. At the very least, extended inflation has shown that the inflationary paradigm is

[7] In fact, by means of a conformal transformation extended inflation can be recast as slow-rollover inflation with an exponential potential and general relativity as the gravity theory.

richer than imagined and that a simple, elegant, and viable model may still await discovery.

Structure formation: In Chapter 9 we focused on two theories of structure formation: hot dark matter (HDM) and cold dark matter (CDM), both with inflation-produced density perturbations. At that time CDM looked very promising and HDM looked very unpromising. After four years and a wealth of new data (especially the COBE DMR result), CDM still looks promising, HDM does not look any more promising, and there are two new contenders.

The first is Peebles' primeval isocurvature baryon (PIB) model [17]. In this model $\Omega_0 = \Omega_B \sim 0.2$ and $h \simeq 0.8$, with *isocurvature* fluctuations in the baryon-number density providing the seed perturbations. PIB is a minimalist model—start with what you see, not what you would like. It is not motivated by early-Universe microphysics (we're not offended), and in fact it is strongly disfavored by inflation and it violates the nucleosynthesis bound to $\Omega_B h^2$ by a factor of six. The model is currently consistent with the bulk of the data on large-scale structure (which motivated its spectrum of perturbations in the first place). However, it seems destined to have difficulty accounting for both the DMR detection and the observed level of CMBR isotropy on angular scales of degrees.

The second encompasses a class of scenarios where the primeval perturbations are isocurvature perturbations arising from defects (cosmic strings, global monopoles, textures, and so on) produced in a very early phase transition (energy scale of about 10^{16} GeV and $t \sim 10^{-38}$ sec) (see Chapter 7 for discussion of topological defects). These scenarios have nonbaryonic dark matter, usually cold, though in the case of cosmic strings, the dark matter could be hot. Structure formation is more difficult to simulate numerically because the seed perturbations are constantly being formed as the network of defects evolves (unlike inflation where the perturbations were put in place in the distant past). To the extent to which numerical simulations have been carried out they are consistent with the observed large-scale structure. CMBR anisotropy measurements are crucial to testing these scenarios: to accommodate the DMR result a high level of biasing is required ($b \sim 4$) and their predictions for anisotropy on the 1° scale are very different than those of cold dark matter [18].

Cold dark matter continues to be the most scrutinized model. In fact, it is probably fair to say that it has helped to spur many of the important observations that are testing theories of structure formation. One of the

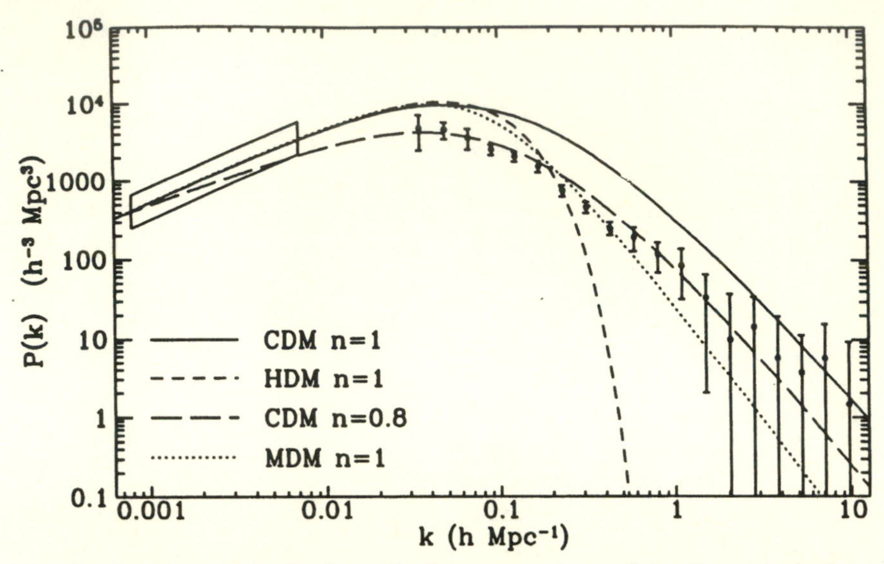

Fig. 4: The empirically determined power spectrum of density perturbations and the (linear-theory) predictions of several models. The power spectrum of density perturbations, cf. Sections 9.2 and 9.3, was inferred from the 1.2Jy IRAS red-shift survey [19] and the COBE DMR measurements. The models shown are cold dark matter; hot dark matter; tilted cold dark matter; and mixed dark matter.

virtues of cold dark matter is its specificity. Before the DMR result, biased cold dark matter with $b \sim 1.5$ was the favored model, though there were indications that fluctuations were too small on large scales to account for the observed structure. Since the shape of the spectrum is specified, only one measurement is needed to determine the entire spectrum (see Sections 9.2 and 9.4). Normalizing to the DMR result leads to a biasing factor of essentially unity, i.e., little or no bias. This is good news and bad news: $b = 1$ is the simplest model and seems to solve the problem of a deficiency of power on large scales; however, such a normalization appears to lead to fluctuations that are too large on small scales, which is illustrated in Fig. 4.

Although part of the attractiveness of cold dark matter has always been its specificity, as the quantity and quality of data have improved, slight variations on the basic theme have been put forth, largely to decrease the level of fluctuations on small scales. These variations involve the composition of the dark matter or the spectrum of perturbations. The oldest variant involves a cosmological constant: $\Omega_\Lambda = 0.8$, $\Omega_{\rm CDM} \simeq 0.15$ and $\Omega_B \simeq 0.05$ [20]. In addition to reducing power on small scales, the presence of a cosmological constant allows for a larger value of the Hubble constant, $h \sim 0.7 - 0.8$. (In a conventional matter-dominated Universe the Hubble constant has to be close to $h = 0.5$ to accommodate other age determinations; see Section 1.4.)

In the mixed dark matter variant a small amount of hot dark matter is added [21]: $\Omega_{\rm HDM} \sim 0.3$, $\Omega_{\rm CDM} \sim 0.65$, and $\Omega_B \sim 0.05$. The "pinch" of hot dark matter, e.g., in the form of neutrinos of mass 7 eV or so, leads to the suppression of fluctuations on small scales because of the free streaming of neutrinos (see Section 9.4).

The third variant involves a modification of the spectrum of perturbations. A generic prediction of inflationary perturbations is "almost scale-invariant" perturbations, the Harrison-Zel'dovich spectrum, characterized by $n = 1$ (or equivalently $\alpha = 0$, cf. Section 9.4). While it was realized from the beginning that there could be deviations from the scale-invariant form ($n = 1$), this fact did not seem important; now it may be [22]. When the spectrum of perturbations is normalized to the COBE DMR result, which fixes the spectrum on a very-large scale ($H_0^{-1} \sim 10^4$ Mpc), a modest amount of "tilt," say $n \simeq 0.8$, can reduce fluctuations on small scales (order $10h^{-1}$ Mpc or so) by the right amount (about a factor of two). If this is the resolution to the small-scale power problem, it provides important information about the inflationary potential, only some potentials predict this much tilt, and makes another prediction, such potentials generally lead to a higher level of gravitational waves [23].

In Fig. 4 the power spectra for the cold dark matter, hot dark matter, tilted cold dark matter, and mixed dark matter models are compared with the empirically determined power spectrum (Λ + cold dark matter is similar to mixed dark matter). All four models have been normalized to the COBE DMR result; by so doing, the resulting biasing factors on the $8h^{-1}$ Mpc scale are respectively: 0.85, 1.4, 1.5, and 1.35. The variants clearly provide better fits to the measured power spectrum; however, unbiased cold dark matter has the great virtue of being the simplest model.[8]

Key tests for CDM models as well as the other contenders are coming soon. For example, the various scenarios make different predictions for the level of anisotropy on the 1° scale, and a number of experiments are now

---

[8] One should keep the words of Francis Crick in mind: A theory that agrees with all the data at any given time is necessarily wrong, as at any given time not all the data are correct.

reporting results on these scales [8]. It is very likely that in the next few years one or more models of structure formation will be falsified. With the great effort focused on the problem of structure formation, this important aspect of the standard cosmology may soon be sorted out.

Electroweak phase transition: Now that almost all the parameters of the electroweak theory are known it is possible in principle to study the dynamics of the electroweak phase transition in some detail. A key word in the above sentence is *almost*. We must remind the reader that the top quark and Higgs masses are still unknown, and it may be that the mechanism for symmetry breaking is more complicated than that in the standard model. In any case, within the minimal electroweak model the problem is essentially well posed.

Throughout almost the entire parameter space (i.e., top quark and Higgs masses) the one-loop calculation predicts that the phase transition should be very weakly first order [24]. Were the transition second order or strongly first order, a number of well known and straightforward theoretical techniques would be applicable [25]. However because of the weakness of the transition the one-loop calculation is not reliable, and calculation of the details of the transition presents a theoretical challenge—one that is important enough to try hard to solve [24,25].

Toward the end of Chapter 6 we discussed baryon number violation within the standard electroweak model due to classical field configurations known as sphalerons. It is now generally believed that $B + L$ is violated rampantly through electroweak interactions at temperatures above the electroweak transition, as well as slightly below through sphalerons. The implications for baryogenesis are manifold: (i) the baryon number could be generated at high temperatures by a GUT process, but with a net $B - L$ so that it cannot be erased by electroweak $B + L$ violation; (ii) the baryon number could be generated at temperatures below the electroweak scale; or (iii) the baryon number could be generated during the electroweak transition itself.

In case (i), there are implications for Majorana neutrino masses and the structure of the neutrino sector. Majorana neutrino masses violate lepton number, and for large enough masses this L violation together with electroweak $B + L$ violation can completely erase any baryon asymmetry [26]. Moreover, it could be that a baryon asymmetry per se is never generated; rather, a lepton asymmetry is produced and is transmuted into a baryon asymmetry by electroweak $B + L$ violation. Possibility (ii) presents challenges, though models for baryogenesis at very low temperatures have been proposed (see Chapter 6). Possibility (iii) is the most intriguing: If the baryon asymmetry is produced at the electroweak scale, the underlying microphysics can probably be tested in laboratory experiments. In fact, a recent paper suggests that a baryon asymmetry of the correct magnitude can be produced within the framework of the standard model [27].

Another Gaze into the Crystal Ball

The Finale at the end of this book reflects upon past accomplishments in cosmology and looks toward the future. Since it was written 4 years ago, it is appropriate to update our view of the future here. (Unlike historians, our view of the past doesn't change!) A view of the future that is unchanged is our belief that we are poised for advances in several areas and that the golden age of cosmology is still ahead.

With the diversity of efforts focused on large-scale structure, it seems very likely that the basic outline of a "standard model" of structure formation will emerge within the next decade, or even sooner. This will occur as the result of a convergence of observation and theory. As mentioned above, there are several very specific and well motivated scenarios for structure formation, which have helped to spur observations and aid in their interpretation. The largest structures in the Universe have now been identified, the CfA voids and walls, and between red-shift surveys and measurements of CMBR anisotropy the spectrum of density perturbations is now being probed on all scales. Larger red-shift surveys and additional CMBR anisotropy measurements should further clarify matters and provide overlapping information that will shed important light on the issue of biasing. In addition, the development of new computational methods, e.g., combining smooth particle hydro and N-body codes, will strengthen the crucial phenomenological link between early-Universe theories and observations of the Universe today.

Not only will a standard model of structure formation fill in the final piece of the hot big-bang cosmology, but it may well open a new window on early-Universe microphysics. For example, if the standard model should turn out to be some variant of cold dark matter, observational data may well allow a "reconstruction" of the inflationary potential [28].

Of course the book on structure formation cannot be closed before the composition of dark matter is discovered. Since luminous matter contributes much less than 1% of the critical density, which is the lower bound to the baryonic contribution based on primordial nucleosynthesis, and there are strong indications that Ω_0 is greater than 0.1, which is the upper bound to the baryonic contribution based on primordial nucleosynthesis, there is evidence for two dark-matter problems, baryonic and nonbaryonic. While nature may prove more elusive than we expect, we are reasonably hopeful of breakthroughs because experimental efforts are underway to directly detect the most promising candidates: neutralinos, through their elastic scattering in cryogenic detectors (see Section 5.6); axions, by means of large-scale Sikivie-type detectors (see Section 10.5.1); and dark stars, through detection of microlensing of stars in the LMC. All three efforts are well underway and could solve the dark-matter riddle by the end of the decade.[9] In addition, a host of experiments will be carried out that bear on the issue of neutrino masses.

Our optimism for the future must be tempered by the realization that some problems in cosmology never seem to reach closure. In a classic paper written in 1961 [30], Sandage discussed how the 200 inch Hale telescope at Palomar could be used to determine our "world model," that is, the basic kinematical parameters which describe the expansion, H_0 and q_0 (see Section 2.3). In one sense, our knowledge of both parameters is about the same today as it was then—poor. However, there is reason to be optimistic. There is a new generation of astronomers attacking the problem of the distance scale (and thereby H_0), armed with new techniques. The Hubble Space Telescope has calibrated the distance to M101 with Cepheid variables, and when the second-generation instruments are installed, Cepheid variables in the Virgo Cluster will be within reach. Most astronomers believe that a reliable distance to Virgo is necessary, if not sufficient, to settle the issue.

Early-Universe cosmologists have a lot at stake in the determination of H_0. If the Hubble constant lies close to $80\,\mathrm{km\,s^{-1}\,Mpc^{-1}}$, then it is hard to see how to avoid an "age crisis" (see Section 3.2). While estimates of the age of the Universe based upon the cooling of white dwarfs, isotopic ratios of radioactive elements, and determination of the age of the oldest stars, all have large inherent uncertainties, it is difficult to imagine that the Universe could be as young as 10 Gyr. And, it is impossible to reconcile $H_0 = 80\,\mathrm{km\,s^{-1}\,Mpc^{-1}}$ with a Universe older than 10 Gyr without either

[9] In fact, as this paperback edition goes to press, three collaborations have reported candidate microlensing events [29]. If these events are due to microlensing by astrophysical objects in the halo of our galaxy, then one of the dark matter problems, the form of the dark baryons, will have been solved.

abandoning $\Omega_0 = 1$ or invoking a cosmological constant, both, equally unpalatable to us. Therefore, we remain steadfast in our prediction that the Hubble constant is $50 \pm 5\,\mathrm{km\,s^{-1}\,Mpc^{-1}}$.[10]

The other parameter that describes our "world model" is q_0, where $2q_0 = \Omega_0(1 + 3p/\rho)$. While the evidence for $\Omega_0 = 1$ is far from conclusive, there is reason to believe that the case will strengthen in the near future. We mentioned earlier the techniques involving peculiar velocities. As larger samples of red shifts become available (e.g., the SDSS will produce four-color CCD measurements of 200 million galaxies) the classic techniques of galaxy number counts and angular size (see Section 2.3) may help determine q_0. In addition, other, new approaches are being pursued, e.g., automated searches for high red-shift supernovae.

The effort focused on looking for physics beyond the standard particle physics model exceeds that focused on structure formation. While everyone is sure that there is new physics beyond the standard model, when and where it will be revealed remains a mystery. The neutrino has a 60-year history of shedding light on new physics; with the diversity of experiments involving neutrinos—four solar-neutrino experiments in operation and several new experiments under construction, a new generation of long-baseline neutrino-oscillation experiments in the planning, and a variety of other experiments sensitive to new physics in the neutrino sector—perhaps history will repeat itself.

Let us now turn to theory. With our present understanding of the standard model of particle physics, there is every reason to believe that both the quark/hadron transition and the electroweak phase transition will soon be understood well enough to extend the "known history" of the Universe back to a time of about 10^{-12} sec or so ($T \sim 1000\,\mathrm{GeV}$).

Some argue that the birth of particle cosmology traces to the development of baryogenesis. While present limits on proton decay seem to push the original GUT scenario of baryogenesis (see Chapter 6) beyond the reach of laboratory verification, new scenarios may be amenable to laboratory tests. For instance, baryogenesis may simply involve electroweak physics, or perhaps neutrino physics, that can be tested in terrestrial laboratories. This issue seems likely to be resolved sooner, rather than later.

Even with unlimited energy, accelerators are the wrong tool to probe the non-perturbative sector of field theories. Early-Universe phase transitions continue to provide the best arena for the study of aspects of particle-

---

[10] We acknowledge a generous contribution to the Chicago Cosmologists' Retirement Fund from Allan Sandage.

physics theories related to coherent, soliton-like objects. The only plausible site for the production of objects such as monopoles, strings, walls, sphalerons, and the like is an early-Universe phase transition. All of these can have very significant implications for the evolution of the Universe. Sphalerons, as well as other solitons produced in the electroweak transition, have some promise of a cosmological payoff. Of course there is an enormous difference between finding a soliton-like solution to the field equations and finding solitons in the Universe. However, even if they are not are found, the techniques developed for their study will be useful additions to the theorist's toolbox.

Of course, disappointments over lack of progress in specific areas are inevitable. We would be remiss if we did not mention them (while reminding the reader that the fact that we do not work in the fields of slow progress is purely coincidental). Any payoff of superstring theory for cosmology seems as remote today as ever. Whatever physics lies beyond the standard-model of particle physics is certain to have cosmological implications, and the standard model has yet to crack under incredible experimental scrutiny. On another front, it is difficult to see identifiable progress in quantum cosmology. As with superstrings, perhaps the mist of early enthusiasm obscured the enormity of the task.

Finally, let us not underestimate the power of a single new idea or discovery to change cosmology. We would be most surprised if the future did not include a revolutionary idea or unexpected discovery. In fact, our biggest disappointment would be if, ten years from now, we did not have to write another book.

Rocky and Mike
Warrenville & Hinsdale, Illinois
September, 1993

References

1. J. C. Mather et al., *Ap. J.* **354**, L37 (1990).
2. J. C. Mather et al., *Ap. J.*, in press (1993).
3. H. Gush, M. Halpern, and E. H. Wishnow, *Phys. Rev. Lett.* **65**, 537 (1990).

4. A. Kogut et al., *Ap. J.*, in press (1993).

5. G. Smoot et al., *Ap. J.* **396**, L1 (1992); E. L. Wright et al., *ibid*, L13 (1992).

6. L. Krauss and M. White, *Phys. Rev. Lett.* **69**, 869 (1992).

7. K. Ganga et al. (FIRS Collaboration), *Ap. J.* **410**, L57 (1993).

8. J. O. Gundersen et al. (MAX Collaboration), *Ap. J.* **413**, L1 (1993); P. R. Meinhold et al. (MAX Collaboration), *ibid* **409**, L1 (1993); E. J. Wollack et al. (BIGPLATE Collaboration), *ibid*, in press (1993); J. Schuster et al., *ibid* **412**, L47 (1993); E. S. Cheng et al. (MSAM Collaboration), *ibid*, in press (1993); M. Dragovan et al. (PYTHON Collaboration), *Ap. J.*, in press (1993); A. Lasenby et al. (Tenerife), *Nature*, in press (1993).

9. M. Rowan-Robinson et al., *Mon. Not. R. astr. Soc.* **247**, 1 (1990); N. Kaiser et al., *ibid* **252**, 1 (1991); M. Strauss et al., *Ap. J.* **385**, 421 (1992); M. Strauss et al., *ibid* **397**, 395 (1992); E. Bertschinger and A. Dekel, *ibid* **336**, L5 (1989); A. Dekel et al., *ibid*, in press (1993).

10. R. Giovanelli and M. Haynes, *Ann. Rev. Astron. Astrophys.* **29**, 499 (1991).

11. J. Loveday, B. A. Peterson, G. Efstathiou, and S. J. Maddox, *Ap. J.* **390**, 338 (1992).

12. T. J. Broadhurst, R. S. Ellis, D. C. Koo, and A. S. Szalay, *Nature* **343**, 726 (1990).

13. The Sloan Digital Sky Survey is a collaboration between The University of Chicago, Fermi National Accelerator Laboratory, the Institute for Advanced Study, the Japan Promotion Group, The Johns Hopkins University, and Princeton University.

14. D. La and P.J. Steinhardt, *Phys. Rev. Lett.* **62**, 376 (1989).

15. E. W. Kolb, *Physica Scripta*, **T36**, 199 (1991); A. Kosowsky, M. S. Turner, and R. Watkins, *Phys. Rev. Lett.* **69**, 2026 (1992).

16. E. J. Weinberg, *Phys. Rev. D* **40**, 3950 (1989).

17. P. J. E. Peebles, *Nature* **327**, 210 (1987); *Ap. J.* **315**, L73 (1987); *Ap. J.* **315**, L73 (1987); R. Cen, J. P. Ostriker, and P. J. E. Peebles, *ibid* **415**, 423 (1993).

18. A. Albrecht and A. Stebbins, *Phys. Rev. Lett.* **69**, 2615 (1992); *ibid* **68**, 2121 (1992); D. Bennett, A. Stebbins, and F. Bouchet, *Ap. J.* **399**, L5 (1992); N. Turok, *Phys. Rev. Lett.* **63**, 2652 (1989); U.-L. Pen, D. N. Spergel, and N. Turok, *Phys. Rev. D*, in press (1993); D. Coulson, U.-L. Pen, and N. Turok, *Phys. Rev. D*, in press (1993).

19. K. B. Fisher, M. Davis, M. A. Strauss, A. Yahil and J. P. Huchra, *Ap. J.* **389**, 188 (1992).

20. M. S. Turner, G. Steigman, and L. Krauss, *Phys. Rev. Lett.* **52**, 2090 (1984); M. S. Turner, *Physica Scripta* **T36**, 167 (1991); P. J. E. Peebles, *Ap. J.* **284**, 439 (1984); G. Efstathiou et al., *Nature* **348**, 705 (1990).

21. Q. Shafi and F. Stecker, *Phys. Rev. Lett.* **53**, 1292 (1984); A. van Dalen and R. K. Schaefer, *Ap. J.* **398**, 33 (1992); M. Davis, F. Summers, and D. Schlegel, *Nature* **359**, 393 (1992); A. Klypin et al., *Ap. J.* **416**, 1 (1993).

22. R. Cen et al., *Ap. J.* **399**, L11 (1992); F. Adams et al., *Phys. Rev. D* **47**, 426 (1993); J. Gelb et al., *Astrophys. J.* **403**, L5 (1993).

23. R. Davis et al., *Phys. Rev. Lett.* **69**, 1856 (1992); F. Lucchin, S. Mattarese, and S. Mollerach, *Ap. J.* **401**, L49 (1992); D. Salopek, *Phys. Rev. Lett.* **69**, 3602 (1992); A. R. Liddle and D. Lyth, *Phys. Lett. B* **291**, 391 (1992); J. E. Lidsey and P. Coles, *Mon. Not. R. astr. Soc.* **358**, 57 (1992); T. Souradeep and V. Sahni, *Mod. Phys. Lett. A* **7**, 3541 (1992).

24. G. W. Anderson and L. J. Hall, *Phys. Rev. D* **45**, 2685 (1992); M. Dine, P. Huet, and R. Singleton, *Nucl. Phys.* **B375**, 625 (1992); M. Dine, R. Leigh, P. Huet, A. Linde, and D. Linde, *Phys. Rev. D* **46**, 550 (1992); M. E. Shaposhnikov, *Phys. Lett.* **B277**, 324 (1992); M. E. Carrington, *Phys. Rev. D* **45**, 2933 (1992); P. Fendley, *Phys. Lett.* **B196**, 175 (1987); J. R. Espinosa, M. Quirós, and F. Zwirner, *Phys. Lett.* **B291**, 115 (1992); P. Arnold and O. Espinosa, *Phys. Rev. D* **47**, 3546 (1993).

25. E. W. Kolb and M. Gleiser, *Phys. Rev. D* **48**, 1560 (1993).

26. M. Fukugita and T. Yanagida, *Phys. Rev. D* **42**, 1285 (1990); J. A. Harvey and M. S. Turner, *ibid* **42**, 3344 (1990).

27. M. E. Shaposhnikov and G. R. Farrar, *Phys. Rev. Lett.* **70**, 2833 (1993). For a review of baryogenesis at the weak scale, see A. Cohen, D. Kaplan, and A. Nelson, *Ann. Rev. Nucl. Part. Sci.* **43**, 27 (1993).

28. H. M. Hodges and G. R. Blumenthal, *Phys. Rev. D* **42**, 3329 (1990); E. J. Copeland, E. W. Kolb, A. R. Liddle, and J. E. Lidsey, *Phys. Rev. Lett.* **71**, 219 (1993) and *Phys. Rev. D* **48**, 2529 (1993); M. S. Turner, *Phys. Rev. D*, in press (1993); E. J. Copeland, E. W. Kolb, A. R. Liddle, and J. E. Lidsey, *Phys. Rev. D*, in press (1994).

29. C. Alcock et al., *Nature* **365**, 621 (1993); E. Aubourg et al., *ibid* **365**, 623 (1993); A. Udalski et al., *Acta Astronomica* **43**, 289 (1993).

30. A. Sandage, *Ap. J.* **133**, 355 (1961).

PREFACE

The past decade has witnessed an explosion of activity and progress in both theoretical and observational cosmology. The catalyst has been the infusion of ideas from modern particle physics. These ideas include the "Standard Model" of the strong and electroweak interactions, which provides an understanding of physics up to the weak scale (about 250 GeV), and attractive, albeit speculative, ideas about grand unification—the unification of the strong and electroweak forces—and about super unification—the unification of all the forces (including gravity). Using these theoretical constructs theorists are able to discuss in a sensible way physics at energies up to the Planck scale (10^{19} GeV) and beyond. Moreover, particle physicists and cosmologists have applied these theoretical ideas to the study of the earliest moments of the Universe. Their speculations have led to very interesting and even compelling scenarios about the events that may have taken place at these early times: baryogenesis, inflation, the production of exotic relics—monopoles, strings, axions, photinos, and so on—and even the possibility of extra spatial dimensions. While these early-Universe scenarios are still untested, it has become clear that the answers to some of the most pressing and fundamental questions facing cosmology today *must* involve events that took place during the first 0.01 sec of the history of the Universe. These fundamental questions include: the origin of the matter–antimatter asymmetry, the nature of the dark matter, the origin of the smoothness and flatness of the Universe, the origin of the density inhomogeneities that initiated structure formation, the origin of the expansion, and even the ultimate fate of the Universe.

A second, equally important, development has been the use of astrophysical and cosmological observations to test and constrain particle physics theories. At present, the most exciting and fundamental ideas in particle theory involve energy scales well beyond the reach of conventional terrestrial accelerators: The highest energies achieved in accelerator laboratories are only 1000 GeV, while the energy scale of grand unification is believed to be in excess of 10^{14} GeV, and that of super unification, in excess of 10^{19} GeV. For this reason, some of the important tests of the most promising theories involve the early Universe or unique contemporary astrophysical environments.

Developments have been taking place in observational cosmology as well: Red shift surveys have begun to reveal the nature of the large-scale structure of the Universe; measurements of the velocity field of the Universe are beginning to shed light upon the distribution of matter (as opposed to just the light); more precise measurements of the spectrum and anisotropies of the microwave background should soon reveal important information about the nature of the primeval inhomogeneities as well as the early history of the Universe; and a number of experiments are being built to search directly for the relic elementary particles that may comprise the dark matter. In brief, observations are beginning to test the very interesting and exciting speculations about the first 0.01 sec that have been put forward in the past decade.

Our purpose in putting together this monograph was to fill the void that now exists between the current standard treatments of cosmology (S. Weinberg, *Gravitation and Cosmology*; C. W. Misner, K. Thorne, and J. A. Wheeler, *Gravitation*; Ya. B. Zel'dovich and I. D. Novikov, *Relativistic Astrophysics Vol. II*; P. J. E. Peebles, *Physical Cosmology*) and the present frontiers of research in early Universe cosmology. While these and other texts discuss the standard hot big-bang cosmology, beginning at the epoch of primordial nucleosynthesis ($t \simeq 0.01$ to 200 sec), they mention only briefly the earliest epoch (the first 0.01 sec)—then referred to as the "hadron era." When these texts were written it was believed that the fundamental particles were leptons and hadrons, and that at a time of about 10^{-5}sec and a temperature of a few hundred MeV the strongly interacting particles should have been so dense that average particle separations would have been less than typical particle sizes, making an extrapolation to earlier times nonsensical. Moreover, the exponential rise with mass in the number of hadronic resonances suggested a maximum temperature for the Universe of only a few hundred MeV. The recognition in the 1970's that the fundamental particles of the strong interaction are point-like quarks and gluons and the discovery of the asymptotic freedom of the strong interaction made it clear that the early Universe should have been a dilute gas of weakly-interacting quarks, leptons, and gauge bosons, and opened the door for the study of the very early Universe.

Our presentation here covers many of the topics of current research in early Universe cosmology. The first three Chapters serve as an introduction to the standard cosmology. In the Fourth Chapter we review primordial nucleosynthesis, discuss recent observations of the light element abundances, and explain how these observations can be used to limit the properties of weakly-interacting particles. The final seven Chapters concern subjects in particle cosmology that were developed after the standard texts were written. In the Fifth Chapter we address the topic of thermodynamics in the expanding Universe, describe the decoupling of massive particles, and discuss neutrino cosmology. Chapter 6 is devoted to baryogenesis—the very attractive theory for the origin of the baryon asymmetry which is based upon grand unification. In Chapter 7 we review cosmological phase transitions and the production of topological defects—domain walls, cosmic string, and massive magnetic monopoles—and discuss in detail the cosmological and astrophysical effects of monopoles. Inflation has revolutionized the way cosmologists view the earliest moments of the Universe and it is the theme of Chapter 8. Chapter 9 addresses the topic of structure formation. Although the basic picture that cosmologists have of structure formation—amplification of small primeval inhomogeneities through the Jeans instability—has not changed significantly, the advent of particle cosmology has led to a renaissance in the subject with the development of the hot and cold dark matter scenarios. The astrophysical and cosmological consequences of axions are the subject of the penultimate Chapter. We devote the final Chapter to recent speculations about the Planck era, including the wave function of the Universe, cosmology with extra dimensions, and superstring cosmology.

The trend of the monograph is clear: The first five Chapters treat traditional big bang cosmology from a modern perspective, while the last six Chapters are devoted to more speculative ideas. The first part of the book reviews areas of cosmology for which we have the most complete observational data base, deals mostly with events that occurred after the first 10^{-2} sec, and is based upon "well-known" low-energy physics. The second part of the book covers the most modern subjects in cosmology for which the observational data base is only now emerging, concerns events that occurred before 10^{-2} sec, and is based upon speculations beyond the standard model of particle physics.

Including all the relevant background material for the wide range of topics covered in the book is impractical, and so we have also prepared a reprint volume, *Early Universe: Reprints* (Addison-Wesley, Redwood City, Calif., 1988), as a companion to this monograph. In the reprint volume we provide a collection of key papers for the important topics that are not thoroughly covered in the monograph—observational cosmology, finite-temperature field theory, light-element abundances, etc. We hope that together these two volumes will provide the graduate student or the interested outsider with a sound introduction to modern cosmology, and even prepare the reader to carry out research in particle cosmology.

This monograph grew out of lectures given by the authors over the years at a number of Summer and Winter Schools and in graduate courses at The University of Chicago. The emphasis of the material covered reflects the judgment (or lack thereof) of the authors. The references are by no means complete. A more thorough review of the literature can be found in the reprint volume.

We are grateful to colleagues, friends, and students who have offered suggestions for improving the book. In particular we would like to thank Marc Davis, Sasha Dolgov, Alan Dressler, George Efstathiou, Margaret Geller, James Hartle, John Huchra, Marc Kamionkowski, Robert Kirshner, Andre Linde, John Preskill, Pierre Sikivie, Sharon Vadas, and Helmut Zaglauer for their careful reading of the manuscript and valuable suggestions. We especially wish to thank Ted Ressell for his preparation of the Index and Richard Holman for his detailed critical comments on several of the Chapters.

As the camera-ready book goes to press, it is completly free of any typographical errors, errors of physics, or errors of judgment. Any errors present in the final product must have crept in during the production process, and are wholly the fault of the publisher.

Edward W. Kolb
Michael S. Turner
Warrenville, Illinois
August, 1989

1

THE UNIVERSE OBSERVED

1.1 Introduction

Our current understanding of the evolution of the Universe is based upon the Friedmann-Robertson-Walker (FRW) cosmological model, or the hot big bang model as it is usually called. The model is so successful that it has become known as the standard cosmology. In this first Chapter we will review the observational basis for the standard cosmology. Direct evidence supporting its validity extends back to the beginning of the epoch of primordial nucleosynthesis, about 10^{-2} sec after the bang. Current speculations about the earliest history of the Universe, the subject of this monograph, derive from an extrapolation of the standard cosmology to very early times. The FRW cosmology is so robust that it is possible to make sensible speculations about the Universe at times as early as 10^{-43} sec after the bang! Of course, such speculations are necessarily based upon some theory of the fundamental interactions at very high energies, energies approaching the Planck scale (10^{19} GeV). At present there exists a standard model of particle physics, the $SU(3)_C\otimes$ $SU(2)_L\otimes$ $U(1)_Y$ gauge theory of the strong and electroweak interactions. It provides a fundamental theory of quarks and leptons and has been tested up to energies approaching 1000 GeV. In addition, the past decade has produced very interesting and important speculations about particle physics at very short distances, e.g., grand unification, supersymmetry, superstring theory, etc. It is these theories of fundamental physics at ultra-high energies which allow us to speculate about the earliest history of the Universe.

Astronomy is a data-starved science. Cosmology is even more so. Observers (and their funding agencies) must pay dearly for each particle detected from distant objects in the Universe. In spite of this handicap, there is indeed a firm observational basis for the standard cosmology, with fossils dating back to about 10^{-2} sec after the bang. Moreover, there is the reasonable expectation that the cosmological data base will grow in the next decades, both in the quantity and in the quality of the observations. At present the cosmological observables include: the expansion of the Universe; the Hubble constant H_0; the deceleration parameter q_0; the age of the Universe t_0; the present mass density ρ_0 and composition of the Universe (ρ_i, i = baryons, radiation, etc.); the cosmic microwave background radiation (CMBR), including its spectrum and spatial structure; other cosmological background radiations (IR, UV, x ray, γ ray, etc.); the abundance of the light elements (particularly D, ^{3}He, ^{4}He, and ^{7}Li); the baryon number of the Universe, quantified as the baryon-to-photon ratio; and the distribution of galaxies and larger structures (clusters of galaxies, superclusters, and voids).

In the companion volume to this monograph, *Early Universe: Reprints* [1], we have reprinted or referred to many of the key papers describing the observed Universe. Here we will briefly summarize the observational evidence that supports the standard cosmology, and in the process describe the present state of the Universe.

Unless explicitly displayed, we will set the fundamental constants $\hbar$, c, and k_B equal to unity. Some handy conversion factors and a list of useful physical parameters for astrophysics and cosmology are given in Appendix A. We will assume that the reader is at least familiar with the basic ideas of the standard model of particle physics and unified gauge theories; we provide a brief primer on modern particle theory in Appendix B. We refer those interested in the standard model and current speculations beyond the standard model to the excellent monographs that exist on these subjects [2].

1.2 The Expansion

A most fundamental feature of the standard cosmology is the expansion of the Universe. The expansion, discovered in the 1920's, plays a most basic role in observational cosmology. Of the almost 28,000 galaxy spectra measured by observers all over the world, all but a handful (those of nearby galaxies) are red shifted, illustrating the universality of the expansion. Many quasi-stellar objects (QSO's) with red shifts in excess of 3 have been observed, and the current record holder has a red shift slightly greater than 4.7 [3]. Many radio galaxies with red shifts in excess of 2 have been

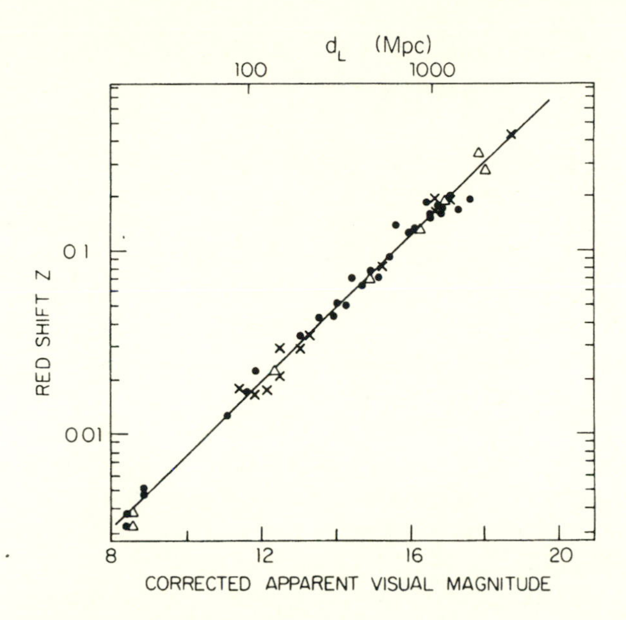

Fig. 1.1: The Hubble diagram. The corrected apparent magnitude is proportional to the logarithm of the luminosity distance. The straight line indicates the theoretical relationship for $q_0 = 1$ (from [7]).

observed, with the current record holder having a red shift of 3.8 [4].[1] The most distant cluster of galaxies observed has a red shift of 0.94 [6]. The light we see today from the most distant objects was emitted when the Universe was only a few billion years old. Thus, galaxies and QSO's provide a probe of the Universe back to times as early as a few billion years after the bang.

The relationship between the luminosity distance, $d_L \equiv (\mathcal{L}/4\pi\mathcal{F})^{1/2}$ ($\mathcal{L}$ = object's luminosity, $\mathcal{F}$ = measured flux),[2] and the red shift of a galaxy z can be written in a power series:

$$H_0 d_L = z + \frac{1}{2}(1 - q_0)z^2 + \cdots \tag{1.1}$$

[1]Few ordinary galaxies with red shifts greater than one have been seen—perhaps because they are more difficult to detect. Very recently a candidate field galaxy has been detected with a purported red shift of 3.38 [5].

[2]The precise definition of d_L and details of the $d_L - z$ relationship will be discussed further in Chapter 2.

or

$$z = H_0 d_L + \frac{1}{2}(q_0 - 1)(H_0 d_L)^2 + \cdots \qquad (1.2)$$

where the Hubble constant is the present expansion rate of the Universe, $H_0 \equiv \dot{R}(t_0)/R(t_0)$, and the deceleration parameter measures the rate of slowing of the expansion, $q_0 \equiv -\ddot{R}(t_0)/R(t_0)H_0^2$ [$R(t)$ is the FRW cosmological scale factor, defined in the next Chapter, and subscript 0 denotes the present value of a quantity]. Red shifts are relatively simple (albeit time consuming) to measure, while determining galaxy distances requires well-established standard candles (i.e., objects with "known" $\mathcal{L}$). Due to the difficulty of calibrating the cosmic distance ladder, the distance scale still has a factor of 2 uncertainty even at modest cosmological distances. Moreover, for the most distant objects in the Universe (red shifts of order unity or greater) one must worry about evolutionary effects: Do the luminosities of the standard candles evolve with time?—after all, some evolution must occur because 20 Gyr ago $\mathcal{L} = 0$.

At modest red shifts, say $z \lesssim 1$, the linear relationship between d_L and z is quite clear and convincing (see Fig. 1.1). Using galaxies at relatively modest red shifts ($z \ll 1$) one may determine H_0: On a $\log z$ vs. $\log d_L$ plot, $\log H_0$ is the intercept on the $\log z$ axis. At present, reported values for H_0 span the range 40 to 100 km sec^{-1} Mpc^{-1}, with many authors quoting standard errors of 10 km sec^{-1} Mpc^{-1} or less! Clearly, systematic uncertainties still dominate in the determination of H_0. [For a historical perspective, compare these values with Hubble's initial determination: $H_0 = 550$ km sec^{-1} Mpc^{-1}.] Without trying to address the complicated issues in determining this most significant number (and risk making even more enemies than we already have!), we summarize the current state of affairs by writing

$$H_0 = 100h \text{ km sec}^{-1} \text{ Mpc}^{-1}, \qquad (1.3)$$

and, like all cosmologists, bury our lack of precise knowledge of H_0 in "little h:"

$$0.4 \lesssim h \lesssim 1.0. \qquad (1.4)$$

It then follows that the Hubble time or Hubble distance H_0^{-1} is

$$\begin{aligned} H_0^{-1} &= 9.78h^{-1} \times 10^9 \text{ yr} \\ &= 3000h^{-1} \text{ Mpc} \\ &= 9.25 \times 10^{27} h^{-1} \text{ cm}. \end{aligned} \qquad (1.5)$$

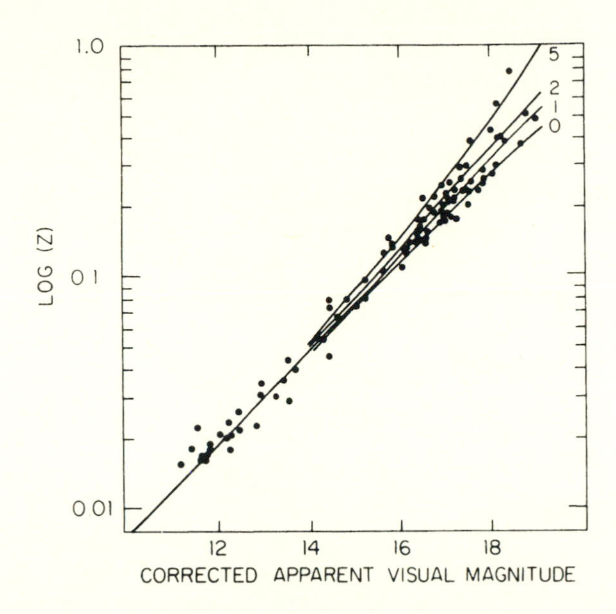

Fig. 1.2: An optical Hubble diagram. The corrected visual magnitude is proportional to the logarithm of the optical luminosity distance. The curves refer to models with $q_0 = 0, 1, 2, 5$. No correction for galactic evolution has been included (from [9]).

With the continued refinement of statistically-based distance indicators,[3] and the advent of the Hubble Space Telescope, there is hope that this frustrating uncertainty in H_0 soon will be be eliminated. For further discussion of the cosmic distance scale see [8] and *Early Universe: Reprints.*

In principle, the deceleration parameter q_0 can be measured without recourse to knowledge of H_0, as it merely measures the deviation of the red shift–distance relationship from the linear "Hubble law," $z = H_0 d_L$. In practice, however, the distance scale is again a problem as one requires reliable distances to objects at moderate to large red shifts. At such red shifts the "look back" times are a significant fraction of the age of the object, and one must worry about evolution. Uncertainties about the effects of galactic evolution on the intrinsic luminosity of galaxies have

[3]Essentially all of the traditional techniques for determining distances to distant galaxies involve the use of only a single object as a standard candle, e.g., the first-ranked galaxy in a cluster or a supernova. Statistical methods like the IR Tully-Fisher, Faber-Jackson, or similar relationships allow one to determine the distance to a cluster by using the properties of a large number of galaxies, e.g., by constructing an IR luminosity vs. rotation speed (or galaxy diameter) diagram.

prevented a definitive (or even near definitive) determination of q_0. In fact, there is not even agreement as to whether the "young" galaxies (i.e., galaxies at large red shift) should be intrinsically brighter or dimmer than the "old" galaxies (i.e., galaxies at small red shift).

An optical Hubble diagram compiled in 1978 [9], and IR and optical Hubble diagrams compiled in a recent review [10] are shown in Figs. 1.2 and 1.3.[4] At present, the Hubble diagram probably only constrains q_0 to be between 0 and a few. As we will discuss in Chapter 3, q_0 is related to Ω_0[5]

$$q_0 = \Omega_0(1+3w)/2, \tag{1.6}$$

where w is the present ratio of the pressure to energy density (for non-relativistic matter, $w \simeq 0$). Thus, current observations only restrict Ω_0 to the interval $[0, few]$.

As we will discuss in Chapter 2, the functional dependence of the angular size of a standard ruler (e.g., the angular size of a galaxy or a cluster) depends upon Ω_0, and can be used to determine Ω_0. The results of such an attempt are shown in Fig. 1.4. Once again, this technique only restricts q_0 to the range 0 to a few.

Finally, we mention a very promising kinematic test of the standard cosmology which can be used to determine q_0 (or Ω_0): the galaxy number count vs. red shift test. The number of galaxies in a volume element comoving with the expansion, defined by the solid area $d\Omega$ and red shift interval dz, depends upon the number density of galaxies (per comoving volume), and the cosmological model. By counting galaxies as a function of red shift it is in principle possible to deduce q_0 (or Ω_0). Loh and Spillar [12] have made a preliminary (and somewhat controversial) attempt to measure Ω_0 based upon this technique and they infer a value of $\Omega_0 = 0.9^{+0.7}_{-0.5}$ (see Fig. 1.5). While this test is also subject to the effects of evolution, it is more sensitive to the evolution of the number density of galaxies than to the evolution of galactic luminosity. We will discuss the galaxy number count test at greater length in Chapter 2.

[4]As emphasized in [10], evolutionary effects may be minimized by using the "IR" luminosity distance. Because of the red shift effect, optical light from distant galaxies comes predominately from hot, young, high-mass stars whose evolutionary time scale is short, while IR light comes from older, low-mass stars whose evolutionary time scale is longer.

[5]The quantity $\Omega_0 \equiv \rho_0/\rho_C$, where ρ_0 is the present mass density of the Universe, and ρ_C is the critical density, $\rho_C \equiv 3H_0^2/8\pi G = 1.88 \times 10^{-29} h^2 \mathrm{g\ cm^{-3}}$.

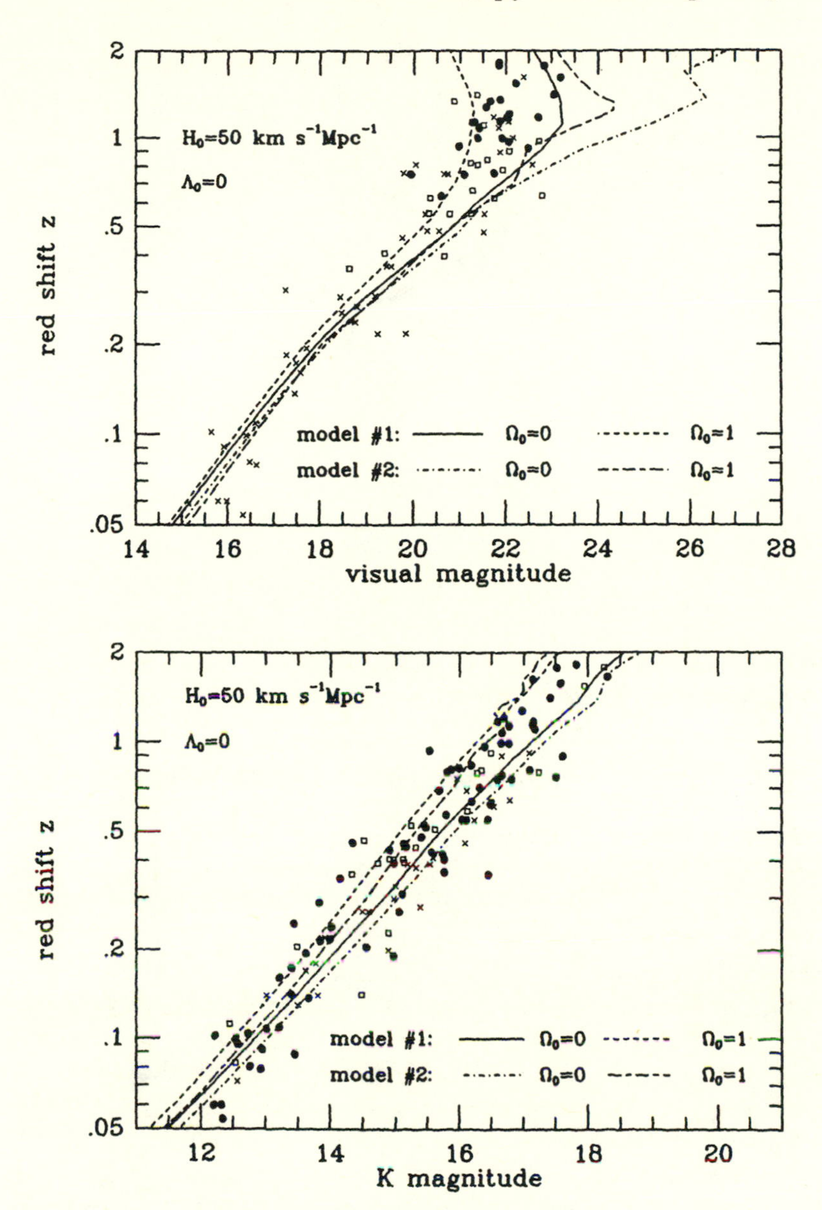

Fig. 1.3: Optical (top) and infrared (bottom) Hubble diagrams. For the infrared diagram, the K magnitude is proportional to the logarithm of the 2 μm luminosity distance. The solid and broken curves in the diagrams indicate effects of galactic evolution for a simple model. Note that evolutionary effects are more severe for the optical case (from [10]).

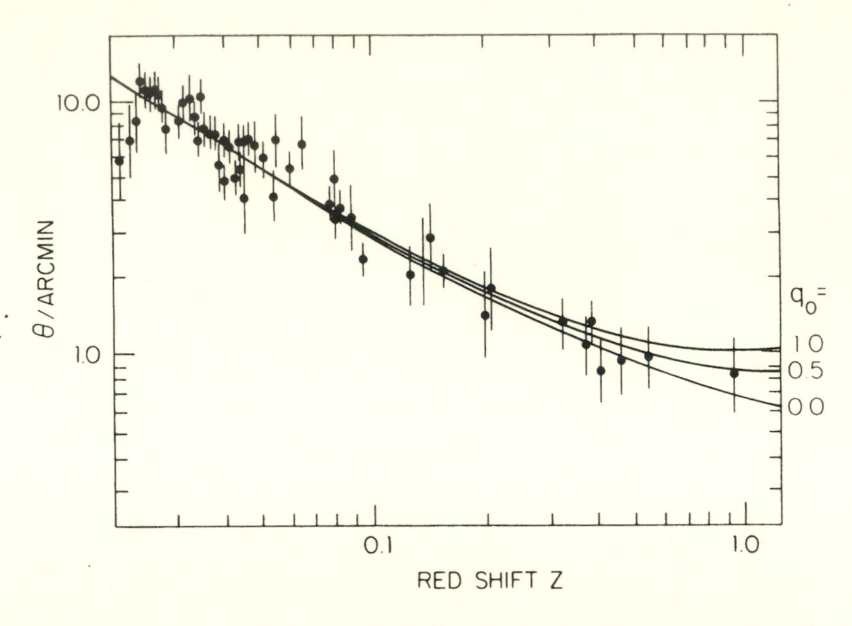

Fig. 1.4: The angle-red shift test. Shown are the angular diameters of galaxy clusters at different red shifts, and the theoretical curves for $q_0 = 0$, 0.5, and 1.0 (from [11]).

1.3 Large-Scale Isotropy and Homogeneity

A cornerstone of the FRW model is the high degree of symmetry of the FRW metric. As a practical matter the simplicity of the metric, which depends upon only one dynamical variable, the cosmic scale factor $R(t)$, makes the theoretical analysis tractable (even for unsophisticated physicists like ourselves!).

The assumption of isotropy and homogeneity dates back to the earliest work of Einstein, who made the assumption not based upon observations, but as theorists often do, to simplify the mathematical analysis. Today there is ample evidence for the isotropy and homogeneity for the part of the Universe we can observe, our present Hubble volume, whose size is characterized by $H_0^{-1} \simeq 3000h^{-1}$ Mpc $\simeq 10^{28}h^{-1}$ cm.

The best evidence for the isotropy of the observed Universe is the uniformity of the temperature of the CMBR: Aside from the observed dipole anisotropy, the temperature difference between two antennas separated by angles ranging from about 10 arc seconds to 180° is smaller than about one part in 10^4 (see Fig. 1.6). The simplest interpretation of the dipole anisotropy is that it is the result of our motion relative to the cosmic rest

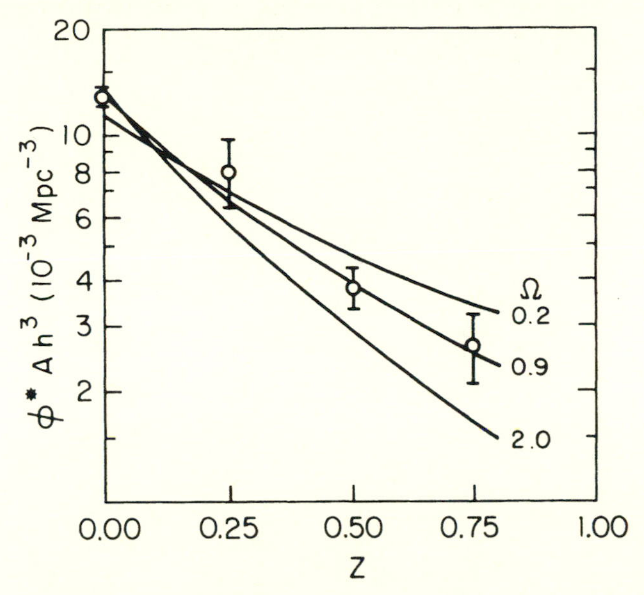

Fig. 1.5: Determination of Ω_0 from the galaxy number count test. The number of galaxies counted in the counting volume $(dz, d\Omega)$ is proportional to $z^2 dz d\Omega H_0^{-3} A(z)\phi^*$, where ϕ^* is proportional to the galaxy number density per comoving volume and $A(z)$ depends upon the cosmological model (from [12]).

frame. If the expansion of the Universe were not isotropic, the expansion anisotropy would lead to a temperature anisotropy in the CMBR of similar magnitude. Likewise, inhomogeneities in the density of the Universe on the last scattering surface would lead to temperature anisotropies. In this regard, the CMBR is a very powerful probe: It is even sensitive to density inhomogeneities on scales larger than our present Hubble volume.[6] The remarkable uniformity of the CMBR indicates that at the epoch of last scattering for the CMBR (about 200,000 yr after the bang) the Universe was to a high degree of precision (order of 10^{-4} or so) isotropic and homogeneous.

There is additional supporting evidence for the isotropy of the Universe: the isotropy of the x-ray background radiation (to about 5%),[7] of

[6]Its sensitivity decreases as $(H_0^{-1}/\lambda)^2$; λ = length scale of the inhomogeneity [14].

[7]In addition, there is weak evidence for a dipole anisotropy of the cosmic x-ray background, with magnitude and direction consistent with that of the CMBR dipole anisotropy [15]. If the x-ray background arises from high-red shift sources, and the x-ray anisotropy indeed coincides with the CMBR anisotropy, this would indicate that

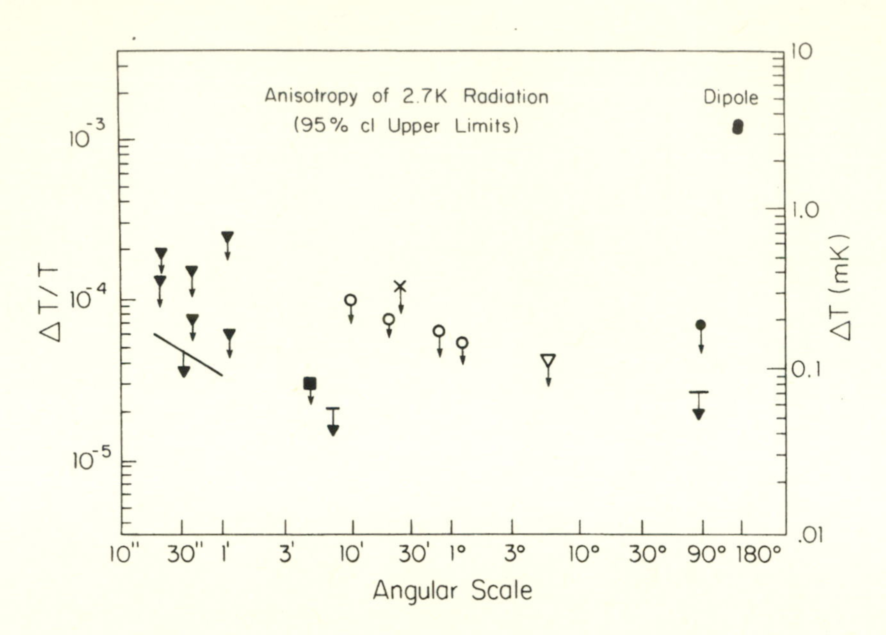

Fig. 1.6: RMS variation of the CMBR temperature as a function of the angular separation of the two antennas (from [13]).

the distribution of faint radio sources, and of galaxies themselves. Some substantial fraction of the x-ray background is believed to be from unresolved sources (e.g., QSO's) at high red shift. Likewise, a substantial fraction of the faint radio sources are radio-bright galaxies at high red shift. Both the Shane-Wirtanen (Lick) catalogue of about a million galaxies with effective depth of about 200 h^{-1} Mpc (see Fig. 1.7) and the IRAS catalogue of infrared selected galaxies with an effective depth of about 60 h^{-1} Mpc provide evidence for the isotropy of the galaxy distribution.

Direct evidence for the homogeneity of the distribution of galaxies is more tenuous. Galaxy counts in deep surveys provide supporting evidence, but their interpretation is not so straightforward. Moreover, since light does not necessarily trace mass, such surveys only determine the distribution of light (i.e., bright galaxies) and not a priori that of mass. In particular, if the mass-to-light ratio varies depending on the local environment (e.g., the local density), the light distribution will not reflect the true distribution of the mass. Moreover, it is somewhat disturbing—although perhaps not totally unexpected—that in the largest red shift surveys com-

---

the frame defined by these distant sources is at rest with respect to the CMBR.

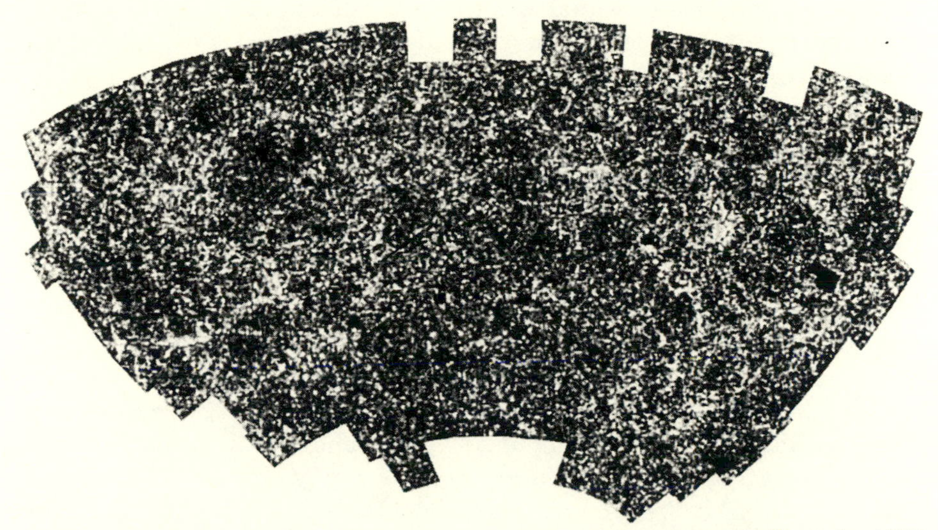

Fig. 1.7: Equal-area projection of a southern sky portion (4080 deg^2) of the APM catalogue. The density of galaxies per pixel (1 pixel = 14 (arc min)^2) is indicated by a grey scale. About 2 million galaxies are represented in this image (from [16]).

pleted, the biggest structures are as large as the limits set by the size of the surveys themselves. As larger red shift surveys are completed (the largest survey completed contains only about 9,000 galaxies), the distribution of light, if not mass, should be better understood.

Finally, we mention the evidence for homogeneity from determination of the "peculiar velocity field of the Universe." Peculiar velocity refers to the motion of an object (e.g., a galaxy) with respect to the cosmic rest frame (defined by the rest frame of the CMBR). In practical terms, it corresponds to a galaxy's velocity after its "expansion velocity" has been subtracted. The inhomogeneous distribution of matter leads to gravitationally-induced peculiar velocities of the order of

$$\delta v/c \simeq \Omega_0^{0.6}(\lambda/H_0^{-1})(\delta\rho/\rho)_\lambda \tag{1.7}$$

on the scale λ, where $(\delta\rho/\rho)_\lambda$ characterizes the amplitude of the mass inhomogeneity on the scale λ. Measured peculiar velocities extend to scales as large as $60h^{-1}$ Mpc or so, where peculiar velocities of order 600 km sec^{-1} have been measured, indicating very roughly that $(\delta\rho/\rho)_\lambda \sim 10^{-1}$ on these scales [17]. Note too, that in principle, peculiar velocity measurements have the attractive feature that galaxies serve only as test particles tracing out the gravitational potential and thus provide direct information about the mass distribution rather than just that of the light. We will discuss the peculiar velocity field in more detail in Chapter 9.

1.4 Age of the Universe

The age of the Universe can be measured in a variety of different ways: by using the expansion rate of the Universe to compute the time back to the bang; by dating the oldest stars in globular clusters; by dating the radioactive elements; by considering the cooling of white dwarf stars; by calculating the cooling time for hot gas in clusters, etc. At present, all techniques yield results consistent with the range 10 to 20 Gyr.

However, the uncertainties, especially the unknown systematics, preclude a definitive determination by any of the above techniques. In principle the present age of the Universe provides a very important test of cosmological models. We remind the reader of the discrepancy that existed until the 1950's between the expansion age, 2 Gyr for the then Hubble constant of $500\,\mathrm{km\,sec^{-1}Mpc^{-1}}$, and the age of the solar system, about 4.5 Gyr. That discrepancy led to the birth of the "ageless" steady state cosmology.

As we will discuss in Chapter 3, the expansion age is set by the Hubble time H_0^{-1} and Ω_0: for $\Omega_0 = 0$, $t_0 = H_0^{-1}$, with the age decreasing with increasing Ω_0 (e.g., for a matter-dominated model, $t_0 = (2/3)H_0^{-1}$ for $\Omega_0 = 1$). For the range of plausible values of H_0 ($h \simeq 0.4$ to 1.0), the Hubble age $H_0^{-1} \simeq 9.8$ to 24.5 Gyr.

Much work has been carried out to determine the ages of the oldest globular clusters, those which contain very metal-poor, pop II stars.[8] By detailed stellar modeling a theoretical Hertzsprung-Russell diagram can be constructed, and then compared to that observed for the cluster. Roughly speaking, the age of the globular cluster is determined by the position of the turn-off point on the main sequence for the red giant phase. The point where this occurs determines the mass of the stars that are just now entering the red giant phase. With recourse to stellar models for such stars one can calculate their ages. Most estimates of the ages of the oldest stars

[8] The oldest generation of stars, those which have low metal abundances (elements with $A > 4$), are referred to as pop II, while younger stars, like our sun, with metal abundances of order 2% (by mass) are referred to as pop I. The hypothetical generation of even older stars (possibly pre-galactic in origin) are referred to as pop III.

in the galaxy determined this way are in the range 10 to 20 Gyr. Systematics inherent to this ingenious and powerful technique (stellar mass loss, convection, metallicity effects, uncertainties in the distance scale, interstellar reddening, etc.) prevent determination of a more precise age or even a definitive estimate of the uncertainty. Since the oldest globular clusters likely formed much less than 1 Gyr after the Galaxy formed, and the Galaxy itself formed less than a few Gyr after the bang, these age determinations also serve to date the Universe.

Since Rutherford, cosmologists have used radioactive clocks to date the Universe (and many other objects within it). For cosmological purposes the most suitable radio isotopes are: ^{232}Th (mean lifetime $\tau = 20.27$ Gyr); ^{235}U ($\tau = 1.015$ Gyr); ^{238}U ($\tau = 6.446$ Gyr); ^{87}Rb ($\tau = 69.2$ Gyr); and ^{187}Re ($\tau = 62.8$ Gyr). In order to use these clocks one must know the relative abundances of these isotopes (or pairs of isotopes) today and at the epoch of their production. All of these isotopes are so-called r-process elements, elements thought to be produced by rapid neutron capture in an early generation of stars. To illustrate this technique, consider the pair ^{235}U and ^{238}U. The production ratio is calculated to be $[^{235}\mathrm{U}/^{238}\mathrm{U}]_P \simeq 1.71$, while the present abundance ratio is $[^{235}\mathrm{U}/^{238}\mathrm{U}]_0 \simeq 0.00723$. The time elapsed since production is then

$$\Delta t = \frac{\ln[^{235}\mathrm{U}/^{238}\mathrm{U}]_P - \ln[^{235}\mathrm{U}/^{238}\mathrm{U}]_0}{\tau_{235}^{-1} - \tau_{238}^{-1}} \simeq 6.6 \text{ Gyr}. \tag{1.8}$$

What does this time interval indicate? If we knew that all the r-process elements were produced shortly after the Galaxy formed, Δt would provide an accurate age for the Galaxy—but we don't! If r-process elements have been formed continuously since the formation of the Galaxy then the age of the Galaxy must be considerably greater. Our simple example illustrates some of the inherent difficulties involved with this technique. Many, much more sophisticated, analyses have been carried out, with derived ages for the Galaxy spanning the range 10 to 20 Gyr.

Recently, Winget, et al. [18] have used the cooling of white dwarf stars to determine the age of the Galaxy. The oldest white dwarf stars are of course the coolest and least luminous. The observed number of white dwarfs drops precipitously below a luminosity of $3 \times 10^{-5} L_\odot$—presumably due to the finite age of the Galaxy. Based upon this observation and models of white dwarf cooling, these authors conclude that the age of the Universe is 10.3 ± 2.2 Gyr.[9]

[9]Since white dwarf stars are only observed in the disk, the authors have actually

To summarize our brief and very pedestrian review of age determinations for the Universe, we can say that all techniques yield an age consistent with the range 10 to 20 Gyr. This in itself is very reassuring, cf. the "age crisis" in cosmology which lasted until the 1950's. Moreover, even this somewhat inprecise dating of the Universe provides an important test of cosmological models. In Chapter 3 we will show that for a matter-dominated model a lower bound to the age of the Universe of 10 Gyr provides an upper bound to Ω_0 of 6.8 (for $h \geq 0.4$) or 3.2 (for $h \geq 0.5$). Further, if we consider a flat ($\Omega_0 = 1$), matter-dominated model, then $t_0 \geq 10$ Gyr, necessarily implies that $h \leq 0.65$. That is, a value of $H_0 \geq 65$ km sec^{-1} Mpc^{-1} would rule out such a cosmological model. A determination of H_0 to a precision of 10% is believed possible with the Hubble Space Telescope; if and when this is done, there could be important cosmological implications.

1.5 Cosmic Microwave Background Radiation

The CMBR provides the fundamental evidence that the Universe began from a hot big bang. As we will discuss in Chapter 3, the surface of last scattering for the CMBR was the Universe at a red shift $z \sim 1100$ and age of $180,000(\Omega_0 h^2)^{-1/2}$ yr. Flux measurements of the CMBR ranging from wavelengths of about 70 cm down to wavelengths of less than 0.1 cm are consistent with that of a black body at temperature 2.75 ± 0.015 K (see Fig. 1.8). Such a temperature corresponds to a present photon number density of 422 cm^{-3}.[10]

As mentioned earlier, the temperature of the CMBR across the sky is remarkably uniform: $\Delta T/T \lesssim 10^{-4}$ on angular scales ranging from 10 arc seconds to 180° (after the dipole anisotropy has been removed); see Fig. 1.6. The observed high degree of isotropy not only provides strong evidence for the present level of large-scale isotropy and homogeneity of our Hubble volume, but also provides an important probe of conditions in the Universe at red shifts of order 1100.

As we will discuss in Chapter 9, the primeval density inhomogeneities necessary to initiate structure formation result in predictable temperature

estimated the age of the disk. If, as some suspect, the disk formed several Gyr after the galaxy, then several Gyr should be added to these estimates for the age of the galaxy.

[10]We mention that recently Lange and his collaborators [19] have reported evidence for a distortion in the CMBR spectrum in the submillimeter region (Wien part of the spectrum). This submillimeter excess corresponds to about 10% of the total energy in the CMBR. If their result stands it will have profound implications for cosmology.

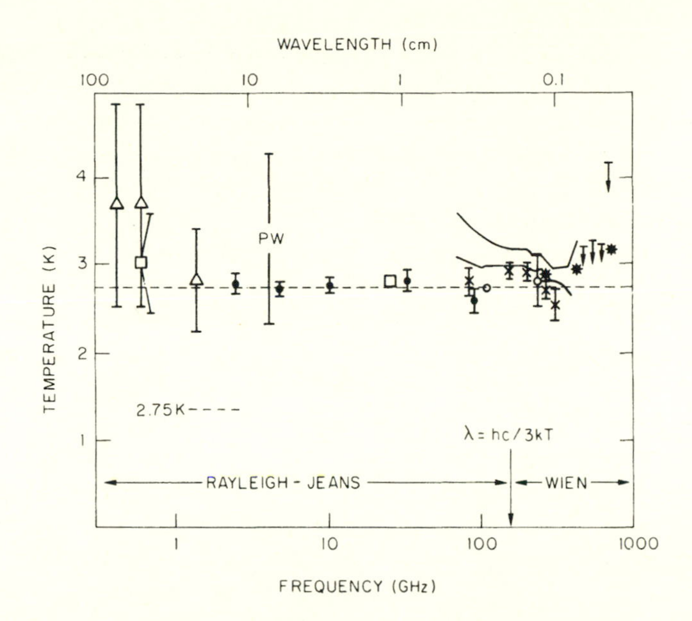

Fig. 1.8: Temperature measurements of the CMBR (from [20]). PW indicates the discovery measurement of Penzias and Wilson, and ⋆ indicates the recent measurements of Lange et al. [19].

fluctuations in the CMBR, and so anisotropies of the CMBR provide a powerful test of theories of structure formation. In fact, the current limits to the anisotropy come within factors of 3 to 10 of the predictions of the most attractive scenarios of structure formation: inflation-produced adiabatic density perturbations with hot or cold dark matter, and cosmic string-induced density perturbations with hot or cold dark matter. Spectral distortions in the CMBR, if they exist, may provide fossil evidence for the early history of galaxy, and possibly even star, formation.

1.6 Light-Element Abundances

Primordial nucleosynthesis is the earliest test of the standard model. Nuclear reactions that took place from $t \simeq 0.01$ to 100 sec ($T \simeq 10$ MeV to 0.1 MeV) resulted in the production of substantial amounts of D (D/H $\simeq$ *few* $\times 10^{-5}$), ^{3}He (^{3}He/H $\simeq$ *few* $\times 10^{-5}$), ^{4}He (mass fraction $Y \simeq 0.25$), and ^{7}Li (^{7}Li/H $\simeq$ 1 to 2 $\times 10^{-10}$). Deuterium and Helium-4 are of particular importance as there are apparently no contemporary astrophysical processes that can account for their observed abundances. While ordinary stars produce ^{4}He, even in regions where there has been significant stellar processing, the stellar contribution is only about $\Delta Y_{\rm stellar} \simeq 0.05$. While the observed Deuterium abundance is very small, even this small amount is difficult, if not impossible, to account for, as almost all astrophysical processes destroy the weakly-bound deuteron which burns at the relatively low temperature of about 0.5×10^6 K.

The comparison of the predicted abundances with "inferred" primordial abundances provides a very powerful test of the standard cosmology. At present there is concordance between the predicted and observed abundances for these four isotopes, provided that the baryon-to-photon ratio η is in the interval $\eta = (4 \text{ to } 7)\times 10^{-10}$, corresponding to $0.015 \leq \Omega_B h^2 \leq 0.026$, or taking $0.4 \leq h \leq 1.0$, $0.014 \leq \Omega_B \leq 0.16$. The standard cosmology passes this very stringent test with flying colors, and further provides important information about the density of baryons in the Universe. In fact, primordial nucleosynthesis provides the most precise determination of the baryon density. Most significantly, primordial nucleosynthesis implies the fraction of critical density in baryons, Ω_B, must be less than one: For $\Omega_B \sim 1$, Deuterium would be severely underproduced, and both ^{4}He and ^{7}Li would be overproduced. If Ω_0 is equal to 1, then primordial nucleosynthesis provides the strong indication that most of the mass density of the Universe is in a form other than baryons.[11]

Finally, we add that primordial nucleosynthesis has also been used as a probe of the early Universe and particle physics. We have already mentioned that it provides an important constraint to the baryon density. It has also been used to constrain the existence of additional hypothetical, light ($\lesssim$ MeV) particle species predicted by some particle theories (e.g., additional light neutrino species). Such species would contribute to the energy density of the Universe, and thereby affect the predicted abundances. Chapter 4 is devoted to a detailed discussion of primordial nucleosynthesis.

1.7 The Matter Density: Dark Matter in the Universe

Previously, we discussed kinematical methods of determining Ω_0; here we discuss dynamical means of measuring Ω_0. Measuring the mass density of the Universe is a challenging task! Figuratively speaking, the average

[11]In Chapter 4 we will discuss two unconventional scenarios of primordial nucleosynthesis which, if correct, would allow $\Omega_B \sim 1$.

density of the Universe ($\langle\rho\rangle$) can be measured by determining the number density of galaxies ($n_{\rm GAL}$) and the average mass per galaxy ($\langle M_{\rm GAL}\rangle$):[12]

$$\langle\rho\rangle = n_{\rm GAL}\langle M_{\rm GAL}\rangle. \tag{1.9}$$

Once $\langle\rho\rangle$ is found, $\Omega_0 = \langle\rho\rangle/\rho_C$ follows easily.

Measuring the mass of a galaxy by dynamical means involves detecting the gravitational effect of the mass in the galaxy in one way or another. The simplest means is the use of Kepler's 3rd Law:

$$GM(r) = v^2 r \tag{1.10}$$

where v is the orbital velocity at a distance r from the center of the galaxy, $M(r)$ is the mass interior to r, and spherical symmetry is assumed. Applying this technique to spiral galaxies (taking the measured rotational velocity to be v), and taking r to be the radius within which most of the light emitted by the galaxy is emitted, one finds that the fraction of critical density directly associated with light is

$$\Omega_{\rm LUM} \simeq 0.01 \text{ or less.} \tag{1.11}$$

Astonishingly, the mass associated with light provides less than 1% of the critical density.

When astronomers extended this technique to distances beyond the point where the light from a galaxy effectively ceases (by observing the rare star, or 21 cm emission from neutral Hydrogen, or HI, gas clouds) they found that $M(r)$ continued to increase. If the mass associated with the light were the whole story, v would decrease as $r^{-1/2}$ beyond the point where the light and mass cut off; rather they found $v \cong$ constant, corresponding to $M(r) \propto r$ (see Fig. 1.9). By definition, the additional mass is "dark," i.e., there is no "detectable" radiation associated with it. Further, there is no convincing evidence for a rotation curve that "turns over" (i.e., $v \propto r^{-1/2}$), indicating the total mass associated with a spiral galaxy has yet to be found! There is additional (weak) evidence that the dark matter is roughly spherically distributed, implying that $\rho_{\rm DARK} \propto r^{-2}$. Rotation curve measurements indicate that virtually all spiral galaxies have a dark,

[12]What astronomers actually do is measure mass-to-light ratios for galaxies (or parts thereof), and then calculate $\langle\rho\rangle$ by: $\langle\rho\rangle = \langle M/L\rangle\mathcal{L}$, where $\mathcal{L}$ is the luminosity density; in the B_T system, $\mathcal{L} \simeq 2.4h \times 10^8 L_{B\odot}\,{\rm Mpc}^{-3}$. In these units, M/L for $\rho = \rho_C$ is $(M/L)_C \simeq 1200hM_\odot/L_\odot$.

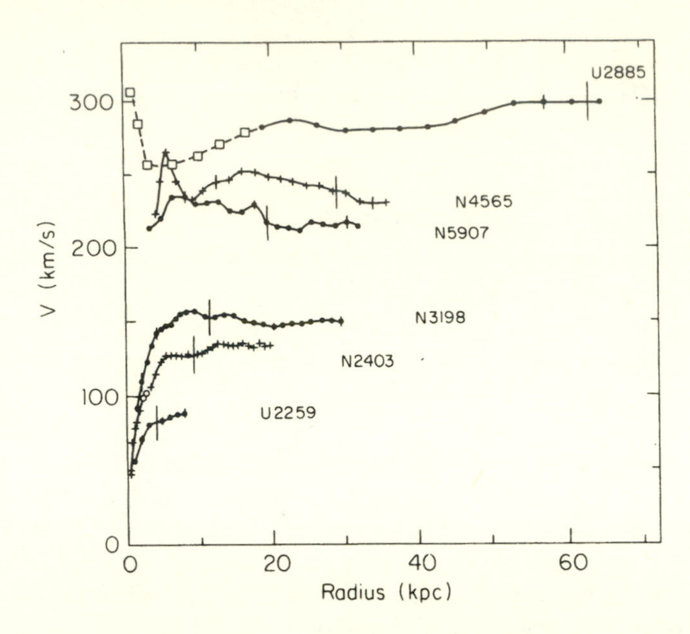

Fig. 1.9: Rotation curves determined from 21 cm observations. Vertical bars indicate the point where the optical light is less than 25 (blue) mag (arc second)$^{-2}$. For reference, this corresponds to about 6% of the surface brightness of the night sky, and less than about 1% of the brightness of the central region of a typical spiral (from [21]).

diffuse "halo" associated with them which contributes at least 3 to 10 times the mass of the "visible matter" (stars and the like). Based upon this we can conclude:

$$\Omega_{\rm HALO} \gtrsim 0.1 \simeq 10\ \Omega_{\rm LUM}. \tag{1.12}$$

This is very strong evidence that dark matter is the dominant component of the mass density of the Universe. Note that a comparison of the lower limit to the baryon density based upon primordial nucleosynthesis, $\Omega_B \geq 0.015$, and $\Omega_{\rm LUM}$ already suggests that there is dark baryonic matter. This is not a great surprise, as there are a variety of forms that baryons can take that are not "luminous," e.g., jupiters, white dwarfs, neutron stars, black holes, etc.

The average mass per galaxy in a cluster can also be determined by dynamical means. Assuming that the cluster in question is a gravitationally-bound and well-relaxed system, the virial theorem applies and

$$GM = \frac{2\langle v^2\rangle}{\langle r^{-1}\rangle} \tag{1.13}$$

where M is the cluster mass, $\langle v^2 \rangle^{1/2}$ is the *rms* velocity of a galaxy, and $\langle r^{-1} \rangle$ is the mean inverse separation between galaxies. The estimates of Ω_0 based upon this technique yield values of the order 0.1 to 0.3, also indicating the presence of substantial amounts of dark matter.[13] There are uncertainties however: Are clusters well-relaxed (i.e., virialized), spherically symmetric objects? Are projection effects important (only projected velocities and positions are measured)? Are some of the galaxies misidentified interlopers—thereby raising $\langle v^2 \rangle$—rather than cluster members? More importantly, only about 10% of galaxies are in clusters, and one may question if the value of $\langle M_{\rm GAL} \rangle$ inferred for cluster galaxies is typical of all galaxies.

The amount of matter in the Universe can be measured by a number of other dynamical methods. For example, the Virgo cluster represents a nearby (about 20 Mpc away) enhancement in the density of galaxies ($\delta n_{\rm GAL}/n_{\rm GAL} \sim$ few) and hence of the mass density, whose presence distorts the local Hubble flow. By modeling the local distortion of the Hubble flow around Virgo, one can determine $\langle M_{\rm GAL} \rangle$ for Virgo and thereby Ω_0. The "Virgo infall" method gives $\Omega_0 = 0.1$ to 0.2. Virgo infall samples Ω_0 on a scale of about 20 Mpc. It is possible to use the infall method on a larger scale, by relating the velocity of the Local Group with respect to the CMBR to the velocity expected from the local cosmic gravitational field that arises due to the inhomogeneous distribution of galaxies. The local gravitational acceleration can be determined from the distribution of matter out to some appropriate distance (see Chapter 9). Using the matter distribution deduced from the IRAS survey, a value of Ω_0 greater than 0.2 and perhaps as large as unity has been inferred [22]. Finally, on even larger scales one can use the cosmic virial and energy theorems to relate the kinetic energies of galaxies relative to the Hubble flow to the gravitational potential energies determined by the mass density. Based upon this technique, values for Ω_0 approaching unity have also been found [23].

With these dynamical determinations of Ω_0, there is a very important caveat that should be kept in mind: All of the aforementioned determinations are only sensitive to material that clusters with bright galaxies. Galaxies and clusters represent large local enhancements in the density of the Universe, $\delta\rho/\rho \simeq 10^5$ (galaxies), $\simeq 10^2$ to 10^3 (clusters), and there-

[13]We mention that many clusters show the presence of baryonic matter that is dark at optical wavelengths, but luminous at x-ray wavelengths, and accounts for an amount of matter comparable to the optically-bright matter, illustrating the fact that "dark" is a relative term.

fore any locally-smooth distribution of matter (i.e., $\delta\rho/\rho \lesssim 1$) makes only a negligible contribution to the mass of these systems whose dynamics is dominated by the large local overdensity (and not the average cosmic density). On the other hand, the kinematical techniques discussed earlier (Hubble diagram, galaxy number count test, angle-red shift test, etc.) measure the average cosmic density (averaged over 1000's of Mpc). While the kinematical techniques have thus far proven inconclusive, the dynamical methods strongly indicate that the material which clusters with bright galaxies on scales less than about 10 to 30 Mpc contributes

$$\Omega_{\lesssim 10-30} \simeq 0.2 \pm 0.1, \tag{1.14}$$

where the $\pm$ is not meant to be a formal error estimate, but rather indicates the spread of values obtained using different techniques (somehow weighted by their reliabilities). The implications for proponents of a flat Universe (including both the authors) are both obvious and very significant. If $\Omega_0 = 1$, and the aforementioned dynamical measurements are correct, then there must be a significant "less-clustered" (or unclustered) component to the energy density of the Universe, contributing

$$\Omega_{\rm SMOOTH} \simeq 0.8 \pm 0.1, \tag{1.15}$$

which is more smoothly distributed on scales less than 10 to 30 Mpc. We will address this important issue, often referred to as "the Ω problem," in more detail in Chapter 9. For purposes of illustration we mention here only three of the possibilities for a smooth component of the matter density: (i) high-velocity particles, such as light ($90h^2$ eV), relic neutrinos, or a sea of undiscovered relativistic particles, which by virtue of their great speeds would not become bound to systems as small as 10 to 30 Mpc; (ii) a relic cosmological term (or vacuum energy) which by definition is spatially constant; (iii) a yet undiscovered (or unidentified) population of very dim galaxies that are significantly less clustered than bright galaxies.

Summarizing our knowledge of Ω_0 based upon dynamical methods: (i) luminous matter contributes only a small fraction of the critical density (less than 0.01); (ii) dark matter dominates the contribution of luminous matter by at least a factor of 10; (iii) the amount of matter that clusters with galaxies on scales of 10 to 30 Mpc contributes 0.2 ± 0.1 of critical density; (iv) dynamical measurements do not preclude a less clustered, or even smooth, component that contributes as much as 0.8 ± 0.1. Taking this information together with our knowledge of Ω_B based upon primordial

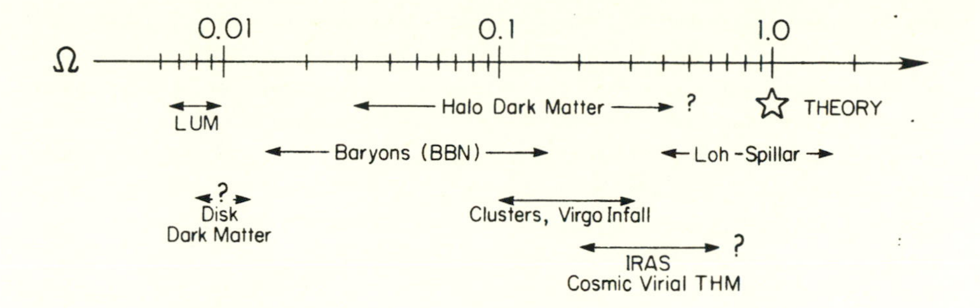

Fig. 1.10: Summary of determinations of Ω_0.

nucleosynthesis ($0.015 \lesssim \Omega_B \lesssim 0.16$), we can infer that some of the dark matter must be baryonic, and if $\Omega_0 < 0.16$, all of it could be baryonic (e.g., in the form of black holes, neutron stars, jupiters, etc.). On the other hand, if $\Omega_0 > 0.16$, then there is strong evidence from primordial nucleosynthesis that the dark component is non-baryonic, e.g., relic, stable Weakly-Interacting Massive Particles (or WIMPs) left over from the earliest moments of the Universe. We will touch upon the subject of WIMPs several times in later Chapters. Fig. 1.10 summarizes the dynamical determinations of Ω_0.

1.8 The Large-Scale Structure of the Universe

To this point we have described the Universe as a fluid of nearly constant density. Such a description for the early Universe, comprised of a soup of elementary particles with short mean free paths, or for the Universe today when viewed on large scales (greater than 100 Mpc), is both well-motivated and quite a good approximation. However, on smaller scales such a description glosses over some of the most salient and conspicuous features of the Universe today—the existence of structures including planets, stars, galaxies, clusters of galaxies, superclusters, voids, etc. The existence of such structures is an important feature of the Universe, and is likely to provide a key to understanding the evolution of the Universe. While an understanding of the origin and evolution of stars and planets is outside the realm of cosmology, the origin and evolution of galaxies and larger structures is definitely not.

Ideally, to develop an understanding of the structure of the Universe one would like to know the distribution of both matter and light in a representative volume of cosmological dimensions, say of order $1000h^{-1}$ Mpc on a side.[14] However, all we can see are galaxies, and at that, only bright ones,[15] like our own. As emphasized previously, it is not a priori true that light faithfully traces the mass distribution.

An ambitious, but more realistic, goal would be to survey a suitably large volume of the Universe, obtaining sky positions, velocities, and distances for a few million galaxies. The largest galaxy catalogue, the APM Galaxy Survey, consists of some 5 million galaxies and has an effective depth of $600h^{-1}$ Mpc [16].[16] The simplest means of obtaining the distance to a galaxy is by determining its red shift: $d_L \simeq H_0^{-1}z$, and of course this necessitates obtaining spectral information. Only a total of 28,000 galaxy red shifts are known, and the largest systematic survey, the CfA slices of the Universe survey [25], contains about 9,000 red shifts. The hang-up is obtaining red shifts: Using traditional techniques, a single red shift determination requires about a half hour of telescope time, and a typical telescope has only about 3,600 useful hours of observing time per year (and more than 10,000 hours of requests for that time!). The present situation then is far from ideal; however it promises to improve dramatically as larger red shift surveys employing automated and multiple object spectrograph techniques are completed in the next decade. In fact, a group of astronomers at Chicago and Princeton have begun a decade-long project to obtain red shifts for a million galaxies—a survey of the northern sky out to a red shift of about 0.1.

We began with the above lengthy preface to place the present observational situation into proper perspective. Stated bluntly, we are just beginning to develop a picture of the distribution of galaxies in the Universe. What then do we know about the distribution of bright galaxies in the Universe and the nature of the large-scale structure? Bright galaxies

---

[14]For reference, using the linear approximation, $d_L = H_0^{-1}z$, it follows that the length ΔL corresponds to the red shift interval $\Delta z \simeq 0.03(\Delta L/100h^{-1}\text{Mpc})$.

[15]Bright galaxies have surface brightnesses which are about a factor of 10 above the surface brightness of the night sky, i.e., the integrated light of all stars and galaxies. By suitable integration techniques galaxies with surface brightnesses of only a few percent of the night sky can be detected. If bright galaxies were a factor of 3 larger in linear size they would fade into the night sky. This fact suggests that there could be substantial numbers of low surface brightness galaxies which have escaped detection.

[16]The Lick (Shane-Wirtanen) Catalogue contains some 1.6 million galaxies and has an effective depth of about $200h^{-1}$ Mpc [24]. This survey is not really a catalogue in that sky positions are not given for individual galaxies, rather just the number of galaxies per 10 arc minute bin on the sky. The Zwicky catalogue of galaxies and clusters contains some 31,000 galaxies.

to a first approximation are distributed uniformly on the sky (see Fig. 1.7), and presumably within our Hubble volume. However, they do show a tendency to cluster, quantified by the galaxy–galaxy correlation function, ξ_{GG}, which measures the probability in excess of random of finding a galaxy at a distance r from another galaxy:

$$\xi_{GG} \simeq (r/5h^{-1}\mathrm{Mpc})^{-1.8}, \tag{1.16}$$

valid for galaxy separations, $0.1h^{-1}\mathrm{Mpc} \lesssim r \lesssim 20h^{-1}\mathrm{Mpc}$.

Many galaxies are found in binary systems or small groups of galaxies. About 10% of galaxies are found in galaxy clusters, bound and often well-virialized groups containing anywhere from tens to thousands of galaxies. The Virgo cluster and Coma clusters are familar nearby clusters. The best-known catalogue of rich clusters is the Abell catalogue, where clusters are categorized by their Abell richness class (classes that roughly correspond to the number of galaxies within the cluster). Abell's combined northern and southern catalogues contain some 4,076 clusters.[17] Not to diminish the importance of this catalogue, we mention that Abell himself warned against the use of his catalogue for statistical purposes. With this in mind, we mention that the clustering properties of Abell clusters have also been determined, quantified by the cluster–cluster correlation function [28],

$$\xi_{CC} \simeq (r/25h^{-1}\mathrm{Mpc})^{-1.8} \tag{1.17}$$

The cluster–cluster correlation function has approximately the same power-law slope as the galaxy–galaxy correlation function (to within the uncertainties), but has a much larger correlation length, $25h^{-1}$ Mpc compared to $5h^{-1}$ Mpc for galaxies. The fact that clusters seem to cluster more strongly is a surprising result—if light faithfully traced mass this would not be true. This suggests that light may be a "biased" tracer of mass; we will return to this issue in Chapter 9. When considering the quantitative difference between the two correlation functions, one should keep in mind

[17]Abell compiled his catalogue in the 1950's using the Palomar all-sky survey plates. He identified clusters and cluster membership by subjective criteria: Clusters were defined visually by an enhancement in the local density of galaxies. Since the exposure time of the plates varied significantly across the sky, and the density of field stars also varied, selection effects (depth of the survey, and the ability to pick out enhancements in the galaxy density) are significant. The Zwicky catalogue of clusters which contains some 4000 clusters suffers from similar shortcomings. A recent analysis has attempted to quantify and correct for selection effects inherent in the Abell catalogue [27].

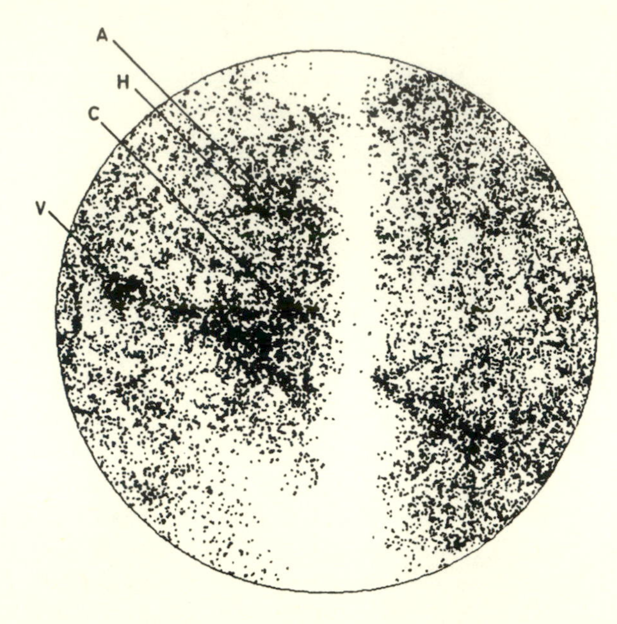

Fig. 1.11: Distribution of UGC (Uppsala General Catalogue) and ESO (European Southern Observatory) galaxies on the sky. The white band is caused by obscuration due to the disk of the Milky Way. The Antlia, Centaurus, Hydra, and Virgo clusters are indicated. The linear feature across the diagonal is referred to as the supergalactic plane (from [26]).

the shortcomings of the Abell catalogue of clusters. Fig. 1.11 shows the distribution of nearby galaxies on the sky, and several nearby clusters can be easily identified.

Even larger structures seem to exist—superclusters, loosely-bound, non-virialized objects with densities about twice the average density of the Universe, containing several to many rich clusters. Nearby superclusters include our own Local Supercluster (centered on the Virgo cluster), Hydra-Centaurus, and Pisces-Cetus. Of order 20 or so such structures have been identified, and attempts have even been made to quantify their clustering properties.

Several surveys have found evidence for the existence of voids in the distribution of bright galaxies. For example, the KOSS survey [29] showed the existence of a void in Boötes of diameter about $50h^{-1}$ Mpc, and the CfA slices of the Universe seem to indicate that voids of size about $20h^{-1}$ Mpc in diameter are quite common. The authors of the CfA survey have even speculated that the overall distribution of galaxies may be "bubbly," with

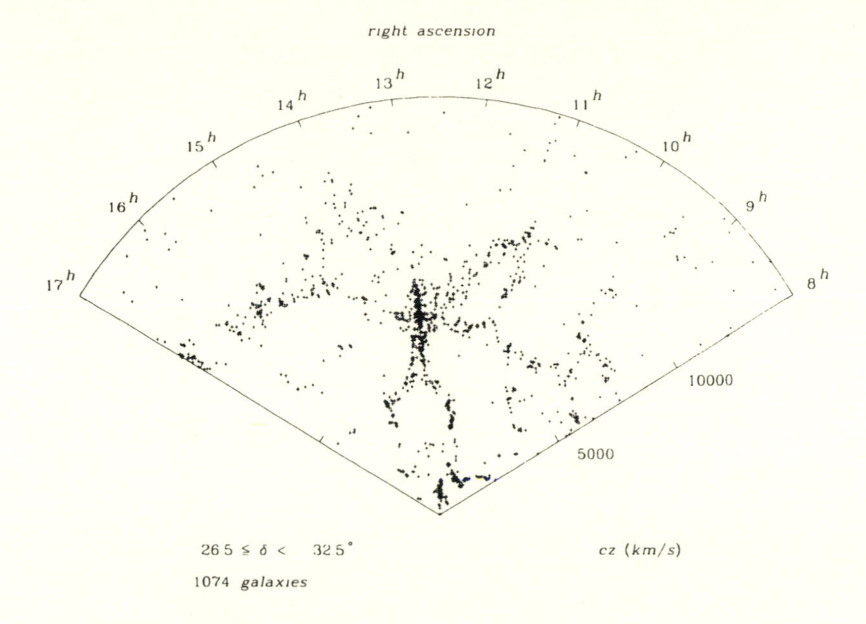

Fig. 1.12: One of the CfA slices of the Universe [25].

galaxies concentrated on sheet-like structures surrounding nearly empty voids (see Fig. 1.12).

Finally, we mention a very interesting probe of the Universe at high red shifts: QSO Absorption Line Systems (QALS's). The spectra of many high-red shift QSO's exhibit series of absorption lines with red shifts less than that of the quasar itself. These absorption lines are thought to be due to absorption of the light from QSO's by intervening objects, some of which are likely to be galactic (or even pre-galactic) disks and/or halos. These absorption line systems are classified by the nature of the absorption lines: Lyman-α systems, damped Lyman-α systems, Lyman-limit systems, and metal systems. The study of QALS's (internal densities, number densities, masses, clustering properties, etc.) will likely prove to be a very important probe of the Universe at large red shift.

It is clear that an understanding of the present distribution of matter in the Universe is crucial to understanding the origin of structure in the Universe and to testing the detailed scenarios of structure formation that have been developed in recent years. In turn, this will also test the theories of the very early Universe that give rise to these scenarios of structure formation. At present, it is probably fair to say that we have an incomplete, but rapidly-emerging, view of the large-scale structure of the Universe. Hopefully, larger red shift surveys will answer pressing questions like: What is the topology of the galaxy distribution and is it "bubbly?" Do clusters cluster significantly more strongly than galaxies? What are the largest coherent structures that exist in the Universe today? How is the nature of the large-scale structure to be quantified? Are the mass and light distributions similar? When did galaxies and clusters of galaxies form? How has the evolution of galaxy clustering proceeded?

1.9 References

1. E. W. Kolb and M. S. Turner, *The Early Universe: Reprints* (Addison-Wesley, Redwood City, Calif. 1988).

2. C. Quigg, *Gauge Theories of the Strong, Weak, and Electromagnetic Interactions* (Addison-Wesley, Redwood City, Calif. 1983); M. B. Green, J. H. Schwarz, and E. Witten, *Superstring Theory, Vols. I and II* (Cambridge Univ. Press, Cambridge, 1987); G. G. Ross, *Grand Unified Theories* (Addison-Wesley, Redwood City, Calif. 1984); J. C. Taylor, *Gauge Theories of Weak Interactions* (Cambridge Univ. Press, Cambridge, 1976); I. J. R. Aitchison and A. J. G. Hey, *Gauge Theories in Particle Physics* (Adam Hilger Ltd., Bristol, 1982).

3. D. P. Schneider, M. Schmidt, and J. E. Gunn, *Astron. J.*, in press (1989). Second place goes to a QSO with a red shift of 4.43; see S. J. Warren, P. C. Hewett, P. S. Osmer, and M. J. Irwin, *Nature* **330**, 453 (1987).

4. K. Chambers, G. Miley, and W. van Breugel, *Ap. J.*, in press (1988).

5. L. L. Cowie and S. J. Lilly, *Ap. J.* **336**, L41 (1989).

6. J. Hoessel, B. Oke, and J. Gunn, *Ap. J.*, in press (1989).

7. A. Sandage, *Physics Today*, Feb. 1970, p. 34.

8. M. Rowan-Robinson, *The Cosmological Distance Ladder* (Freeman, San Francisco, 1985).

9. J. Kristian, A. Sandage, and J. Westphal, *Ap. J.* **221**, 383 (1978).

10. H. Spinrad and S. Djorgovski, in *Observational Cosmology* (IAU Symposium 124), eds. A. Hewitt, G. Burbidge, and Li-Zhi Fang (Reidel, Dordrecht, 1987), p. 129.

11. G. Bruzual and H. Spinrad, *Ap. J.* **220**, 1 (1978).

12. E. Loh and E. Spillar, *Ap. J.* **307**, L1 (1986); however see S. Bahcall and S. Tremaine, *Ap. J.* **326**, L1 (1988) for a discussion of the shortcomings of this determination.

13. D. T. Wilkinson, in *Inner Space/Outer Space*, eds. E. W. Kolb, M. S. Turner, D. Lindley, K. Olive, and D. Seckel (Univ. Chicago Press, Chicago, 1986), p. 126; D. T. Wilkinson, in *13th Texas Symposium on Relativistic Astrophysics*, ed. M. P. Ulmer (World Scientific, Singapore, 1987), p. 209; R. B. Partridge, *Rep. Prog. Phys.* **51**, 647 (1988). Not included in Fig. 1.6 are the recent measurements of R. D. Davis, et al., *Nature* **326**, 462 (1987) and A. C. S. Readhead, et al., *Ap. J.*, in press (1989).

14. L. P. Grischuk and Ya. B. Zel'dovich, *Sov. Astron.* **22**, 125 (1978).

15. E. Boldt, *Phys. Rep.* **146**, 215 (1987).

16. S. J. Maddox, W. J. Sutherland, G. Efstathiou, and J. Loveday, *Mon. Not. Roy. Astron. Soc.*, in press (1989).

17. *Galaxy Distances and Deviations from Universal Expansion*, eds. B. F. Madore and R. B. Tully (Reidel, Dordrecht, 1986); A. Dressler, et al., *Ap. J.* **313**, L37 (1987); C. A. Collins, et al., *Nature* **320**, 506 (1986); M. Aaronson, et al., *Ap. J.* **302**, 536 (1986).

18. D. E. Winget, et al., *Ap. J.* **315**, L77 (1987).

19. T. Matsumoto, et al., *Ap. J.* **329**, 567 (1988).

20. J. B. Peterson, et al., in *Inner Space/Outer Space*, eds. E. W. Kolb, M. S. Turner, D. Lindley, K. Olive, and D. Seckel (Univ. Chicago Press, Chicago, 1986), p. 119; D. T. Wilkinson, in *13th Texas Symposium on Relativistic Astrophysics*, ed. M. P. Ulmer (World Scientific, Singapore, 1987), p. 209. Not included in Fig. 1.8 are the recent measurements of M. Bersanelli, et al., *Ap. J.* **339**, 632 (1989).

21. R. Sancisi and T. S. van Albada, in *Dark Matter in the Universe*, eds. J. Kormendy and G. Knapp (Reidel, Dordrecht, 1987), p. 67.

22. M. A. Strauss and M. Davis, in *Proceedings of the Vatican Study Week on Large-Scale Motions in the Universe*, eds. G. Coyne and V. Rubin (Pontifical Scientific Academy, Vatican City, 1988); A. Yahil, *ibid.*

23. M. Davis and P. J. E. Peebles, *Ap. J.* **267**, 465 (1983); M. Davis, M. J. Geller, and J. Huchra, *Ap. J.* **221**, 1 (1978).

24. M. Seldner, B. Siebers, E. Groth, and P. J. E. Peebles, *Astron. J.* **82**, 249 (1977).

25. V. de Lapparent, M. J. Geller, and J. Huchra, *Ap. J.* **302**, L1 (1986); *ibid*, **332**, 44 (1988); M. J. Geller, J. Huchra, and V. de Lapparent, in *Observational Cosmology* (IAU Symposium 124), eds. A. Hewitt, G. Burbidge, and Li-Zhi Fang (Reidel, Dordrecht, 1987), p. 301.

26. D. Lynden-Bell, *Q. Jl. R. astr. Soc.* **28**, 187 (1987); *ibid.* **27**, 319 (1986).

27. A. Dekel, G. R. Blumenthal, J. R. Primack, and S. Oliver, *Ap. J. (Letter)*, in press, (1989).

28. For a recent review of the clustering of rich clusters, see, N. A. Bahcall, *Ann. Rev. Astron. Astrophys.* **26**, 631 (1988).

29. R. P. Kirshner, A. Oemler Jr., P. Schechter, and S. Shectman, *Ap. J.* **248**, L57 (1981).

2

ROBERTSON-WALKER METRIC

2.1 Open, Closed, and Flat Spatial Models

As discussed in the last Chapter, the distribution of matter and radiation in the observable Universe is homogeneous and isotropic. While this by no means guarantees that the *entire* Universe is smooth, it does imply that a region at least as large as our present Hubble volume is smooth. (We will return to the smoothness of the Universe on scales larger than the Hubble volume when we discuss the topic of inflation in Chapter 8.) So long as the Universe is spatially homogeneous and isotropic on scales as large as the Hubble volume, for purposes of description of our local Hubble volume we may assume the entire Universe is homogeneous and isotropic. While a homogeneous and isotropic region within an otherwise inhomogeneous and anisotropic Universe will not remain so forever, causality implies that such a region will remain smooth for a time comparable to its light-crossing time. This time corresponds to the Hubble time, about 10 Gyr.

The metric for a space with homogeneous and isotropic spatial sections is the maximally-symmetric Robertson-Walker (RW) metric, which can be written in the form[1]

$$ds^2 = dt^2 - R^2(t)\left\{\frac{dr^2}{1-kr^2} + r^2 d\theta^2 + r^2 \sin^2\theta d\phi^2\right\} \qquad (2.1)$$

---

[1]The sign conventions are the same as Landau and Lifshitz [4]: $ds^2 = dt^2 - \vec{\mathbf{dl}}^2$; $G_{\mu\nu} = +8\pi G T_{\mu\nu}$; and $R^\mu{}_{\nu\alpha\beta} = \partial_\alpha \Gamma^\mu{}_{\nu\beta} - \partial_\beta \Gamma^\mu{}_{\nu\alpha} + \Gamma^\mu{}_{\sigma\alpha}\Gamma^\sigma{}_{\nu\beta} - \Gamma^\mu{}_{\sigma\beta}\Gamma^\sigma{}_{\nu\alpha}$, and $\Gamma^\mu{}_{\alpha\beta} = (1/2)g^{\mu\sigma}(\partial_\beta g_{\sigma\alpha} + \partial_\alpha g_{\sigma\beta} - \partial_\sigma g_{\alpha\beta})$. Greek indices run from 0 to 3, while Latin indices run over the spatial indices 1 to 3. We assume that the reader is familiar with the rudiments of general relativity; for those interested in textbook treatments of general relativity we suggest [1–4].

where $(t,\ r,\ \theta,\ \phi)$ are coordinates (referred to as comoving coordinates), $R(t)$ is the cosmic scale factor, and with an appropriate rescaling of the coordinates, k can be chosen to be $+1$, -1, or 0 for spaces of constant positive, negative, or zero spatial curvature, respectively.[2] The coordinate r in (2.1) is dimensionless, i.e., $R(t)$ has dimensions of length, and r ranges from 0 to 1 for $k = +1$. Notice that for $k = +1$ the circumference of a one sphere of coordinate radius r in the $\phi =$ const plane is just $2\pi R(t)r$, and that the area of a two sphere of coordinate radius r is just $4\pi R^2(t)r^2$; however, the physical radius of such one and two spheres is $R(t)\int_0^r dr/(1-kr^2)^{1/2}$, and not $R(t)r$.

The time coordinate in (2.1) is just the proper (or clock) time measured by an observer at rest in the comoving frame, i.e., $(r,\ \theta,\ \phi)$=const. As we shall discover shortly, the term *comoving* is well chosen: Observers at rest in the comoving frame remain at rest, i.e., $(r,\ \theta,\ \phi)$ remain unchanged, and observers initially moving with respect to this frame will eventually come to rest in it. Thus, if one introduces a homogeneous, isotropic fluid initially at rest in this frame, the t =const hypersurfaces will always be orthogonal to the fluid flow, and will always coincide with the hypersurfaces of both spatial homogeneity and constant fluid density.

The spatial part of the metric will be denoted as

$$d\vec{\mathbf{l}}^{\,2} = h_{ij}dx^i dx^j \tag{2.2}$$

where $i,\ j = 1,\ 2,\ 3$. Note that $h_{ij} = -g_{ij}$. The three-dimensional Riemann tensor, Ricci tensor, and Ricci scalar will be denoted as ${}^3R_{ijkl}$, ${}^3R_{ij}$, and ${}^3\mathcal{R}$. Because the spatial metric is maximally symmetric, ${}^3R_{ijkl}$ and ${}^3R_{ij}$ can be expressed in terms of the spatial metric and $R(t)$,

$$\begin{aligned} {}^3R_{ijkl} &= \frac{k}{R^2(t)}(h_{ik}h_{jl} - h_{il}h_{kj}) \\ {}^3R_{ij} &= \frac{2k}{R^2(t)}h_{ij} \\ {}^3\mathcal{R} &= \frac{6k}{R^2(t)}. \end{aligned} \tag{2.3}$$

Because of the symmetry of the RW metric, calculation of the affine

[2]The scale factor $R(t)$ should not be confused with the Ricci scalar, $\mathcal{R}$.

connection is particularly simple. The non-zero components are:

$$\begin{aligned}
\Gamma^i{}_{jk} &= \frac{1}{2}h^{il}\left(\frac{\partial h_{lj}}{\partial x^k} + \frac{\partial h_{lk}}{\partial x^j} - \frac{\partial h_{jk}}{\partial x^l}\right) \\
\Gamma^0{}_{ij} &= \frac{\dot{R}}{R}h_{ij} \\
\Gamma^i{}_{0j} &= \frac{\dot{R}}{R}\delta^i{}_j.
\end{aligned} \tag{2.4}$$

For completeness, we also mention the non-zero components of the Ricci tensor:

$$\begin{aligned}
R_{00} &= -3\frac{\ddot{R}}{R} \\
R_{ij} &= -\left[\frac{\ddot{R}}{R} + 2\frac{\dot{R}^2}{R^2} + \frac{2k}{R^2}\right] g_{ij},
\end{aligned} \tag{2.5}$$

and the Ricci scalar:

$$\mathcal{R} = -6\left[\frac{\ddot{R}}{R} + \frac{\dot{R}^2}{R^2} + \frac{k}{R^2}\right]. \tag{2.6}$$

In order to illustrate the construction of the metric, consider the simpler case of two spatial dimensions. Examples of two-dimensional spaces that are homogeneous and isotropic are: (i) the flat "$x - y$" plane (R^2); (ii) the positively-curved two sphere (S^2); and (iii) the negatively-curved hyperbolic plane (H^2).

First consider the two sphere. To visualize the two sphere it is convenient to introduce an extra fictitious spatial dimension and to embed this two-dimensional curved space in a three-dimensional Euclidean space with cartesian coordinates x_1, x_2, x_3. The equation of the two sphere of radius R is

$$x_1^2 + x_2^2 + x_3^2 = R^2. \tag{2.7}$$

The element of length in the three-dimensional Euclidean space is

$$\vec{\mathrm{dl}}^{\,2} = dx_1^2 + dx_2^2 + dx_3^2. \tag{2.8}$$

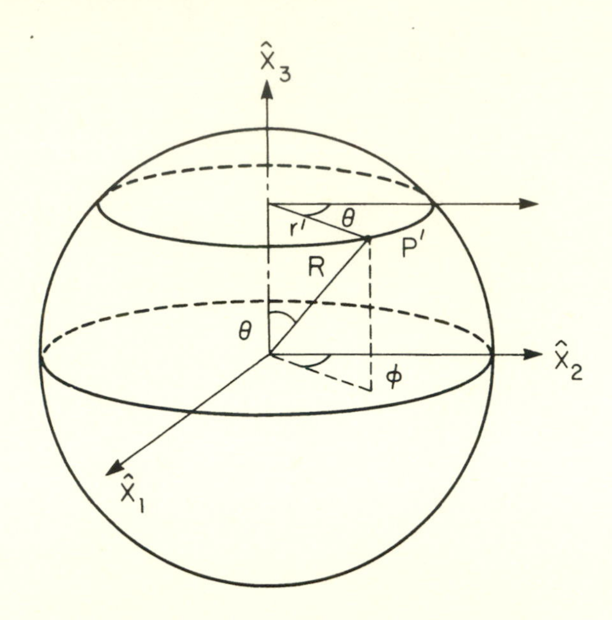

Fig. 2.1: A point on the two sphere specified by coordinates (r', θ) or (θ, ϕ).

If x_3 is taken as the fictitious third spatial coordinate, it can be eliminated from $\vec{\mathbf{dl}}^{\,2}$ by use of (2.7)

$$\vec{\mathbf{dl}}^{\,2} = dx_1^2 + dx_2^2 + \frac{(x_1 dx_1 + x_2 dx_2)^2}{R^2 - x_1^2 - x_2^2}. \tag{2.9}$$

Now introduce the coordinates r' and θ defined in terms of x_1 and x_2 by

$$x_1 = r' \cos\theta, \qquad x_2 = r' \sin\theta. \tag{2.10}$$

Physically, r' and θ correspond to polar coordinates in the x_3-plane ($x_3^2 = R^2 - r'^2$). In terms of the new coordinates, (2.9) becomes

$$\vec{\mathbf{dl}}^{\,2} = \frac{R^2 dr'^2}{R^2 - r'^2} + r'^2 d\theta^2. \tag{2.11}$$

Finally, with the definition of a dimensionless coordinate $r \equiv r'/R$ ($0 \leq$

$r \leq 1$), the spatial metric becomes

$$\vec{\mathrm{dl}}^{\,2} = R^2 \left\{ \frac{dr^2}{1-r^2} + r^2 d\theta^2 \right\}. \tag{2.12}$$

Note the similarity between this metric and the $k = 1$ RW metric. It should also now be clear that $R(t)$ is the radius of the space.

The two sphere is shown in Fig. 2.1. The poles are at $r = 0$, the equator is at $r = 1$. The locus of points of constant r sweep out the latitudes of the sphere, while the locus of points of constant θ sweep out the longitudes of the sphere.

Another convenient coordinate system for the two sphere is that specified by the usual polar and azimuthal angles $(\theta,\ \phi)$ of spherical coordinates, related to the x_i by

$$x_1 = R\sin\theta\cos\phi, \qquad x_2 = R\sin\theta\sin\phi, \qquad x_3 = R\cos\theta. \tag{2.13}$$

In terms of these coordinates, (2.8) becomes

$$\vec{\mathrm{dl}}^{\,2} = R^2 \left[d\theta^2 + \sin^2\theta d\phi^2 \right]. \tag{2.14}$$

This form makes manifest the fact that the space is the two sphere of radius R. The volume of the two sphere is easily calculated:

$$V = \int d^2x\sqrt{h} = \int_0^{2\pi} d\phi \int_0^{\pi} d\theta R^2 \sin\theta = 4\pi R^2 \tag{2.15}$$

as expected (h is the determinant of the spatial metric).

The two sphere is homogeneous and isotropic. Every point in the space is equivalent to every other point, and there is no preferred direction. In other words, the space embodies the *Cosmological Principle*, i.e., no observer (especially us) occupies a preferred position in the Universe. Note that this space is unbounded; there are no edges on the two sphere. It is possible to circumnavigate the two sphere, but it is impossible to fall off. Although the space is unbounded, the volume is finite.

The expansion (or contraction) of this 2-dimensional Universe is equivalent to an increase (or decrease) in the radius of the two sphere, R. Since the Universe is *spatially* homogeneous and isotropic, the scale factor (radius R) can only be a function of time. As the two sphere expands or contracts, the coordinates (r and θ in the case of the two sphere) remain unchanged; they are "comoving." Also note that the physical distance between any two comoving points in the space scales with R (hence the name scale factor).

The equivalent formulas for a space of constant negative curvature can be obtained with the replacement $R \to iR$ in (2.7). In this case the embedding is in a three-dimensional Minkowski space. The metric corresponding to the form of (2.12) for the negative curvature case is

$$\vec{\mathrm{dl}}^{\,2} = R^2 \left\{ \frac{dr^2}{1+r^2} + r^2 d\theta^2 \right\}, \tag{2.16}$$

and the metric in the form corresponding to (2.14) is

$$\vec{\mathrm{dl}}^{\,2} = R^2 \left[d\theta^2 + \sinh^2 \theta d\phi^2 \right]. \tag{2.17}$$

The hyperbolic plane is unbounded with infinite volume since $0 \leq \theta \leq \infty$. The embedding of H^2 in a Euclidean space requires three fictitious extra dimensions, and such an embedding is of little use in visualizing the geometry.[3]

The spatially-flat model can be obtained from either of the above examples by taking the radius R to infinity. The flat model is unbounded with infinite volume. For the flat model the scale factor does not represent any physical radius as in the closed case, or an imaginary radius as in the open case, but merely represents how the physical distance between comoving points scales as the space expands or contracts.

The generalization of the two-dimensional models discussed above to three spatial dimensions is trivial. For the three sphere a fictitious fourth spatial dimension is introduced, and in cartesian coordinates the three sphere is defined by: $R^2 = x_1^2 + x_2^2 + x_3^2 + x_4^2$. The spatial metric is $\vec{\mathrm{dl}}^{\,2} = dx_1^2 + dx_2^2 + dx_3^2 + dx_4^2$. The fictitious coordinate can be removed to give

$$\vec{\mathrm{dl}}^{\,2} = dx_1^2 + dx_2^2 + dx_3^2 + \frac{(x_1 dx_1 + x_2 dx_2 + x_3 dx_3)^2}{R^2 - x_1^2 - x_2^2 - x_3^2}. \tag{2.18}$$

[3]While the hyperbolic plane cannot be globally embedded in R^3, it can be partially represented by the psuedosphere, a 2-dimensional space of constant negative curvature which has a cusp. The pseudosphere is the surface of revolution of the tractrix, $z = \ln[\tan(\phi/2 + \pi/4)] - \sin\phi$, $x = \cos\phi$ $(-\pi/2 \leq \phi \leq \pi/2)$, about the z-axis [6].

In terms of coordinates $x_1 = r' \sin\theta \cos\phi$, $x_2 = r' \sin\theta \sin\phi$, $x_3 = r' \cos\theta$, the metric is given by (2.1) with $k = +1$ and $r = r'/R$, cf. (2.12). In the coordinate system that employs the 3 angular coordinates $(\chi,\ \theta,\ \phi)$ of a four-dimensional spherical coordinate system, $x_1 = R \sin\chi \sin\theta \cos\phi$, $x_2 = R \sin\chi \sin\theta \sin\phi$, $x_3 = R \sin\chi \cos\theta$, $x_4 = R \cos\chi$, the metric is given by

$$\vec{\mathrm{dl}}^{\,2} = R^2 \left[d\chi^2 + \sin^2\chi (d\theta^2 + \sin^2\theta d\phi^2) \right] . \tag{2.19}$$

The volume of the three sphere is $V = \int d^3x \sqrt{h} = 2\pi^2 R^3$.

As in the two-dimensional example, the three-dimensional open model is obtained by the replacement $R \to iR$, which gives the metric in the form (2.1) with $k = -1$, cf. (2.16), or in the form (2.19) with $\sin\chi \to \sinh\chi$, cf. (2.17). Again the space is unbounded with infinite volume and $R(t)$ sets the curvature scale. Embedding H^3 in a Euclidean space requires four fictitious extra dimensions.

Finally, it is often convenient to express the metric in terms of "conformal time" η, defined by $d\eta = dt/R(t)$:

$$ds^2 = R^2(\eta) \left\{ d\eta^2 - \frac{dr^2}{1 - kr^2} - r^2 d\theta^2 - r^2 \sin^2\theta d\phi^2 \right\} . \tag{2.20}$$

It then directly follows that the flat RW model is conformal to Minkowski, i.e., the line element is equal to the Minkowski line element times a conformal factor. In fact, all the RW models are locally conformal to Minkowski space, by which we mean that for a suitable local choice of coordinates, the metric can be written as the Minkowski line element times a conformal factor.[4]

It should be noted that the assumption of local homogeneity and isotropy only implies that the spatial metric is *locally* S^3, H^3, or R^3, and the space can have different *global* properties. For instance, for the spatially flat case the global properties of the space might be that of the three torus, T^3, rather than R^3; this is accomplished by identifying the opposite sides of a fundamental spatial volume element. Such non-trivial topologies may be relevant in light of recent work on theories with extra dimensions, such

---

[4]The RW line element has vanishing Weyl curvature, which is both a necessary and sufficient condition for conformal flatness. The Weyl curvature tensor, which corresponds to the traceless part of the Riemann curvature tensor, is given by $C^{\alpha\beta}{}_{\gamma\delta} = R^{\alpha\beta}{}_{\gamma\delta} - 2\delta^{[\alpha}{}_{[\gamma} R^{\beta]}{}_{\delta]} + (1/3)\delta^{[\alpha}{}_{[\gamma} \delta^{\beta]}{}_{\delta]} \mathcal{R}$, where $[\cdots]$ indicates that the indicies are to be antisymmetrized.

as superstrings. In many such theories the internal space (of the extra dimensions) is compact, but with non-trivial topology, e.g., containing topological defects such as holes, handles, etc. If the internal space is not simply connected, it suggests that the external space may also be non-trivial, and the global properties of our 3-space might be much richer than the simple S^3, H^3, or R^3 topologies.

2.2 Particle Kinematics

A fundamental question in cosmology that one might ask is: What fraction of the Universe is in causal contact? More precisely, for a comoving observer with coordinates (r_0, θ_0, ϕ_0), for what values of (r, θ, ϕ) would a light signal emitted at $t = 0$ reach the observer at, or before, time t? This can be calculated directly in terms of the metric (2.1). A light signal satisfies the geodesic equation $ds^2 = 0$. Because of the homogeneity of space, without loss of generality we may choose $r_0 = 0$. Geodesics passing through $r_0 = 0$ are lines of constant θ and ϕ, just as great circles eminating from the poles of a two sphere are lines of constant θ (i.e., constant longitude), so $d\theta = d\phi = 0$. And, of course, the isotropy of space makes the choice of direction (θ_0, ϕ_0) irrelevant. Thus, a light signal emitted from coordinate position (r_H, θ_0, ϕ_0) at time $t = 0$ will reach $r_0 = 0$ in a time t determined by

$$\int_0^t \frac{dt'}{R(t')} = \int_0^{r_H} \frac{dr}{\sqrt{1-kr^2}}. \tag{2.21}$$

The proper distance to the horizon measured at time t,

$$d_H(t) = \int_0^{r_H} \sqrt{g_{rr}}dr, \tag{2.22}$$

is simply $R(t)$ times (2.21)

$$d_H(t) = R(t) \int_0^t \frac{dt'}{R(t')}. \tag{2.23}$$

If $d_H(t)$ is finite, then our past light cone is limited by a particle horizon, which is the boundary between the visible Universe and the part of the Universe from which light signals have not reached us. The behavior of $R(t)$ near the singularity will determine whether or not d_H is finite. In the next Chapter we will see that in the standard cosmology $d_H(t) \sim t$ is finite.

Using the conformal form for the metric, (2.20), the causal structure of the space time is clear since light propagates on 45° lines in the η-space plane, and the distance to the horizon is just

$$d_H(t) = R(t)\left[\eta(t) - \eta(t=0)\right]. \tag{2.24}$$

Next, consider the geodesic motion of a particle that is not necessarily massless. The four-velocity u^μ of a particle with respect to the comoving frame is referred to as its *peculiar* velocity. The equation of geodesic motion is

$$\frac{du^\mu}{d\lambda} + \Gamma^\mu{}_{\nu\alpha} u^\nu \frac{dx^\alpha}{d\lambda} = 0, \tag{2.25}$$

where $u^\mu \equiv dx^\mu/ds$, and λ is some affine parameter. The four velocity may be expressed in terms of the ordinary three velocity $v^i \equiv dx^i/dt$ by the familiar relation $u^\mu = (u^0,\ u^i) = (\gamma,\ \gamma v^i)$, where $|\vec{v}|^2 = h_{ij}v^i v^j$, and $\gamma \equiv (1 - |\vec{v}|^2)^{-1/2}$. If we choose the affine parameter to be the proper length, $ds^2 = g_{\mu\nu}dx^\mu dx^\nu$, the $\mu = 0$ component of the geodesic equation is

$$\frac{du^0}{ds} + \Gamma^0{}_{\nu\alpha} u^\nu u^\alpha = 0. \tag{2.26}$$

For the Robertson-Walker metric, the only non-vanishing component of $\Gamma^0{}_{\nu\alpha}$ is $\Gamma^0{}_{ij} = (\dot{R}/R)h_{ij}$, and using the fact that $h_{ij}u^i u^j = |\vec{u}|^2$, the geodesic equation becomes

$$\frac{du^0}{ds} + \frac{\dot{R}}{R}|\vec{u}|^2 = 0. \tag{2.27}$$

Since $(u^0)^2 - |\vec{u}|^2 = 1$, it follows that $u^0 du^0 = |\vec{u}|d|\vec{u}|$, and the geodesic equation can be written as

$$\frac{1}{u^0}\frac{d|\vec{u}|}{ds} + \frac{\dot{R}}{R}|\vec{u}| = 0. \tag{2:28}$$

Finally, since $u^0 \equiv dt/ds$, this equation reduces to $|\dot{\vec{u}}|/|\vec{u}| = -\dot{R}/R$, which implies that $|\vec{u}| \propto R^{-1}$. Recalling that the four-momentum is $p^\mu = mu^\mu$, we see that the magnitude of the three-momentum of a freely-propagating particle "red shifts" as R^{-1}.[5] Note that in (2.28) the factors of ds cancel.

[5] Since we are dealing with $|\vec{u}|$ or $|\vec{p}|$, we don't have to worry about the fact that the RW metric is not orthonormal. In a basis that is not orthonormal, neither p^i nor p_i corresponds to the "physical" momentum. What most physicists (as opposed to

That implies that the above discussion also applies to massless particles, where $ds = 0$ (formally by the choice of a different affine parameter).

In terms of the ordinary three-velocity $|\vec{\mathbf{v}}|$,

$$|\vec{\mathbf{u}}| = \frac{|\vec{\mathbf{v}}|}{\sqrt{1 - |\vec{\mathbf{v}}|^2}} \propto R^{-1}. \tag{2.29}$$

We again see why the comoving frame is the natural frame. Consider an observer initially ($t = t_1$) moving non-relativistically with respect to the comoving frame with physical three velocity of magnitude $|\vec{\mathbf{v}}_1|$. At a later time t_0, the magnitude of the observer's physical three velocity, $|\vec{\mathbf{v}}_0|$, will be

$$|\vec{\mathbf{v}}_0| = |\vec{\mathbf{v}}_1| \frac{R(t_1)}{R(t_0)}. \tag{2.30}$$

In an expanding Universe, the freely-falling observer is destined to come to rest in the comoving frame even if he has some initial velocity with respect to it.

A more formal derivation of the evolution of the momentum of a particle propagating in a Robertson-Walker background may prove useful. Choose the origin of the coordinate system at any point along the trajectory of the particle. With the metric in the form

$$ds^2 = R^2(\eta)\left[d\eta^2 - d\chi^2 - \sin^2\chi(d\theta^2 + \sin^2\theta d\phi^2)\right], \tag{2.31}$$

choose the trajectory of the particle along $d\theta = d\phi = 0$. The Hamilton–Jacobi equation of motion is

$$g^{\mu\nu}\frac{\partial S}{\partial x^\mu}\frac{\partial S}{\partial x^\nu} - m^2 = 0, \tag{2.32}$$

where S is the action of a free particle, $S \propto \int ds$. With the metric in the above form, the Hamilton-Jacobi equation is particularly simple:

$$\left(\frac{\partial S}{\partial \chi}\right)^2 - \left(\frac{\partial S}{\partial \eta}\right)^2 + m^2 R^2(\eta) = 0. \tag{2.33}$$

The three-momentum $|\vec{\mathbf{p}}|$ of a particle is given by $|\vec{\mathbf{p}}| = \partial S/\partial|\vec{\mathbf{dl}}\,|$, and with $d\theta = d\phi = 0$, $|\vec{\mathbf{dl}}\,| = R(\eta)d\chi$. Therefore $|\vec{\mathbf{p}}| = R^{-1}\partial S/\partial\chi$. The

---

mathematicians) would refer to as the physical three-momentum is that measured in an orthonormal frame, which is also equivalent to $|\vec{\mathbf{p}}| = (p^i p_i)^{1/2}$.

energy of a particle is given by $E \equiv (E^0 E_0)^{1/2} = \partial S/\partial \tau = R^{-1}\partial S/\partial \eta$. Thus we recognize that the Hamilton-Jacobi equation corresponds to the familiar relation $E^2 - |\vec{\mathbf{p}}|^2 = p^\mu p_\mu = m^2$. Since χ is a cyclic coordinate, $dS/d\chi$=const, from which it follows that $\partial S/\partial \chi = R|\vec{\mathbf{p}}|$=const, or $|\vec{\mathbf{p}}| \propto R^{-1}$.

2.3 Kinematics of the RW Metric

Without explicitly solving Einstein's equations for the dynamics of the expansion, it is still possible to understand many of the kinematic effects of the expansion upon light from distant galaxies. The light emitted by a distant object can be viewed quantum mechanically as freely-propagating photons, or classically as propagating plane waves.

In the quantum mechanical description, the wavelength of light is inversely proportional to the photon momentum ($\lambda = h/p$). If the momentum changes, the wavelength of the light must change. It was shown in the previous section that the momentum of a photon changes in proportion to R^{-1}. Since the wavelength of a photon is inversely proportional to its momentum, the wavelength at time t_0, denoted as λ_0, will differ from that at time t_1, denoted as λ_1, by

$$\frac{\lambda_1}{\lambda_0} = \frac{R(t_1)}{R(t_0)}. \tag{2.34}$$

As the Universe expands, the wavelength of a freely-propagating photon increases—just as all physical distances increase with the expansion. This means that the red shift of the wavelength of a photon is due to the fact that the Universe was smaller when the photon was emitted!

It is also possible to derive the same result by considering the propagation of light from a distant galaxy as a classical wave phenomenon. Suppose a wave is emitted from a source at coordinate $r = r_1$ at time t_1 and arrives at a detector at time t_0 at coordinate $r = 0$. The massless wave will travel on a geodesic ($ds^2 = 0$), and the coordinate distance and time will be related by

$$\int_{t_1}^{t_0} \frac{dt}{R(t)} = \int_0^{r_1} \frac{dr}{(1-kr^2)^{1/2}} \equiv f(r_1). \tag{2.35}$$

The wavecrest emitted at a time $t_1 + \delta t_1$ will arrive at the detector at a time $t_0 + \delta t_0$. The equation of motion will be the same as (2.35) with

$t_1 \to t_1 + \delta t_1$ and $t_0 \to t_0 + \delta t_0$. Since $f(r_1)$ is constant (the source is fixed in the comoving coordinate system)

$$\int_{t_1}^{t_0} \frac{dt}{R(t)} = \int_{t_1+\delta t_1}^{t_0+\delta t_0} \frac{dt}{R(t)}. \tag{2.36}$$

By simple rearrangement of the limits of integration we have

$$\int_{t_1}^{t_1+\delta t_1} \frac{dt}{R(t)} = \int_{t_0}^{t_0+\delta t_0} \frac{dt}{R(t)}. \tag{2.37}$$

If δt is sufficiently small, corresponding to wavelengths $\lambda = \delta t \ll t$, $R(t)$ can be taken to be constant over the integration time of (2.37), and

$$\frac{\delta t_1}{R(t_1)} = \frac{\delta t_0}{R(t_0)}. \tag{2.38}$$

Since δt_1 (δt_0) is the time between successive wave crests of the emitted (detected) light, δt_1 (δt_0) is the wavelength of the emitted (detected) light, and

$$\frac{\lambda_1}{\lambda_0} = \frac{R(t_1)}{R(t_0)}, \tag{2.39}$$

just the result obtained above.

Astronomers define the *red shift* of an object, z, in terms of the ratio of the detected wavelength to the emitted wavelength:

$$1 + z \equiv \frac{\lambda_0}{\lambda_1} = \frac{R(t_0)}{R(t_1)}. \tag{2.40}$$

Any increase (decrease) in $R(t)$ leads to a red shift (blue shift) of the light from distant sources. Since today astronomers observe distant galaxies to have red shifted spectra, we can conclude that the Universe is expanding.

Hubble's Law, the linear relationship between the "distance" to a galaxy and its observed red shift, may be deduced directly from the RW metric without specific knowledge of the dynamics of the expansion. Suppose a source, e.g., a galaxy has an absolute luminosity $\mathcal{L}$ (the energy per time produced by the source in its rest frame), its luminosity distance is defined in terms of the measured flux $\mathcal{F}$ (the energy per time per area measured by a detector) by

$$d_L^2 \equiv \frac{\mathcal{L}}{4\pi\mathcal{F}}. \tag{2.41}$$

The motivation for this *definition* of d_L is clear: In the absence of expansion, the detected flux is simply equal to the fraction of the area of the two sphere surrounding the source covered by the detector ($dA/4\pi d_L^2$, where dA is the area of the detector) times the luminosity $\mathcal{L}$. In this case d_L is simply the distance to the source. The expansion of the Universe complicates matters: Is d_L the distance to the source at the time of emission, or at the time of detection, etc? Of course, it is neither. If a source at comoving coordinate $r = r_1$ emits light at time t_1, and a detector at comoving coordinate $r = 0$ detects the light at $t = t_0$, conservation of energy ($T^{\mu\nu}{}_{;\nu} = 0$) implies

$$\mathcal{F} = \frac{\mathcal{L}}{4\pi R^2(t_0) r_1^2 (1+z)^2}, \tag{2.42}$$

or

$$d_L^2 = R^2(t_0) r_1^2 (1+z)^2. \tag{2.43}$$

The rhs of (2.42) may be understood as follows: At the detection time t_0, the fraction of the area of a two sphere surrounding the source covered by the detector is $dA/4\pi R^2(t_0) r_1^2$. In addition, the expansion decreases the energy per unit time crossing the two sphere surrounding the source by a factor of $(1+z)^2$: one factor of $(1+z)$ arising from the decrease in total energy due to the red shift of the energy of the individual photons, and the other factor of $(1+z)$ arising from the increased time interval between the detection of incoming photons due to the fact that two photons separated by a time δt at emission, will be separated by a time $\delta t(1+z)$ at the time of detection [as in (2.38)]. Therefore $d_L^2 = [R(t_0)r_1]^2(1+z)^2$ as written in (2.43).

In order to express d_L in terms of the red shift z, the explicit dependence upon r_1 must be removed. With explicit knowledge of $R(t)$ (which requires solving the Einstein equations) this is straightforward; without explicitly specifying $R(t)$ this formally can be accomplished by expanding $R(t)$ (and other quantities) in a Taylor series about the present epoch (i.e., for small $H_0(t-t_0)$):

$$\frac{R(t)}{R(t_0)} = 1 + H_0(t-t_0) - \frac{1}{2} q_0 H_0^2 (t-t_0)^2 + \cdots \tag{2.44}$$

where

$$H_0 \equiv \frac{\dot{R}(t_0)}{R(t_0)} \qquad q_0 \equiv \frac{-\ddot{R}(t_0)}{\dot{R}^2(t_0)} R(t_0) = -\frac{\ddot{R}}{R H_0^2} \tag{2.45}$$

are the present values of the expansion rate (or the Hubble constant) and the so-called deceleration parameter. Now, remembering that $R(t_0)/R(t) = 1+z$, (2.44) can be inverted for small $H_0(t_0-t)$ to give

$$z = H_0(t_0 - t) + \left(1 + \frac{q_0}{2}\right) H_0^2(t_0 - t)^2 + \cdots \tag{2.46}$$

Eq. (2.46) yields $(t_0 - t)$ in terms of z:

$$(t_0 - t) = H_0^{-1}\left[z - \left(1 + \frac{q_0}{2}\right) z^2 + \cdots\right]. \tag{2.47}$$

It is also possible to expand $f(r_1)$, defined in (2.35), in a power series

$$f(r_1) = \begin{cases} \sin^{-1} r_1 & = r_1 + r_1^3/6 + \cdots \quad (k = +1) \\ r_1 & \qquad\qquad\qquad\quad (k = 0) \\ \sinh^{-1} r_1 & = r_1 - r_1^3/6 + \cdots \quad (k = -1). \end{cases} \tag{2.48}$$

Using the expansion (2.44) on the lhs of (2.35) gives

$$r_1 = R^{-1}(t_0)\left[(t_0 - t_1) + \frac{1}{2}H_0(t_0 - t_1)^2 + \cdots\right]. \tag{2.49}$$

Substituting expression (2.47) for $(t_0 - t_1)$ in terms of z gives

$$r_1 = R(t_0)^{-1}H_0^{-1}\left[z - \frac{1}{2}(1 + q_0)z^2 + \cdots\right]. \tag{2.50}$$

Finally, using this expression in (2.43) yields Hubble's law:

$$H_0 d_L = z + \frac{1}{2}(1 - q_0)z^2 + \cdots. \tag{2.51}$$

Note that the relationship between d_L and z deviates from linear for moderate z if $q_0 \neq 1$.

In the next Chapter we will solve the Einstein equations for $R(t)$ to obtain r_1 as a function of z, cf. (3.112). Using the *exact* result for $r_1(z)$ in a matter-dominated Universe gives

$$H_0 d_L = q_0^{-2}\left[zq_0 + (q_0 - 1)\left(\sqrt{2q_0 z + 1} - 1\right)\right]. \tag{2.52}$$

Of course, expansion of (2.52) for small z yields (2.51).

Note that the relationship between the observable quantities d_L and z depends upon the cosmological model, and thus, *in principle*, we can use the Hubble diagram (plot of z vs. d_L) to infer q_0. As discussed in Chapter 1, the difficulty lies in finding appropriate standard candles.[6]

Another relationship whose form also depends upon the cosmological model is the the galaxy count–red shift relationship. As mentioned in Chapter 1, this relationship holds great promise as an observational tool to determine our cosmological model. Consider a comoving volume element dV_c, containing $dN_{\rm gal}$ galaxies. In terms of the number density of galaxies per comoving volume element $n_c(t)$

$$dN_{\rm gal} = n_c(t)dV_c = n_c(t)\frac{r^2}{(1-kr^2)^{1/2}}drd\Omega. \qquad (2.53)$$

Of course, r is not an observable, and we must express r in terms of z. Using (2.50)

$$\frac{r^2dr}{(1-kr^2)^{1/2}} = H_0^{-3}R_0^{-3}z^2dz[1-2(q_0+1)z+\cdots], \qquad (2.54)$$

and we may express the galaxy count as a function of red shift:

$$\frac{1}{z^2}\frac{dN_{\rm gal}}{dzd\Omega} = (H_0R_0)^{-3}n_c(z)[1-2(q_0+1)z+\cdots]. \qquad (2.55)$$

The exact solution for $r(z)$ found in the next chapter gives

$$\frac{dN_{\rm gal}}{dzd\Omega} = \frac{n_c(z)}{H_0^3R_0^3(1+z)^3q_0^4}\frac{\left[zq_0+(q_0-1)(\sqrt{2q_0z+1}-1)\right]^2}{[1-2q_0+2q_0(1+z)]^{1/2}}. \qquad (2.56)$$

Expansion of the exact result for small z of course leads to (2.55).

If one makes the reasonable assumption that $n_c(z)$ is constant (galaxies are neither created or destroyed—probably a good assumption at red shifts

[6]The luminosity distance, $d_L \equiv (4\pi\mathcal{L}/\mathcal{F})^{1/2}$, is an observable *provided* that the absolute luminosity $\mathcal{L}$ is known. Determining absolute luminosities involves the use of standard candles, i.e., classes of objects that are easily identifiable and are bright enough to be seen at cosmological distances, whose absolute luminosities have been determined and have very little variance.

less than a few), then the differential galaxy count can be used to determine q_0.

Finally, we mention one other relationship that depends upon the cosmological model, the angular diameter–red shift relation. Let us assume that there is an object (say, a galaxy or a cluster) of proper diameter D at coordinate $r = r_1$, which emitted light at $t = t_1$, detected by the observer at $t = t_0$. As usual, we may choose the coodinate system such that the observer is at the coordinate $r = 0$. From the metric (2.1) it follows that the observed angular diameter of the source, δ, is related to D by

$$\delta = \frac{D}{R(t_1)r_1}. \tag{2.57}$$

The angular diameter distance, d_A, is defined as

$$d_A \equiv \frac{D}{\delta} = R(t_1)r_1. \tag{2.58}$$

In a static Euclidean geometry $d_A = d_L$ is simply the "distance" to the source.

It is easy to express d_A in terms of d_L and z using (2.43): $d_A = d_L(1+z)^{-2}$. For small $H_0(t_1 - t_0)$ the approximate relationship between d_A and z is, cf. (2.51),

$$H_0 d_A = z - \frac{1}{2}(3 + q_0)z^2 + \cdots \tag{2.59}$$

while the exact relationship is, cf. (2.52)

$$H_0 d_A = q_0^{-2}(1+z)^{-2}\left[zq_0 + (q_0 - 1)\left(\sqrt{2q_0 z + 1} - 1\right)\right]. \tag{2.60}$$

Again, the form of the relationship depends upon the cosmological model (q_0), and thus one may in principle use "standard rulers" to infer q_0.

We have discussed three relationships (d_L vs z, $dN_{\rm gal}$ vs z, and d_A vs z) which all depend upon the cosmological model through q_0. There are also relationships between observables which *do not* depend upon the cosmological model. The simplest example is the measured surface brightness of an object, say a galaxy or a cluster. The surface brightness, SB, of an object is the energy flux per projected solid angle (on the sky): $\mathrm{SB} \equiv d\mathcal{F}/d\Omega$. The projected solid angle $d\Omega$ is related to the projected surface area on the object by $dA = R^2(t_1)r_1^2 d\Omega$, from which it follows that $d\mathcal{F}/d\Omega = \mathrm{SB}_{\rm intrinsic}/(1+z)^4$, where $\mathrm{SB}_{\rm intrinsic}$ is the intrinsic surface brightness of the object (erg sec^{-1} cm^{-2}). Model-independent relationships are useful because in principle they can be used to test whether spurious astrophysical or observational problems affect the model-dependent relationships. For instance, a measured deviation of $d\mathcal{F}/d\Omega$ from a $(1+z)^{-4}$ dependence would indicate that galactic evolution (and/or other effects) must be taken into account when using standard candles. While earth-based telescopes probably lack the resolving power to test this relationship, the Hubble Space Telescope with its expected gain in resolution, may well be able to exploit this relationship.

In the next Chapter we will derive the dynamical relationship between the physical parameters of the cosmological model (curvature radius, energy density, pressure, etc.) and q_0 and H_0.

2.4 References

Further discussion of the Robertson-Walker metric can be found in:

1. S. Weinberg, *Gravitation and Cosmology* (Wiley, New York, 1972).

2. C. W. Misner, K. Thorne, and J. A. Wheeler, *Gravitation* (Freeman, San Francisco, 1973).

3. R. M. Wald, *General Relativity* (Univ. Chicago Press, Chicago, 1984).

4. L. D. Landau and E. M. Lifshitz, *Classical Theory of Fields* (Pergamon Press, Oxford, 1975).

5. Ya. B. Zel'dovich and I. D. Novikov, *Relativistic Astrophysics, Vol. II* (Univ. Chicago Press, Chicago, 1983).

There are a multitude of theorems concerning the isometric (i.e., metric preserving) embedding of Riemannian manifolds into Euclidean space; for example, any n-dimensional manifold can be locally embedded into R^N for $N = n(n+1)/2$. Global embeddings are more difficult and the number of dimensions required depends upon the smoothness of the embedding. Any n-dimensional manifold can be C^1 globally embedded in a Euclidean space of at most $2n+1$ dimensions; further, if the manifold is compact, then at most $2n$ dimensions are required. (A C^k ($k \geq 3$) global embedding requires at most $n(n+1)(3n+11)/2$ dimensions, $n(3n+11)/2$ for a compact manifold.) In addition, it is known that the n-dimensional surface of constant negative curvature cannot be globally embedded in R^{n+1}. Since the two sphere is compact, at most 4 dimensions are required for a global

Euclidean embedding, and our example shows that 3 suffice. On the other hand, these theorems guarantee that the hyperbolic plane can be embedded in R^5, but not in R^3. For further discussion, see the following reference and references therein:

6. *Encyclopedic Dictionary of Mathematics*, eds. S. Iyanaga and Y. Kawada (MIT Press, Cambridge, 1980), Secs. **96H, 114, 177, 211, 283, 360**.

The proof of the conformal equivalence of Robertson-Walker and Minkowski spaces can be found in:

7. A. P. Lightman, W. H. Press, R. H. Price, and S. L. Teukolsky, *Problem Book in Relativity and Gravitation* (Princeton Univ. Press, Princeton, 1975), problem 19.8.

References to some historical papers can be found in *Early Universe: Reprints*. Also reprinted there is a paper of M. A. MacCallum that discusses anisotropic and inhomogeneous cosmologies:

8. M. A. MacCallum, in *General Relativity: An Einstein Centenary Survey*, eds. S. W. Hawking and W. Israel (Cambridge Univ. Press, Cambridge, 1979).

3

STANDARD COSMOLOGY

3.1 The Friedmann Equation

All of the discussions in the previous Chapter concerned the "kinematics" of a Universe described by a Robertson-Walker metric. The dynamics of the expanding Universe only appeared implicitly in the time dependence of the scale factor $R(t)$. To make this time dependence explicit, one must solve for the evolution of the scale factor using the Einstein equations:

$$R_{\mu\nu} - \frac{1}{2}\mathcal{R}g_{\mu\nu} \equiv G_{\mu\nu} = 8\pi G T_{\mu\nu} + \Lambda g_{\mu\nu}, \tag{3.1}$$

where $G_{\mu\nu}$ is the Einstein tensor, $T_{\mu\nu}$ is the stress-energy tensor for all the fields present—matter, radiation, etc., and Λ is a cosmological constant.[1]

The Einstein equations follow directly from an action principle: The action is stationary under small variations of the metric tensor. The action that leads to the Einstein field equations is $S = S_{\text{E-H}} + S_{\text{M}}$, where $S_{\text{E-H}}$ is the Einstein-Hilbert (gravitational) action and S_{M} is the "matter" action:

$$\begin{aligned} S_{\text{E-H}} &= -\frac{1}{16\pi G}\int d^4x\sqrt{-g}\,(\mathcal{R} + 2\Lambda) \\ S_{\text{M}} &= \sum_{\text{fields}} \int d^4x\sqrt{-g}\,\mathcal{L}_{\text{fields}}, \end{aligned} \tag{3.2}$$

where the sum in S_{M} runs over the Lagrangian densities for all the funda-

[1] From (3.1) it follows that $\mathcal{R} = -8\pi G T$, where $T \equiv T^{\mu}_{\ \mu}$ is the trace of $T_{\mu\nu}$, and the Einstein equations can also be written in the form $R_{\mu\nu} = 8\pi G\,[T_{\mu\nu} - (1/2)T g_{\mu\nu}]$.

mental fields. The variation of the matter and gravitational actions are:

$$\delta S_{\text{E-H}} = \frac{1}{16\pi G}\int d^4x\sqrt{-g}\left(R^{\mu\nu} - \frac{1}{2}\mathcal{R}g^{\mu\nu} - \Lambda g^{\mu\nu}\right)\delta g_{\mu\nu}$$

$$\delta S_{\text{M}} = -\frac{1}{2}\sum_{\text{fields}}\int d^4x\sqrt{-g}\,T^{\mu\nu}_{\text{fields}}\delta g_{\mu\nu}, \tag{3.3}$$

and so the principle that $\delta S/\delta g_{\mu\nu} = 0$ yields the Einstein equations.

With very minimal assumptions about the dread right-hand side of the Einstein equations, it is possible to proceed without detailed knowledge of the properties of the fundamental fields that contribute to the stress tensor $T_{\mu\nu}$. To be consistent with the symmetries of the metric, the total stress-energy tensor $T_{\mu\nu}$ must be diagonal, and by isotropy the spatial components must be equal. The simplest realization of such a stress-energy tensor is that of a perfect fluid characterized by a time-dependent energy density $\rho(t)$ and pressure $p(t)$:[2]

$$T^{\mu}_{\ \nu} = \text{diag}(\rho, -p, -p, -p). \tag{3.4}$$

The $\mu = 0$ component of the conservation of stress energy ($T^{\mu\nu}_{\ \ ;\nu} = 0$) gives the 1st law of thermodynamics in the familiar form

$$d(\rho R^3) = -pd(R^3), \tag{3.5}$$

or equivalently, in the unfamiliar form

$$d\left[R^3(\rho + p)\right] = R^3 dp. \tag{3.6}$$

The physical significance of (3.5) is clear: The change in energy in a comoving volume element, $d(\rho R^3)$, is equal to minus the pressure times the change in volume, $-pd(R^3)$. For the simple equation of state $p = w\rho$, where w is independent of time,[3] the energy density evolves as $\rho \propto R^{-3(1+w)}$. Examples of interest include

$$\text{RADIATION} \quad (p = \frac{1}{3}\rho) \quad \Longrightarrow \quad \rho \propto R^{-4}$$

[2]Note that an imperfect fluid with bulk viscosity would also satisfy the symmetry requirements.

[3]Note that p can always be expressed as $w\rho$; the assumption here is that the constant of proportionality is independent of time.

$$\begin{aligned} &\text{MATTER} && (p=0) \implies \rho \propto R^{-3} \\ &\text{VACUUM ENERGY} && (p=-\rho) \implies \rho \propto \text{const.} \end{aligned} \qquad (3.7)$$

The "early" Universe was radiation dominated, the "adolescent" Universe was matter dominated, and in the absence of vacuum energy, the "late" Universe will continue to be be matter dominated (or curvature dominated if $k \neq 0$). If the Universe underwent inflation, there was a "very early" period when the stress-energy was dominated by vacuum energy. As we shall see next, once we know the evolution of ρ and p in terms of the scale factor $R(t)$, it is straightforward to solve for $R(t)$.

Before going on, we want to emphasize the utility of describing the stress energy in the Universe by the simple equation of state $p = w\rho$. This is the most general form for the stress energy in a RW space-time—and the observational evidence indicates that on large scales the RW metric is quite a good approximation to the space-time within our Hubble volume. Throughout this monograph we use the equation of state $p = w\rho$ (with w constant) to describe a variety of contributions to the right-hand side of the Einstein equations—matter, radiation, vacuum energy, coherent scalar fields, cosmic strings, domain walls, and so on. This simple, but often very accurate, approximation will allow us to explore many early-Universe phenomena with a single parameter!

The dynamical equations that describe the evolution of the scale factor $R(t)$ follow from the Einstein field equations. The non-zero components of the Ricci tensor for the Robertson-Walker metric are

$$\begin{aligned} R_{00} &= -3\frac{\ddot{R}}{R} \\ R_{ij} &= -\left[\frac{\ddot{R}}{R} + 2\frac{\dot{R}^2}{R^2} + \frac{2k}{R^2}\right] g_{ij}, \end{aligned} \qquad (3.8)$$

and the Ricci scalar $\mathcal{R}$ is

$$\mathcal{R} = -6\left[\frac{\ddot{R}}{R} + \frac{\dot{R}^2}{R^2} + \frac{k}{R^2}\right]. \qquad (3.9)$$

The $0-0$ component of the Einstein equation, $R_{\mu\nu} - \frac{1}{2}g_{\mu\nu}\mathcal{R} = 8\pi G T_{\mu\nu}$, gives the so-called Friedmann equation

$$\frac{\dot{R}^2}{R^2} + \frac{k}{R^2} = \frac{8\pi G}{3}\rho, \qquad (3.10)$$

while the $i-i$ component gives

$$2\frac{\ddot{R}}{R} + \frac{\dot{R}^2}{R^2} + \frac{k}{R^2} = -8\pi G p. \tag{3.11}$$

The three field equations, (3.5), (3.10), and (3.11) are related by the Bianchi identities and only two are independent. The two independent equations are usually taken as (3.5) and (3.10). The difference of (3.11) and (3.10) provides an equation for the acceleration $\ddot{R}$ alone,

$$\frac{\ddot{R}}{R} = -\frac{4\pi G}{3}(\rho + 3p). \tag{3.12}$$

Today $\dot{R} \geq 0$; if in the past $\rho + 3p$ was always positive, then $\ddot{R}$ was always negative, and thus at some finite time in the past R must have been equal to zero. This event, referred to as the big bang, is usually identified as time zero. At $R = 0$ there is a singularity; extrapolation past the singularity is not possible in the framework of classical general relativity.

The expansion rate of the Universe is determined by the Hubble *parameter* $H \equiv \dot{R}/R$. The Hubble parameter is not constant, and in general varies as t^{-1}. The Hubble time (or Hubble radius) H^{-1} sets the time scale for the expansion: R roughly doubles in a Hubble time. The Hubble *constant*, H_0, is the present value of the expansion rate. The Friedmann equation can be recast as

$$\frac{k}{H^2R^2} = \frac{\rho}{3H^2/8\pi G} - 1 \equiv \Omega - 1, \tag{3.13}$$

where Ω is the ratio of the density to the critical density ρ_C:

$$\Omega \equiv \frac{\rho}{\rho_C} \qquad \rho_C \equiv \frac{3H^2}{8\pi G}. \tag{3.14}$$

Since $H^2R^2 \geq 0$, there is a correspondence between the sign of k, and the sign of $\Omega - 1$

$$\begin{aligned} k = +1 &\implies \Omega > 1 \qquad \text{CLOSED} \\ k = 0 &\implies \Omega = 1 \qquad \text{FLAT} \\ k = -1 &\implies \Omega < 1 \qquad \text{OPEN.} \end{aligned} \tag{3.15}$$

In terms of Ω, the Friedmann equation for the expansion rate is

$$\frac{k}{H^2R^2} = \Omega - 1. \tag{3.16}$$

Equation (3.16) is valid for all times; note however that $\Omega - 1$ and ρ_C are not constant, but change as the Universe expands. At early times when the curvature term is negligible, $H^2 \propto \rho$ which gives $H^2 \propto R^{-3}$ for a matter-dominated Universe, and $H^2 \propto R^{-4}$ for a radiation-dominated Universe. Since $|\Omega - 1|$ is at most of order unity today, then at early times

$$|\Omega - 1| \simeq \begin{cases} R/R_0 = (1+z)^{-1} & \text{(MD)} \\ (R_{EQ}/R_0)(R/R_{EQ})^2 \simeq 10^4(1+z)^{-2} & \text{(RD)}, \end{cases} \tag{3.17}$$

where $R_{EQ} \simeq 10^{-4}R_0$ is the value of R at the transition between matter domination and radiation domination. Apparently, at earlier epochs the Universe was *very* nearly critical.

Recall from (2.3) that the *spatial* curvature of the FRW model is ${}^3\mathcal{R} = 6k/R^2(t)$. Using (3.13), we may also express the spatial curvature as ${}^3\mathcal{R} = 6H^2(\Omega - 1)$. From the form of the RW metric, it is clear that the effects of spatial curvature become very significant for $r \sim |k|^{-1/2}$, so we define a physical "radius of curvature" of the Universe, $R_{\rm curv} \equiv R(t)|k|^{-1/2} = (6/|{}^3\mathcal{R}|)^{1/2}$, related to the Hubble radius, H^{-1}, by

$$R_{\rm curv} = \frac{H^{-1}}{|\Omega - 1|^{1/2}}. \tag{3.18}$$

When $|\Omega - 1|$ is of order unity the two are comparable; when $|\Omega - 1|$ is very small, $R_{\rm curv} \gg H^{-1}$. A Universe close to critical density is very flat. Since $|\Omega - 1|$ must have been very small at early epochs, it is safe to ignore spatial curvature in the early Universe.

Note that for the closed models ($k > 0$), $R_{\rm curv}$ is just the physical radius of the three sphere. Further, while we have scaled r such that $k = +1, -1$, or 0, there are an infinity of RW models, characterized by the curvature radius, $R_{\rm curv} = R(t)|k|^{-1/2}$, at some specified epoch.

Recall the definition of the deceleration parameter q_0 from (2.45): $q_0 \equiv -(\ddot{R}(t_0)/R(t_0))/H_0^2$. Here and below, sub-0 will denote the present value of a quantity. By taking the ratio of Eqns. (3.12) and (3.10), and using the definition of Ω_0, it follows that

$$q_0 = \Omega_0[1 + 3p/\rho]/2 \equiv \Omega_0(1 + 3w)/2. \tag{3.19}$$

For a matter-dominated model $q_0 = \Omega_0/2$; for a radiation-dominated model $q_0 = \Omega_0$; and for a vacuum-dominated model $q_0 = -\Omega_0$ (in a vacuum-dominated model the expansion is accelerating, $\ddot{R} > 0$).

3.2 The Expansion Age of the Universe

The Friedmann equation may be integrated to give the age of the Universe in terms of present cosmological parameters. The energy density scales as $\rho/\rho_0 = (R/R_0)^{-3}$ for a matter-dominated (MD) Universe, and scales as $\rho/\rho_0 = (R/R_0)^{-4}$ for a radiation-dominated (RD) Universe. The Friedmann equation becomes

$$\begin{aligned} \left(\frac{\dot{R}}{R_0}\right)^2 + \frac{k}{R_0^2} &= \frac{8\pi G}{3}\rho_0 \frac{R_0}{R} \qquad \text{(MD)} \\ \left(\frac{\dot{R}}{R_0}\right)^2 + \frac{k}{R_0^2} &= \frac{8\pi G}{3}\rho_0 \left(\frac{R_0}{R}\right)^2 \qquad \text{(RD)}. \end{aligned} \tag{3.20}$$

Using the fact that $k/R_0^2 \equiv H_0^2(\Omega_0 - 1)$, the age of the Universe as a function of $R_0/R = 1 + z$ is given by

$$\begin{aligned} t &\equiv \int_0^{R(t)} \frac{dR'}{\dot{R}'} \\ &= H_0^{-1} \int_0^{(1+z)^{-1}} \frac{dx}{[1 - \Omega_0 + \Omega_0 x^{-1}]^{1/2}} \qquad \text{(MD)} \\ &= H_0^{-1} \int_0^{(1+z)^{-1}} \frac{dx}{[1 - \Omega_0 + \Omega_0 x^{-2}]^{1/2}} \qquad \text{(RD)}. \end{aligned} \tag{3.21}$$

As expected, the time scale for the age of the Universe is set by the Hubble time H_0^{-1}.

As noted above, it is conventional to define the zero of time to be that time when the scale factor R extrapolates to zero. Of course, this is arbitrary as all the equations governing the Universe are invariant under time translation. As we will see, in the calculation of physical quantities, the timescale enters as H^{-1}, which is independent of the zero of time. In calculating the age of the Universe, a potential problem arises: Earlier than some time, say $t_?$ (more accurately, for R less than some $R_?$), our knowledge of the Universe is uncertain, and so the time elapsed from $R = 0$

to $R = R_?$ cannot be reliably calculated. However, the contribution to the age of the Universe from $R = 0$ to $R = R_?$ is very small, providing $R \sim t^n$ ($n < 1$) during this period: $\int_0^{R_?} dR/\dot{R} \sim H_?^{-1}$. Put another way, so long as $R \sim t^n$ ($n < 1$), most of the time elapsed since $R = 0$ accumulated during the most recent few Hubble times. Unless there was some very long early epoch where $R \sim t^n$ ($n > 1$),[4] the "error" in calculating the age of the Universe at late times is negligible, and of course the precise age of the Universe is irrelevant for any microphysical calculation since the only timescale that enters such calculations is the Hubble time at the epoch of interest.

The integrals in (3.21) are easily evaluated. Consider the matter-dominated case first. In terms of Ω_0 and z,

$$t = H_0^{-1}\frac{\Omega_0}{2(\Omega_0 - 1)^{3/2}} \times \left[\cos^{-1}\left(\frac{\Omega_0 z - \Omega_0 + 2}{\Omega_0 z + \Omega_0}\right) - \frac{2(\Omega_0 - 1)^{1/2}(\Omega_0 z + 1)^{1/2}}{\Omega_0(1+z)}\right] \tag{3.22}$$

for $\Omega_0 > 1$, and

$$t = H_0^{-1}\frac{\Omega_0}{2(1 - \Omega_0)^{3/2}} \times \left[-\cosh^{-1}\left(\frac{\Omega_0 z - \Omega_0 + 2}{\Omega_0 z + \Omega_0}\right) + \frac{2(1 - \Omega_0)^{1/2}(\Omega_0 z + 1)^{1/2}}{\Omega_0(1+z)}\right] \tag{3.23}$$

for $\Omega_0 < 1$. For $\Omega_0 = 1$, $t = (2/3)H_0^{-1}(1+z)^{-3/2}$. The present age of a matter dominated Universe, t_0, is given by the above expressions with $z = 0$

$$t_0 = H_0^{-1}\frac{\Omega_0}{2(\Omega_0 - 1)^{3/2}}\left[\cos^{-1}(2\Omega_0^{-1} - 1) - \frac{2}{\Omega_0}(\Omega_0 - 1)^{1/2}\right] \tag{3.24}$$

for $\Omega_0 > 1$, and

$$t_0 = H_0^{-1}\frac{\Omega_0}{2(1 - \Omega_0)^{3/2}}\left[\frac{2}{\Omega_0}(1 - \Omega_0)^{1/2} - \cosh^{-1}(2\Omega_0^{-1} - 1)\right] \tag{3.25}$$

[4]As an example, consider a Universe that "began" in a vacuum-dominated phase; then early on $R \sim \exp(Ht)$, and the time back to the singularity is infinite.

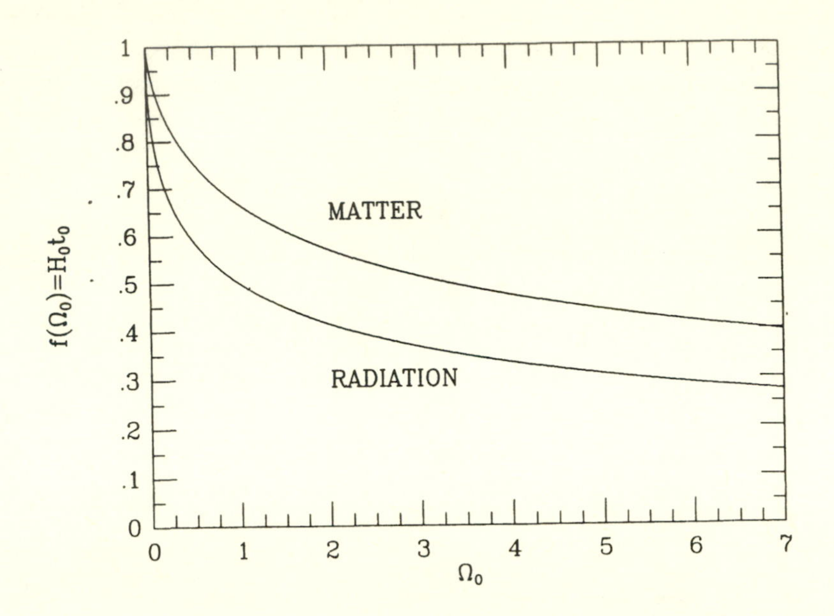

Fig. 3.1: The present age of the Universe in units of H_0^{-1} as a function of Ω_0.

for $\Omega_0 < 1$. If $\Omega_0 = 1$, $t_0 = (2/3)H_0^{-1}$.

The age of the Universe is a decreasing function of Ω_0: Larger Ω_0 implies faster deceleration, which in turn corresponds to a more rapidly expanding Universe early on. In the limit $\Omega_0 \to 0$, $t \to H_0^{-1}(1+z)^{-1} = 9.78\times 10^9 h^{-1}(1+z)^{-1}$ years. Expressions (3.24) and (3.25) can be expanded about $\Omega_0 = 1$:

$$t_0 \simeq \frac{2}{3}H_0^{-1}\left[1 - \frac{1}{5}(\Omega_0 - 1) + \cdots\right]. \tag{3.26}$$

At early times, $(1+z) \gg \Omega_0^{-1}$, both (3.22) and (3.23) reduce to

$$t \simeq \frac{2}{3}(1+z)^{-3/2}H_0^{-1}\Omega_0^{-1/2}. \tag{3.27}$$

For large $1+z$ the dependence of t upon Ω_0 simplifies greatly; this occurs because for $(1+z) \gg \Omega_0^{-1}$ the k/R^2 term becomes negligible compared to the matter density term, and $t \to (2/3)H^{-1} \sim \rho^{-1/2} \propto z^{-3/2}H_0^{-1}\Omega_0^{-1/2}$.

The present age of a matter-dominated, $\Omega_0 = 1$ Universe is $6.52 \times 10^9 h^{-1}$ years. If h is not too much larger than 1/2, this age is consistent with the lower end of estimates for the age of the Universe based upon stellar evolution and nucleocosmochronology.

Now consider the case of a radiation-dominated Universe. In terms of Ω_0 and z,

$$t = H_0^{-1} \frac{\sqrt{\Omega_0(1+z)^2} - \sqrt{\Omega_0(1+z)^2 - \Omega_0 + 1}}{(\Omega_0 - 1)(1+z)}. \tag{3.28}$$

The present age of a radiation-dominated Universe is given by the above expression with $z = 0$

$$t_0 = H_0^{-1} \frac{\sqrt{\Omega_0} - 1}{\Omega_0 - 1}. \tag{3.29}$$

For Ω_0 not too different from 1, (3.28) can be expanded to give

$$t_0 \simeq \frac{1}{2} H_0^{-1} \left[1 - \frac{1}{4}(\Omega_0 - 1) + \cdots \right]. \tag{3.30}$$

For $(1+z) \gg \Omega_0^{-1}$, (3.28) reduces to

$$t \simeq \frac{1}{2}(1+z)^{-2} H_0^{-1} \Omega_0^{-1/2}. \tag{3.31}$$

For both the matter-dominated and radiation-dominated cases, the present age of the Universe is shown in Fig. 3.1 as a function of Ω_0. In both cases, $H_0 t_0 \equiv f(\Omega_0)$ is a decreasing function of Ω_0. As $\Omega_0 \to 0$, $f(\Omega_0) \to 1$. If $\Omega_0 = 1$, $f(\Omega_0) = 2/3$ for the matter-dominated case and $f(\Omega_0) = 1/2$ for the radiation-dominated case. As $\Omega_0 \to \infty$, $f(\Omega_0) \to (\pi/2)\Omega_0^{-1/2}$ for the matter-dominated case and $f(\Omega_0) \to \Omega_0^{-1/2}$ for the radiation-dominated case.

Finally, consider the age of a model Universe that is flat, and contains both matter and (positive) vacuum energy (equivalently, a cosmological constant). The present age of such a model is also easily computed,

$$t_0 = \frac{2}{3} H_0^{-1} \Omega_{\rm VAC}^{-1/2} \ln \left[\frac{1 + \Omega_{\rm VAC}^{1/2}}{(1 - \Omega_{\rm VAC})^{1/2}} \right] \tag{3.32}$$

where $\Omega_{\rm VAC} = \rho_{\rm VAC}/\rho_C$ and by assumption $\Omega_{\rm VAC} + \Omega_{\rm MATTER} = 1$. The present age of such a model universe is shown in Fig. 3.2. It is interesting to note that unlike previous models, a model Universe with $\Omega_{\rm VAC} \gtrsim 0.74$

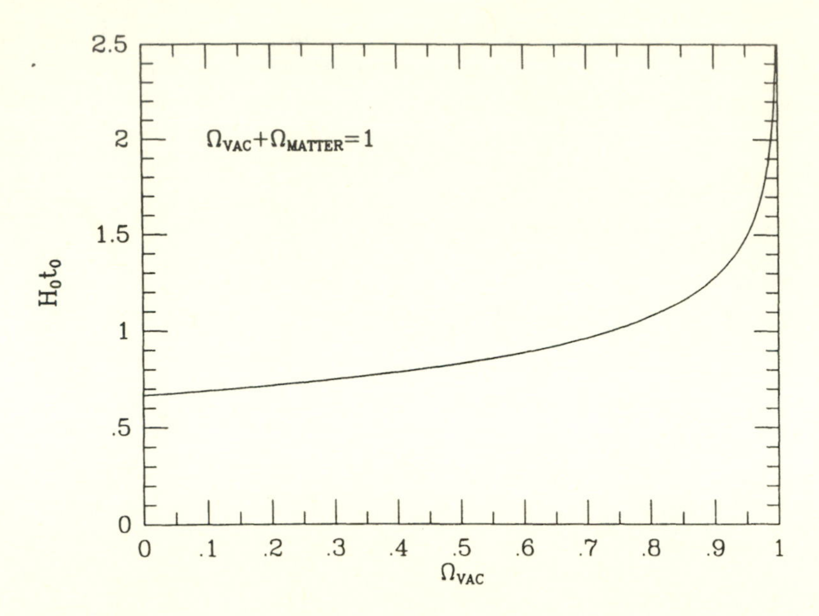

Fig. 3.2: The age of a $k = 0$ Universe with both matter and vacuum energy.

is older than H_0^{-1}; this occurs because the expansion rate is accelerating. And of course in the limit $\Omega_{VAC} \to 1$, $t_0 \to \infty$.

As we will discuss in greater length in Chapters 7 and 8, there is no understanding at present for the absence of a cosmological constant; moreover, the naive expectation for such is $\rho_{VAC} \sim m_{Pl}^4$, a value which would result in an expansion rate which is about a factor of 10^{61} larger than that observed today. The problem of reconciling a "youthful" expansion age with other independent age determinations has at several times in the past led cosmologists to invoke a cosmological constant (and may again in the future). Lacking a fundamental understanding of what its value should be, we should probably keep an open mind to the possibility that $\rho_{VAC} \neq 0$.

As we discussed in Chapter 1, attempts to directly measure Ω_0 probably only restrict it to the range zero to a few. The age of the present Universe provides a very powerful constraint to the value of Ω_0, and to the present energy density of the Universe. To see this, consider a matter-dominated model. Its present age is $t_0 = H_0^{-1}f(\Omega_0) = 9.78h^{-1}f(\Omega_0)$ Gyr. The function $f(\Omega_0)$ decreases monontonically with Ω_0 as discussed above (and shown in Fig. 3.1). For fixed h, a model with larger Ω_0 is younger. Using this fact and an independent age determination one can then derive an upper limit to Ω_0. Independent measures of the age of the Universe suggest that $t_0 = 10$ to 20 Gyr (see Chapter 1). Since a smaller value for $f(\Omega_0)$ (larger Ω_0) can be compensated for by a lower value of h, we also need a lower limit to h in order to constrain Ω_0.

Taking $t_0 \geq 10$ Gyr and $h \geq 0.4$ (0.5), it follows from Fig. 3.1 that Ω_0 must be less than 6.4 (3.1). And of course a larger lower bound to t_0, say 12 Gyr, results in a more stringent upper bound to Ω_0: $\Omega_0 \leq 3.5$ ($h \geq 0.4$), or $\Omega_0 \leq 1.5$ ($h \geq 0.5$).

In constraining the mass density contributed by a relic particle species,

$$\rho_X = \Omega_X \rho_C = \Omega_X h^2 1.88 \times 10^{-29} \mathrm{g\,cm}^{-3}, \tag{3.33}$$

it proves useful to have a constraint to $\Omega_0 h^2$. By considering the age of the Universe once again, one can derive a more stringent constraint than the one that simply follows by using the individual constraints to Ω_0 and h. Using again the fact that $t_0 = 9.78h^{-1}f(\Omega_0)$ Gyr, and defining t_{10} by $t_0 = t_{10} \times 10^{10}$ yr, we have

$$\Omega_0 f^2(\Omega_0) = \Omega_0 h^2 (t_0/9.78\mathrm{Gyr})^2 \tag{3.34}$$

$$0.956\Omega_0 f^2(\Omega_0)/t_{10}^2 \geq \Omega_0 h^2. \tag{3.35}$$

The function $\Omega_0 f^2(\Omega_0)$ increases monotonically, and is bounded from above by its value at $\Omega_0 = \infty$: $\Omega_0 f^2(\Omega_0) \to \pi^2/4 \simeq 2.47$ as $\Omega_0 \to \infty$. Irrespective of the value of h, if $t_{10} \geq 1$, then $\Omega_0 h^2 \leq 2.4$. Using our limited knowledge of h, we can do considerably better. Taking h to be greater than 0.4 (0.5), we find that $\Omega_0 h^2 \leq 1$ (0.8); see Fig. 3.3.

One of the routine duties of an early-Universe cosmologist is to compute the present mass density contributed by some massive stable particle species hypothesized by a particle physicist down the hall, and to determine whether such a particle is at odds with the standard cosmology. In so doing, it is very convenient to express the mass density contributed by the particle species "X" in terms of ρ_C: $\rho_X = 1.88 \times 10^{-29}(\Omega_X h^2)$ g cm^{-3}. The best limit to ρ_X follows from the age constraint discussed above: $\Omega_X h^2 \leq \Omega_0 h^2 \lesssim 1$. If the computed value of $\Omega_X h^2$ exceeds unity, it is often—and incorrectly—stated that the relic is forbidden because it would "overclose" the Universe. The existence of a cosmological relic cannot, of course, modify the geometry of the Universe, i.e., change an open, infinite Universe with $\Omega < 1$ to a closed, finite Universe with $\Omega > 1$. Rather, the

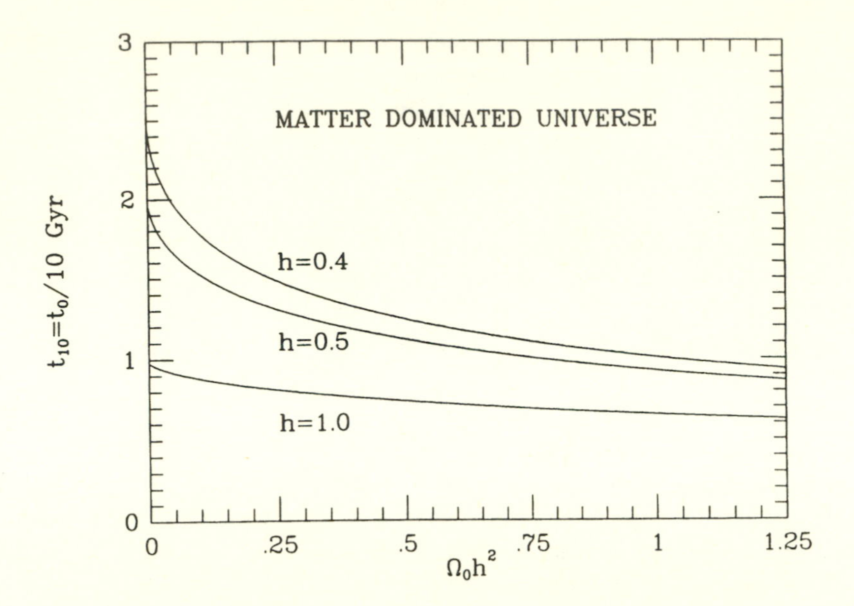

Fig. 3.3: The age of a matter-dominated Universe as a function of $\Omega_0 h^2$.

existence of a relic with $\Omega_X h^2 \gtrsim 1$ would lead to a Universe that, at a temperature of 2.75 K, would have a higher mass density, larger expansion rate (Hubble constant), and smaller age than a Universe without relic X's, because it would have become matter dominated at an earlier epoch. To illustrate, consider a flat model where the relic density of X's at present is such that $\Omega_X h^2 = \beta^2 \gg 1$. Since a flat model is characterized by $\Omega = 1$ whether X's are present or not, such a Universe must have $h^2 = \beta^2 \gg 1$. This would lead to a Hubble constant of 100β km sec^{-1}Mpc^{-1} $\gg 100$ km sec^{-1}Mpc^{-1}, in contradiction to the observed value.

The equations previously derived for $t(z)$ can be inverted to give the scale factor R as a function of t. To begin, ignore spatial curvature ($k = 0$) and consider the simple case in which the rhs of the Friedmann equation is dominated by a fluid whose pressure is given by $p = w\rho$. Then it follows that

$$\rho \propto R^{-3(1+w)} \tag{3.36}$$

$$R \propto t^{2/3(1+w)} \tag{3.37}$$

which leads to the familiar results: $R \propto t^{1/2}$ for $w = 1/3$ (RD); $R \propto t^{2/3}$

for $w = 0$ (MD); $R \propto \exp(H_0 t)$ for $w = -1$ (vacuum dominated); and $R \propto t$ for $w = -1/3$, which corresponds to a curvature-dominated model ($H^2 \propto R^{-2}$).

Now consider a matter-dominated model with arbitrary $\Omega_0 > 1$. While $R(t)$ cannot be given in closed form, R and t can be represented parametrically in terms of the "development" angle θ:

$$\frac{R(t)}{R_0} = (1 - \cos\theta)\frac{\Omega_0}{2(\Omega_0 - 1)} \tag{3.38}$$

$$H_0 t = (\theta - \sin\theta)\frac{\Omega_0}{2(\Omega_0 - 1)^{3/2}}. \tag{3.39}$$

The scale factor $R(t)$ increases from zero (at $t = \theta = 0$) to its maximum value

$$\frac{R_{MAX}}{R_0} = \frac{\Omega_0}{(\Omega_0 - 1)} \tag{3.40}$$

$$H_0 t_{MAX} = \frac{\pi}{2}\frac{\Omega_0}{(\Omega_0 - 1)^{3/2}} \tag{3.41}$$

$$\theta_{MAX} = \pi \tag{3.42}$$

and then back to zero (at $\theta = 2\pi$, $t = 2t_{MAX}$). The maximum value of R obtains when $8\pi G\rho/3 = k/R^2$ and the expansion rate H vanishes. Further expansion would result in a negative value for H^2, so recollapse must follow.

Now consider a matter-dominated model with $\Omega_0 < 1$. In this case the development angle is imaginary, $\theta = i\psi$:

$$\frac{R(t)}{R_0} = (\cosh\psi - 1)\frac{\Omega_0}{2(1 - \Omega_0)^{3/2}}, \tag{3.43}$$

$$H_0 t = (\sinh\psi - \psi)\frac{\Omega_0}{2(1 - \Omega_0)^{3/2}}. \tag{3.44}$$

Here, of course, $R(t)$ increases without limit.

The evolution of $R(t)$ for both the open, closed, and flat FRW models is shown in Fig. 3.4.

Finally, consider a model with comparable contributions to the energy density from both matter and radiation, but with negligible curvature: To

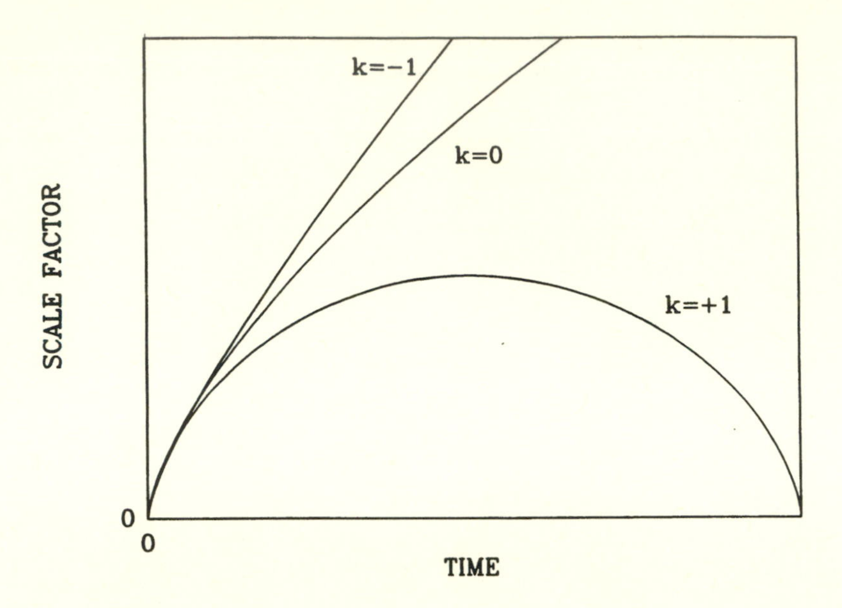

Fig. 3.4: The evolution of $R(t)$ for closed ($k = +1$), open ($k = -1$), and flat ($k = 0$) FRW models.

a very good approximation this corresponds to the Universe around the time of matter–radiation equality ($R \equiv R_{EQ} \simeq 4 \times 10^{-5}(\Omega_0 h^2)^{-1} R_0$). It is simple to obtain the following expression for the time as a function of the scale factor (inverting this expression is not so simple):

$$t/t_{EQ} = [(R/R_{EQ} - 2)(R/R_{EQ} + 1)^{1/2} + 2]/(2 - \sqrt{2}) \tag{3.45}$$

where the epoch of equal matter and radiation densities is denoted by subscript EQ, and $t_{EQ} = 4(\sqrt{2} - 1)H_{EQ}^{-1}/3$. As they must be, the limiting forms of (3.45) are: $R \propto t^{1/2}$ ($t \ll t_{EQ}$), and $R \propto t^{2/3}$ ($t \gg t_{EQ}$).

3.3 Equilibrium Thermodynamics

Today the radiation, or relativistic particles, in the Universe is comprised of the 2.75 K microwave photons, and the 3 cosmic seas of 1.96 K relic neutrinos. Because the early Universe was to a good approximation in thermal equilibrium, there should have been other relativistic particles present, with comparable abundances. Before going on to discuss the early radiation-dominated phase, we will quickly review some basic thermodynamics.

The number density n, energy density ρ, and pressure p of a dilute, weakly-interacting gas of particles with g internal degrees of freedom is given in terms of its phase space distribution (or occupancy) function $f(\vec{\mathbf{p}})$:

$$n = \frac{g}{(2\pi)^3} \int f(\vec{\mathbf{p}}) d^3p \tag{3.46}$$

$$\rho = \frac{g}{(2\pi)^3} \int E(\vec{\mathbf{p}}) f(\vec{\mathbf{p}}) d^3p \tag{3.47}$$

$$p = \frac{g}{(2\pi)^3} \int \frac{|\vec{\mathbf{p}}|^2}{3E} f(\vec{\mathbf{p}}) d^3p \tag{3.48}$$

where $E^2 = |\vec{\mathbf{p}}|^2 + m^2$. For a species in *kinetic equilibrium* the phase space occupancy f is given by the familiar Fermi-Dirac or Bose-Einstein distributions,

$$f(\vec{\mathbf{p}}) = [\exp((E - \mu)/T) \pm 1]^{-1} \tag{3.49}$$

where μ is the chemical potential of the species, and here and throughout $+1$ pertains to Fermi-Dirac species and -1 to Bose-Einstein species. Moreover, if the species is in *chemical equilibrium*, then its chemical potential μ is related to the chemical potentials of other species with which it interacts. For example, if the species (denoted by i) interacts with species j, k, l,

$$i \ + \ j \ \longleftrightarrow \ k \ + \ l \tag{3.50}$$

then $\mu_i + \mu_j = \mu_k + \mu_l$, whenever chemical equilibrium holds. We will return to the question of when local thermodynamic equilibrium (or, LTE) holds later.

From the equilibrium distributions, it follows that the number density n, energy density ρ, and pressure p of a species of mass m with chemical potential μ at temperature T is

$$\rho = \frac{g}{2\pi^2} \int_m^\infty \frac{(E^2 - m^2)^{1/2}}{\exp[(E - \mu)/T] \pm 1} E^2 dE$$

$$n = \frac{g}{2\pi^2} \int_m^\infty \frac{(E^2 - m^2)^{1/2}}{\exp[(E - \mu)/T] \pm 1} E dE,$$

$$p = \frac{g}{6\pi^2}\int_m^\infty \frac{(E^2-m^2)^{3/2}}{\exp[(E-\mu)/T]\pm 1}dE. \tag{3.51}$$

In the relativistic limit ($T \gg m$), for $T \gg \mu$,

$$\begin{aligned} \rho &= \begin{cases} (\pi^2/30)gT^4 & \text{(BOSE)} \\ (7/8)(\pi^2/30)gT^4 & \text{(FERMI)} \end{cases} \\ n &= \begin{cases} (\zeta(3)/\pi^2)gT^3 & \text{(BOSE)} \\ (3/4)(\zeta(3)/\pi^2)gT^3 & \text{(FERMI)}, \end{cases} \\ p &= \rho/3 \end{aligned} \tag{3.52}$$

while for degenerate fermions ($\mu \gg T$)

$$\begin{aligned} \rho &= (1/8\pi^2)g\mu^4 \\ n &= (1/6\pi^2)g\mu^3 \\ p &= (1/24\pi^2)g\mu^4. \end{aligned} \tag{3.53}$$

Here $\zeta(3) = 1.20206...$ is the Riemann zeta function of 3. For a Bose-Einstein species $\mu > 0$ indicates the presence of a Bose condensate, which must be treated separately from the other modes. For relativistic bosons or fermions with $\mu < 0$ and $|\mu| < T$, it follows that

$$\begin{aligned} n &= \exp(\mu/T)(g/\pi^2)T^3 \\ \rho &= \exp(\mu/T)(3g/\pi^2)T^4 \\ p &= \exp(\mu/T)(g/\pi^2)T^4. \end{aligned} \tag{3.54}$$

In the non-relativistic limit ($m \gg T$) the number density, energy density and pressure are the same for Bose and Fermi species

$$\begin{aligned} n &= g\left(\frac{mT}{2\pi}\right)^{3/2}\exp[-(m-\mu)/T] \\ \rho &= mn \\ p &= nT \ll \rho. \end{aligned} \tag{3.55}$$

For a nondegenerate, relativistic species, the average energy per particle is

$$\langle E\rangle \equiv \frac{\rho}{n} = [\pi^4/30\zeta(3)]T \simeq 2.701T \quad \text{(BOSE)}$$
$$\langle E\rangle \equiv \frac{\rho}{n} = [7\pi^4/180\zeta(3)]T \simeq 3.151T \quad \text{(FERMI)}. \quad (3.56)$$

For a degenerate, relativistic fermion species,

$$\langle E\rangle \equiv \frac{\rho}{n} = \frac{3}{4}\mu. \quad (3.57)$$

For a non-relativistic species, $\langle E\rangle = m + (3/2)T$.

The excess of a fermion species over its antiparticle is often of interest, and is straightforward to compute in the relativistic and non-relativistic limits. Assuming that $\mu_+ = -\mu_-$ (true if reactions like *particle* + *antiparticle* $\leftrightarrow \gamma + \gamma$ are occurring rapidly), the net fermion number density is

$$\begin{aligned} n_+ - n_- &= \frac{g}{2\pi^2}\int_m^\infty E(E^2 - m^2)^{1/2}dE \\ &\quad \times \left[\frac{1}{1+\exp[(E-\mu)/T]} - \frac{1}{1+\exp[(E+\mu)/T]}\right] \\ &= \begin{cases} \dfrac{gT^3}{6\pi^2}\left[\pi^2\left(\dfrac{\mu}{T}\right) + \left(\dfrac{\mu}{T}\right)^3\right] & (T \gg m) \\ 2g(mT/2\pi)^{3/2}\sinh(\mu/T) & \\ \times\exp(-m/T) \quad (T \ll m). & \end{cases} \end{aligned} \quad (3.58)$$

The total energy density and pressure of all species in equilibrium can be expressed in terms of the photon temperature T

$$\rho_R = T^4 \sum_{i=all species}\left(\frac{T_i}{T}\right)^4 \frac{g_i}{2\pi^2}\int_{x_i}^\infty \frac{(u^2 - x_i^2)^{1/2}u^2du}{\exp(u - y_i) \pm 1} \quad (3.59)$$

$$p_R = T^4 \sum_{i=all species}\left(\frac{T_i}{T}\right)^4 \frac{g_i}{6\pi^2}\int_{x_i}^\infty \frac{(u^2 - x_i^2)^{3/2}du}{\exp(u - y_i) \pm 1} \quad (3.60)$$

where $x_i \equiv m_i/T$, $y_i \equiv \mu_i/T$, and we have taken into account the possibility that the species i may have a thermal distribution, but with a different temperature than that of the photons.

Since the energy density and pressure of a non-relativistic species (i.e., one with mass $m \gg T$) is exponentially smaller than that of a relativistic species (i.e., one with mass $m \ll T$), it is a very convenient and good approximation to include only the relativistic species in the sums for ρ_R and p_R, in which case the above expressions greatly simplify:

$$\begin{aligned} \rho_R &= \frac{\pi^2}{30} g_* T^4, \\ p_R &= \rho_R/3 = \frac{\pi^2}{90} g_* T^4, \end{aligned} \tag{3.61}$$

where g_* counts the total number of effectively massless degrees of freedom (those species with mass $m_i \ll T$), and

$$g_* = \sum_{i=bosons} g_i \left(\frac{T_i}{T}\right)^4 + \frac{7}{8} \sum_{i=fermions} g_i \left(\frac{T_i}{T}\right)^4 . \tag{3.62}$$

The relative factor of 7/8 accounts for the difference in Fermi and Bose statistics. Of course, it is a straightforward matter to obtain an exact expression for $g_*(T)$ from (3.59).[5] Note also that g_* is a function of T since the sum runs over only those species with mass $m_i \ll T$. For $T \ll$ MeV, the only relativistic species are the 3 neutrino species (assuming that they are very light) and the photon; since $T_\nu = (4/11)^{1/3} T_\gamma$ (see below), $g_*(\ll \text{MeV}) = 3.36$. For 100 MeV $\gtrsim T \gtrsim$ 1 MeV, the electron and positron are additional relativistic degrees of freedom and $T_\nu = T_\gamma$; $g_* = 10.75$. For $T \gtrsim 300$ GeV, all the species in the standard model—8 gluons, $W^\pm Z^o$, 3 generations of quarks and leptons, and 1 complex Higgs doublet—should have been relativistic; $g_* = 106.75$. The dependence of $g_*(T)$ upon T is shown in Fig. 3.5.

During the early radiation-dominated epoch ($t \lesssim 4 \times 10^{10}$ sec) $\rho \simeq \rho_R$; and further, when $g_* \simeq$ const, $p_R = \rho_R/3$ (i.e., $w = 1/3$) and $R(t) \propto t^{1/2}$. From this it follows

$$\begin{aligned} H &= 1.66 g_*^{1/2} \frac{T^2}{m_{Pl}} \\ t &= 0.301 g_*^{-1/2} \frac{m_{Pl}}{T^2} \sim \left(\frac{T}{\text{MeV}}\right)^{-2} \text{ sec.} \end{aligned} \tag{3.63}$$

[5]If the contribution to ρ and p of non-relativistic or semi-relativistic species are significant the $g_*(T)$ for ρ and p are not equal; see Fig. 3.5.

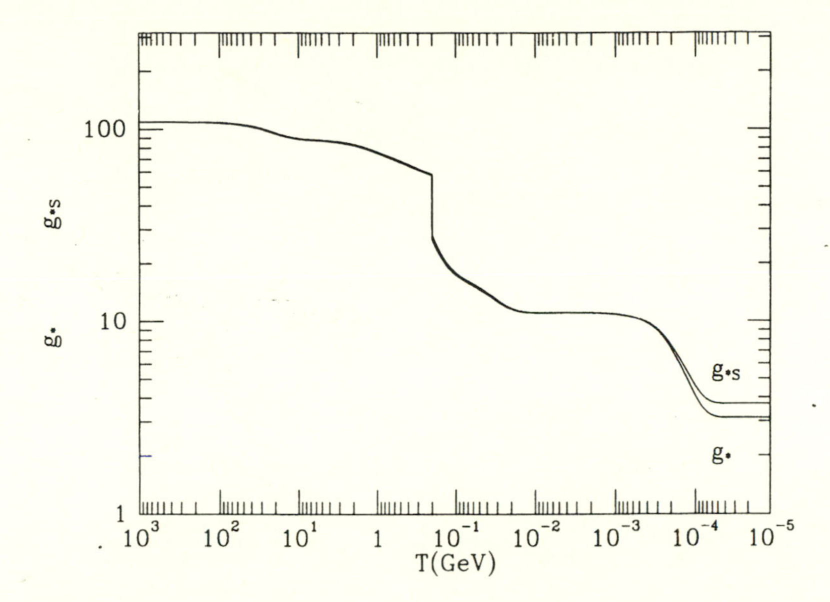

Fig. 3.5: The evolution of $g_*(T)$ as a function of temperature in the $SU(3)_C \otimes SU(2)_L \otimes U(1)_Y$ theory.

3.4 Entropy

Throughout most of the history of the Universe (in particular the early Universe) the reaction rates of particles in the thermal bath, Γ_{int}, were much greater than the expansion rate, H, and local thermal equilibrium (LTE) should have been maintained. In this case the entropy per comoving volume element remains constant. The entropy in a comoving volume provides a very useful fiducial quantity during the expansion of the Universe.

In the expanding Universe, the second law of thermodynamics, as applied to a comoving volume element of unit coordinate volume[6] and physical volume $V = R^3$, implies that

$$TdS = d(\rho V) + pdV = d[(\rho + p)V] - Vdp, \tag{3.64}$$

where ρ and p are the equilibrium energy density and pressure. Moreover,

---

[6]For simplicity, unless otherwise noted, we will take our comoving volume to be of unit coordinate volume.

the integrability condition,

$$\frac{\partial^2 S}{\partial T \partial V} = \frac{\partial^2 S}{\partial V \partial T} \tag{3.65}$$

relates the energy density and pressure:

$$T\frac{dp}{dT} = \rho + p, \tag{3.66}$$

or equivalently[7]

$$dp = \frac{\rho + p}{T} dT. \tag{3.67}$$

If we substitute (3.67) into (3.64), it follows that

$$dS = \frac{1}{T} d[(\rho + p)V] - (\rho + p)V\frac{dT}{T^2} = d\left[\frac{(\rho + p)V}{T} + \text{const}\right]. \tag{3.68}$$

That is, up to an additive constant, the entropy per comoving volume is $S = R^3(\rho + p)/T$. Recall that the first law (energy conservation) can be written as

$$d[(\rho + p)V] = V dp. \tag{3.69}$$

Substituting (3.67) into (3.69), it follows that

$$d\left[\frac{(\rho + p)V}{T}\right] = 0. \tag{3.70}$$

This result implies that in thermal equilibrium, the entropy per comoving volume, S, is conserved.[8]

It is useful to define the entropy density s

$$s \equiv \frac{S}{V} = \frac{\rho + p}{T}. \tag{3.71}$$

---

[7]Note, as it must, (3.67) follows directly from the equilibrium expressions for the pressure and energy density.

[8]Here we have assumed that all chemical potentials are zero—a very good approximation, as all evidence indicates that $|\mu| \ll T$. It is straightforward to include a species with a chemical potential; in this case $TdS = d(\rho V) + pdV - \mu d(nV)$, and $S = R^3(\rho + p - \mu n)/T$.

The entropy density is dominated by the contribution of relativistic particles, so that to a very good approximation,

$$s = \frac{2\pi^2}{45} g_{*S} T^3, \tag{3.72}$$

where

$$g_{*S} = \sum_{i=bosons} g_i \left(\frac{T_i}{T}\right)^3 + \frac{7}{8} \sum_{i=fermions} g_i \left(\frac{T_i}{T}\right)^3. \tag{3.73}$$

For most of the history of the Universe all particle species had a common temperature, and g_{*S} can be replaced by g_*. Note also that s is proportional to the number density of relativistic particles, and that in particular s is related to the photon number density, $s = 1.80 g_{*S} n_\gamma$, where n_γ is the number density of photons. Today $s = 7.04 n_\gamma$. Since g_{*S} is a function of temperature, s and n_γ cannot always be used interchangeably.

Conservation of S implies that $s \propto R^{-3}$, and therefore that $g_{*S} T^3 R^3$ remains constant as the Universe expands. The first fact, that $s \propto R^{-3}$, implies that the physical size of a comoving volume element $\propto R^3 \propto s^{-1}$. Thus the number of some species in a comoving volume, $N \equiv R^3 n$, is equal to the number density of that species divided by s:

$$N \equiv n/s. \tag{3.74}$$

For a species in thermal equilibrium

$$\begin{aligned} N &= \frac{45\zeta(3) g}{2\pi^4 g_{*S}} \qquad T \gg m,\ \mu \\ &= \frac{45 g}{4\sqrt{2}\pi^5 g_{*S}} (m/T)^{3/2} \exp(-m/T + \mu/T) \qquad T \ll m. \end{aligned} \tag{3.75}$$

If the number of a given species in a comoving volume is not changing, i.e., particles of that species are not being created or destroyed, then $N = n/s$ remains constant. The number of a species in thermal equilibrium in a comoving volume is shown in Fig. 3.6 for $\mu = 0$.

As an example of the utility of the ratio n/s, consider baryon number; the baryon number in a comoving volume is

$$\frac{n_B}{s} \equiv \frac{n_b - n_{\bar{b}}}{s}. \tag{3.76}$$

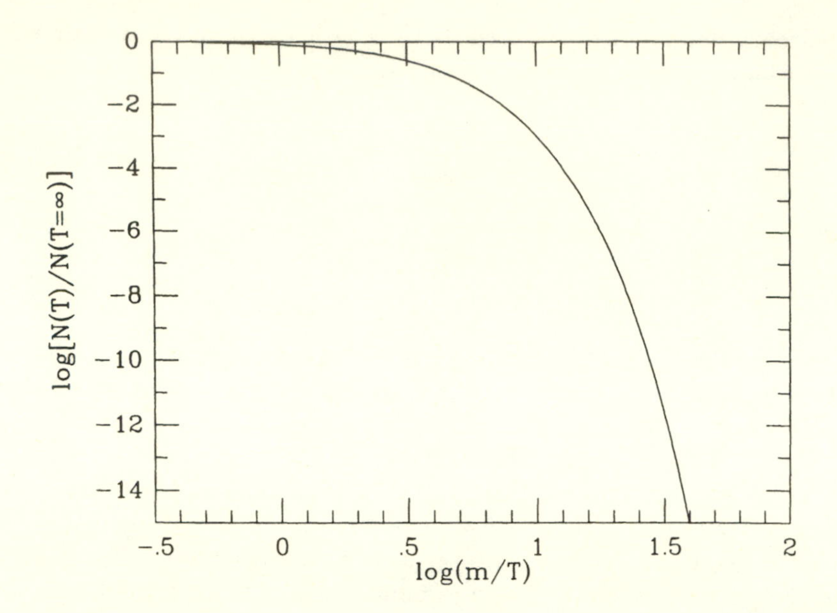

Fig. 3.6: The equilibrium abundance of a species in a comoving volume element, $N = n/s$. Since both n_γ and s vary as T^3, N is also proportional to n/n_γ.

So long as baryon number nonconserving interactions (if such exist in nature) are occurring very slowly, the baryon number in a comoving volume, n_B/s, is conserved. Although $\eta = n_B/n_\gamma = 1.8g_{*S}(n_B/s)$, the baryon number-to-photon ratio does not remain constant with time because g_{*S} changes. During the era of $e^\pm$ annihilation, the number of photons per comoving volume, $N_\gamma = R^3 n_\gamma$, increases by a factor of 11/4, so that η decreases by the same factor. After the time of $e^\pm$ annihilations, however, g_* is constant, and $\eta \simeq 7n_B/s$ and n_B/s can be used interchangeably.

The second fact, that $S = g_{*S}T^3R^3 = \text{const}$, implies that the temperature of the Universe evolves as

$$T \propto g_{*S}^{-1/3} R^{-1}. \tag{3.77}$$

Whenever g_{*S} is constant, the familiar result, $T \propto R^{-1}$, obtains. The factor of $g_{*S}^{-1/3}$ enters because whenever a particle species becomes nonrelativistic and disappears (see Fig. 3.6), its entropy is transferred to the other relativistic particle species still present in the thermal plasma, causing T to *decrease slightly less slowly.*[9]

Massless particles that are decoupled from the heat bath will not share in the entropy transfer as the temperature drops below the mass threshold of a species; instead, the temperature of a massless decoupled species scales as $T \propto R^{-1}$ as we will now show. Consider a massless particle species initially in LTE which decouples at time t_D, temperature T_D, when the scale factor was R_D. The phase-space distribution at decoupling is given by the equilibrium distribution

$$f(\vec{p}, t_D) = [\exp(E/T_D) \pm 1]^{-1}. \tag{3.78}$$

After decoupling, the energy of each massless particle is red shifted by the expansion of the Universe: $E(t) = E(t_D)(R(t_D)/R(t))$. In addition, the number density of particles decreases due to the expansion of the Universe: $n \propto R^{-3}$. As a result, the phase space distribution function, $f(\vec{p}) = d^3n/d^3p$, at time t will be precisely that of a species in LTE with temperature $T(t) = T_D R_D / R(t)$:

$$\begin{aligned} f(\vec{p}, t) &= f(\vec{p}\frac{R}{R_D}, t_D) = \left[\exp\left(\frac{ER}{R_D T_D}\right) \pm 1\right]^{-1} \\ &= [\exp(E/T) \pm 1]^{-1}. \end{aligned} \tag{3.79}$$

Thus the distribution function for a massless particle species remains self-similar as the Universe expands, with the temperature red shifting as R^{-1}

$$\boxed{T = T_D \frac{R_D}{R} \propto R^{-1}} \qquad \text{DECOUPLED} - \text{MASSLESS}, \tag{3.80}$$

and not as $g_{*S}^{-1/3} R^{-1}$ which holds for the particle species remaining in equilibrium.

Next, consider the evolution of the phase space distribution of a massive, non-relativistic ($m \gg T_D$) particle species that was in LTE, and then decouples (when $t = t_D$, $T = T_D$, and $R = R_D$). The momentum of each particle red shifts as the Universe expands: $|\vec{p}(t)| = |\vec{p}_D| R_D / R$; from which it follows that the kinetic energy of each particle red shifts as R^{-2}: $E_K(t) = E_K(t_D) R_D^2 / R^2$. Further, the number density of particles

[9] We have been careful to distinguish between g_* and g_{*S} in this Section. While today $g_* \neq g_{*S}$, earlier than 1 sec the difference should have been small, and henceforth, we will not distinguish between the two.

decreases as R^{-3}. Owing to both of these effects, such a decoupled species will have precisely an equilibrium phase space distribution characterized by temperature T,

$$T = T_D \left(\frac{R_D}{R}\right)^2 \propto R^{-2} \qquad \text{DECOUPLED} - \text{MASSIVE}, \tag{3.81}$$

and chemical potential

$$\mu(t) = m + (\mu_D - m)T(t)/T_D. \tag{3.82}$$

The chemical potential must vary in this way to insure that the number density of particles scales with the expansion as R^{-3}.

So we see that a decoupled species that is either highly relativistic ($T_D \gg m$) or highly non-relativistic ($m \gg T_D$) at decoupling maintains an equilibrium distribution; the former characterized by a temperature $\propto R^{-1}$, and the latter by a temperature $\propto R^{-2}$. For a species that decouples when it is semi-relativistic, $T_D \sim m$, the phase space distribution *does not* maintain an equilibrium distribution in the absence of interactions.

3.5 Brief Thermal History of the Universe

In the strictest mathematical sense it is not possible for the Universe to be in thermal equilibrium, as the FRW cosmological model does not possess a time-like Killing vector. For practical purposes, however, the Universe has for much of its history been very nearly in thermal equilibrium. Needless to say the departures from equilibrium have been very important—without them, the past history of the Universe would be irrelevant, as the present state would be merely that of a system at 2.75 K, very uninteresting indeed! The key to understanding the thermal history of the Universe is the comparison of the particle interaction rates and the expansion rate. Ignoring the temperature variation of g_* for this discussion, $T \propto R^{-1}$ and the rate of change of the temperature $\dot{T}/T$ is just set by the expansion rate: $\dot{T}/T = -H$. So long as the interactions necessary for particle distribution functions to adjust to the changing temperature are rapid compared to the expansion rate, the Universe will, to a good approximation, evolve through a succession of nearly thermal states with temperature decreasing as R^{-1}. A useful rule of thumb is that a reaction is occurring rapidly enough to maintain thermal distributions when $\Gamma \gtrsim H$, where Γ is the interaction rate per particle, $\Gamma \equiv n\sigma|v|$. Here n is the number density of target particles and $\sigma|v|$ is the cross section for interaction times relative velocity (appropriately averaged).

This criterion is simple to understand. Suppose, as is very often the case, $\Gamma \propto T^n$; then the number of interactions a species has from time t onward is given by

$$N_{int} = \int_t^\infty \Gamma(t')dt'. \tag{3.83}$$

Taking the Universe to be radiation dominated, it follows that $N_{int} = (\Gamma/H)|_t/(n-2)$: For $n > 2$, a particle interacts less than one time subsequent to the time when $\Gamma \simeq H$.

We remind the reader that $\Gamma < H$ is not a sufficient condition for a departure from thermal equilibrium to occur: A massless, non-interacting species once in thermal equilibrium will forever maintain an equilibrium distribution with $T \propto R^{-1}$. In order for a departure from equilibrium to develop, the rate for some reaction crucial to maintaining equilibrium must remain less than H.

The correct way to evolve particle distributions is to integrate the Boltzmann equation; this approach will be used in Chapter 5 to precisely calculate the relic density of a stable particle species which decouples, and in Chapter 6 to compute the baryon asymmetry of the Universe. For the moment we will use $\Gamma > H$ ($\Gamma < H$) as the criterion for whether or not a species is coupled to (decoupled from) the thermal plasma in the Universe.

To get a rough understanding of the decoupling of a particle species in the expanding Universe, consider two types of interactions: (i) interactions mediated by a massless gauge boson, e.g., the photon; (ii) interactions mediated by a massive gauge boson, e.g., a $W^\pm$ or Z^o boson below the scale of electroweak symmetry breaking ($T \lesssim 300$ GeV). In the first case the cross section for a $2 \leftrightarrow 2$ scattering of relativistic particles with significant momentum transfer is $\sigma \sim \alpha^2/T^2$ ($g = \sqrt{4\pi\alpha}$ = gauge coupling strength). In the second case, for $T \lesssim m_X$, the corresponding cross section is $\sigma \sim G_X^2 T^2$ where m_X is the mass of the gauge boson, $G_X \sim \alpha/m_X^2$, and the m_X^{-2} factor results from the propagator of the massive gauge boson. For $T \gg m_X$, the cross section is the same as that for massless gauge boson exchange.

For interactions mediated by massless gauge bosons $\Gamma \sim n\sigma|v| \sim \alpha^2 T$; during the radiation-dominated epoch $H \sim T^2/m_{Pl}$, so that $\Gamma/H \sim \alpha^2 m_{Pl}/T$.[10] Therefore, for $T \lesssim \alpha^2 m_{Pl} \sim 10^{16}$ GeV or so, such reactions

[10]We will assume the target particles are relativistic and in LTE, so that $n \sim T^3$.

are occurring rapidly, while for $T \gtrsim \alpha^2 m_{Pl} \sim 10^{16}$ GeV, such reactions are effectively "frozen out." Now consider interactions mediated by massive gauge bosons: $\Gamma \sim n\sigma|v| \sim G_X^2 T^5$ and $\Gamma/H \sim G_X^2 m_{Pl} T^3$. Thus for $m_X \gtrsim T \gtrsim G_X^{-2/3} m_{Pl}^{-1/3} \sim (m_X/100\,\text{GeV})^{4/3}$ MeV, such reactions are occurring rapidly, while for $T \lesssim (m_X/100\,\text{GeV})^{4/3}$ MeV such reactions have effectively frozen out.

We should emphasize that for $T \gtrsim \alpha^2 m_{Pl} \sim 10^{16}$ GeV all perturbative interactions should be frozen out and ineffective in maintaining or establishing thermal equilibrium. Thus the known interactions—plus any new interactions arising from grand unification—are not capable of thermalizing the Universe at temperatures greater than 10^{16} GeV, corresponding to times earlier than about 10^{-38} sec. Perhaps there exist other, as of yet unknown, interactions that can thermalize the Universe at these earliest times (e.g., "strong" gravitational interactions). We should keep in mind the fact that the Universe may not have been in thermal equilibrium during its earliest epoch ($T \gtrsim 10^{16}$GeV).

Fig. 3.7 provides a brief summary of the thermal history of the Universe, based upon extrapolating our present knowledge of the Universe and particle physics back to the Planck epoch ($t \sim 10^{-43}$ sec and $T \sim 10^{19}$ GeV), the point at which quantum corrections to general relativity should render it invalid. At the earliest times the Universe was a plasma of relativistic particles, including the quarks, leptons, gauge bosons, and Higgs bosons. If current ideas are correct, a number of spontaneous symmetry breaking (or SSB) phase transitions should take place during the course of the early history of the Universe. They include the grand unification (or GUT) phase transition at a temperature of 10^{14} to 10^{16} GeV, and the electroweak SSB phase transition at a temperature of about 300 GeV. During these SSB phase transitions some of the gauge bosons and other particles acquire mass via the Higgs mechanism and the full symmetry of the theory is broken to a lower symmetry. Subsequent to the phase transition the interactions mediated by the X bosons which acquire mass will be characterized by a coupling strength G_X, and particles which only interact via such interactions will decouple from the thermal plasma at $T \sim G_X^{-2/3} m_{Pl}^{-1/3}$. SSB phase transitions will be discussed in Chapter 7.

At a temperature of about 100 to 300 MeV ($t \sim 10^{-5}$ sec) the Universe should undergo a transition associated with chiral symmetry breaking and color confinement, after which the strongly-interacting particles are color-singlet–quark-triplet states (baryons) and color-singlet–quark-antiquark states (mesons). While we will not discuss the quark/hadron

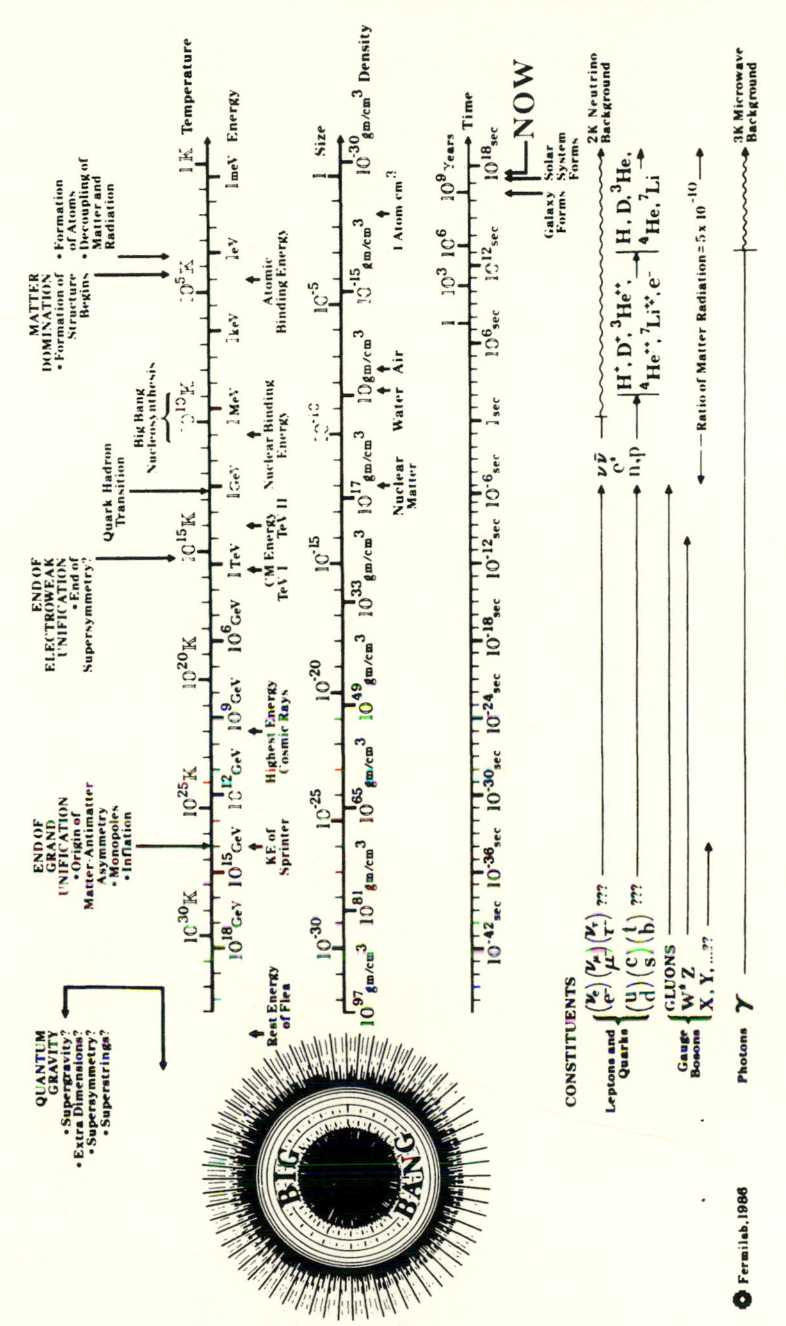

Fig. 3.7: The complete history of the Universe.

transition, the details and the nature (1st order, 2nd order, etc.) of this transition are of some cosmological interest, as local inhomogeneities in the baryon number density may be produced and could possibly affect the outcome of primordial nucleosynthesis (as discussed in the next Chapter).

The epoch of primordial nucleosynthesis follows when $t \sim 10^{-2}$ to 10^2 sec and $T \sim 10$ to 0.1 MeV; the next Chapter will be devoted to a detailed discussion of this very important subject. At present, primordial nucleosynthesis is the earliest test of the standard cosmology. At a time of about 10^{11} sec the matter density becomes equal to that of the radiation. This marks the beginning of the current matter-dominated epoch and the start of structure formation. Structure formation will be addressed in Chapter 9. Finally, at time of about 10^{13}sec, the ions and electrons combine to form atoms, and the matter and radiation decouple, ending the long epoch of near thermal equilibrium that existed in the early Universe. The surface of last scattering for the microwave background radiation is the Universe itself at decoupling. We will now discuss some of these events in more detail.

• *Neutrino Decoupling:* The phenomenon of decoupling of a massless species is nicely illustrated by considering the decoupling of massless neutrinos. In the early Universe neutrinos are kept in equilibrium via reactions of the sort $\bar{\nu}\nu \leftrightarrow e^+e^-$, $\nu e \leftrightarrow \nu e$, etc. The cross section for these weak interaction processes is given by $\sigma \simeq G_F^2 T^2$, where G_F is the Fermi constant. The number density of massless particles is $n \simeq T^3$, and so the interaction rate (per neutrino) is

$$\Gamma_{int} = n\sigma|v| \simeq G_F^2 T^5. \tag{3.84}$$

The ratio of the interaction rate to the expansion rate is

$$\frac{\Gamma_{int}}{H} \simeq \frac{G_F^2 T^5}{T^2/m_{Pl}} \simeq \left(\frac{T}{1\ \mathrm{MeV}}\right)^3. \tag{3.85}$$

At temperatures above 1 MeV, the interaction rate is greater than the expansion rate and neutrinos are in good thermal contact with the plasma. At temperatures below 1 MeV the interaction rate is less than the expansion rate and neutrino interactions are too weak to keep them in equilibrium: Thus at a temperature of order 1 MeV light neutrino species decouple from the plasma. Below 1 MeV the neutrino temperature T_ν scales as R^{-1}. Shortly after neutrino decoupling the temperature drops below the mass of the electron, and the entropy in $e^\pm$ pairs is transferred to the photons, but not to the decoupled neutrinos. For $T \gtrsim m_e$, the

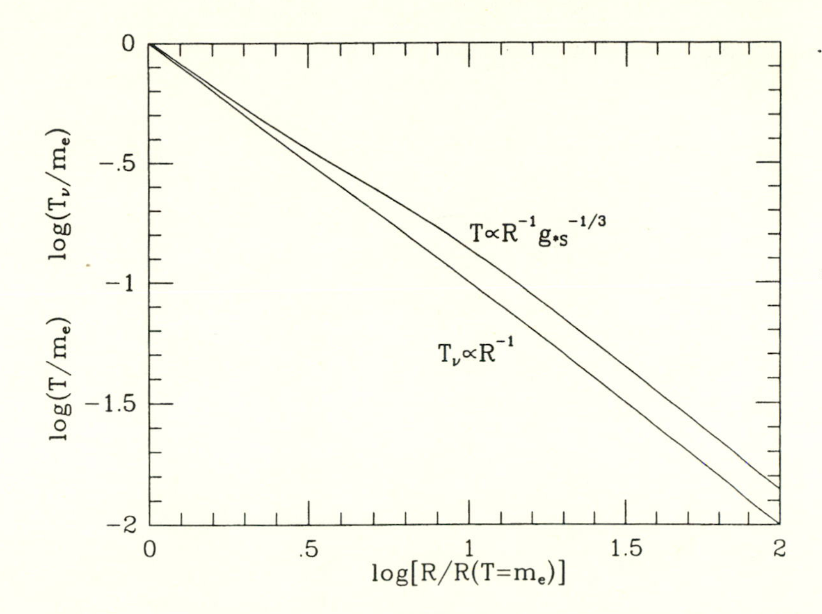

Fig. 3.8: The evolution of T and T_ν through the epoch of $e^\pm$ annihilation.

particle species in thermal equilibrium with photons include the photon ($g = 2$) and $e^\pm$ pairs ($g = 4$), for a value of $g_* = 11/2$. For $T \ll m_e$, only the photons are in equilibrium for a value of $g_* = 2$. For the particles in thermal equilibrium with the photons $g_*(RT)^3$ remains constant; therefore the value of RT after $e^\pm$ annihilation must be larger than that before $e^\pm$ annihilation by a factor of the third-root of the ratio of g_* before $e^\pm$ annihilation (=11/2) to g_* after $e^\pm$ annihilation (=2). Thus the $e^\pm$ entropy transfer increases (RT_γ) by a factor of $(11/4)^{1/3}$, while (RT_ν) remains constant. Therefore today the ratio of T and T_ν should be

$$\frac{T}{T_\nu} = \left(\frac{11}{4}\right)^{1/3} = 1.40 \tag{3.86}$$

which gives $T_\nu = 1.96$ K. The increase of T relative to T_ν is shown in Fig. 3.8. Note that the decrease in g_* does not lead to an actual increase in T, but rather causes T to decrease less slowly than R^{-1}.

Using this result we can now compute the values of g_* and g_{*S} today

(assuming 3 massless neutrino species)

$$g_*(today) = 2 + \frac{7}{8} \times 2 \times 3 \times \left(\frac{4}{11}\right)^{4/3} = 3.36, \quad (3.87)$$

$$g_{*S}(today) = 2 + \frac{7}{8} \times 2 \times 3 \times \frac{4}{11} = 3.91. \quad (3.88)$$

Note that since $T_\nu \neq T$, $g_* \neq g_{*S}$. Also note that since the photon and neutrino species are decoupled their entropies are separately conserved (a fact which we implicitly used above). Using these results we can compute the present energy density and entropy density

$$\begin{aligned}
\rho_R &= \frac{\pi^2}{30} g_* T^4 = 8.09 \times 10^{-34} \text{g cm}^{-3} \\
\Omega_R h^2 &= 4.31 \times 10^{-5} \\
s &= \frac{2\pi^2}{45} g_{*S} T^3 \simeq 2970 \text{ cm}^{-3} \\
n_\gamma &= \frac{2\zeta(3)}{\pi^2} T^3 = 422 \text{ cm}^{-3}.
\end{aligned} \quad (3.89)$$

assuming that $T_0 = 2.75$K.

• *Graviton Decoupling:* Another example of decoupling is that of gravitons. On purely dimensional grounds, the interaction rate for particles with only gravitational strength interactions should be $\Gamma_{int} = n\sigma|v| \simeq G^2T^5 \simeq T^5/m_{Pl}^4$ (which follows by taking $G_X = G_N$ and $\alpha_G \sim 1$). This will become less than the expansion rate, $H \simeq T^2/m_{Pl}$, at temperatures less than about m_{Pl}. At the Planck time, the contribution to g_* from the particles of the standard model is 106.75; of course, if current ideas about unification are correct, there are likely many more relativistic degrees of freedom at $T \sim m_{Pl}$. In any case, if gravitons were in thermal equilibrium at the Planck epoch (or before) and then decoupled, their present temperature should be at most $(3.91/106.75)^{1/3}T \simeq 0.91$ K, corresponding to a number density of less than about 15 cm^{-3}.

• *Matter–Radiation Equality:* If we define ρ_M as the total energy density in "matter" (i.e., in non-relativistic particles: baryons and whatever else), then today $\rho_M = 1.88 \times 10^{-29}\Omega_0 h^2$ g cm^{-3}, where Ω_0 is the fraction of the critical density contributed by matter. Using (3.89), and the fact that $\rho_R/\rho_M \propto R_0/R = 1 + z$, it then follows that the red shift, time, and temperature of equal matter and radiation energy densities are given by

$$\begin{aligned} 1 + z_{EQ} &\equiv R_0/R_{EQ} = 2.32 \times 10^4\ \Omega_0 h^2 \\ T_{EQ} &= T_0(1 + z_{EQ}) = 5.50\ \Omega_0 h^2\ \text{eV} \\ t_{EQ} &\simeq \frac{2}{3} H_0^{-1} \Omega_0^{-1/2} (1 + z_{EQ})^{-3/2} \\ &= 1.4 \times 10^3 (\Omega_0 h^2)^{-2}\ \text{years.} \end{aligned} \tag{3.90}$$

Note that the exact relationship for t_{EQ} [cf. (3.45)] results in a slightly different value: $t_{EQ} = 0.39 H_0^{-1} \Omega_0^{-1/2} (1 + z_{EQ})^{-3/2}$.

• *Photon Decoupling and Recombination:* In the early Universe the matter and radiation were in good thermal contact, because of rapid interactions between the photons and electrons. However, eventually the density of free electrons became too low to maintain thermal contact and matter and radiation decoupled. Roughly speaking this occurs when $\Gamma_\gamma \simeq H$, or equivalently when the mean free path of the photons, $\lambda_\gamma \simeq \Gamma_\gamma^{-1}$, became larger than the Hubble distance, H^{-1}.

The interaction rate of the photons is given by

$$\Gamma_\gamma = n_e \sigma_T, \tag{3.91}$$

where n_e is the number density of free electrons, and σ_T is the Thomson cross section, $\sigma_T = 6.65 \times 10^{-25} \text{cm}^2$. The *equilibrium* abundance of free electrons is determined by the Saha equation. The derivation of the Saha equation will be a useful warm-up exercise for the calculation of elemental abundances in nuclear statistical equilibrium, to be done in the next Chapter.

Let n_H, n_p, and n_e denote the number density of hydrogen, free protons, and free electrons, respectively. For simplicity we will ignore the one ^{4}He nucleus per 10 protons and assume that all the baryons in the Universe are in the form of protons. The charge neutrality of the Universe implies $n_p = n_e$, and baryon number conservation implies that $n_B = n_p + n_H$. In thermal equilibrium, at temperatures less than m_i

$$n_i = g_i \left(\frac{m_i T}{2\pi} \right)^{3/2} \exp\left(\frac{\mu_i - m_i}{T} \right), \tag{3.92}$$

where $i = e,\ p,\ H$, m_i is the mass of species i, and μ_i is the chemical potential of i. In chemical equilibrium, the process $p + e \to H + \gamma$ guarantees that $\mu_p + \mu_e = \mu_H$. The factor of μ_H in n_H can be expressed in terms of μ_e and μ_p, which, in turn, can be expressed in terms of n_p and n_e:

$$n_H = \frac{g_H}{g_p g_e} n_p n_e \left(\frac{m_e T}{2\pi}\right)^{-3/2} \exp(B/T), \tag{3.93}$$

where B is the binding energy of hydrogen, $B \equiv m_p + m_e - m_H = 13.6$ eV. In the pre-exponential factor we have set $m_H = m_p$. In terms of the total baryon number density the fractional ionization is

$$X_e \equiv \frac{n_p}{n_B}. \tag{3.94}$$

Using $g_p = g_e = 2$ and $g_H = 4$, and $n_B = \eta n_\gamma$, the equation for n_H gives the equilibrium fractional ionization

$$\frac{1 - X_e^{eq}}{(X_e^{eq})^2} = \frac{4\sqrt{2}\zeta(3)}{\sqrt{\pi}} \eta \left(\frac{T}{m_e}\right)^{3/2} \exp(B/T). \tag{3.95}$$

This is the Saha equation for the equilibrium ionization fraction.

The baryon-to-photon ratio η is related to $\Omega_B h^2$ by $\eta = (\Omega_B h^2) 2.68 \times 10^{-8}$, and the temperature T is related to the red shift z by $T = (1+z)\, 2.75$ K. Using these relations we have solved (3.95) for the equilibrium ionization for three values of $\Omega_B h^2$, with the results shown in Fig. 3.9. If we define recombination as the point when 90% of the electrons have combined with protons, we see from Fig. 3.9 that recombination occurred at a red shift between 1200 and 1400, depending on the value of $\Omega_B h^2$. Taking $1 + z = 1300$ for the red shift of recombination, the temperature at recombination is

$$T_{rec} = T_0(1 + z_{rec}) = 3575 \text{ K} = 0.308 \text{ eV}. \tag{3.96}$$

Assuming that the Universe was matter dominated at recombination, the age of the Universe at recombination is

$$\begin{aligned} t_{rec} &= \frac{2}{3} H_0^{-1} \Omega_0^{-1/2} (1 + z_{rec})^{-3/2} \\ &= 4.39 \times 10^{12} \, (\Omega_0 h^2)^{-1/2} \text{ sec.} \end{aligned} \tag{3.97}$$

Note that recombination occurs at a temperature of about 0.3 eV, not at $T \sim B \sim 13.6$ eV. This is due to the small value of the prefactors

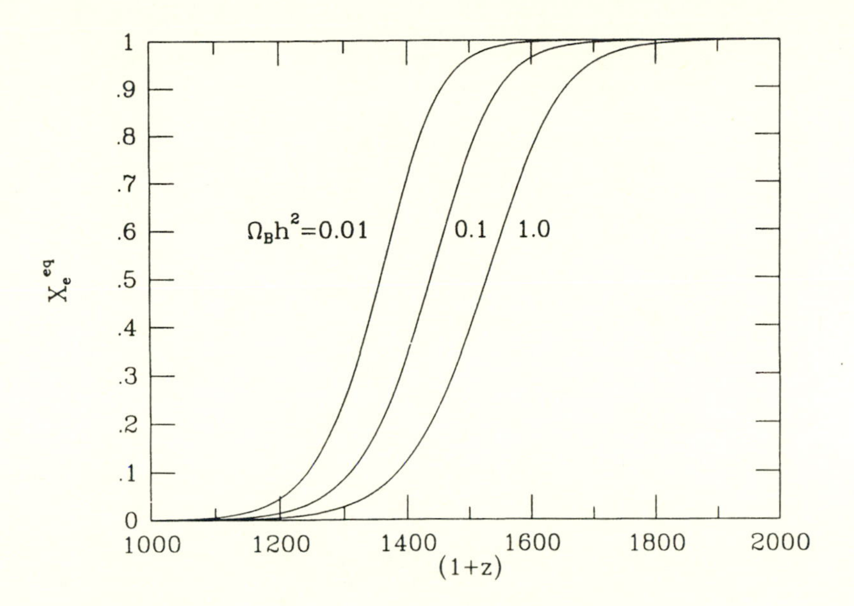

Fig. 3.9: The equilibrium ionization fraction as a function of $(1+z)$.

to $\exp(B/T)$ in (3.95): The large entropy (small η) and $(T/m_e)^{3/2}$ factor result in T_{rec} being much less than the binding energy of hydrogen.

The equilibrium ionization is only appropriate to use when equilibrium is maintained, i.e., when the reaction rate for $e + p \leftrightarrow H + \gamma$ is greater than the expansion rate. This question will be treated in detail in Chapter 5, where it will be shown that the equilibrium ionization is maintained for $(1+z) > 1100$, and that a residual ionization fraction of

$$X_\infty \simeq 3 \times 10^{-5} \Omega_0 / \Omega_B h \tag{3.98}$$

remains for $(1+z) \lesssim 1100$.

Using the equilibrium ionization, and the fact that the density of free electrons is $n_e = X_e n_B = X_e \eta n_\gamma \simeq X_e(\Omega_B h^2)(1+z)^3 1.13 \times 10^{-5}\ \mathrm{cm}^{-3}$, the photon mean free path can be found from (3.91), and compared to the age of the Universe, $t = (2/3)(1+z)^{-3/2} H_0^{-1} \Omega_0^{-1/2}$. The mean free path of the photons and the age of the Universe as a function of $(1+z)$ are shown in Fig. 3.10. The age of the Universe depends upon Ω_0, while the mean free path depends on Ω_B. Decoupling occurs when $\lambda_\gamma \sim \Gamma_\gamma^{-1} \simeq t \sim H^{-1}$.

The red shift at decoupling depends upon Ω_0 and Ω_B (remember they

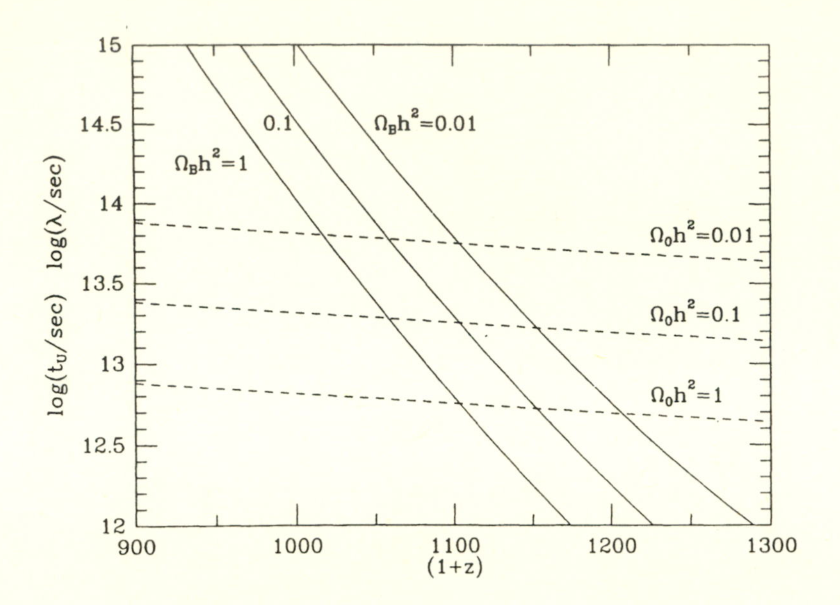

Fig. 3.10: The mean free path of the microwave photons (solid line), and the age of the Universe (dashed line) as a function of $(1 + z)$.

might well be different), but is somewhere in the range $1 + z = 1100$ to 1200. If decoupling occurred at a red shift of $1+z \simeq 1100$, the temperature at decoupling was

$$T_{dec} = T_0(1 + z_{rec}) = 3030 \text{ K} = 0.26 \text{ eV}. \tag{3.99}$$

If the Universe was matter dominated at decoupling, the age of the Universe at decoupling was

$$\begin{aligned} t_{dec} &= \frac{2}{3} H_0^{-1} \Omega_0^{-1/2} (1 + z_{dec})^{-3/2} \\ &= 5.64 \times 10^{12} \, (\Omega_0 h^2)^{-1/2} \text{ sec}. \end{aligned} \tag{3.100}$$

In finishing our discussion of "decoupling," we want to emphasize that there are at least three distinct events occurring at this epoch: the recombination of matter, the decoupling of radiation, and the "freeze in" of a residual ionization. Fitting the results shown in Figs. 3.9 and 3.10 to simple analytic formulae, and stating the results obtained for the freeze in of the residual ionization derived in Section 5.3, we find that these events occur at red shifts of

$$\begin{aligned} 1+z_{dec} &\simeq 1100(\Omega_0/\Omega_B)^{0.018} \simeq 1100-1200 \\ 1+z_{rec} &\simeq 1380(\Omega_B h^2)^{0.023} \simeq 1240-1380 \\ 1+z_{\text{freeze in}} &\simeq \frac{1180}{1+0.021\ln(\Omega_0/\Omega_B)} \simeq 1080-1180, \end{aligned} \quad (3.101)$$

where the red shift range corresponds to $\Omega_B h^2$ =0.01 to 1 and Ω_0/Ω_B =1 to 100, and we have assumed that the Universe is matter dominated at this epoch. For simplicity, and somewhat arbitrarily, we shall define the red shift of "decoupling/recombination/freeze in" to be $1+z_{dec} \equiv 1100$, corresponding to a temperature of 0.26 eV. We shall consistently use this value throughout the monograph.

• *The Baryon Number of the Universe:* The baryon number density is $n_B \equiv n_b - n_{\bar{b}}$, where n_b ($n_{\bar{b}}$) is the baryon (antibaryon) number density. Today, all evidence suggests that there are very few antibaryons in the Universe, and the only baryons present are nucleons, so

$$n_B = n_N = 1.13 \times 10^{-5}\ (\Omega_B h^2)\ \text{cm}^{-3}. \quad (3.102)$$

As discussed earlier, the ratio of the net baryon number to the entropy, n_B/s, corresponds to the net baryon number in a comoving volume, a number that is conserved in the absence of baryon number violating reactions. Today that ratio is also the nucleon number in a comoving volume. Since that ratio serves to quantify the baryon number, we define it to be the *baryon number of the Universe*:

$$B \equiv \frac{n_B}{s} = 3.81 \times 10^{-9}\ (\Omega_B h^2). \quad (3.103)$$

Since the epoch of $e^{\pm}$ annihilation the entropy density s and the photon number density n_γ have been related by a constant factor, $s \simeq 7.04 n_\gamma$, so that

$$\eta \simeq 7B \simeq 2.68 \times 10^{-8} (\Omega_B h^2). \quad (3.104)$$

As we will discuss in the next Chapter, primordial nucleosynthesis constrains η to the interval (4 to 7)$\times 10^{-10}$, corresponding to $B \simeq$ (6 to 10)

$\times 10^{-11}$.[11] The inverse of B, $s/n_B \simeq$ (1 to 2) $\times 10^{10}$, is the entropy per baryon; we live in a Universe with very high entropy.

3.6 Horizons

In Section 2.2 it was shown that the proper distance to the horizon in a Robertson-Walker space is given by

$$d_H(t) = R(t)\int_0^t \frac{dt'}{R(t')}. \tag{3.105}$$

If $R(t) \propto t^n$, then for $n < 1$, $d_H(t)$ is finite and equal to $t/(1-n)$. That is, in spite of the fact that all physical distances approach zero as $R \to 0$, the expansion of the Universe precludes all but a tiny fraction of the Universe from being in causal contact. This is a vexing feature of the standard cosmology to which we will return again. If curvature effects can be neglected (which is a good approximation in the early Universe), the Friedmann equation ($\dot{R}/R = (8\pi G\rho/3)^{1/2}$) implies $R(t) \propto t^{2/3(1+w)}$ for $p = w\rho$. For a radiation-dominated Universe, $R \propto t^{1/2}$, and $d_H(t) = 2t$, while for a matter-dominated Universe $R \propto t^{2/3}$, and $d_H(t) = 3t$.

It is straightforward to find a more general expression for d_H that includes the effect of curvature. The integral for d_H can be written as

$$d_H(t) = R(t)\int_0^t \frac{dt'}{R(t')} = R(t)\int_0^{R(t)} \frac{dR(t')}{\dot{R}(t')R(t')}. \tag{3.106}$$

Generalizing (3.20) for $\rho \propto R^{-3(1+w)}$,

$$\dot{R}^2 = R_0^2 H_0^2 \left[1 - \Omega_0 + \Omega_0 \left(\frac{R_0}{R}\right)^{1+3w}\right], \tag{3.107}$$

the expression for $d_H(t)$ becomes

$$d_H(t) = \frac{1}{H_0(1+z)}\int_0^{(1+z)^{-1}} \frac{dx}{\left[x^2(1-\Omega_0) + \Omega_0 x^{(1-3w)}\right]^{1/2}}. \tag{3.108}$$

---

[11] Here, and throughout, η is defined as the value of the baryon-to-photon ratio at the present epoch.

From this expression for d_H we see that if $w < -1/3$, the integral will diverge and the horizon distance will be infinite, i.e., no particle horizon exists.

For a matter-dominated Universe ($w = 0$)

$$\begin{aligned} d_H(t) &= \frac{1}{H_0(1+z)\sqrt{\Omega_0 - 1}} \cos^{-1}\left[1 - \frac{2(\Omega_0 - 1)}{\Omega_0(1+z)}\right] \qquad \Omega_0 > 1 \\ &= 2H_0^{-1}(1+z)^{-3/2} \qquad \Omega_0 = 1 \qquad\qquad (3.109) \\ &= \frac{1}{H_0(1+z)\sqrt{1-\Omega_0}} \cosh^{-1}\left[1 + \frac{2(1-\Omega_0)}{\Omega_0(1+z)}\right] \qquad \Omega_0 < 1 \end{aligned}$$

For $1 + z \gg \Omega_0^{-1}$, $d_H(t) \to 2/H_0\sqrt{\Omega_0}(1+z)^{3/2} = 3t$ as expected.

It is interesting to calculate the fraction of a matter-dominated Universe that is visible as a function of time. For $k = 0$ or $k = -1$, space is infinite, so the question only makes sense for $k = +1$. Consider a matter-dominated, $k = +1$ RW Universe. The value of the scale factor $R(t)$ can formally be related to its present value R_0 by $R(t) = R_0/(1+z)$ in both the past ($z > 0$) and the future ($z < 0$), so that we can use the above expressions to compute $d_H(t)$. At the moment of recollapse the scale factor achieves its maximum value: $R_{\max}/R_0 = \Omega_0/(\Omega_0 - 1)$, corresponding to a minimum (negative) value of z: $z_{\min} = -\Omega_0^{-1}$. The circumference of the 3-sphere at time t is $2\pi R(t) = 2\pi R_0/(1+z)$, and so the ratio of the horizon distance to the circumference is

$$\frac{d_H(t)}{2\pi R(t)} = \frac{1}{2\pi} \cos^{-1}\left[1 - \frac{2(\Omega_0 - 1)}{\Omega_0(1+z)}\right]. \qquad (3.110)$$

As $(1+z) \to \infty$, $d_H(t)/2\pi R(t) \to \sqrt{(\Omega_0 - 1)/\pi^2\Omega_0(1+z)} \to 0$. As expected, the fraction of the Universe that is visible vanishes at early times. At the epoch of maximum expansion, $z = z_{\min}$, the ratio $d_H(t)/2\pi R(t) = 1/2$. At the point of maximum expansion it is possible to see half way around the Universe. Since the subsequent contracting phase is just the time reverse of the expanding phase, we can infer that by the time of the crunch $d_H(t)/2\pi R(t) \to 1$, i.e., the entire Universe will become visible.

Returning again to the expressions (3.105)–(3.109) for $d_H(t)$, by changing the limits of integration we can obtain a relationship between the coordinate r_1 of a source (e.g., a galaxy) whose radiation suffers a red shift

z in propagating to us (at the present epoch):

$$\int_0^{r_1} \frac{dr}{(1-kr^2)^{1/2}} = \int_{R_1}^{R_0} \frac{dR(t')}{\dot{R}(t')R(t')}. \tag{3.111}$$

For $w = 0$, $k = -1$, 0, or $+1$, the solution is

$$r_1 = \frac{2\Omega_0 z + (2\Omega_0 - 4)(\sqrt{\Omega_0 z + 1} - 1)}{H_0 R_0 \Omega_0^2 (1+z)}, \tag{3.112}$$

and the very useful quantity $dr/(1-kr^2)^{1/2}$ is related to z and dz by

$$\frac{dr_1}{(1-kr_1^2)^{1/2}} = \frac{dz}{R_0 H_0 (1+z)[1 - \Omega_0 + \Omega_0(1+z)]^{1/2}}. \tag{3.113}$$

It is useful to know the entropy and number of baryons contained within the horizon. During the radiation-dominated era $d_H(t) = 2t$ and $s = 2\pi^2 g_* T^3/45$, so that

$$\begin{aligned} S_{\rm HOR} &= \frac{4\pi}{3} t^3 s = 0.050 g_*^{-1/2} \left(\frac{m_{Pl}}{T}\right)^3 \\ (N_B)_{\rm HOR} &= B S_{\rm HOR} = 1.9 \times 10^{-10} (\Omega_B h^2) g_*^{-1/2} \left(\frac{m_{Pl}}{T}\right)^3 \\ (M_B)_{\rm HOR} &= 0.29 M_\odot (\Omega_B h^2) g_*^{-1/2} \left(\frac{\rm MeV}{T}\right)^3 . \end{aligned} \tag{3.114}$$

During the matter-dominated epoch, $d_H(t) \simeq 3t$ and $s = s_0(1+z)^3 = 2970(1+z)^3\ {\rm cm}^{-3}$, so that

$$\begin{aligned} S_{\rm HOR} &\simeq \frac{4\pi}{3} t^3 s = 2.9 \times 10^{87} (\Omega_0 h^2)^{-3/2} (1+z)^{-3/2} \\ (N_B)_{\rm HOR} &= B S_{\rm HOR} \simeq 1.1 \times 10^{79} (\Omega_B/\Omega_0^{3/2} h)(1+z)^{-3/2} \\ (M_B)_{\rm HOR} &\simeq 9.4 \times 10^{21} M_\odot (\Omega_B/\Omega_0^{3/2} h)(1+z)^{-3/2}. \end{aligned} \tag{3.115}$$

We see that today the horizon contains an entropy of order 10^{88}, and a baryon number of order 10^{79}, while at very early times the horizon only contained an entropy of order $(m_{Pl}/T)^3$ and a baryon number of order

$10^{-10}(m_{Pl}/T)^3$. The impact of particle horizons in the standard model is very clear.

We will end this Chapter with an often asked question: How big is our Universe? A simple question with a complicated answer! For the closed model the answer is easy: The circumference of our three sphere is just $2\pi R_{\rm curv}(t) = 2\pi H^{-1}(t)/|\Omega(t)-1|^{1/2}$. The negatively curved and flat models are now, and were always, infinite in spatial extent. For these models, a more sensible question to ask is: How large was the presently observable Universe at some earlier epoch? For simplicity, let us consider the flat model. As just shown, the distance to the horizon today is $d_H = 3t_0 = 2H_0^{-1}$—the Universe observable today has a diameter $D_0(t_0) = 4H_0^{-1}$. At some earlier time, this diameter corresponded to a distance[12]

$$D_0(t) = \frac{R(t)}{R_0} D_0(t_0) = \frac{R(t)}{R_0} 4H_0^{-1}. \tag{3.116}$$

Assuming no significant entropy production since time t, $R(t)$ and R_0 are related through the constancy of S: $g_{*S}(t_0)R_0^3T_0^3 = g_{*S}(t)R^3(t)T^3(t)$. Using $g_{*S}(t_0) = 3.91$, it follows that

$$D_0(t) = 4\left[\frac{3.91}{g_{*S}(t)}\right]^{1/3} \frac{T_0}{T(t)} H_0^{-1}. \tag{3.117}$$

As an example, let us calculate the diameter of the presently observable Universe at the Planck time. Assuming that the Universe is radiation dominated at the Planck era, the temperature at the Planck time, T_{Pl}, and the Planck time, $t_{Pl} = m_{Pl}^{-1}$, are related by

$$\begin{aligned} t_{Pl} &= 0.301 g_*^{-1/2} m_{Pl}/T_{Pl}^2 \\ T_{Pl} &= 0.55 g_*^{-1/4} m_{Pl}. \end{aligned} \tag{3.118}$$

(Note that the temperature at the Planck time is not equal to the Planck energy.) Substituting into (3.117), we find

$$D_0(t_{Pl}) = \frac{11.5}{g_{*S}^{1/12}(t_{Pl})} \frac{T_0}{m_{Pl}} H_0^{-1} = 1.4h^{-1} \times 10^{-3}\text{cm}, \tag{3.119}$$

[12]It is important to note that $D_0(t)$ does *not* correspond to the diameter of the observable Universe at time t, but rather the diameter at time t of the region that eventually grows to become the presently observable Universe.

o

Fig. 3.11: The presently observable Universe at the Planck time, assuming $h = 0.4$ (100 × magnification).

where the second expression follows from taking $g_{*S} = 106.75$—the value that follows by including just the degrees of freedom in the $SU(3)_C \otimes SU(2)_L \otimes U(1)_Y$ model. The presently observable Universe at the Planck time is shown in Fig. 3.11.

3.7 References

References to other developments of the standard cosmology were given in the previous Chapter, cf. Sec. 2.4. For a discussion of the determination of the cosmological parameters, H_0, q_0, Ω_0, Ω_B, t_0, etc., we refer the reader to the papers reprinted in Chapters 1 and 3 of *The Early Universe: Reprints*, and references therein.

4

BIG-BANG NUCLEOSYNTHESIS

4.1 Nuclear Statistical Equilibrium

As a first step to understanding primordial nucleosynthesis we will consider the consequences of nuclear statistical equilibrium (NSE) among the light nuclear species. In *kinetic* equilibrium, the number density of a very nonrelativistic nuclear species $A(Z)$ with mass number A and charge Z is given by

$$n_A = g_A \left(\frac{m_A T}{2\pi}\right)^{3/2} \exp\left(\frac{\mu_A - m_A}{T}\right), \tag{4.1}$$

where μ_A is the chemical potential of the species. If the nuclear reactions that produce nucleus A out of Z protons and $A - Z$ neutrons occur rapidly compared to the expansion rate, *chemical* equilibrium also obtains. In chemical equilibrium, the chemical potential of the species $A(Z)$ is related to the neutron and proton chemical potentials by

$$\mu_A = Z\mu_p + (A - Z)\mu_n. \tag{4.2}$$

Equation (4.1) also applies to the neutron and proton. Using this fact, when chemical equilibrium pertains we can express $\exp(\mu_A/T)$ in terms of the neutron and proton number densities:[1]

$$\begin{aligned} \exp(\mu_A/T) &= \exp\left[(Z\mu_p + (A - Z)\mu_n)/T\right] \\ &= n_p^Z n_n^{A-Z} \left(\frac{2\pi}{m_N T}\right)^{3A/2} 2^{-A} \exp\left[(Zm_p + (A - Z)m_n)/T\right]. \end{aligned} \tag{4.3}$$

[1] In all pre-exponential factors, the difference between m_n, m_p, and m_A/A is not important, and all will be taken to be equal to a common mass, the nucleon mass, m_N.

^{A}Z	B_A	g_A
^{2}H	2.22 MeV	3
^{3}H	6.92 MeV	2
^{3}He	7.72 MeV	2
^{4}He	28.3 MeV	1
^{12}C	92.2 MeV	1

Table 4.1: The binding energies of some light nuclei.

Recalling the definition of the binding energy of the nuclear species $A(Z)$,

$$B_A \equiv Zm_p + (A-Z)m_n - m_A, \tag{4.4}$$

and substituting (4.3) into (4.1), the abundance of species $A(Z)$ is

$$n_A = g_A A^{3/2} 2^{-A} \left(\frac{2\pi}{m_N T}\right)^{3(A-1)/2} n_p^Z n_n^{A-Z} \exp(B_A/T). \tag{4.5}$$

A list of binding energies of some light nuclei is given in Table 4.1.

Since particle number densities in the expanding Universe decrease as R^{-3} (for constant number per comoving volume), it is useful to use the total nucleon density, $n_N = n_n + n_p + \sum_i (An_A)_i$, as a fiducial quantity and to consider the mass fraction contributed by nuclear species $A(Z)$,

$$\begin{aligned} X_A &\equiv \frac{n_A A}{n_N} \\ \sum_i X_i &= 1. \end{aligned} \tag{4.6}$$

Using this definition we find that in NSE the mass fraction of species $A(Z)$ is given by

$$\begin{aligned} X_A &= g_A[\zeta(3)^{A-1}\pi^{(1-A)/2}2^{(3A-5)/2}]A^{5/2}(T/m_N)^{3(A-1)/2} \\ &\quad \times \eta^{A-1} X_p^Z X_n^{A-Z} \exp(B_A/T), \end{aligned} \tag{4.7}$$

where as usual,

$$\eta \equiv \frac{n_N}{n_\gamma} = 2.68 \times 10^{-8}\ (\Omega_B h^2) \tag{4.8}$$

is the present baryon-to-photon ratio. The fact that the Universe is "hot" ($\eta \ll 1$, i.e., very high entropy per baryon) is of the utmost significance to primordial nucleosynthesis.[2] After considering the "initial conditions" for primordial nucleosynthesis we will describe nucleosynthesis in three simple steps.

4.2 Initial Conditions ($T \gg 1$ MeV, $t \ll 1$ sec)

The ratio of neutrons to protons is of particular importance to the outcome of primordial nucleosynthesis, as essentially all the neutrons in the Universe become incorporated into ^{4}He. The balance between neutrons and protons is maintained by the weak interactions (here $\nu \equiv \nu_e$):

$$\begin{aligned} n &\longleftrightarrow p + e^- + \bar{\nu}, \\ \nu + n &\longleftrightarrow p + e^- \\ e^+ + n &\longleftrightarrow p + \bar{\nu}. \end{aligned} \tag{4.9}$$

When the rates for these interactions are rapid compared to the expansion rate H, chemical equilibrium obtains,

$$\mu_n + \mu_\nu = \mu_p + \mu_e, \tag{4.10}$$

from which it follows that in chemical equilibrium

$$n/p \equiv n_n/n_p = X_n/X_p = \exp[-Q/T + (\mu_e - \mu_\nu)/T] \tag{4.11}$$

where $Q \equiv m_n - m_p = 1.293$ MeV. Based upon the charge neutrality of the Universe we can infer that $\mu_e/T \sim (n_e/n_\gamma) = (n_p/n_\gamma) \sim \eta$, from which it follows that $\mu_e/T \sim 10^{-10}$, cf. (3.55). The electron neutrino number of the Universe is similarly related to μ_ν/T; however, since the relic neutrino background has not been detected, none of the neutrino lepton numbers (e, μ, or τ) is known. For the moment we will assume that the lepton numbers, like the baryon number, are small ($\ll 1$), so that $|\mu_\nu|/T \ll 1$. In our discussion of non-standard scenarios of nucleosynthesis we will return to the possibility that they might be of order unity (or larger). Having

---

[2]For purposes of comparison, in a star like our sun, $n_\gamma/n_N \sim 10^{-2}$; even in the post-collapse core of a supernova n_N/n_γ is only a few. Indeed, the entropy of the Universe is enormous.

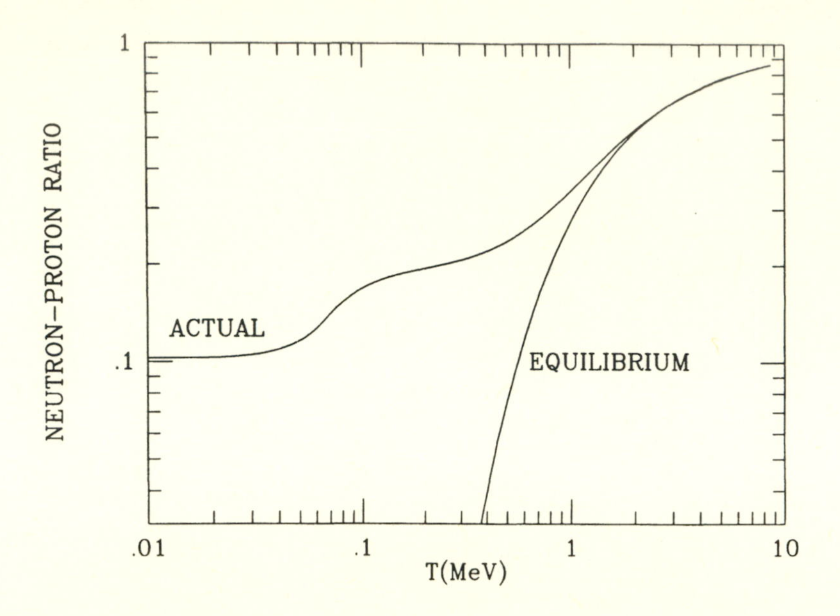

Fig. 4.1: The equilibrium and actual values of the neutron to proton ratio.

made this assumption, the equilibrium value of the neutron-to-proton ratio is

$$\left(\frac{n}{p}\right)_{EQ} = \exp(-Q/T) \tag{4.12}$$

which is shown in Fig. 4.1.

Now consider the rates for the weak interactions that interconvert neutrons and protons; the rates (per nucleon per time) for these reactions are found by integrating the square of the matrix element for a given process, weighted by the available phase-space densities of particles (other than the initial nucleon), while enforcing four-momentum conservation. As an example, the rate for $pe \to \nu n$ is given by

$$\begin{aligned}\Gamma_{pe\to\nu n} &= \int f_e(E_e)[1 - f_\nu(E_\nu)]|\mathcal{M}|^2_{pe\to\nu n}(2\pi)^{-5}\delta^4(p+e-\nu-n) \\ &\times \frac{d^3p_e}{2E_e}\frac{d^3p_\nu}{2E_\nu}\frac{d^3p_n}{2E_n}.\end{aligned} \tag{4.13}$$

All of these processes have in common a factor from the nuclear matrix element for the β-decay of the neutron,

$$|\mathcal{M}|^2 \propto G_F^2(1+3g_A^2), \tag{4.14}$$

where $g_A \simeq 1.26$ is the axial-vector coupling of the nucleon. This factor can be expressed in terms of the mean neutron lifetime τ_n (or the neutron half life $\tau_{1/2}(n) = \ln(2)\tau_n$) as

$$\tau_n^{-1} = \Gamma_{n\to pe\nu} = \frac{G_F^2}{2\pi^3}(1+3g_A^2)m_e^5\lambda_0 \tag{4.15}$$

where

$$\lambda_0 \equiv \int_1^q d\epsilon\, \epsilon(\epsilon-q)^2(\epsilon^2-1)^{1/2} \simeq 1.636 \tag{4.16}$$

simply represents a numerical factor from the phase space integral for neutron decay [9].

The neutron–proton mass difference and the electron mass determine the limits of integration for these rates. In terms of the dimensionless quantities $q = Q/m_e$, $\epsilon = E_e/m_e$, $z = m_e/T$, and $z_\nu = m_e/T_\nu$,

$$\Gamma_{pe\to\nu n} = (\tau_n\lambda_0)^{-1}\int_q^\infty d\epsilon \frac{\epsilon(\epsilon-q)^2(\epsilon^2-1)^{1/2}}{[1+\exp(\epsilon z)][1+\exp((q-\epsilon)z_\nu)]}, \tag{4.17}$$

In the high-temperature and low-temperature limits

$$\Gamma_{pe\to\nu n} \longrightarrow \begin{cases} \tau_n^{-1}(T/m_e)^3\exp(-Q/T) & T \ll Q,\ m_e \\ \frac{7}{60}\pi(1+3g_A^2)G_F^2T^5 \simeq G_F^2T^5 & T \gg Q,\ m_e. \end{cases} \tag{4.18}$$

By comparing Γ to the expansion rate of the Universe, $H \simeq 1.66 g_*^{1/2}T^2/m_{Pl}$ $\simeq 5.5T^2/m_{Pl}$, we find that

$$\Gamma/H \sim (T/0.8\,\mathrm{MeV})^3 \tag{4.19}$$

for $T \gtrsim m_e$. Thus at temperatures greater than about 0.8 MeV one expects the neutron-to-proton ratio to be equal to its equilibrium value, which at temperatures much greater than an MeV implies $X_n \simeq X_p$.

At temperatures greater than about an MeV, not only are the rates for the weak interactions more rapid than the expansion rate, but so are the rates for the nuclear reactions that build up the light elements, and so NSE should obtain. For purposes of illustration, consider the following system of light elements: neutrons, protons, deuterons, ^{3}He nuclei, ^{4}He nuclei, and ^{12}C nuclei. We choose to represent mass 3 with the more tightly-bound ^{3}He nucleus, and "metals" with ^{12}C. In NSE the mass fractions of the various nuclear species are:

$$X_n/X_p = \exp(-Q/T) \tag{4.20}$$

$$X_2 = 16.3(T/m_N)^{3/2}\eta\exp(B_2/T)X_nX_p \tag{4.21}$$

$$X_3 = 57.4(T/m_N)^3\eta^2\exp(B_3/T)X_nX_p^2 \tag{4.22}$$

$$X_4 = 113(T/m_N)^{9/2}\eta^3\exp(B_4/T)X_n^2X_p^2 \tag{4.23}$$

$$X_{12} = 3.22\times 10^5(T/m_N)^{33/2}\eta^{11}\exp(B_{12}/T)X_n^6X_p^6 \tag{4.24}$$

$$1 = X_n + X_p + X_2 + X_3 + X_4 + X_{12}. \tag{4.25}$$

In Fig. 4.2 the NSE abundances for this system are displayed. It is significant to note that although the binding energies per nucleon are of the order of 1 to 8 MeV, the equilibrium abundance of nuclear species does not become of order unity until a temperature of order 0.3 MeV. This is due to the high entropy of the Universe, that is, the very small value of η. Although for temperatures less than a few MeV nuclei are favored on energetic grounds, entropy considerations favor free nucleons, and the entropy of the Universe is very high. As a rough estimate of when a nuclear species A becomes thermodynamically favored, let us solve for the temperature T_{NUC} when $X_A \sim 1$ (assuming $X_n \sim X_p \sim 1$):

$$T_{NUC} \simeq \frac{B_A/(A-1)}{\ln(\eta^{-1}) + 1.5\ln(m_N/T)}. \tag{4.26}$$

For deuterium, $T_{NUC} \simeq 0.07$ MeV, for ^{3}He $T_{NUC} \simeq 0.11$ MeV, for ^{4}He $T_{NUC} \simeq 0.28$ MeV, and for ^{12}C $T_{NUC} \simeq 0.25$ MeV. The fact that the abundances of the light elements did not begin to build up until temperatures of much less than an MeV is often blamed on the very small binding energy of the deuteron—the so called "deuterium bottleneck." In fact, the NSE abundances of ^{4}He and ^{12}C (nuclei with large binding energies) are very small until temperatures that are less than about 0.3 MeV—a fact that traces to the high entropy of the Universe and not the small binding energy of the deuteron. At a somewhat lower temperature ($T \sim 0.1$ MeV), the low abundances of D and ^{3}He delay nucleosynthesis briefly; the myth of the "deuterium bottleneck" is just that.

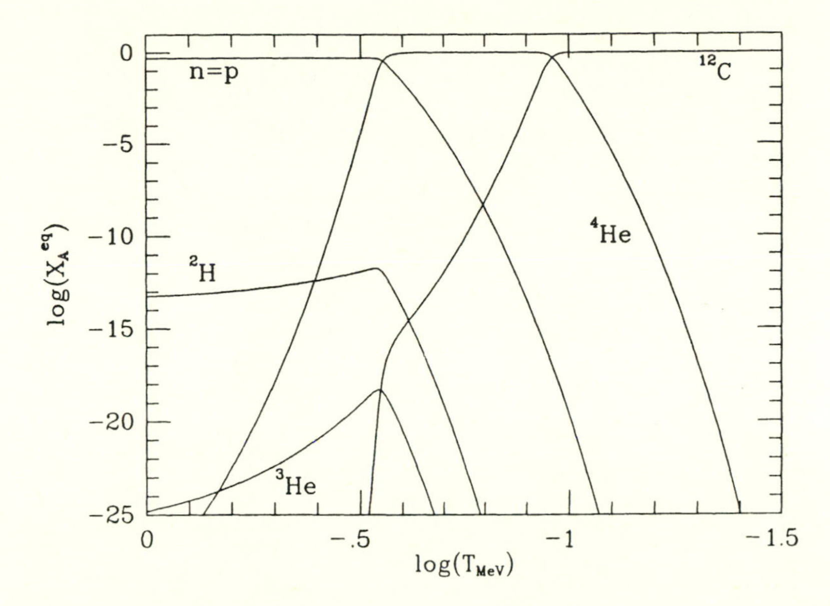

Fig. 4.2: The NSE mass fractions for the system of n, p, D, ^{3}He, ^{4}He, and ^{12}C as a function of temperature. For simplicity we have taken $X_n = X_p$.

4.3 Production of the Light Elements: 1–2–3

• *Step 1* ($t = 10^{-2}$sec, $T = 10$ MeV): At this epoch, the energy density of the Universe is dominated by radiation, and the relativistic degrees of freedom are: $e^{\pm}$, γ, and 3 neutrino species, so that $g_* = 10.75$.[3] All the weak rates are much larger than the expansion rate H, so $(n/p) = (n/p)_{EQ} \simeq 1$ and $T_\nu = T$. The light elements are in NSE, but they have very small abundances due to the fact that η is so small. For example, with $\eta = 10^{-9}$

$$
\begin{aligned}
X_n,\, X_p &= 0.5 \\
X_2 &= 4.1(T/m_N)^{3/2}\eta \exp(2.22/T_{\rm MeV}) \simeq 6\times 10^{-12}
\end{aligned}
$$

[3]These are the light ($\lesssim$ MeV) particle species known to exist. The upper limits to the masses of the 3 neutrino species are about: 13 eV (electron neutrino); 0.25 MeV (muon neutrino); and 35 MeV (tau neutrino). The electron and muon neutrinos are clearly "light;" we will also take the tau neutrino to be light. We will use $g_* = 10.75$ as the value for the standard scenario; later, we consider the possibility of additional, hypothetical light species. If the tau neutrino is heavy ($\gg$ MeV), then $g_* = 9$.

$$
\begin{aligned}
X_3 &= 7.2(T/m_N)^3\eta^2\exp(7.72/T_{\rm MeV}) \simeq 2\times 10^{-23} \\
X_4 &= 7.1(T/m_N)^{9/2}\eta^3\exp(28.3/T_{\rm MeV}) \simeq 2\times 10^{-34} \\
X_{12} &= 79(T/m_N)^{33/2}\eta^{11}\exp(92.2/T_{\rm MeV}) \simeq 2\times 10^{-126} \quad (4.27)
\end{aligned}
$$

where $T_{\rm MeV} = T/{\rm MeV}$.

• *Step 2* ($t \simeq 1$ sec, $T = T_F \simeq 1$ MeV): Shortly before this epoch, the 3 neutrino species decouple from the plasma, and as we have discussed in the previous Chapter, a little later ($T \simeq m_e/3$) the $e^\pm$ pairs annihilate, transferring their entropy to the photons alone and thereby raising the photon temperature relative to that of the neutrinos by a factor of $(11/4)^{1/3}$. At about this time the weak interactions that interconvert neutrons and protons freeze out (Γ becomes smaller than H). When this occurs, the neutron-to-proton ratio is given approximately by its equilibrium value,

$$
\left(\frac{n}{p}\right)_{freeze-out} = \exp(-Q/T_F) \simeq \frac{1}{6}. \quad (4.28)
$$

After freeze-out the neutron-to-proton ratio does not remain truly constant, but actually slowly decreases due to occassional weak interactions (eventually dominated by free neutron decays). The deviation of n/p from its equilibrium value becomes significant by the time nucleosynthesis begins (see Fig. 4.1). At this time the light nuclear species are still in NSE, with very small abundances:

$$
\begin{aligned}
X_n &\simeq 1/7 \\
X_p &\simeq 6/7 \\
X_2 &\simeq 10^{-12} \\
X_3 &\simeq 10^{-23} \\
X_4 &\simeq 10^{-28} \\
X_{12} &\simeq 10^{-108} \quad (4.29)
\end{aligned}
$$

• *Step 3* ($t = 1$ to 3 minutes, $T = 0.3$ to 0.1 MeV): At about this time g_* has decreased to its value today, 3.36, because the $e^\pm$ pairs have disappeared and transferred their entropy to the photons. The neutron-to-proton ratio has decreased from $\sim 1/6$ to $\sim 1/7$ due to occassional weak interactions; compare this to the equilibrium value for $T = 0.3$

MeV, $(n/p)_{EQ} = 1/74$. At a temperature of 0.3 MeV the NSE value of the mass fraction of ^{4}He rapidly approaches unity. Shortly before this ($T \sim 0.5$ MeV), the actual amount of ^{4}He present first falls below its NSE value. This occurs because the rates for the processes that synthesize ^{4}He [D(D,n)^{3}He(D,p)^{4}He, D(D,p)^{3}H(D,n)^{4}He, and D(D,γ)^{4}He] are not fast enough to keep up with the rapidly increasing "NSE demand" for ^{4}He. The reaction rates, Γ, which are proportional to $n_A\langle\sigma|v|\rangle$, are not fast enough for two reasons: (1) While the abundances of D, ^{3}He, and ^{3}H are actually beginning to exceed their NSE values, the NSE abundances are still very small, $X_i = 10^{-12}$, 2×10^{-19}, 5×10^{-19}, respectively. For this reason the number densities of these fuels, $n_A = (X_A/A)\eta n_\gamma$, are small. (2) Coulomb-barrier suppression is beginning to become significant: The thermal average of the barrier-penetration factor is given by

$$\langle\sigma|v|\rangle \propto \exp[-2\bar{A}^{1/3}(Z_1Z_2)^{2/3}T_{\rm MeV}^{-1/3}], \tag{4.30}$$

where $\bar{A} = A_1A_2/(A_1+A_2)$, and $\langle\sigma|v|\rangle$ indicates the thermally-averaged cross section times relative velocity. Until the abundances of D, ^{3}He, and ^{3}H become of order unity at $T \simeq T_{NUC} \sim 0.1$ MeV, these reactions cannot produce sufficient ^{4}He to establish its NSE abundance. Once these abundances build up, essentially all the available neutrons are quickly bound into ^{4}He, the most tightly bound light nuclear species. Assuming that all the neutrons wind up in ^{4}He, the resulting mass fraction of ^{4}He is easy to estimate:

$$X_4 \simeq \frac{4n_4}{n_N} = \frac{4(n_n/2)}{n_n+n_p} = \frac{2(n/p)_{NUC}}{1+(n/p)_{NUC}} \tag{4.31}$$

where $(n/p)_{NUC} \simeq 1/7$ is the ratio of neutrons to protons at the time ^{4}He synthesis finally takes place ($T \simeq 0.1$ MeV).

While the binding energies per nucleon of ^{12}C, ^{16}O, etc. are larger than that of ^{4}He, by the time "the light-element bottleneck" is finally broken and ^{4}He is produced, Coulomb-barrier suppression is very significant. This fact, together with the absence of tightly-bound isotopes with mass 5 and 8, prevents significant nucleosynthesis beyond ^{4}He. Moreover, the low nucleon density supresses the triple-alpha reaction by which stellar nuclear burning bridges these mass gaps.

Some ^{7}Li is synthesized, ^{7}Li/H $\sim 10^{-10}$ to 10^{-9}, produced for $\eta \lesssim 3\times 10^{-10}$ by the process ^{4}He(^{3}H,γ)^{7}Li, and for $\eta \gtrsim 3\times 10^{-10}$ by ^{4}He(^{3}He,γ)^{7}Be (followed by the eventual β-decay of ^{7}Be to ^{7}Li by electron capture). As we shall soon see, even this trace amount of ^{7}Li proves to be a very valuable probe of primordial nucleosynthesis. In addition, substantial amounts of both D and ^{3}He are left unburnt, D, ^{3}He/H $\sim 10^{-5}$ to 10^{-4}, as the rates for the reactions that burn them to ^{4}He, $\Gamma \propto X_{2,3}(\eta n_\gamma)\langle\sigma|v|\rangle$, become small as X_2, X_3 become small and the reactions freeze out. Since these rates are proportional to η, it is clear that the amounts of D, ^{3}He left unburnt should decrease with increasing η.

4.4 Primordial Abundances: Predictions

The idea of primordial nucleosynthesis dates back to Gamow [2] in 1946.[4] In the 1950's Alpher, Follin, and Herman [3] all but wrote a code to calculate the synthesis of ^{4}He. Shortly before the discovery of the CMBR, Hoyle and Tayler estimated the amount of ^{4}He that would be synthesized in the early stages of a hot big bang [4]. Almost immediately after the discovery of the CMBR Peebles [5] wrote a very simple code to follow ^{4}He synthesis, and in 1967 Wagoner, Fowler, and Hoyle [6] wrote a very detailed reaction network to follow primordial nucleosynthesis all the way up the periodic table. Other independent nucleosynthesis codes have been written since [7]. Wagoner's 1973 version of the code [8] has become the "standard code" for primordial nucleosynthesis. The nuclear reaction rates have been updated periodically and the weak rates corrected for finite temperature and radiative/Coulomb corrections [9]. The code has recently been "modernized" (faster integration procedures, more user-friendly operation, better documentation, etc.) by Kawano [10], and FORTRAN versions of the code and documentation are available. All of the results presented here were calculated using the most up to date version of the code.

How accurate are the predicted abundances? The numerical accuracy of the code is better than 1%. In principle the abundances are sensitive to the input nuclear physics data. In practice, the relevant cross sections are known to sufficient accuracy that, with a few important exceptions, the theoretical uncertainties are irrelevant. The predicted ^{4}He abundance essentially only depends upon the weak interaction rates (which determine the neutron-to-proton ratio at freeze out). In turn, all of these rates depend upon the same matrix element, which also sets the neutron half life, $\tau_{1/2}(n)$.

[4]Gamow originally proposed that all of the elements in the periodic table be built up during big bang nucleosynthesis. Of course, it is now generally believed that the heavy elements ($A \geq 12$) are produced by massive stars, and that the light elements B, Be, and ^{6}Li are produced by cosmic ray spallation processes.

At present, the uncertainty in the neutron half life is surprisingly large:

$$\tau_{1/2}(n) = 10.5 \pm 0.2\,\text{min}. \tag{4.32}$$

In addition, some recent determinations suggest an even lower half life, $\tau_{1/2}(n)$ = 10.1 to 10.3 min [11]. Shortly, we will discuss the sensitivity of ^{4}He production to the neutron half life. The only other significant sensitivity of the predicted abundances to input parameters involves ^{7}Li: due to uncertainties in the cross sections for several reactions that both destroy and produce ^{7}Li, there is about a 50% uncertainty in the predicted abundance of ^{7}Li.

In Fig. 4.3 we display the time evolution of the light-element abundances for 3 neutrino species, $\eta = 3 \times 10^{-10}$, and $\tau_{1/2}(n) = 10.6$ min, and in Fig. 4.4 their final abundances as a function of the present baryon-to-photon ratio, η. We note that the predicted abundances are a function of η alone; this is simple to understand. During primordial nucleosynthesis the Universe is radiation dominated, so that the expansion rate $H = H(T)$. The various reaction rates are proportional to thermally-averaged cross sections, $\langle \sigma |v| \rangle = f(T)$, times the number densities of various nuclear species, $n_A = (X_A/A)\eta n_\gamma = n_A(\eta, T)$. Thus, all reaction rates are only a function of the density parameter η and the temperature T. Of course, η can be expressed in terms of Ω_B (at the expense of the unknown factor h), or in terms of ρ_B (by specifying the present photon temperature, which is now rather well determined).

Before comparing the predicted abundances with the observed abundances, it is useful to consider the sensitivity of the abundances to the one free cosmological parameter, η, and the two physical parameters $g_*(T \sim \text{MeV})$ and $\tau_{1/2}(n)$.

• $\tau_{1/2}(n)$: As discussed previously, all the weak rates are proportional to $G_F^2(1 + 3g_A^2)$, or in terms of the neutron half life, $\Gamma \propto T^5/\tau_{1/2}(n)$. An increase in the input value of $\tau_{1/2}(n)$ decreases all of the weak rates that interconvert neutrons and protons, and thereby leads to a freeze out of the neutron-to-proton ratio at a higher temperature, $T_F \propto \tau_{1/2}(n)^{1/3}$, and larger value of (n/p). Since the final ^{4}He abundance depends upon the value of $(n/p)_{freeze-out}$, cf. (4.28), this leads to an increase in the predicted ^{4}He abundance. The abundances of the other light elements change also, but since their present abundances are known far less precisely than that of ^{4}He, the changes are not yet of great interest.

• g_*: Since $H \propto g_*^{1/2}T^2$, an increase in the input value of g_* leads to a faster expansion rate (for the same temperature); this too leads to an

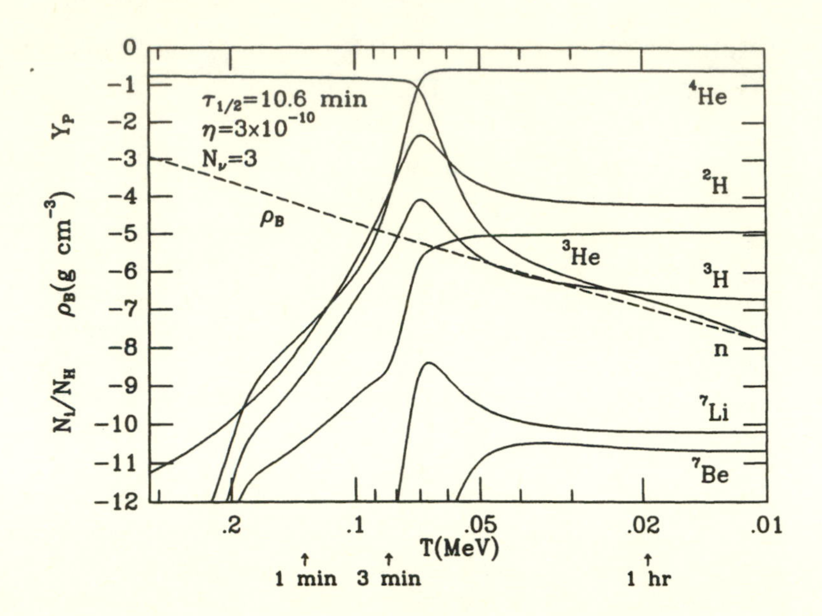

Fig. 4.3: The development of primordial nucleosynthesis. The dashed line is the baryon density, and the solid lines are the mass fraction of ^{4}He, and the number abundance (relative to H) for the other light elements.

earlier freeze out of the neutron-to-proton ratio: $T_F \propto g_*^{1/6}$, at a higher value, and hence more ^{4}He. Later we will use the dependence of T_F upon g_* to study the effects of the possible existence of additional light particle species (e.g., additional light neutrino species).

• η: In NSE the abundances of species $A(Z)$, $X_A \propto \eta^{A-1}$. For a larger value of η, the abundances of D, ^{3}He, and ^{3}H build up slightly earlier, cf. (4.26), and thus ^{4}He synthesis commences earlier, when the neutron-to-proton ratio is larger, resulting in more ^{4}He. Around the time ^{4}He synthesis begins in earnest ($T \simeq 0.1$ MeV), the neutron-to-proton ratio is only slowly decreasing (due to neutron decays), and the sensitivity of ^{4}He production to η is only slight. As mentioned earlier, the amount of D and ^{3}He left unburnt depends upon η (decreasing with increasing η). This sensitivity to η is much more significant, with the yields of D and ^{3}He decreasing as η^{-n} ($n \sim 1$ to 2). The ^{7}Li "trough" at a value of $\eta \simeq 3 \times 10^{-10}$ results because of the two different production processes, one which dominates at small η and one which dominates at large η.

An accurate analytic fit to the primordial mass fraction of ^{4}He is $Y_P = 0.230 + 0.025\log(\eta/10^{-10}) + 0.0075(g_* - 10.75) + 0.014[\tau_{1/2}(n) - 10.6 \text{ min}]$.

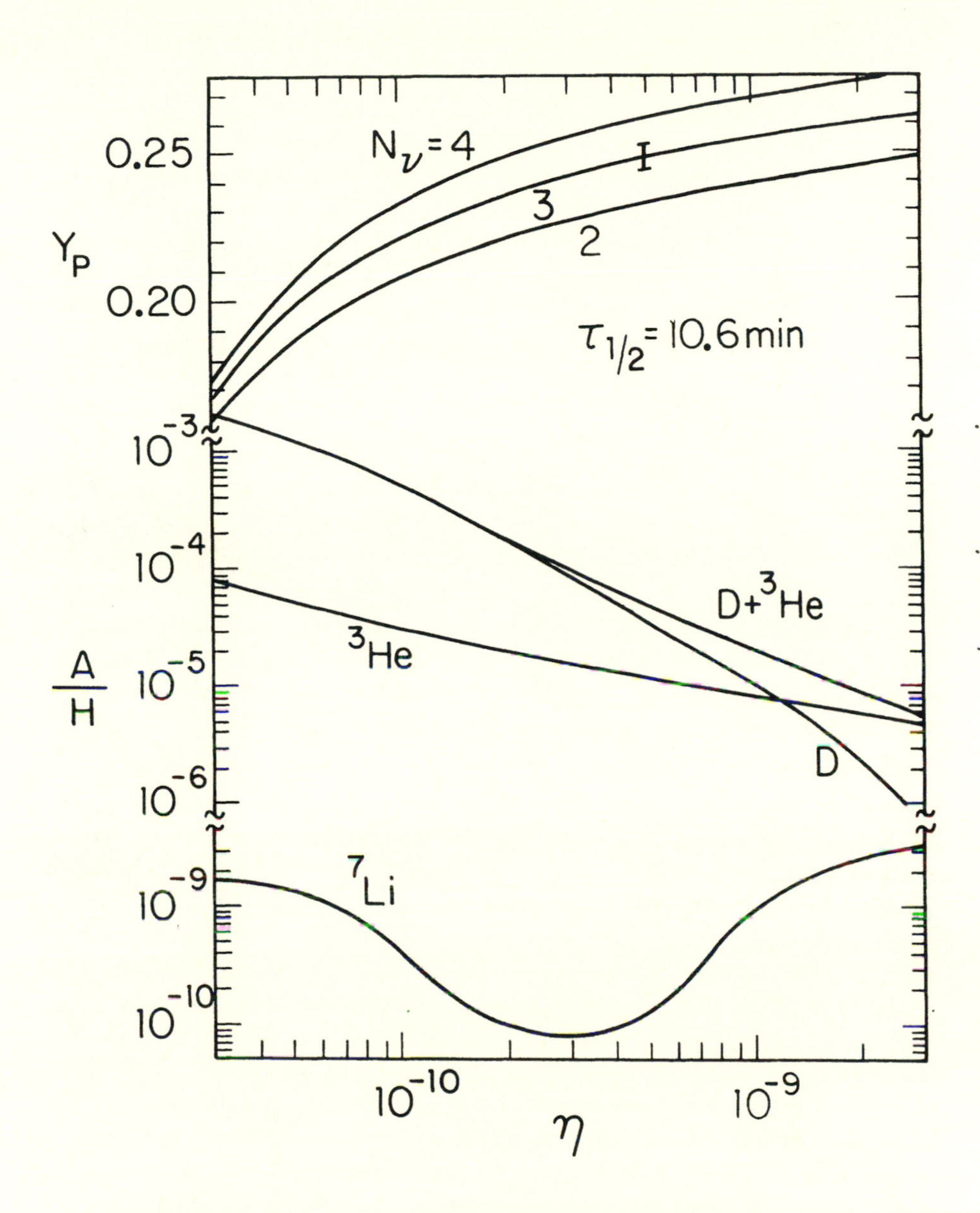

Fig. 4.4: The predicted primordial abundances of the light elements as a function of η. The error bar indicates the change in Y_P for $\Delta\tau_{1/2} = \pm 0.2$ min.

4.5 Primordial Abundances: Observations

Unlike the predicted primordial abundances, where within the context of the standard cosmology the theoretical uncertainties are well defined—and small, the uncertainties in the "observed" primordial abundances are far less certain—and large. As the reader by now must appreciate, to a first approximation, all observables in cosmology are impossible to measure! The problems with the primordial abundances of the light elements are manifold. What we would like to measure are the primordial, cosmic abundances, i.e., abundances at an epoch before other astrophysical processes, e.g., stellar production and destruction, became important. What we can measure are present-day abundances in selected astrophysical sites. From these we must try to infer primordial abundances.

We will not review the important work of numerous astrophysicists which led to the modern view that the present abundances of the light elements D, ^{3}He, ^{4}He, and ^{7}Li are predominantly cosmological in origin (for such a review see, e.g., [12] and [13]). However, it is interesting to note that even in the mid–1960's it was not generally accepted that ^{4}He had a primordial component, and it wasn't until the early 1970's that it was realized that no contemporary astrophysical process could account for the observed D abundance. And, as we shall mention below, the primordial component of ^{7}Li was not determined until the early 1980's.

• *Deuterium:* The abundance of D has been measured in solar system studies, in UV absorption studies of the local interstellar medium (ISM), and in studies of deuterated molecules (DCO, DHO) in the ISM. The solar system determinations are based upon measuring the abundances of deuterated molecules in the atmosphere of Jupiter: $D/H \simeq (1 \text{ to } 4) \times 10^{-5}$, and inferring the pre-solar (i.e., at the time of the formation of the solar system) D/H ratio from meteoritic and solar data on the abundance of ^{3}He: $D/H \simeq (1.5 \text{ to } 2.9) \times 10^{-5}$. These determinations are consistent with a pre-solar value of $(D/H) \simeq (2 \pm 1) \times 10^{-5}$. An average ISM value for $(D/H) \simeq 2 \times 10^{-5}$ has been derived from UV absorption studies of the local ISM (distances less than a few hundred pc), with individual measurements spanning the range $(1 \text{ to } 4) \times 10^{-5}$. Note that these measurements are consistent with the solar system determinations of D/H. The studies of deuterated molecules in the ISM that $D/H \sim 1 \times 10^{-5}$ with about a factor of 2 uncertainty. While isotopic chemical effects have been taken into account, it should be noted that they can be significant, cf. the abundance of deuterium in the ocean, $D/H \simeq 1.5 \times 10^{-4}$, is about a factor of 10 above cosmic,[5] and the abundance of D in the atmosphere of Venus is 10^{-2}, a factor of 10^3 above cosmic. Very recently, studies of QSO absorption line systems, clouds of gas backlit by QSO's, have been used to try to determine the extragalactic deuterium abundance. An absorption system with red shift $z_{abs} = 0.03$ associated with Mrk 509 yielded an upper limit to the D abundance of: $D/H \lesssim 10^{-4}$, and a tentative detection in an absorption system with red shift $z_{abs} = 3.09$ associated with Q0420-388 gave D/H $\sim 4 \times 10^{-5}$ [14].

Since the deuteron is very weakly-bound, it is easily destroyed (burned at temperatures greater than about 0.5×10^6 K) and hard to produce. Thus, it has been difficult to find a contemporary astrophysical site to produce even the small cosmic abundance of D.[6] Therefore, it is seems highly plausible that the presently-observed deuterium abundance provides a *lower* bound to the primordial abundance. Using $(D/H)_P \gtrsim 1 \times 10^{-5}$ it follows that η must be less than about 10^{-9} in order for primordial nucleosynthesis to account for the observed abundance of D.[7] [Note: because of the rapid variation of $(D/H)_P$ with η, this upper bound to η is rather insensitive to the precise lower bound to $(D/H)_P$ assumed.] Since $\eta = 2.68 \times 10^{-8} \Omega_B h^2$, an upper bound to Ω_B also follows: $\Omega_B \lesssim 0.037 h^{-2} \leq 0.20$, indicating that baryons alone cannot close the Universe. One is also tempted to exploit the sensitive dependence of $(D/H)_P$ upon η to derive a *lower* bound to η; this is not possible because D is so easily destroyed, and the present abundance could be significantly lower than the primordial abundance. As we will see this end can be accomplished by instead using the abundances of both D and ^{3}He.

• *Helium-3:* The abundance of ^{3}He has been measured in solar system studies and by observations of the $^3He^+$ hyperfine line in galactic HII regions[8] (the analog of the 21 cm line of H). The abundance of ^{3}He in the oldest meteorites, carbonaceous chondrites, has been determined to be

---

[5]The D abundance in some ocean-class lakes (e.g., Lake Michigan!) is about a factor of 2 higher than the "ocean" value.

[6]One might wonder why D can be produced in the big bang if it is so easily destroyed elsewhere. As is apparent from Fig. 4.2, the primordial NSE value of D/H should never exceed about 10^{-13}. However, because reaction rates cannot keep pace with the changing temperature, large departures from NSE occur. In stars, densities and time scales are much larger, and NSE is more closely tracked.

[7]Here and throughout the subscript P refers to the *primordial* abundance of an element.

[8]An HII region is a cloud of ionized hydrogen gas; typical dimensions are 10's of pc and typical temperatures are order 10,000 to 30,000 K. The energy input is usually provided by several young O or B stars at the center.

$^3He/H = 1.4 \pm 0.4 \times 10^{-5}$. Since these objects are believed to have formed at about the same time as the solar system, they provide a sample of pre-solar material. The abundance of ^3He in the solar wind has been determined by analyzing gas-rich meteorites, lunar soil, and the foil placed upon the surface of the moon by the Apollo astronauts. Since D is burned to ^3He during the sun's approach to the main sequence, these measurements represent the pre-solar sum of D and ^3He. These determinations of D + ^3He are all consistent with a pre-solar $[(D + ^3He)/H] \simeq (3.6 \pm 0.6) \times 10^{-5}$. The 3.46 cm hyperfine transition of $^3He^+$ is difficult to detect. Bania, Rood, and Wilson [15] have searched 17 galactic HII regions for this line, and have 9 detections with $^3He^+/H \simeq (1.2 \text{ to } 15) \times 10^{-5}$, and 8 upper limits (i.e., no detection) with $^3He^+/H \lesssim (0.4 \text{ to } 6.2) \times 10^{-5}$.

^3He is much more difficult to destroy than D. It is very hard to efficiently dispose of ^3He without also producing heavy elements or large amounts of ^4He (environments hot enough to burn ^3He are usually hot enough to burn protons to ^4He). When the kind of stars we see today process material they return more than 50% of their original ^3He to the ISM. In the absence of a generation of very exotic Pop III stars that process essentially all the material in the Universe and in so doing destroy most of the ^3He without overproducing ^4He or heavy elements, ^3He can have been astrated (i.e. reduced by stellar burning) by a factor of no more than $f_a \simeq 2$.[9] Using this argument, the inequality

$$\begin{aligned} \left[\frac{(D + ^3He)}{H}\right]_P &\leq \left(\frac{D}{H}\right)_{pre-\odot} + f_a \left(\frac{^3He}{H}\right)_{pre-\odot} \\ &\leq (1 - f_a)\left(\frac{D}{H}\right)_{pre-\odot} + f_a \left[\frac{(D + ^3He)}{H}\right]_{pre-\odot} \end{aligned} \tag{4.33}$$

and the presolar abundances of D and D+^3He, we can derive an upper bound to the primordial abundance of D+^3He:

$$\left[\frac{(D + ^3He)}{H}\right]_P \lesssim 8 \times 10^{-5}. \tag{4.34}$$

For a very conservative astration factor, $f_a \simeq 4$, the upper limit becomes 13×10^{-5}. Using 8×10^{-5} as an upper bound to the primordial D+^3He

[9]The youngest stars, e.g., our sun, are called Pop I; the oldest observed stars are called Pop II. Pop III refers to a yet to be discovered, hypothetical first generation of stars.

production implies that for concordance, η must be greater than 4×10^{-10} (for the very conservative upper bound of 13×10^{-5}, η must be greater than 3×10^{-10}). To summarize, consistency between the predicted big bang abundances of D and ^{3}He and their abundances observed today requires η to lie in the range $\simeq$ (4 to 10) $\times 10^{-10}$.

• *Lithium-7:* Until very recently, our knowledge of the ^{7}Li abundance was limited to observations of meteorites, the local ISM, and Pop I stars, with a derived present abundance of $^7\mathrm{Li/H} \simeq 10^{-9}$ (to within a factor of 2), which is about a factor of 10 greater than that predicted by primordial nucleosynthesis. Given that ^{7}Li is produced by cosmic ray spallation and some stellar processes (e.g., in novae outbursts), and is easily destroyed in environments where $T \gtrsim 2 \times 10^6$ K (e.g., $^7\mathrm{Li/H} \simeq 10^{-11}$ in the atmosphere of our sun), there was not the slightest reason to suspect (or even hope!) that this value accurately reflects the primordial abundance. In 1982 Spite and Spite [16] attempted to observe ^{7}Li lines in the atmospheres of 13 unevolved halo and old disk stars with very low metal abundances ($Z = Z_\odot/12$ to $Z_\odot/250$), and masses in the range (0.6 to $1.1)M_\odot$. To the surprise of many they were successful. For these objects they saw an interesting correlation in the ^{7}Li abundance with stellar mass (actually surface temperature): a plateau in the ^{7}Li abundance for the highest mass stars. While there was clear evidence that the lower mass stars had astrated their ^{7}Li (rapidly decreasing lithium abundance with decreasing mass), the plateau provided very good evidence that the more massive stars had not. Further observations of other very metal-poor, old stars have confirmed their initial results. While a fundamental quantitative understanding of the plateau and the mass where the ^{7}Li abundance starts to decrease is lacking, a qualitative explaination does exist: lower mass stars have deeper surface convective zones and convect their ^{7}Li deep enough to burn it. The observational evidence for this qualitative explanation is very convincing. From the plateau a primordial ^{7}Li abundance follows: $^7\mathrm{Li/H} \simeq (1.1 \pm 0.4) \times 10^{-10}$. This measured abundance represents the pre-pop II ^{7}Li abundance, which, barring the existence of a generation of Pop III stars that very efficiently destroyed (or produced) ^{7}Li, is then the primordial abundance.[10]

Remarkably, this is the predicted big bang production of ^{7}Li for η in the range (2 to 5) $\times 10^{-10}$. If we take this to be the primordial ^{7}Li abundance,

---

[10]Recently, SN1987A provided the opportunity to determine the ^{7}Li abundance in the metal-poor ISM of the Large Magellanic Cloud. Absorption studies of the light from SN1987A indicate $(^7\mathrm{Li/H})_{\mathrm{LMC}} \lesssim 1 \times 10^{-10}$, consistent with the pop II abundance in our own galaxy [17].

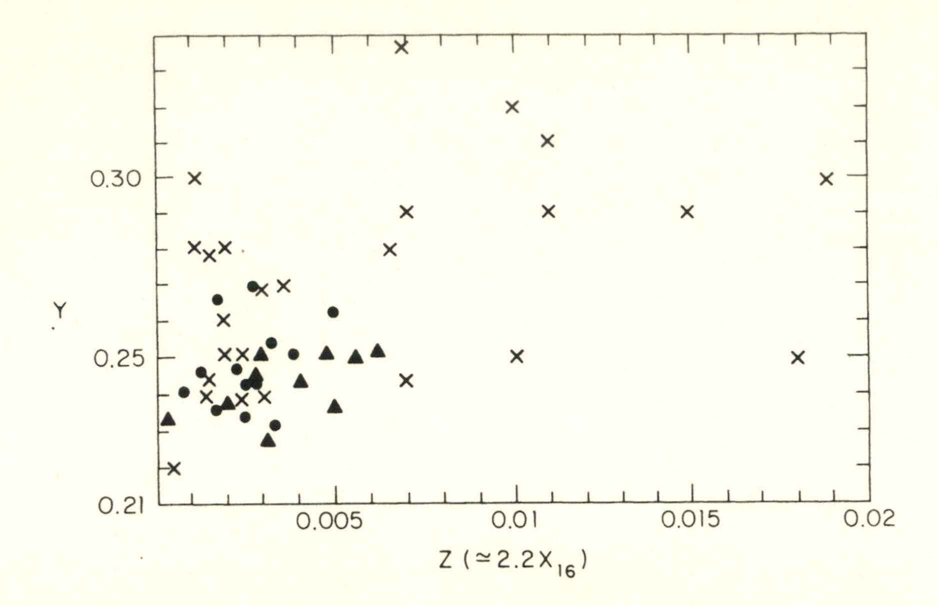

Fig. 4.5: Summary of ^{4}He determinations for galactic and extragalactic HII regions. The filled circles and triangles are two carefully studied sets of metal-poor extragalactic HII regions.

and allow for a possible 50% uncertainty in the predicted abundance of ^{7}Li (due to estimated uncertainties in some of the reaction rates that affect ^{7}Li), then consistency for ^{7}Li restricts η to the range (1 to 7) $\times 10^{-10}$. It is interesting to note that this range of values for η lies at the bottom of the ^{7}Li trough, making ^{7}Li a very powerful probe of the nucleon density: Any value of η away from the trough results in the overproduction of ^{7}Li.

To summarize thus far, concordance of the big bang nucleosynthesis predictions with the derived abundances of D and ^{3}He requires $\eta \simeq$ (4 to 10) $\times 10^{-10}$ (or being more conservative, (3 to 10) $\times 10^{-10}$); moreover, concordance for D, ^{3}He, and ^{7}Li together further restricts η: $\eta \simeq$ (4 to 7) $\times 10^{-10}$, with the more conservative range being (3 to 10) $\times 10^{-10}$.

• *Helium-4:* In the past ten years the quality and quantity of ^{4}He observations has increased markedly. In Fig. 4.5 all the ^{4}He abundance determinations derived from observations of recombination lines in HII regions (galactic and extragalactic) are shown as a function of metallicity Z (more precisely, 2.2 times the mass fraction of ^{16}O).

Since ^{4}He is also synthesized in stars, some of the observed ^{4}He is most certainly *not* primordial. Since stars also produce metals, one would expect some correlation between Y and Z, or at least a trend: lower Y where Z is lower. Such a trend is apparent in Fig. 4.5. From Fig. 4.5 it is also clear that there is a large primordial component to ^{4}He: $Y_P \simeq 0.22$ to 0.26. Is it possible to pin down the value of Y_P more precisely?

There are many steps in going from the line strengths (what the observer actually measures), to a mass fraction of ^{4}He (e.g., corrections for neutral ^{4}He, ^{4}He^{++}, reddening, etc.). In galactic HII regions where abundances can be determined for various positions within a region, variations are seen within a given region. Observations of extragalactic HII regions are actually observations of a superposition of *several* HII regions. Although observers have quoted statistical uncertainties of $\Delta Y \simeq \pm 0.01$ (or smaller), from the scatter in Fig. 4.5 it is clear that the systematic uncertainties must be larger. For example, different observers have derived ^{4}He abundances of between 0.22 and 0.25 for I Zw18, an extremely metal-poor dwarf emission line galaxy.[11]

Perhaps the safest way to estimate Y_P is to concentrate on the ^{4}He determinations for metal-poor objects. From Fig. 4.5 $Y_P \simeq 0.23$ to 0.25 appears to be consistent with all the data (although Y_P as low as 0.22 or high as 0.26 could not be ruled out). Kunth and Sargent have studied 13 metal-poor ($Z \lesssim Z_\odot/5$) Blue Compact galaxies (the solid circles in Fig. 4.5). From a weighted average for their sample they derive a *primordial abundance* $Y_P \simeq 0.245 \pm 0.003$; allowing for a 3σ variation this suggests $0.236 \leq Y_P \leq 0.254$.

For the range deduced from D, ^{3}He, and ^{7}Li: $\eta \geq 4 \times 10^{-10}$ (and $\tau_{1/2}(n) \geq 10.3$ min), the predicted ^{4}He abundance is (see Fig. 4.6)

$$Y_P \geq \begin{cases} 0.227 & \text{for } N_\nu = 2 \\ 0.242 & \text{for } N_\nu = 3 \\ 0.254 & \text{for } N_\nu = 4. \end{cases} \tag{4.35}$$

[Using the more conservative lower bound, $\eta \geq 3 \times 10^{-10}$, the above numbers become 0.224, 0.238, and 0.251, respectively.] Since $Y_P \simeq 0.23$ to 0.25 there are values of η, N_ν, and $\tau_{1/2}(n)$ for which there is agreement between the abundances predicted by big bang nucleosynthesis and the primordial abundances of D, ^{3}He, ^{4}He, and ^{7}Li derived from observational data.

• *Summary of the Confrontation Between Theory and Observation:* The only isotopes that are predicted to be produced in significant amounts

[11] An excellent discussion of the difficulties and uncertainties associated with ^{4}He determinations in HII regions (using I Zw18 as a case study) is given in [18].

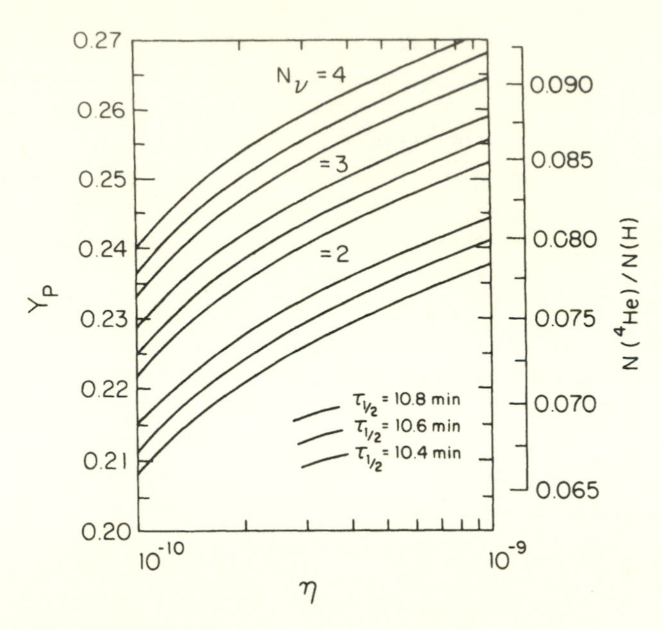

Fig. 4.6: Predicted ^{4}He abundance.

(A/H $> 10^{-12}$) during the epoch of primordial nucleosynthesis are: D, ^{3}He, ^{4}He, and ^{7}Li, and their predicted abundances span some 9 orders of magnitude. At present there is good agreement between the predicted primordial abundances of all 4 of these elements and their observed abundances for values of $N_\nu, \tau_{1/2}(n)$, and η in the following intervals: $2 \leq N_\nu \leq 4$; 10.3 min $\leq \tau_{1/2}(n) \leq$ 10.7 min; and $4 \times 10^{-10} \leq \eta \leq 7 \times 10^{-10}$ (or, with the more conservative range, (3 to 10) $\times 10^{-10}$). This is a truly remarkable achievement, and strong evidence that the standard model is valid at times as early as 10^{-2} sec after the bang.

The existence of a concordant range for η is a spectacular success for the standard cosmology. The narrowness of the range is both reassuring—in the end there can be but one set of concordant values—and a cause for continued close scrutiny. For example, focussing on the ^{4}He abundance; should the primordial abundance be unambiguously determined to be 0.22 or less, then the present agreement would disappear. What recourses exist if $Y_P \leq 0.22$? If a generation of Pop III stars that efficiently destroyed ^{3}He and ^{7}Li existed, then the lower bound to η based upon D, ^{3}He (and ^{7}Li) no longer exists. The only solid lower bound to η would then be that based upon the amount of luminous matter in galaxies: $\eta \geq 0.3 \times 10^{-10}$.

In this case the predicted Y_P could be as low as 0.15 or 0.16. Although small amounts of anisotropy increase the primordial production of ^{4}He, recent work suggests that larger amounts could *decrease* the primordial production of ^{4}He [19]. Another possibility is neutrino degeneracy; as mentioned earlier, a large electron lepton number, $|\mu_\nu| \gtrsim T$, modifies the equilibrium value of the neutron-to-proton ratio,

$$\left[\left(\frac{n}{p}\right)_{\mu_\nu \neq 0}\right]_{EQ} = \exp(-\mu_\nu/T)\left[\left(\frac{n}{p}\right)_{\mu_\nu = 0}\right]_{EQ} \tag{4.36}$$

and therefore the value of (n/p) at freeze out. Since the yield of ^{4}He is determined by this value, it is possible to dial in the desired ^{4}He abundance.[12] Finally, one might have to discard the standard cosmology altogether.

4.6 Primordial Nucleosynthesis as a Probe

If, based upon its apparent success, we accept the validity of the standard model, we can use primordial nucleosynthesis as a probe of conditions in the early Universe, and thereby of cosmology and particle physics. For example, concordance requires: $4 \times 10^{-10} \lesssim \eta \lesssim 7 \times 10^{-10}$ (or being more conservative, (3 to 10) $\times 10^{-10}$). This is the most precise determination we have of the present baryon density

$$\begin{aligned} 4(3) \times 10^{-10} &\leq \eta \leq 7(10) \times 10^{-10} \\ 0.015(0.011) &\leq \Omega_B h^2 \leq 0.026(0.037) \\ 0.015(0.011) &\leq \Omega_B \leq 0.16(0.21) \\ 6(4) \times 10^{-11} \leq n_B/s &\simeq \eta/7 \lesssim 1(1.4) \times 10^{-10}, \end{aligned} \tag{4.37}$$

where the more conservative bounds are given in parentheses.

Recall that luminous matter contributes less than about 0.01 of critical density, and that the dynamical determinations of the cosmic density suggest that, $\Omega_0 = 0.2 \pm 0.1$. First, since primordial nucleosynthesis indicates that $\Omega_B \gtrsim 0.015h^{-2} \gtrsim 0.015$, there must be baryonic matter that is not luminous—not much of a surprise. Second, if, as some dynamical studies suggest, $\Omega_0 > 0.15$ to 0.20, then some other, non-baryonic, form of matter

[12] Large chemical potentials for any, or several, of the 3 neutrino species also affects the energy density in the neutrino species, and thereby the expansion rate.

must account for the difference between Ω_0 and Ω_B. Numerous candidates have been proposed for the dark matter, including primordial black holes, axions, quark nuggets, photinos, gravitinos, higgsinos, relativistic debris, massive neutrinos, sneutrinos, monopoles, pyrgons, maximons, etc. We will return to this intriguing possibility in the next Chapter.

Let us turn now to probing particle physics; the best known constraint from primordial nucleosynthesis is the limit to the number of light neutrino flavors: $N_\nu \leq 4$, which we will now discuss. In their 1964 paper, Hoyle and Tayler realized from their estimates of primordial ^{4}He production that an additional neutrino species would lead to greater ^{4}He production [4]. Wagoner, Fowler, and Hoyle were more quantitative: They calculated primordial nucleosynthesis with an arbitrary "speed-up" factor, which is equivalent to changing g_* [6]. Schvartsman [21] and Peebles [22] also emphasized the dependence of the yield of ^{4}He on the expansion rate of the Universe during nucleosynthesis: ^{4}He production increases with increasing expansion rate, or g_*. Steigman, Schramm, and Gunn used this fact to place a limit to the number of light neutrino species (originally $N_\nu \leq 7$) [20]. As discussed earlier, the reason for increased ^{4}He involves the freeze out of the neutron-to-proton ratio, which occurs at a temperature of ~ 0.8 MeV. At this epoch the known relativistic degrees of freedom are: γ, 3 species of $\nu\bar{\nu}$'s, and $e^\pm$ pairs, and of course there may be additional, as of yet unknown, light (mass less than about a MeV) particle species. The possiblities include, additional neutrino species, axions, majorons, right-handed neutrinos, etc.

Now let us be more quantitative. Recall that Y_P increases with increasing values of η, $\tau_{1/2}(n)$, and $g_*(T)$, shown in Fig. 4.6. An upper limit to $g_*(T \sim \text{MeV})$ can be obtained from the following: (i) a lower limit to η—based upon the present D + ^{3}He abundance $\eta \geq (3 \text{ to } 4) \times 10^{-10}$; (ii) a lower limit to $\tau_{1/2}(n)$—present results suggest $\tau_{1/2}(n) \geq 10.3$ min (1σ); (iii) an upper bound to Y_P—present observations suggest that $Y_P \leq 0.25$. The resulting limit, stated in terms of g_*, or the equivalent number of neutrino species, is

$$N_\nu \leq 4 \qquad \text{or} \qquad g_*(T \sim \text{MeV}) \leq 12.5. \tag{4.38}$$

Recall the definition of g_*,

$$\begin{aligned} g_* &= 10.75 \text{ (std model; } N_\nu = 3) \\ &\quad + \sum_{\text{new bosons}} g_i (T_i/T)^4 + \frac{7}{8} \sum_{\text{new fermions}} g_i (T_i/T)^4. \end{aligned} \tag{4.39}$$

Then $N_\nu \leq 4$, $g_* \leq 12.5$, implies that

$$\begin{aligned} 1.75 \geq\ & 1.75(N_\nu - 3) \\ & + \sum_{\text{new bosons}} g_i(T_i/T)^4 + \frac{7}{8} \sum_{\text{new fermions}} g_i(T_i/T)^4. \qquad (4.40) \end{aligned}$$

While at most 1 additional light ($m \lesssim 1$ MeV) neutrino species can be tolerated, many more additional species can be tolerated if their temperatures T_i are less than the photon temperature T. For example, consider a particle that decoupled at a temperature $T_D \gtrsim 300$ GeV, when $g_* \gtrsim 106.75$. Mimicking the calculation we carried out for neutrino decoupling in Chapter 3, we find that $(T_i/T)^4 \leq (10.75/106.75)^{4/3} \simeq 0.047$, so that such a species makes a very small contribution to $g_*(T \sim \text{MeV})$.

The number of neutrino species can also be determined by measuring the width of the Z^0 boson: Each neutrino flavor less massive than $m_Z/2$ contributes about 190 MeV to the width of the Z^0. Present UA(1) and CDF collider data on the production of $W^\pm$ and Z^0 bosons and e^+e^- experiments that look for the process $e^+ + e^- \rightarrow \text{nothing} + \gamma$ (which includes $e^+ + e^- \rightarrow \nu\bar{\nu} + \gamma$) have placed limits to N_ν which are of the order of 4 to 7 [23]. When the width of the Z^0 boson is measured to high precision at SLC and LEP, N_ν should be determined very accurately, $\Delta N_\nu \lesssim 0.2$, and the big bang constraint put to the test.[13]

4.7 Concluding Remarks

Primordial nucleosynthesis is both the earliest and most stringent test of the standard cosmology, and an important probe of cosmology and particle physics. The present state of agreement between theory and observation indicates that the standard cosmology is a valid description of the Universe at least back to times as early as 10^{-2} sec after the bang and temperatures as high as 10 MeV. As a probe it provides us with the best determination of Ω_B, a very stringent limit to N_ν, and numerous other constraints on particle physics and cosmology. Given the important role occupied by

[13]While big bang nucleosynthesis and the width of the Z^0 both provide information about the number of neutrino flavors, they "measure" slightly different quantities. Big bang nucleosynthesis is sensitive to the number of light ($m \lesssim 1$ MeV) neutrino species and all other light degrees of freedom, whereas the width of the Z^0 is determined by the number of particles less massive than about 46 GeV that couple to the Z^0 (neutrinos among them). As we go to press, SLC reports $N_\nu = 2.79 \pm 0.63$, and the four groups at LEP report $N_\nu = 3.27 \pm 0.30$, 3.42 ± 0.48, 3.1 ± 0.4, and $2.4 \pm 0.4 \pm 0.5$ [24].

big bang nucleosynthesis, it is clear that continued scrutiny is in order. The importance of new observational data cannot be overemphasized: extragalactic D abundance determinations (Is the D abundance universal? What is its value?); more measurements of the ^{3}He abundance (What is its primordial value?); continued improvement in the accuracy of ^{4}He abundances in very metal poor HII regions (recall, a difference between $Y_P = 0.22$ and $Y_P = 0.23$ is crucial); and further study of the ^{7}Li abundance in very old stellar populations (Has the primordial abundance of ^{7}Li actually been measured?). Data from particle physics will prove useful too: A high precision determination of $\tau_{1/2}(n)$ (i.e., $\Delta\tau_{1/2}(n) \leq \pm 0.05$ min) will all but eliminate the uncertainty in the predicted ^{4}He primordial abundance; an accurate measurement of the width of the Z^0 will determine the total number of neutrino species (less massive than about 46 GeV) and thereby bound the total number of light neutrino species. All these data will not only make primordial nucleosynthesis a more stringent test of the standard cosmology, but they will also make primordial nucleosynthesis a more powerful probe of the early Universe.

On the theoretical side, one should not let the success of the standard scenario keep us from at least exploring other possibilities. For example, standard nucleosynthesis precludes the possibility of $\Omega_B = 1$. If our theoretical prejudice is correct and $\Omega_0 = 1$, primordial nucleosynthesis then makes a very bold prediction: Most of the material in the cosmos today is non-baryonic. Rather than blindly embracing this profound prediction, one should keep an open mind to variants of the standard scenario of nucleosynthesis. Recently, two non-standard scenarios have been suggested as a means to allow $\Omega_B = 1$. (1) A relic, decaying particle species with mass $\gtrsim$ *few* GeV and lifetime $\tau \sim 10^4$ to 10^6 sec. The hadronic decays of this relic, which occur well after the standard epoch of nucleosynthesis, initiate a second epoch of nucleosynthesis, resetting the light-element abundances—to acceptable values the authors hope [25]. (2) Large, local inhomogeneities in the baryon-to-photon ratio produced by a strongly, first order quark/hadron phase transition. In this scenario, η is highly inhomogeneous (varying by a factor of greater than 30 or so); in addition, due to the ability of neutrons to diffuse through the cosmic plasma more easily than protons, the high density regions become proton rich and the low density regions neutron rich [26]. Because of the inhomogeneties, the cosmic light-element abundances represent the average over the two very different zones of nucleosynthesis, and the authors hope that the resulting abundances are compatible with the observations and $\Omega_0 = \Omega_B = 1$. At present both scenarios have "serious lithium problems;" in the first scenario ^{7}Li tends to be underproduced (by a factor of 2 to 3) and ^{6}Li overproduced[14] (by a factor of about 10). In the second scenario, the ^{7}Li yield is (10 to 1000) $\times 10^{-10}$, compared to the observed abundance in old pop II stars of about 10^{-10}. (It also appears that in addition to the lithium problem, ^{4}He may be overproduced). An interesting aside is that both scenarios probably would have been viable 6 years ago prior to determination of the primordial ^{7}Li abundance. The lithium abundance is proving to be a very powerful probe of the early Universe. The success of the very simple, standard scenario of nucleosynthesis is all that more impressive in the light of the apparent inability of scenarios with adjustable parameters to achieve the same success.

4.8 References

References to some historical papers on primordial nucleosynthesis are given in *Early Universe: Reprints.*

1. A review of primordial nucleosynthesis stressing observational data is given in A. M. Boesgaard and G. Steigman, *Ann. Rev. Astron. Astro.* **23**, 319 (1985). This paper is reprinted in *Early Universe: Reprints.*

2. G. Gamow, *Phys. Rev.* **70**, 527 (1946); R. A. Alpher, H. Bethe, and G. Gamow, *Phys. Rev.* **73**, 803 (1948).

3. R. A. Alpher, J. W. Follin, R. C. Herman, *Phys. Rev.* **92**, 1347 (1953).

4. F. Hoyle and R. J. Tayler, *Nature* **203**, 1108 (1964).

5. P. J. E. Peebles, *Ap. J.* **146**, 542 (1966).

6. R. V. Wagoner, W. A. Fowler, and F. Hoyle, *Ap. J.* **148**, 3 (1967).

7. A. Yahil and G. Beaudet, *Ap. J.* **206**, 26 (1976); B. V. Vainer, et al., *Sov. Astron.* **22**, 1 (1976); Y. David and H. Reeves, in *Physical Cosmology*, eds., R. Balian, J. Audouze, and D. N. Schramm (North-Holland, Amsterdam, 1980); N. Terasawa and K. Sato, *Ap. J.* **294**, 9 (1985).

8. R. V. Wagoner, *Ap. J.* **179**, 343 (1973).

[14]At present there is no convincing evidence for a detection of ^{6}Li; in the ISM the upper limit is $^6\mathrm{Li/H} \lesssim 10^{-10}$ and in old pop II stars $^6\mathrm{Li/H} \lesssim 10^{-11}$.

9. D. A. Dicus, et al., *Phys. Rev. D* **26**, 2694 (1982).

10. L. Kawano, *Let's Go: Early Universe*, Fermilab preprint PUB-88/34-A (unpublished).

11. P. Bopp, et al., *Phys. Rev. Lett.* **56**, 919 (1986); *ibid.* **57**, 1192E (1986); J. Last, et al., *ibid.* **60**, 995 (1988); E. Klemt, et al., *Z. Phys.* **C37**, 179 (1988).

12. D. N. Schramm and R. V. Wagoner, *Ann. Rev. Nucl. Part. Sci.* **27**, 37 (1979).

13. S. M. Austin, *Prog. Part. Nucl. Phys.* **7**, 1 (1981).

14. D. York, et al., *Ap. J.* **276**, 92 (1984); R. Carswell, et al., *Ap. J.*, in press (1988).

15. T. M. Bania, R. T. Rood, and T. L. Wilson, *Ap. J.* **323**, 30 (1987).

16. F. Spite and M. Spite, *Astron. Astrophys.* **115**, 357 (1982); for a recent review of ^{7}Li determinations, see L. Hobbs, in *13th Texas Symposium on Relativistic Astrophysics*, ed. M. Ulmer, (World Scientific Press, Singapore, 1987), p. 185.

17. D. Baade and P. Magain, *Astron. Astrophys.* **194**, 237 (1988); A. Vidal-Madjar, et al., *ibid* **177**, L17 (1987); K. C. Sahu and M. Sahu, *ibid*, in press (1989); R. A. Malaney and C. R. Alcock, *Ap. J.*, in press (1989).

18. K. Davidson and T. D. Kinman, *Ap. J. Suppl.* **50**, 321 (1985).

19. R. Matzner and T. Rothman, *Phys. Rev. Lett.* **48**, 1565 (1982), and references therein.

20. G. Steigman, D. N. Schramm, and J. Gunn, *Phys. Lett.* **66B**, 202 (1977).

21. V. F. Shvartsman, *JETP Lett.* **9**, 184 (1969).

22. P. J. E. Peebles, *Physical Cosmology* (Princeton University Press, Princeton, 1971), p. 267.

23. D. DeNegri, B. Saudolet, and M. Spiro, CERN preprint CERN-EP-88, LBL preprint 88-26014, *Rev. Mod. Phys.*, in press (1989); F. Abe, et al., *Phys. Rev. Lett.* **63**, 720 (1989); G. Abrams, et al., *ibid* **63**, 724 (1989).

24. G. Abrams, et al. (SLC), *Phys. Rev. Lett.*, in press (1989); M. Z. Akrawy, et al. (OPAL), *Phys. Lett.* **B**, in press (1989); B. Adeva, et al. (L3), *Phys. Lett.* **B**, in press (1989); D. Decamp, et al. (ALEPH), *Phys. Lett.* **B**, in press (1989); P. Aarino, et al. (DELPHI), *Phys. Lett.* **B**, in press (1989).

25. S. Dimopoulos, R. Esmailzadeh, L. J. Hall, and G. D. Starkman, *Phys. Rev. Lett.* **60**, 7 (1988); *Ap. J.* **330**, 545 (1988).

26. J. Applegate, C. Hogan, and R. J. Scherrer, *Phys. Rev. D* **35**, 1151 (1987); C. Alcock, G. Fuller, and G. J. Mathews, *Ap. J.* **320**, 439 (1987); H. Kurki-Suonio, R. A. Matzner, J. M. Centrella, T. Rothman, and J. R. Wilson, *Phys. Rev. D* **38**, 1091 (1988); M. S. Turner, *Phys. Rev. D* **37**, 304 (1988).

5

THERMODYNAMICS IN THE EXPANDING UNIVERSE

5.1 The Boltzmann Equation

For much of its early history, most of the constituents of the Universe were in thermal equilibrium, thereby making an equilibrium description a good approximation. However, there have been a number of very notable departures from thermal equilibrium—neutrino decoupling, decoupling of the background radiation, primordial nucleosynthesis, and on the more speculative side, inflation, baryogenesis, decoupling of relic WIMPs, etc. As we have previously emphasized, if not for such departures from thermal equilibrium, the present state of the Universe would be completely specified by the present temperature. The departures from equilibrium have led to important relics—the light elements, the neutrino backgrounds, a net baryon number, relic WIMPs (weakly-interacting massive particles), relic cosmologists, and so on.

As discussed in Chapters 2 and 3, once a species totally decouples from the plasma its evolution is very simple: particle number density decreasing as R^{-3} and particle momenta decreasing as R^{-1}. The evolution of the phase space distribution of a species which is in LTE or is completely decoupled is simple. It is the evolution of particle distributions around the epoch of decoupling that is challenging. Recall that the rough criterion for a particle species to be either coupled or decoupled involves the comparison of the interaction rate of the particle, Γ, with the expansion rate of the Universe, H:

$$\begin{array}{ll} \Gamma \gtrsim H & \text{(coupled)} \\ \Gamma \lesssim H & \text{(decoupled)} \end{array} \qquad (5.1)$$

where Γ is the interaction rate (per particle) for the reaction(s) that keep the species in thermal equilibrium. The units of Γ, of course, are time^{-1}.

While this is a very useful rule of thumb which is usually surprisingly accurate, in order to properly treat decoupling one must follow the microscopic evolution of the particle's phase space distribution function $f(p^\mu,\ x^\mu)$. This of course is governed by the Boltzmann equation, which can be written as

$$\hat{\mathbf{L}}[f] = \mathbf{C}[f] \tag{5.2}$$

where $\mathbf{C}$ is the collision operator and $\hat{\mathbf{L}}$ is the Liouville operator. The familiar non-relativistic Liouville operator for the phase space density $f(\vec{\mathbf{v}},\ \vec{\mathbf{x}})$ of a particle species of mass m subject to a force $\vec{\mathbf{F}} = d\vec{\mathbf{p}}/dt$ is

$$\hat{\mathbf{L}}_{\rm NR} = \frac{d}{dt} + \frac{d\vec{\mathbf{x}}}{dt}\cdot\vec{\nabla}_x + \frac{d\vec{\mathbf{v}}}{dt}\cdot\vec{\nabla}_v = \frac{d}{dt} + \vec{\mathbf{v}}\cdot\vec{\nabla}_x + \frac{\vec{\mathbf{F}}}{m}\cdot\vec{\nabla}_v. \tag{5.3}$$

The covariant, relativistic generalization of the Liouville operator is

$$\hat{\mathbf{L}} = p^\alpha\frac{\partial}{\partial x^\alpha} - \Gamma^\alpha_{\beta\gamma}p^\beta p^\gamma\frac{\partial}{\partial p^\alpha}. \tag{5.4}$$

Note, as expected, gravitational effects enter the equation only through the affine connection. For the FRW model the phase space density is spatially homogeneous and isotropic: $f = f(|\vec{\mathbf{p}}|,\ t)$ [or equivalently $f(E,\ t)$]. For the Robertson-Walker metric the Liouville operator is

$$\hat{\mathbf{L}}[f(E,t)] = E\frac{\partial f}{\partial t} - \frac{\dot{R}}{R}|\vec{\mathbf{p}}|^2\frac{\partial f}{\partial E}. \tag{5.5}$$

Using the definition of the number density in terms of the phase space density

$$n(t) = \frac{g}{(2\pi)^3}\int d^3p f(E,\ t), \tag{5.6}$$

and upon integration by parts, the Boltzmann equation can be written in the form

$$\frac{dn}{dt} + 3\frac{\dot{R}}{R}n = \frac{g}{(2\pi)^3}\int \mathbf{C}[f]\frac{d^3p}{E}. \tag{5.7}$$

The collision term for the process $\psi + a + b + \cdots \longleftrightarrow i + j + \cdots$ is given by

$$\frac{g}{(2\pi)^3}\int \mathbf{C}[f]\frac{d^3p_\psi}{E_\psi} = -\int d\Pi_\psi d\Pi_a d\Pi_b\cdots d\Pi_i d\Pi_j\cdots$$

$$\times(2\pi)^4\delta^4(p_\psi + p_a + p_b \cdots - p_i - p_j \cdots)$$
$$\times \Big[|\mathcal{M}|^2_{\psi+a+b+\cdots\rightarrow i+j+\cdots} f_a f_b \cdots f_\psi (1\pm f_i)(1\pm f_j)\cdots$$
$$-|\mathcal{M}|^2_{i+j+\cdots\rightarrow\psi+a+b+\cdots} f_i f_j \cdots (1\pm f_a)(1\pm f_b)\cdots(1\pm f_\psi)\Big] \quad (5.8)$$

where f_i, f_j, , f_a, f_b, $\cdots$ are the phase space densities of species i, j, $\cdots$, a, b, $\cdots$; f_ψ is the phase space density of ψ (the species whose evolution we are focusing on); $(+)$ applies to bosons; $(-)$ applies to fermions; and

$$d\Pi \equiv g\frac{1}{(2\pi)^3}\frac{d^3p}{2E} \quad (5.9)$$

where g counts the internal degrees of freedom. The 4-dimensional delta function enforces energy and momentum conservation, and the matrix element squared, $|\mathcal{M}|^2_{i+j+\cdots\rightarrow\psi+a+b+\cdots}$, for the process $i+j+\cdots\rightarrow\psi+a+b+\cdots$, is averaged over initial and final spins, and includes the appropriate symmetry factors for identical particles in the initial or final states.[1] The topic of statistical mechanics in the expanding Universe is discussed by Wagoner [2], and in the monograph of Bernstein [2].

In the most general case, the Boltzmann equations are a coupled set of integral-partial differential equations for the phase space distributions of all the species present! Fortunately, in problems of interest to us, all but one (or two) species will have equilibrium phase space distribution functions because of their rapid interactions with other species, reducing the problem to a single integral-partial differential equation for the one species of interest, denoted by ψ.

There are two well motivated approximations that greatly simplify (5.8). The first is the assumption of T (or CP) invariance,[2] which implies

$$|\mathcal{M}|^2_{i+j+\cdots\rightarrow\psi+a+b+\cdots} = |\mathcal{M}|^2_{\psi+a+b+\cdots\rightarrow i+j+\cdots} \equiv |\mathcal{M}|^2. \quad (5.10)$$

The second simplification is the use of Maxwell-Boltzmann statistics for all species instead of Fermi-Dirac for fermions and Bose-Einstein for bosons.[3]

[1] For n identical particles of a given species in the initial or final state, a factor of $1/n!$. The rules for calculating $|\mathcal{M}|^2$ are given in many standard texts, including Quigg, and Bjorken and Drell [1].

[2] In Chapter 6 we will relax the assumption of CP invariance when we discuss baryogenesis. Since the only observation of CP violation is in the K^0–$\bar{K}^0$ system, this assumption is well justified.

[3] In the absence of a degenerate (i.e., $\mu_i \gtrsim T$) Fermi species or a Bose condensate,

In the absence of Bose condensation or Fermi degeneracy, the blocking and stimulated emission factors can be ignored, $1 \pm f \simeq 1$, and $f_i(E_i) = \exp[-(E_i - \mu_i)/T]$ for all species in kinetic equilibrium. With these two assumptions the Boltzmann equation may be cast in the familiar form

$$\dot{n}_\psi + 3Hn_\psi = -\int d\Pi_\psi d\Pi_a d\Pi_b \cdots d\Pi_i d\Pi_j \cdots (2\pi)^4 |\mathcal{M}|^2 \times \delta^4(p_i + p_j \cdots - p_\psi - p_a - p_b \cdots) [f_a f_b \cdots f_\psi - f_i f_j \cdots] \quad (5.11)$$

where, as usual, $H \equiv \dot{R}/R$.

The significance of the individual terms is manifest: The $3Hn_\psi$ term accounts for the dilution effect of the expansion of the Universe, and the right hand side of (5.11) accounts for interactions that change the number of ψ's present. In the absence of interactions the solution to (5.11) is $n_\psi \propto R^{-3}$.

Finally, it is usually useful to scale out the effect of the expansion of the Universe by considering the evolution of the number of particles in a comoving volume. This is done by using the entropy density, s, as a fiducial quantity, and by defining as the dependent variable

$$Y \equiv \frac{n_\psi}{s}. \quad (5.12)$$

Using the conservation of entropy per comoving volume ($sR^3 =$ constant), it follows that

$$\dot{n}_\psi + 3Hn_\psi = s\dot{Y}. \quad (5.13)$$

Furthermore, since the interaction term will usually depend explicitly upon temperature rather than time, it is useful to introduce as the independent variable

$$x \equiv m/T, \quad (5.14)$$

where m is any convenient mass scale (usually taken as the mass of the particle of interest). During the radiation-dominated epoch x and t are related by

$$t = 0.301 g_*^{-1/2} \frac{m_{Pl}}{T^2} = 0.301 g_*^{-1/2} \frac{m_{Pl}}{m^2} x^2 \quad (5.15)$$

---

the use of Maxwell-Boltzmann statistics introduces only a small quantitative change, as all three distribution functions are very similar (and much less than one) for momenta near the peak of the distribution. Moreover, for any non-relativistic species, Maxwell-Boltzmann statistics becomes exact in the limit $(m_i - \mu_i)/T \gg 1$.

so that the Boltzmann equation can be rewritten as

$$\frac{dY}{dx} = -\frac{x}{H(m)s}\int d\Pi_\psi d\Pi_a d\Pi_b \cdots d\Pi_i d\Pi_j \cdots (2\pi^4)|\mathcal{M}|^2$$
$$\times \delta^4(p_i + p_j \cdots - p_\psi - p_a - p_b \cdots)[f_a f_b \cdots f_\psi - f_i f_j \cdots] \quad (5.16)$$

where $H(m) = 1.67 g_*^{1/2} m^2/m_{Pl}$, and $H(x) = H(m)x^{-2}$.

We will now consider some specific applications of the formalism we have developed here to treat non-equilibrium thermodynamics.

5.2 Freeze Out: Origin of Species

If a massive particle species remained in thermal equilibrium until the present, its abundance, $n/s \sim (m/T)^{3/2}\exp(-m/T)$, would be absolutely negligible because of the exponential factor. If the interactions of the species freeze out (i.e., $\Gamma < H$) at a temperature such that m/T is not much greater than 1, the species can have a significant relic abundance today. We will now calculate that relic abundance.

First, suppose that the species is stable (or very long-lived compared to the age of the Universe when its interactions freeze out). In the next section we will consider the case where the species is unstable. Given that it is stable, only annihilation and inverse annihilation processes, e.g.,

$$\psi\bar{\psi} \longleftrightarrow X\bar{X}, \quad (5.17)$$

can change the number of ψ's and $\bar{\psi}$'s in a comoving volume.[4] Here X generically denotes all the species into which ψ's can annihilate. In addition, we assume that there is no asymmetry between ψ's and $\bar{\psi}$'s.[5]

We will also assume that all the species X, $\bar{X}$ into which ψ, $\bar{\psi}$ annihilate have thermal distributions with zero chemical potential. Because these particles will usually have additional interactions which are "stronger" than their interactions with ψ's, the assumption of equilibrium for the X's is almost always a good one. For example, let ψ, $\bar{\psi} = \nu$, $\bar{\nu}$ and X, $\bar{X} = e^-$, e^+; while the neutrinos only have weak interactions, the $e^\pm$'s have weak and electromagnetic interactions.

[4]For simplicity we will only consider $2 \leftrightarrow 2$ annihilation and creation processes; it is straightforward to generalize from here.

[5]Again, it is straightforward to generalize this to include the possibility of an excess (or deficit) of ψ's over $\bar{\psi}$'s.

Now consider the factor $[f_\psi f_{\bar\psi} - f_X f_{\bar X}]$ in the collision term in the Boltzmann equation. Since X, $\bar X$ are in thermal equilibrium (and for simplicity we assume they have zero chemical potential)[6]

$$
\begin{aligned}
f_X &= \exp(-E_X/T), \\
f_{\bar X} &= \exp(-E_{\bar X}/T).
\end{aligned} \tag{5.18}
$$

The energy part of the δ-function enforces $E_\psi + E_{\bar\psi} = E_X + E_{\bar X}$, so that

$$
f_X f_{\bar X} = \exp[-(E_X + E_{\bar X})/T] = \exp[-(E_\psi + E_{\bar\psi})/T] = f_\psi^{\rm EQ} f_{\bar\psi}^{\rm EQ}, \tag{5.19}
$$

since $f_\psi^{\rm EQ} \equiv \exp(-E_\psi/T)$ and $f_{\bar\psi}^{\rm EQ} \equiv \exp(-E_{\bar\psi}/T)$. Therefore, it follows that

$$
[f_\psi f_{\bar\psi} - f_X f_{\bar X}] = [f_\psi f_{\bar\psi} - f_\psi^{\rm EQ} f_{\bar\psi}^{\rm EQ}]. \tag{5.20}
$$

Now the interaction term can be written in terms of n_ψ, the actual number density of ψ's, and $n_\psi^{\rm EQ}$, the equilibrium number density of ψ's, as

$$
\frac{dn_\psi}{dt} + 3Hn_\psi = -\langle\sigma_{\psi\bar\psi\to X\bar X}|v|\rangle[n_\psi^2 - (n_\psi^{\rm EQ})^2], \tag{5.21}
$$

or,

$$
\frac{dY}{dx} = \frac{-x\langle\sigma_{\psi\bar\psi\to X\bar X}|v|\rangle s}{H(m)}(Y^2 - Y_{\rm EQ}^2), \tag{5.22}
$$

where $Y \equiv n_\psi/s = n_{\bar\psi}/s$ is the *actual* number of ψ, $\bar\psi$'s per comoving volume, and $Y_{\rm EQ} \equiv n_\psi^{\rm EQ}/s = n_{\bar\psi}^{\rm EQ}/s$ is the *equilibrium* number of ψ, $\bar\psi$'s per comoving volume. The thermally-averaged annihilation cross section times velocity is given by

$$
\begin{aligned}
\langle\sigma_{\psi\bar\psi\to X\bar X}|v|\rangle \equiv \left(n_\psi^{\rm EQ}\right)^{-2} \int & d\Pi_\psi d\Pi_{\bar\psi} d\Pi_X d\Pi_{\bar X} (2\pi)^4 \\
& \times\delta^4(p_\psi + p_{\bar\psi} - p_X - p_{\bar X})|\mathcal{M}|^2 \exp(-E_\psi/T)\exp(-E_{\bar\psi}/T).
\end{aligned} \tag{5.23}
$$

Now if we consider other annihilation channels for $\psi\bar\psi$, say $\psi\bar\psi$ to some final state F (not necessarily a two-body final state), there is an additional term in $\dot n_\psi$, which is simply (5.21) with $\langle\sigma_{\psi\bar\psi\to X\bar X}|v|\rangle$ replaced by

[6]While the Boltzmann equation is explicitly covariant, specifying the particle distribution functions singles out a frame—the comoving frame—and breaks the covariance. Once this is done, all quantities must be evaluated in this frame.

$\langle\sigma_{\psi\bar{\psi}\to F}|v|\rangle$. Summing over all annihilation channels yields the final result in terms of the *total* annihilation cross section $\langle\sigma_A|v|\rangle$

$$\begin{aligned}\frac{dn_\psi}{dt}+3Hn_\psi &= -\langle\sigma_A|v|\rangle[n_\psi^2-(n_\psi^{\rm EQ})^2]\\ \frac{dY}{dx} &= \frac{-x\langle\sigma_A|v|\rangle s}{H(m)}(Y^2-Y_{\rm EQ}^2).\end{aligned}\tag{5.24}$$

In the non-relativistic ($x\gg 3$) and in the extreme relativistic ($x\ll 3$) regimes the equilibrium value of the number of ψ's per comoving volume has the following simple limiting forms,

$$\begin{aligned}Y_{EQ}(x) &= \frac{45}{2\pi^4}\left(\frac{\pi}{8}\right)^{1/2}\frac{g}{g_{*S}}x^{3/2}e^{-x}=0.145\frac{g}{g_{*S}}x^{3/2}e^{-x}\quad (x\gg 3)\\ Y_{EQ}(x) &= \frac{45\zeta(3)}{2\pi^4}\frac{g_{\rm eff}}{g_{*S}}=0.278\frac{g_{\rm eff}}{g_{*S}}\quad (x\ll 3)\end{aligned}\tag{5.25}$$

where $g_{\rm eff}=g$ (bosons) and $g_{\rm eff}=3g/4$ (fermions).

This very convenient form of the Boltzmann equation that we have just derived has all the "features" that we would have expected: The destruction rate of ψ, $\bar{\psi}$'s per comoving volume is just proportional to the annihilation rate of ψ, $\bar{\psi}$'s, and the net rate of destruction is just balanced by inverse (creation) processes when $n_\psi=n_\psi^{\rm EQ}$. Moreover, creation processes are Boltzmann suppressed for $T\ll m$, because only a small fraction of $X\bar{X}$ pairs have sufficient KE to create a $\psi\bar{\psi}$ pair. In fact, we could have forgone the sleight-of-hand involved in rewriting f_X, $f_{\bar{X}}$ in terms of $f_\psi^{\rm EQ}$, $f_{\bar{\psi}}^{\rm EQ}$, and just derived (5.24) by invoking the *principle of detailed balance*—however, we would do so at the expense of pedagogy. Whenever all species except the one of interest are in thermal equilibrium, the creation processes can be obtained from the destruction processes by substituting $n\to n_{\rm EQ}$ for the species of interest.

Remembering that $H\propto x^{-2}$, so that $H(T)=x^{-2}H(m)$, we see that (5.24) can be cast in a very suggestive way,

$$\begin{aligned}\frac{x}{Y_{\rm EQ}}\frac{dY}{dx} &= -\frac{\Gamma_A}{H}\left[\left(\frac{Y}{Y_{\rm EQ}}\right)^2-1\right],\\ \Gamma_A &\equiv n_{\rm EQ}\langle\sigma_A|v|\rangle.\end{aligned}\tag{5.26}$$

In this form, we see that the change of ψ's per comoving volume is controlled by the *effectiveness of annihilations*, the usual Γ/H factor, times a measure of the deviation from equilibrium. It is then clear that when Γ/H is less than order unity, the relative change in the number of ψ's in a comoving volume becomes small, $-\Delta Y/Y \sim -(xdY/dx)/Y_{\rm EQ} \sim (\Gamma/H) \lesssim 1$, annihilations freeze out, and the number of ψ's in a comoving volume "freezes in."

The Boltzmann equation for the evolution of the abundance of a species is a particular form of the Riccati equation, for which there are no general, closed-form solutions. Before we solve the equation by approximate methods, let's consider the qualitative behavior of the solution. The annihilation rate Γ_A varies as n_{EQ} times the thermally-averaged annihilation cross section $\langle\sigma_A|v|\rangle$. In the relativistic regime, $n_{EQ} \sim T^3$, and like other rates, Γ_A will vary as some power of T. In the non-relativistic regime, $n_{EQ} \sim (mT)^{3/2}\exp(-m/T)$, cf. (3.52), so that Γ_A decreases exponentially. In either regime, Γ_A decreases as T decreases, and so eventually annihilations become impotent, roughly when $\Gamma_A \simeq H$, which for definiteness, say occurs for $x = x_f$ ("freeze out"). Thus, we expect that for $x \lesssim x_f$, $Y \simeq Y_{\rm EQ}$, while for $x \gtrsim x_f$ the abundance "freezes in:" $Y(x \gtrsim x_f) = Y_{\rm EQ}(x_f)$.

• *Hot Relics:* First consider the case of a particle species for which $x_f \lesssim 3$. In this case, freeze out occurs when the species is still relativistic and $Y_{\rm EQ}$ is not changing with time. Since $Y_{\rm EQ}$ is constant, the final value of Y is very insensitive to the details of freeze out (i.e., the precise value of x_f), and the asymptotic value of Y, $Y(x \to \infty) \equiv Y_\infty$, is just the equilibrium value at freeze out:

$$Y_\infty = Y_{\rm EQ}(x_f) = 0.278 g_{\rm eff}/g_{*S}(x_f) \quad (x_f \lesssim 3). \tag{5.27}$$

Thus the species freezes out with order unity abundance relative to s (or the number density of photons). Assuming the expansion remains isentropic thereafter (constant entropy per comoving volume), the abundance of ψ's today is (s_0 is the present entropy density)

$$\begin{aligned} n_{\psi 0} &= s_0 Y_\infty = 2970 Y_\infty \ {\rm cm}^{-3} & (5.28)\\ &= 825[g_{\rm eff}/g_*(x_f)]\ {\rm cm}^{-3}. & (5.29)\end{aligned}$$

If, after freeze out, the entropy per comoving volume of the Universe should increase, say by a factor of γ, the present abundance of ψ's in a comoving volume would be diminished by γ: $Y_\infty = Y(x_f)/\gamma$. In the next Section we will discuss an example of entropy production.

A species that decouples when it is relativistic is often called a *hot relic.* The present relic mass density contributed by a hot relic of mass m is simple to compute:

$$\rho_{\psi 0} = s_0 Y_\infty m = 2.97 \times 10^3 Y_\infty (m/\text{eV}) \text{ eV cm}^{-3} \tag{5.30}$$

$$\Omega_\psi h^2 = 7.83 \times 10^{-2} [g_{\text{eff}}/g_{*S}(x_f)](m/\text{eV}). \tag{5.31}$$

Based upon the present age of the Universe we know that $\Omega_0 h^2 \lesssim 1$; applying this bound to the contribution of the species ψ to $\Omega_0 h^2$ we obtain a cosmological bound to the mass of the ψ:

$$m \lesssim 12.8 \text{ eV}[g_{*S}(x_f)/g_{\text{eff}}]. \tag{5.32}$$

Light (mass $\lesssim$ MeV) neutrinos decouple when $T \sim$ few MeV, and $g_{*S} = g_* = 10.75$. For a single, 2-component neutrino species $g_{\text{eff}} = 2 \times (3/4) = 1.5$, so that $g_{\text{eff}}/g_{*S} = 0.140$. This implies that

$$\Omega_{\nu\bar{\nu}} h^2 = \frac{m_\nu}{91.5 \text{ eV}}, \tag{5.33}$$

$$m_\nu \lesssim 91.5 \text{ eV}. \tag{5.34}$$

This cosmological bound to the mass of a stable, light neutrino species is often referred to as the Cowsik-McClelland bound.[7]

Note that the present density of ψ's depends upon $g_{*S}(x_f)$. If a species decouples very early on, when g_{*S} is large, its present number density is proportionally smaller. As an example, consider a species with $g_{\text{eff}} = 1.5$ which decouples at a temperature $T \gtrsim 300$ GeV, when $g_{*S} \simeq g_* \gtrsim 106.75$. For such a species the present contribution to the energy density is

$$\Omega_\psi h^2 = \frac{m}{910 \text{ eV}}, \tag{5.35}$$

about a factor of 10 less than that of a conventional neutrino species. We see here that a species that decouples when $g_{*S} \gg 1$ has a present abundance much less than that of the microwave photons, and if the species

[7]In their original paper, Coswik and McClelland [6] consider a 4-component neutrino species ($g = 4$), and took $\Omega < 3.8$, $h = 1/2$ and $T_\nu = T$, which resulted in the bound $m \lesssim 8$ eV.

is massless, a temperature much less than the photon temperature, $T_\psi \simeq (3.91/g_{*S})^{1/3}T$. For the latter reason, such a relic is often referred to as a *warm relic*. Examples of possible warm relics include a light gravitino, or a light photino (here, "light" means mass less than about a keV).

• *Cold Relics:* Now consider the more difficult case where freeze out occurs when the species is non-relativistic ($x_f \gtrsim 3$), and $Y_{\rm EQ}$ is decreasing exponentially with x. In this case the precise details of freeze out are important.

It is useful to parameterize the temperature dependence of the annihilation cross section. On general theoretical grounds the annihilation cross section should have the velocity dependence $\sigma_A|v| \propto v^p$, where $p = 0$ corresponds to s-wave annihilation, $p = 2$ to p-wave annihilation, etc. Since $\langle v\rangle \sim T^{1/2}$, $\langle\sigma_A|v|\rangle \propto T^n$, $n = 0$ for s-wave annihilation, $n = 1$ for p-wave annihilation, etc. Therefore we parameterize $\langle\sigma_A|v|\rangle$ as

$$\langle\sigma_A|v|\rangle \equiv \sigma_0(T/m)^n = \sigma_0 x^{-n} \quad (for\ x \gtrsim 3). \tag{5.36}$$

With this parameterization, the Boltzmann equation for the abundance of ψ's becomes,

$$dY/dx = -\lambda x^{-n-2}(Y^2 - Y_{\rm EQ}^2), \tag{5.37}$$

where

$$\begin{aligned} \lambda &= \left[\frac{x\langle\sigma_A|v|\rangle s}{H(m)}\right]_{x=1} = 0.264(g_{*S}/g_*^{1/2})m_{Pl}\, m\, \sigma_0, \\ Y_{\rm EQ} &= 0.145(g/g_{*S})x^{3/2}e^{-x}. \end{aligned} \tag{5.38}$$

As we will now describe, this differential equation can be solved approximately to very good accuracy (better than 5%). To begin, consider the differential equation for $\Delta \equiv Y - Y_{\rm EQ}$, the departure from equilibrium,

$$\Delta' = -Y'_{\rm EQ} - \lambda x^{-n-2}\Delta(2Y_{\rm EQ} + \Delta), \tag{5.39}$$

where prime denotes d/dx. At early times ($1 < x \ll x_f$), Y tracks $Y_{\rm EQ}$ very closely, and both Δ and $|\Delta'|$ are small, so that an approximate solution is obtained by setting $\Delta' = 0$:

$$\begin{aligned} \Delta &\simeq -\lambda^{-1}x^{n+2}Y'_{\rm EQ}/(2Y_{\rm EQ} + \Delta) \\ &\simeq x^{n+2}/2\lambda. \end{aligned} \tag{5.40}$$

At late times ($x \gg x_f$), Y tracks $Y_{\rm EQ}$ very poorly: $\Delta \simeq Y \gg Y_{\rm EQ}$, and the terms involving $Y'_{\rm EQ}$ and $Y_{\rm EQ}$ can be safely neglected, so that

$$\Delta' = -\lambda x^{-n-2}\Delta^2. \tag{5.41}$$

Upon integration of (5.41) from $x = x_f$ to $x = \infty$, we obtain

$$Y_\infty = \Delta_\infty = \frac{n+1}{\lambda} x_f^{n+1}. \tag{5.42}$$

Now we must determine x_f. Recall $x = x_f$ is the time when Y ceases to track Y_{EQ}, or equivalently, when Δ becomes of order Y_{EQ}. Defining x_f by the criterion: $\Delta(x_f) = c\,Y_{\rm EQ}(x_f)$, c = numerical constant of order unity, the early time solution of (5.41) becomes $\Delta(x_f) \simeq x_f^{n+2}/\lambda(2+c)$, and the freeze-out criterion gives

$$x_f \cong \ln\left[(2+c)\lambda ac\right] - \left(n + \frac{1}{2}\right) \ln\left\{\ln\left[(2+c)\lambda ac\right]\right\} \tag{5.43}$$

where $a = 0.145(g/g_{*S})$. Note that x_f depends only logarithmically upon the numerical criterion for freeze out, i.e., the value of c, as does the final abundance. The results of a numerical integration of the Boltzmann equation are shown in Fig. 5.1.

Choosing $c(c+2) = n+1$ gives the best fit to the numerical results for the final abundance Y_∞ (to better than 5% for any $x_f \gtrsim 3$). With this choice

$$\begin{aligned} x_f &= \ln[0.038(n+1)(g/g_*^{1/2})m_{Pl}\, m\, \sigma_0] \\ &\quad - \left(n+\frac{1}{2}\right)\ln\left\{\ln\left[0.038(n+1)(g/g_*^{1/2})m_{Pl}\, m\, \sigma_0\right]\right\} \end{aligned} \tag{5.44}$$

$$Y_\infty = \frac{3.79(n+1)x_f^{n+1}}{(g_{*S}/g_*^{1/2})m_{Pl}\, m\, \sigma_0}. \tag{5.45}$$

We mention in passing that one could have obtained a very similar result to (5.45) by estimating x_f by the freeze-out criterion $\Gamma(x_f) \simeq H(x_f)$, and setting $Y_\infty = Y(x_f)$. The formulae for x_f and Y_∞ obtained this way differ very little; for x_f, the coefficient of the lnln term is $(-n+1/2)$ rather than $(-n-1/2)$, and for Y_∞, a factor of 5 instead of $3.79(n+1)$.

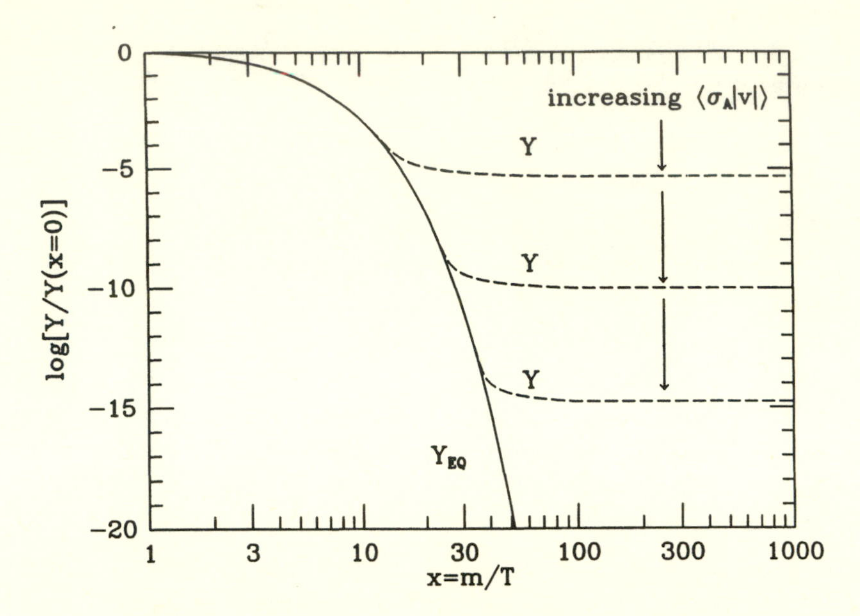

Fig. 5.1: The freeze out of a massive particle species. The dashed line is the actual abundance, and the solid line is the equilibrium abundance.

In some circumstances the annihilation cross section may be better approximated by $\langle\sigma_A|v|\rangle = \sigma_0 x^{-n}(1+bx^{-m})$, e.g., if both s-wave and p-wave annihilation processes are important. The modification to (5.45) is straightforward to compute: $Y_\infty \to Y_\infty/[1+(n+1)bx_f^{-m}/(m+n+1)]$, and $x_f \to x_f + \ln[1+b\{\ln(0.038(g/g_*^{1/2})m_{Pl}m\sigma_0)\}^{-m}]$.

As with a hot relic, the present number density and mass density of relic ψ's is easy to compute,

$$\begin{aligned} n_{\psi 0} &= s_0 Y_\infty = 2970 Y_\infty \ \mathrm{cm}^{-3} \\ &= 1.13\times 10^4 \frac{(n+1)x_f^{n+1}}{(g_{*S}/g_*^{1/2})m_{Pl}\, m\, \sigma_0}\ \mathrm{cm}^{-3} \end{aligned} \tag{5.46}$$

$$\Omega_\psi h^2 = 1.07\times 10^9 \frac{(n+1)x_f^{n+1}\ \mathrm{GeV}^{-1}}{(g_{*S}/g_*^{1/2})m_{Pl}\sigma_0}. \tag{5.47}$$

1.9733×10^{-14} cm = 1 GeV^{-1}
$g_* = 8$
$g_{*S} = 100$
$x_f^{n+1} = x_f \sim 20-40$

It is very interesting to note that the relic density of ψ's is inversely

$\therefore \Omega_\psi h^2 = \frac{3.783\times10^{-11}\ \mathrm{GeV}^{-2}}{\sigma_0} = \frac{3.783\times10^{-11}\times(1.9733\times10^{-14})^2\ \mathrm{cm}^2}{\sigma_0} = \frac{1.47\times10^{-38}\ \mathrm{cm}^2}{\sigma_0}$

$\therefore$ need $\sigma_0 \sim 10^{-37}$ cm^2

proportional to the annihilation cross section and mass of the particle

$$Y_\infty = \frac{3.79(n+1)(g_*^{1/2}/g_{*S})x_f}{m\, m_{Pl}\langle\sigma_A|v|\rangle}. \tag{5.48}$$

The smaller the annihilation cross section, the greater the relic abundance—the weak prevail. Moreover, the present mass density ($\rho_{\psi 0} \propto mY_\infty$) only depends upon the annihilation cross section at freeze out, which for $n = 0$ (s-wave annihilation) is independent of temperature (and energy).

• *Two Examples of Cold Relics:* Now let's apply this formalism to two simple examples. First, consider a baryon symmetric Universe (not ours!), that is equal numbers of nucleons and antinucleons. Using the above machinery we can calculate the surviving relic abundance of nucleons and antinucleons. Taking the nucleon–antinucleon annihilation cross section to be $\langle\sigma_A|v|\rangle = c_1 m_\pi^{-2}$, where $m_\pi = 135$ MeV is the pion mass and c_1 is a numerical constant of order unity, we find that

$$\begin{aligned} x_f &\simeq 42 + \ln\, c_1, \\ T_f &\simeq 22 \text{ MeV}, \\ Y_\infty &\simeq 7 \times 10^{-20} c_1^{-1}. \end{aligned} \tag{5.49}$$

Today, we know that the abundance of nucleons is $Y_\infty = n_B/s \simeq \eta/7 \sim$ (6 to 10) $\times 10^{-11}$, some 9 orders of magnitude larger than the relic abundance of nucleons in a baryon-symmetric Universe. As we will discuss later, this calculation provides strong evidence for the necessity of a baryon asymmetry—to prevent the so-called annihilation catastrophe in a baryon symmetric Universe. The presence of an excess of nucleons over antinucleons, of course, precludes this catastrophe, as all the excess nucleons necessarily survive annihilation. While the formalism developed here does not allow for a ψ–$\bar{\psi}$ asymmetry, it is straightforward to extend the formalism. In the case of the nucleon–antinucleon asymmetric Universe (like ours), the relic abundance of antinucleons can be computed (i.e., those that survive annihilation); it is

$$Y_\infty(\bar{N}) = 10^{18} \exp(-9 \times 10^5). \tag{5.50}$$

This very, very small number illustrates an interesting feature of freeze out. In the symmetric case, as time goes on the annihilation rate falls exponentially:

$$\Gamma_A = n_{\rm EQ}\langle\sigma_A|v|\rangle \quad\propto\quad (m/T)^{3/2}\exp(-m/T)\langle\sigma_A|v|\rangle, \tag{5.51}$$

and annihilations necessarily quench as particles and antiparticles become very rare.[8] In the case of antinucleons in a baryon asymmetric Universe, the annihilation rate of antinucleons is: $\Gamma_A = n_N\langle\sigma_A|v|\rangle$, which does not decrease exponentially, as the nucleon abundance levels off at the value set by the asymmetry, $n_N \simeq n_B$. We will return to the origin of the all important baryon asymmetry in the next Chapter.

Next, consider the relic abundance of a hypothetical heavy, stable neutrino species of mass $m \gg$ MeV. Because of its large mass, such a neutrino species will decouple when it is non-relativistic (though not necessarily at the canonical $T \sim$ few MeV), and the formulae for a cold relic pertain.

Annihilation for such a species proceeds through Z^0 exchange to final states $i\bar{i}$; where $i = \nu_L, e, \mu, \tau, u, d, s, \cdots$ (ν_L denotes any lighter neutrino species). The annihilation cross section depends upon whether the heavy neutrino is of the Dirac or Majorana type; for $T \lesssim m \lesssim M_Z$, the annihilation cross section is

$$\begin{aligned}
\langle\sigma_A|v|\rangle_{\rm Dirac} &= \frac{G_F^2 m^2}{2\pi}\sum_i (1-z_i^2)^{1/2} \\
&\quad \times[(C_{V_i}^2 + C_{A_i}^2) + \frac{1}{2}z_i^2(C_{V_i}^2 + C_{A_i}^2)] \\
\langle\sigma_A|v|\rangle_{\rm Majorana} &= \frac{G_F^2 m^2}{2\pi}\sum_i (1-z_i^2)^{1/2} \\
&\quad \times[(C_{V_i}^2 + C_{A_i}^2)8\beta_i^2/3 + C_{A_i}^2 2z_i^2],
\end{aligned} \tag{5.52}$$

where $z_i = m_i/m$, β is the relative velocity, and C_V and C_A are given in terms of the weak isospin j_3, the electric charge q, and the Weinberg angle θ_W by $C_A = j_3$, $C_V = j_3 - 2q\sin^2\theta_W$.[9] The sum is over all quark and lepton species lighter than m.

---

[8]While annihilations cease to significantly affect the abundance of $\psi\bar{\psi}$'s after freeze out, annihilations do occur at a rate per comoving volume per Hubble time proportional to T^r ($r = 1 + n$, RD; $r = 1.5 + n$, MD)—and may have interesting consequences [7].

[9]We have assumed that the neutrino is less massive than M_Z. If the neutrino is more massive than the Z^0 the annihilation cross section will be $\sigma_A \sim \alpha^2/M^2$, and the calculation of freeze out will be modified.

In the Dirac case, annihilations proceed through the s-wave and $\langle\sigma_A|v|\rangle$ is velocity independent:

$$\sigma_0 \simeq c_2\, G_F^2 m^2/2\pi \tag{5.53}$$

where $c_2 \sim 5$. Taking $g = 2$ and $g_* \simeq 60$, from our formulae for x_f and Y_∞ we find

$$\begin{aligned} x_f &\simeq 15 + 3\ln(m/\mathrm{GeV}) + \ln(c_2/5) \\ Y_\infty &\simeq 6\times 10^{-9}\left(\frac{m}{\mathrm{GeV}}\right)^{-3}\left[1 + \frac{3\ln(m/\mathrm{GeV})}{15} + \frac{\ln(c_2/5)}{15}\right] \end{aligned} \tag{5.54}$$

from which we compute that

$$\Omega_{\nu\bar{\nu}}h^2 = 3(m/\mathrm{GeV})^{-2}\left[1 + \frac{3\ln(m/\mathrm{GeV})}{15}\right], \tag{5.55}$$

where we have included the identical relic abundance of the antineutrino species ($\Omega_{\nu\bar{\nu}} = 2\Omega_\nu$). Note that freeze out takes place at $T_F \simeq m/15 \simeq 70\ \mathrm{MeV}(m/\mathrm{GeV})$—before the interactions of light neutrinos freeze out. This is because as neutrinos annihilate and become rare, the annihilation process quenches. Requiring $\Omega_{\nu\bar{\nu}}h^2 \lesssim 1$ we obtain the so-called Lee-Weinberg bound:

$$m \gtrsim 2\ \mathrm{GeV}. \tag{5.56}$$

Although it is often called the Lee-Weinberg bound [8], it was discovered independently by a number of people.

For the Majorana case, annihilation proceeds through both the s-wave and p-wave; however the formulae for x_f, Y_∞ and $\Omega_{\nu\bar{\nu}}h^2$ are similar. In Fig. 5.2 we show the contribution to $\Omega_0 h^2$ for a stable, massive neutrino species. For $m \lesssim$ MeV, $\Omega_{\nu\bar{\nu}}h^2 \propto m$ as the relic abundance is constant. For $m \gtrsim$ MeV, $\Omega_{\nu\bar{\nu}}h^2 \propto m^{-2}$ as the relic abundance decreases as m^{-3}. The relic mass density achieves its maximum for $m \sim$ MeV; neutrino masses less than about $92h^2$ eV, or more than about 2 GeV (Dirac) or about 5 GeV (Majorana) are cosmologically acceptable.

The calculation of the relic abundance of some hypothetical, massive stable particle species that was once in thermal equilibrium in the early Universe (*Origin of Species*, if you will) is one of the routine chores of an early-Universe cosmologist. Among the numerous potential relics venerated by cosmologists are hot or warm relics such as a light neutrino,

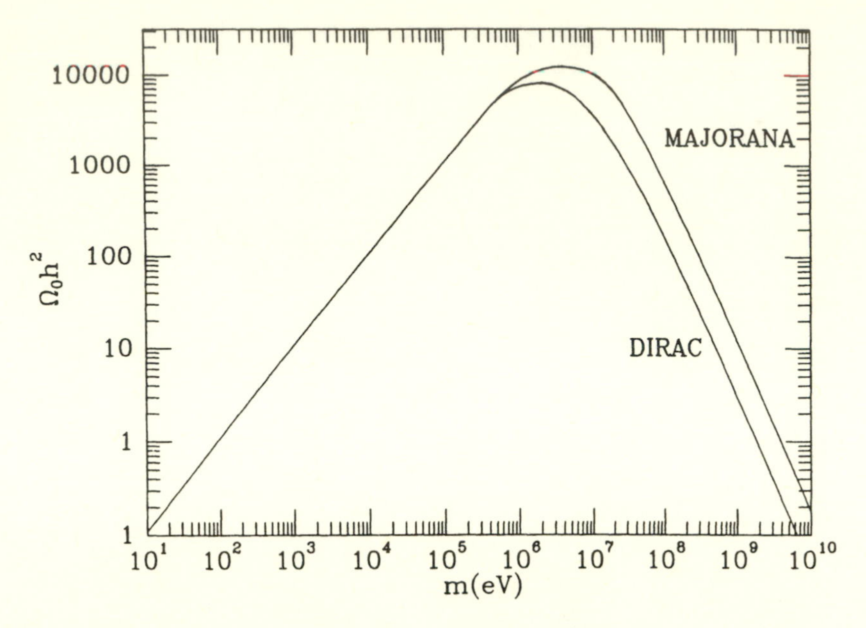

Fig. 5.2: The contribution to $\Omega_0 h^2$ for a stable neutrino species of mass m.

gravitino, photino, right-handed neutrino, etc., or cold relics such as a heavy neutrino, photino, sneutrino, higgsino, pyrgon, etc. In Chapters 7 and 10 we will discuss examples of non-thermal relics, monopoles and axions, respectively. The production of non-thermal relics differs from that of their thermal counterparts in that they were never in thermal equilibrium. In Chapter 9 we will stress the importance of early-Universe relics as candidates for the dark matter.

5.3 Out–of–Equilibrium Decay

We will now consider another kind of non-equilibrium process, the decay of a massive particle species which occurs out of equilibrium (i.e., after the particle has decoupled and has an abundance $Y \gg Y_{\rm EQ}$). As we shall see, such a process can produce considerable entropy.

Consider a non-relativistic and relatively long-lived particle species ψ that is decoupled and has a pre-decay abundance $Y_i = n_\psi/s$, (e.g., a 100 GeV gravitino whose lifetime is about 10^6 sec). Since the ψ is non-relativistic, its contribution to the energy density decreases as R^{-3}, but grows relative to the radiation energy density as R. If it is sufficiently long lived, it decays while dominating the energy density of the Universe, and thereby releases considerable entropy. Let's make a simple estimate of the entropy it releases. Suppose all the decays occur when $t \sim \tau$ (τ is the mean lifetime of the ψ). To make things interesting let us assume that its energy density dominates that of the radiation present before it decays.

The ψ decays at a time $t \sim H^{-1} \sim \tau$, when the temperature of the Universe is $T = T_D$, and the energy density of the Universe is $\rho \sim \rho_\psi = sY_i m$. Just before the ψ's decay T_D and τ are related by:

$$H^2(T_D) \equiv H_D^2 \sim G\rho \sim Y_i T_D^3 m/m_{Pl}^2 \sim \tau^{-2}. \tag{5.57}$$

Suppose that the ψ's decay into relativistic particles that rapidly (compared to the expansion timescale) thermalize, yielding a post-decay radiation density ρ_R. Then after the ψ's decay,

$$\rho_R \sim g_* T_{RH}^4, \tag{5.58}$$

which, by energy conservation, must also equal the energy density in ψ's just before their demise, $H_D^2 m_{Pl}^2$. Comparing (5.57) and (5.58) we find that the ratio of the entropy per comoving volume after decay to that before decay is

$$\frac{S_{\rm after}}{S_{\rm before}} = \frac{g_* R^3 T_{RH}^3}{g_* R^3 T_D^3} \sim g_*^{1/4} \frac{Y_i m \tau^{1/2}}{m_{Pl}^{1/2}}. \tag{5.59}$$

In addition, it appears that the Universe has been heated up by the ψ decays:

$$T_{\rm after}/T_{\rm before} = T_{RH}/T_D = (S_{\rm after}/S_{\rm before})^{1/3}. \tag{5.60}$$

As we will now see, the estimate for the entropy increase is quite accurate; however, the temperature of the Universe never increases—rather, it just decreases more slowly than it would in the absence of ψ decays.

Now let's treat the problem more carefully. First consider the energy density in ψ particles: Due to ψ decays, the number of ψ's in a comoving volume ($\equiv R^3 n_\psi$) decreases according to the usual exponential decay law, $d(R^3 n_\psi)/dt = -\tau^{-1}(R^3 n_\psi)$, which gives

$$\dot{n}_\psi + 3Hn_\psi = -\tau^{-1} n_\psi. \tag{5.61}$$

Since the ψ's are non-relativistic, their energy density is given by $\rho_\psi = m\, n_\psi$, and

$$\dot{\rho}_\psi + 3H\rho_\psi = -\tau^{-1}\rho_\psi. \tag{5.62}$$

$\frac{dn}{dt} + 3Hn = -\Gamma n$

The solution is easily found:

$$\rho_\psi(R) = \rho_\psi(R_i)\left(\frac{R}{R_i}\right)^{-3} \exp(-t/\tau). \tag{5.63}$$

Let us assume that the energy released from ψ decays is rapidly (i.e., on a timescale less than the expansion timescale) converted into relativistic particles which are then thermalized. In this case the energy density of the Universe resides in two components: non-relativistic ψ particles and radiation. The decay of ψ's transfers energy from the former to the latter. The second law of thermodynamics applied to a comoving volume element implies that

$$dS = \frac{dQ}{T} = -d(R^3\rho_\psi)/T = \frac{R^3\rho_\psi}{T}(dt/\tau). \tag{5.64}$$

Using the fact that $S = (2\pi^2/45)g_* T^3 R^3$, (5.64) can be rewritten as

$$S^{1/3}\dot{S} = \left(\frac{2\pi^2}{45}g_*\right)^{1/3} R^4\rho_\psi/\tau. \tag{5.65}$$

A formal solution to this equation is easily obtained:

$$S^{4/3} = S_i^{4/3} + \frac{4}{3}\rho_\psi(R_i)R_i^4\tau^{-1}\int_{t_i}^{t}\left(\frac{2\pi^2 g_*}{45}\right)^{1/3}\frac{R(t')}{R_i}\exp(-t'/\tau)dt'. \tag{5.66}$$

Note that in the limit that g_* is constant, the first law of thermodynamics provides a simple means for obtaining the evolution equation for the radiation energy density ρ_R:

$$\begin{aligned} d(R^3\rho_R) &= -p_R d(R^3) - d(R^3\rho_\psi) = -\frac{\rho_R}{3}d(R^3) + (R^3\rho_\psi)dt/\tau \\ \dot{\rho}_R + 4H\rho_R &= \tau^{-1}\rho_\psi. \end{aligned} \tag{5.67}$$

This equation is equivalent to (5.65) if g_* is constant. If g_* is not constant, then p_R is not simply $\rho_R/3$, (5.67) is not valid, and (5.65) must be used. The physics of (5.67) is manifest: The $4H\rho_R$ term represents the usual red shift of the radiation energy density, and the ρ_ψ/τ term accounts for the energy input from ψ decays. In the absence of the ρ_ψ/τ term, the solution to (5.67) is just $\rho_R \propto R^{-4}$.

The expansion rate of the Universe is governed as usual by the Friedmann equation,

$$H^2 = (\dot{R}/R)^2 = \frac{8\pi}{3m_{Pl}^2}(\rho_\psi + \rho_R). \tag{5.68}$$

The energy density in radiation is related to the entropy per comoving volume by

$$\rho_R = \frac{3}{4}\Big(\frac{45}{2\pi^2 g_*}\Big)^{1/3} S^{4/3} R^{-4}. \tag{5.69}$$

Equations (5.68), (5.65) [or (5.67)], and (5.62) form a closed set of differential equations governing the evolution of R, S [or ρ_R], and ρ_ψ.

We will consider their solution in the interesting limit of significant entropy generation: Physically, this means that the ψ's come to dominate the energy density of Universe at some time before $t = \tau$. Suppose that this occurs at time $t = t_\psi$. Roughly speaking then, until the time $t = \tau$ when the energy density in ψ's begins to exponentially decrease, the Universe is matter dominated, and $R \propto t^{2/3}$. During this time $\rho_\psi \simeq \rho_\psi(R_i)(R_i^3/R^3) \propto t^{-2}$, and it is simple to solve (5.67) for the energy density in radiation

$$\rho_R = \rho_R(R_i)\left(\frac{R_i}{R}\right)^4 + \frac{5}{3}\rho_\psi(R_i)\frac{t_i^2}{t\tau}, \tag{5.70}$$

where $\rho(R_i)$ is the energy density at some initial epoch $t = t_i$, and for convenience we have taken t_i to be some time shortly after the ψ's begin to dominate the energy density. The physical significance of this solution is manifest: The first term represents the "primeval radiation" and the second that produced by ψ decays. From this we see that the ψ-produced radiation starts to be the dominant component when $t \simeq t_\psi(\tau/t_\psi)^{3/5}$. This means that until this time $\rho_R \propto R^{-4}$, and thereafter $\rho_R \propto t^{-1} \propto R^{-3/2}$. For $t \gtrsim \tau$, the "source term" ρ_ψ/τ dies away exponentially and ρ_R once again decreases as R^{-4}. Thus we see that ρ_R always *decreases*, albeit at a much slower rate when ψ decays are producing significant energy density. This fact is due to two things: (i) the exponential decay law—ψ's don't suddenly and simultaneously decay at $t = \tau$; (ii) the expansion of the Universe which red shifts the decay-produced radiation energy density ($\propto R^{-4}$) after it is produced. During the time interval from $t \simeq t_\psi(\tau/t_\psi)^{3/5}$ to $t \simeq \tau$, when the ψ produced radiation is the dominant radiation component, the entropy per comoving volume is also growing: $S \propto R^3\rho_R^{3/4} \propto R^{15/8} \propto t^{5/4}$.

Finally, let us compute the total entropy increase due to ψ decays, in the case that ψ's come to dominate the energy density of the Universe well before they decay. In general it follows from (5.67) that the ratio of the final ($t \gg \tau$) to initial ($t \ll \tau$) entropy per comoving volume can be written as

$$\frac{S_f}{S_i} = \left[1+\frac{4}{3}\left(\frac{45}{2\pi^2 g_*(T_i)}\right)^{1/3}\frac{mY_i}{T_i}I\right]^{3/4},$$

$$I = \tau^{-1}\int_0^\infty \left(\frac{2\pi^2 g_*}{45}\right)^{1/3}\frac{R(t)}{R_i}\exp(-t/\tau)dt. \qquad (5.71)$$

The integral I depends upon the functional form of the time dependence of the scale factor $R(t)$. In the case where ψ's dominate the energy density of the Universe before they decay, I can be evaluated numerically:

$$I = 1.09\left(\frac{8\pi\rho_{\psi i}}{3m_{Pl}^2}\right)^{1/3}\tau^{2/3}\left(\frac{2\pi^2}{45}\right)^{1/3}\langle g_*^{1/3}\rangle, \qquad (5.72)$$

where the brackets indicate the appropriately-averaged value of $g_*^{1/3}$ over the decay interval. Bringing together all the numerical factors we find

$$\frac{S_f}{S_i} \simeq 1.83\langle g_*^{1/3}\rangle^{3/4}\frac{mY_i\tau^{1/2}}{m_{Pl}^{1/2}}, \qquad (5.73)$$

which differs from our simple-minded estimate by only a numerical factor of order unity. Note that the temperature at the end of the decay epoch follows directly from the Friedmann equation:

$$H^2(t=\tau) \simeq \frac{1}{4}\tau^{-2} \simeq \frac{8\pi}{3m_{Pl}^2}\frac{\pi^2 g_*}{30}T_{RH}^4 \qquad (5.74)$$

$$T_{RH} = T(\tau) \simeq 0.55 g_*^{-1/4}(m_{Pl}/\tau)^{1/2}. \qquad (5.75)$$

And as it should, this estimate for T_{RH} agrees with that obtained from (5.60).

The numerical solution of the system of equations (5.68), (5.65) [or (5.67)], and (5.62) through the epoch of out of equilibrium decay of ψ is shown in Fig. 5.3. The dashed lines indicate the evolution of the energy densities in the absence of ψ decay. Also indicated is the increase in the entropy S due to ψ decay. For this example, $Y_i = 3.2\times10^{-5}$,

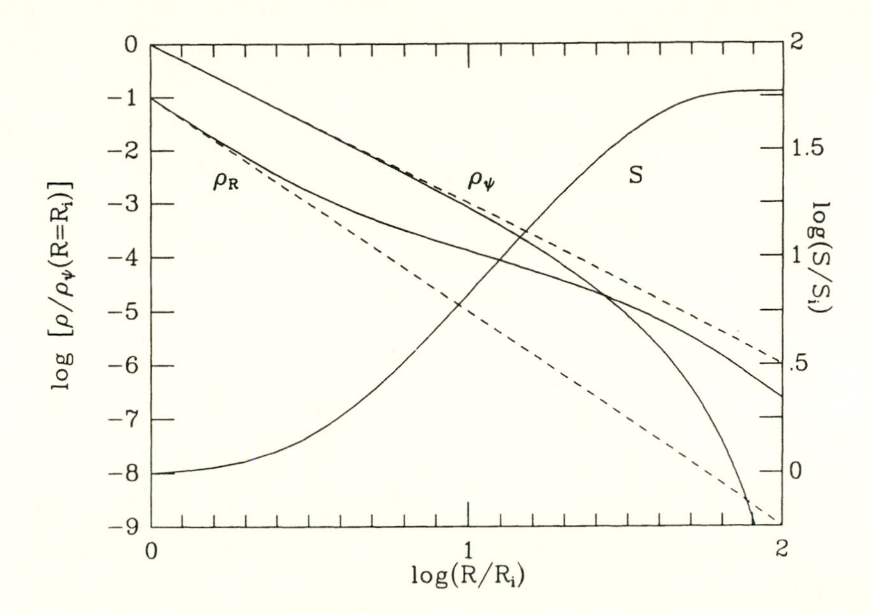

Fig. 5.3: The evolution of the radiation and ψ energy densities and entropy S through the epoch of out-of-equilibrium ψ decay (solid lines). Broken lines indicate the evolution of ρ_R and ρ_ψ in the absence of ψ decays.

$\rho_\psi(R_i)/\rho_R(R_i) = 10$, $H_i\tau = 100$, and $\tau/m_{Pl} = 10^{12}/m^2 g_*^{1/2}$. Equation (5.73) predicts $S_f/S_i = 59$, in excellent agreement with the numerical result shown in Fig. 5.3. Likewise, the analytic expression for the epoch when ψ-produced radiation starts to dominate the primeval radiation, $R/R_i \sim 3$, is also in agreement with the numerical results.

We will return to these equations when discussing the reheating of an inflationary Universe, as they also describe that process; in that case ψ is the scalar field that drives inflation. Before going on, let's consider an example of a hypothetical, long-lived particle that decays out of equilibrium, and produces significant entropy—the gravitino. The gravitino is the supersymmetric partner of the graviton. Its mass and lifetime,

$$m \simeq \mu^2/m_{Pl}, \tag{5.76}$$

$$\tau \simeq m_{Pl}^2/m^3 \simeq 9.8 \times 10^{13} m_{\rm GeV}^{-3}\ \rm sec \tag{5.77}$$

are determined by the scale of supersymmetry breaking μ. Like the graviton its interactions are so weak that it decouples very early on, while it is still relativistic. Thus we expect its initial abundance to be

$$Y_i = 0.278(g_{\rm eff}/g_{*S}) \sim 10^{-3}. \tag{5.78}$$

Using the formulae just derived, we find that

$$T_{RH} \sim m^{3/2}/m_{Pl}^{1/2} \sim 3\times 10^{-10} m_{\rm GeV}^{3/2}\ {\rm GeV} \tag{5.79}$$

$$S_f/S_i \sim 10^{-3}(m_{Pl}/m)^{1/2} \sim 10^7 m_{\rm GeV}^{-1/2}. \tag{5.80}$$

For the interesting value of the gravitino mass $m \sim 10^2$ to 10^3 GeV, gravitino decays produce an enormous amount of entropy after nucleosynthesis, seemingly excluding such a gravitino mass. This problem can be circumvented if the Universe inflated after gravitinos decoupled; then $Y_i \ll 10^{-3}$, as the only gravitinos present are those produced during the reheating process, a number which is much, much less than the equilibrium value. We will return to gravitinos and inflation again.

5.4 Recombination Revisited

Armed with our Boltzmann technology we will now calculate more precisely the residual ionization of the Universe. Recall from our earlier discussion in Chapter 3, that $X_e = n_p/(n_H + n_p)$ is the ionization fraction, where n_e, n_p and n_H are the number densities of free electrons, free protons and hydrogen atoms, and that for simplicity we have ignored the small abundances of the other light elements. The equilibrium ionization fraction satisfies

$$1 - X_e^{EQ} = (X_e^{EQ})^2 \frac{4\sqrt{2}\zeta(3)}{\sqrt{\pi}} \eta \left(\frac{T}{m_e}\right)^{3/2} \exp(B/T), \tag{5.81}$$

where η is the baryon-to-photon ratio and $B = 13.6$ eV is the binding energy of hydrogen. In the "post-recombination" era, where $X_e^{EQ} \ll 1$, the expression for X_e^{EQ} reduces to

$$X_e^{EQ} \simeq 0.51\eta^{-1/2}\left(\frac{m_e}{T}\right)^{3/4} \exp(-B/2T). \tag{5.82}$$

Following the evolution of the ionization fraction X_e is analogous to following the evolution of the abundance of a stable, massive particle species.

The key reaction there is annihilation ($\psi\bar{\psi} \longleftrightarrow X\bar{X}$), while here the key process is recombination ($e + p \longleftrightarrow H + \gamma$). By inspection we can write the Boltzmann equation for n_e:

$$\dot{n}_e + 3Hn_e = -\langle\sigma_{\rm rec}|v|\rangle \left(n_e^2 - (n_e^{\rm EQ})^2\right), \tag{5.83}$$

where $\langle\sigma_{\rm rec}|v|\rangle$ is the thermally-averaged recombination cross section. This cross section is given by

$$\langle\sigma_n|v|\rangle = \frac{4\pi^2\alpha}{m_e^2}\frac{B/n}{(3m_eT)^{1/2}}, \tag{5.84}$$

where $\langle\sigma_n|v|\rangle$ is the thermally-averaged cross section for $p + e \rightarrow H^*$ (nth excited level), and $E_n = 13.6 \text{ eV}/n^2$ is the binding energy of the nth level. For simplicity, we will consider only the ground state ($n = 1$), so that

$$\begin{aligned} \langle\sigma_{\rm rec}|v|\rangle &= \frac{4\pi^2\alpha}{m_e^2}\frac{B}{(3m_eT)^{1/2}} \\ &= 4.7 \times 10^{-24} \text{ cm}^2 T_{\rm eV}^{-1/2} \end{aligned} \tag{5.85}$$

where $T_{\rm eV} = T/\text{eV}$.

Taking the Universe to be matter-dominated, and defining $x = T_{\rm eV}^{-1}$, (5.83) can be rewritten as

$$X_e' = -\lambda x^{-2}(X_e^2 - (X_e^{EQ})^2), \tag{5.86}$$

$$\lambda = \left[\frac{n_B\langle\sigma_{\rm rec}|v|\rangle}{xH}\right]_{x=1} = 1.4 \times 10^5(\Omega_B h/\Omega_0^{1/2}), \tag{5.87}$$

$$X_e^{EQ} = 5.95 \times 10^7(\Omega_B h^2)^{-1/2}x^{3/4}\exp(-6.8x). \tag{5.88}$$

In the same way we did for the relic abundance of a massive particle species we can obtain an approximate solution for X_e^∞, the residual ionization fraction. Again, consider $\Delta = X_e - X_e^{EQ}$, the deviation from ionization equilibrium. The equation governing the evolution of Δ is

$$\Delta' = -X_e^{EQ\prime} - \lambda x^{-2}(\Delta + 2X_e^{EQ})\Delta. \tag{5.89}$$

For $x \lesssim x_f$ an approximate solution is obtained by setting $\Delta' = 0$,

$$\Delta \simeq -\frac{X_e^{EQ'} x^2}{\lambda(\Delta + 2X_e^{EQ})} \simeq \frac{3.4x^2}{\lambda} \tag{5.90}$$

On the other hand, for $x \gtrsim x_f$, $\Delta \simeq X_e \gg X_e^{EQ}$, $X_e^{EQ'}$, so that X_e^{EQ}, $X_e^{EQ'}$ can be neglected, and

$$\Delta' \simeq -\lambda x^{-2}\Delta^2. \tag{5.91}$$

Integrating from $x = x_f$ to $x = \infty$ we obtain

$$X_e^{\infty} = \frac{x_f}{\lambda}. \tag{5.92}$$

As before, "freeze out" ($x = x_f$) is found by using the approximate solution for $\Delta(x)$ to determine when the deviation from equilibrium ionization becomes of order unity, $\Delta \simeq X_e^{EQ}$:

$$\frac{3.4}{\lambda} x_f^2 \simeq 5.95 \times 10^7 (\Omega_B h^2)^{-1/2} x_f^{3/4} \exp(-6.8 x_f). \tag{5.93}$$

The value of x at freeze out is

$$\begin{aligned} x_f &\simeq \frac{1}{6.8} \ln\left[2.5 \times 10^{12} (\Omega_B/\Omega_0)^{1/2}\right] \\ &\quad - \frac{5/4}{6.8} \ln\left\{\ln\left[2.5 \times 10^{12} (\Omega_B/\Omega_0)^{1/2}\right]\right\} \\ &\simeq 3.6 - 0.074 \ln(\Omega_B/\Omega_0). \end{aligned} \tag{5.94}$$

Taking $\Omega_B/\Omega_0 \sim 0.1$, the freeze in of the residual ionization occurs for $x_f \simeq 3.8$, or

$$T_F = \frac{1}{3.8}\ \text{eV} \simeq 0.26\ \text{eV}, \tag{5.95}$$

and the residual ionization fraction is

$$X_e^{\infty} \simeq \frac{x_f}{\lambda} \simeq 2.7 \times 10^{-5} (\Omega_0^{1/2}/\Omega_B h), \tag{5.96}$$

about comparable to the abundance of D or ^{3}He in the Universe. We note that using the poor man's criterion for freeze out, $(\Gamma_{rec}/H)|_{x_f} \simeq 1$, and setting $X_e^\infty = X_e^{EQ}(x_f)$, we would obtain a numerical factor of 2.9×10^{-5} rather than 2.7×10^{-5}.

5.5 Neutrino Cosmology

The fact that neutrinos decouple relatively early in the evolution of the Universe ($T \gtrsim$ few MeV) guarantees that neutrinos should have a substantial relic abundance today. In turn, this means that they may have important cosmological consequences. Moreover, their cosmological consequences can be used to place very significant constraints on their properties, constraints which go far beyond the reach of conventional terrestrial laboratories.

For a stable neutrino species, we have seen that requiring their relic mass density to be not so great as to lead to a Universe that today is younger than 10 Gyr (i.e., $\Omega_0 h^2 \lesssim 1$) restricts any neutrino mass to be either less than about $92h^2$ eV, or greater than a few GeV. The limit of $92h^2$ eV is quite impressive when compared with the laboratory limits to the mass of the μ and τ neutrinos, 250 keV and 35 MeV respectively.

Now consider the possibility of an unstable neutrino species whose decay products are relativistic, even at the present epoch. It is clear that the mass density bound for such a species must be less stringent: from the epoch at which they decay (say, $z = z_D$) until the present, the mass density of the relativistic neutrino decay products decreases as R^{-4}, as opposed to the R^{-3} had the neutrinos not decayed. Roughly speaking then, the mass density today of the decay products is a factor of $(1+z_D)^{-1}$ less than that of a stable neutrino species.

The precise abundance of the neutrino decay products is very easy to compute. Denote the energy density of the relativistic decay products by ρ_D, and for simplicity assume that they do not thermalize. The equations governing the evolution of the daughter products are essentially the same equations we discussed earlier for a decaying particle species:

$$\begin{aligned} \dot{\rho}_D &+ 4H\rho_D = \rho_\nu/\tau, \\ \rho_\nu(R) &= \rho_\nu(R_i)\left(\frac{R}{R_i}\right)^{-3} \exp(-t/\tau), \end{aligned} \qquad (5.97)$$

where R_i, t_i is some convenient epoch prior to decay, $t_i \ll \tau$. The relic density of the decay products is obtained by integrating (5.97):

$$\rho_D(t) = \rho_{\nu i}\tau^{-1}\left(\frac{R_i}{R}\right)^4 \int_i^t \frac{R(t')}{R_i}\exp(-t'/\tau)dt'. \tag{5.98}$$

Assuming that around the time the neutrinos decay ($t \sim \tau$) the scale factor $R \propto t^n$ ($n = 1/2$ radiation dominated; $n = 2/3$ matter dominated) we can evaluate this integral directly, and find that the present density of relic, relativistic particles from neutrino decays is

$$\rho_D(t_0) = n!\rho_\nu(t_0)\frac{R(\tau)}{R_0} \tag{5.99}$$

where $\rho_\nu(t_0)$ is the present density that neutrinos and antineutrinos would have had they not decayed, and $R(\tau)$ is the value of the scale factor at the time $t = \tau$. As expected, the present energy density of the decay products is less than that of a stable neutrino species, by a factor of $n!R(\tau)/R_0 \sim (1 + z_D)^{-1}$.[10] During the matter-dominated epoch ($t \gtrsim 4.4 \times 10^{10}(\Omega_0 h^2)^{-2}$ sec), $R(t)/R_0 = 2.9 \times 10^{-12}(\Omega_0 h^2)^{1/3} t_{\rm sec}^{2/3}$, so that the reduction factor is

$$n!R(\tau)/R_0 = 2.6 \times 10^{-12}(\Omega_0 h^2)^{1/3}\tau_{\rm sec}^{2/3}. \tag{5.100}$$

During the radiation-dominated epoch, $R(t)/R_0 = 2.4 g_*^{-1/12} \times 10^{-10} t_{\rm sec}^{1/2}$, so that the reduction factor is

$$n!R(\tau)/R_0 = 2.1 \times 10^{-10} g_*^{-1/12}\tau_{\rm sec}^{1/2}. \tag{5.101}$$

Using the results of our earlier calculations for $\Omega_{\nu\bar\nu}h^2$, we obtain the following constraint to the epoch of decay (for neutrino masses which fall in the previously disallowed range)

$$\begin{aligned}
&m \lesssim 4 \times 10^{11}\text{eV} g_*^{1/12}\tau_{\rm sec}^{-1/2} && (\text{light}, \tau \lesssim t_{\rm EQ})\\
&m \lesssim 4 \times 10^{13}\text{eV}(\Omega_0 h^2)^{-1/3}\tau_{\rm sec}^{-2/3} && (\text{light}, \tau \gtrsim t_{\rm EQ})\\
\\
&m \gtrsim 3 \times 10^{-5}\text{GeV} g_*^{-1/24}\tau_{\rm sec}^{1/4} && (\text{heavy Dirac}, \tau \lesssim t_{\rm EQ})\\
&m \gtrsim 3 \times 10^{-6}\text{GeV}(\Omega_0 h^2)^{1/6}\tau_{\rm sec}^{1/3} && (\text{heavy Dirac}, \tau \gtrsim t_{\rm EQ})\\
\\
&m \gtrsim 7 \times 10^{-5}\text{GeV} g_*^{-1/24}\tau_{\rm sec}^{1/4} && (\text{heavy Majorana}, \tau \lesssim t_{\rm EQ})\\
&m \gtrsim 8 \times 10^{-6}\text{GeV}(\Omega_0 h^2)^{1/6}\tau_{\rm sec}^{1/3} && (\text{heavy Majorana}, \tau \gtrsim t_{\rm EQ})
\end{aligned} \tag{5.102}$$

[10]For reference, $(1/2)! = \sqrt{\pi}/2 \simeq 0.886$ and $(2/3)! \simeq 0.903$.

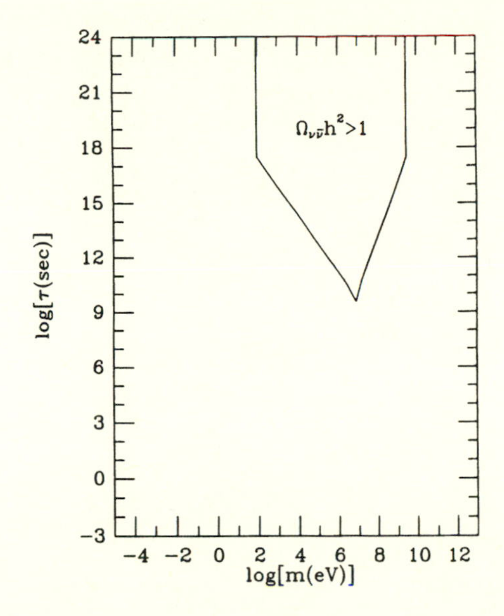

Fig. 5.4: The forbidden region of the neutrino mass–lifetime plane based upon the requirement that $\Omega_{\nu\bar{\nu}}h^2 \lesssim 1$.

This requirement excludes a region of the neutrino mass–lifetime plane as shown in Fig. 5.4.[11]

The limits just discussed apply irrespective of the nature of the decay products (so long as they are relativistic). If the decay products include "visible" particles, e.g., photons, $e^\pm$ pairs, pions, etc, much more stringent limits can be obtained. We will now consider the additional constraints which apply when the decay products include a photon.[12] The limits that follow depend both qualitatively and quantitatively upon the decay epoch, and we will consider five distinct epochs.

Before discussing these limits, it is useful to calculate the time at which the energy density of the massive neutrino species would dominate the energy density in photons. The energy density in photons is $\rho_\gamma = (\pi^2/15)T^4$, and assuming the neutrinos are NR, their energy density is $\rho_\nu = Y_\infty ms$.

[11]Consideration of the formation of structure in the Universe leads to a significantly more stringent constraint to the mass density of the relativistic decay products; as we shall discuss in Chapter 9, structure cannot grow in a radiation-dominated Universe. For a discussion of these constraints see [12].

[12]For the most part these same limits also apply if the decay products include $e^\pm$ pairs.

Taking $g_{*S} \simeq 4$, the energy densities are equal when $T \simeq 3Y_\infty m$. For heavy neutrinos Y_∞ is given by (5.54), and for light neutrinos, $Y_\infty \simeq 0.04$. Thus we find that the relic neutrino energy density will exceed the photon energy density at $T/m \lesssim 0.1$ for light neutrinos, and $T/m \lesssim 2 \times 10^{-8} m_{\rm GeV}^{-3}$ for heavy neutrinos. Using $t \simeq 1\ {\rm sec}/T_{\rm MeV}^2$ for the age of the Universe, the epoch of matter domination (by massive neutrinos) is given by[13]

$$t({\rm sec}) \simeq \begin{cases} 10^{14}(m/1\ {\rm eV})^{-2} & \text{light neutrinos} \\ 3 \times 10^9 m_{\rm GeV}^4 & \text{heavy neutrinos.} \end{cases} \tag{5.103}$$

• $t_U \simeq 3 \times 10^{17}{\rm sec} \leq \tau$: If the neutrino lifetime is greater than the age of the Universe, neutrinos will still be decaying at the present and decay-produced photons will contribute to the diffuse photon background. Assuming that the neutrinos are unclustered (the most conservative assumption), the differential number flux of decay-produced photons (per cm^2 sr sec erg) is

$$\frac{d\mathcal{F}_\gamma}{dEd\Omega} = \frac{n_\nu c}{4\pi\tau H_0} E^{-1}\left(\frac{E}{m/2}\right)^{3/2} \qquad (E \leq m/2) \tag{5.104}$$

where for simplicity we have assumed that each decay produces one photon of energy $m/2$ and that $\Omega_0 = 1$. Taking the number flux to be $d\mathcal{F}_\gamma/d\Omega \simeq E d\mathcal{F}_\gamma/dEd\Omega$ and $H_0 = 50$ km sec^{-1}Mpc^{-1}, we find

$$\begin{aligned} \frac{d\mathcal{F}_\gamma}{d\Omega} &\simeq 10^{29}\tau_{\rm sec}^{-1}\ {\rm cm}^{-2}\,{\rm sr}^{-1}\,{\rm sec}^{-1}\ \text{light neutrinos} \\ &\simeq 3 \times 10^{22}\tau_{\rm sec}^{-1} m_{\rm GeV}^{-3}\ {\rm cm}^{-2}\,{\rm sr}^{-1}\,{\rm sec}^{-1}\ \text{heavy neutrinos} \end{aligned} \tag{5.105}$$

A summary of the observations of the diffuse photon background is shown in Fig. 5.5. The differential energy flux, $d\mathcal{F}/dEd\Omega$, is shown as a function of energy and wavelength. From this data, a very rough limit of

$$\frac{d\mathcal{F}_\gamma}{d\Omega} \lesssim \left(\frac{1\ {\rm MeV}}{E}\right) {\rm cm}^{-2}{\rm sr}^{-1}{\rm sec}^{-1}, \tag{5.106}$$

can be placed to the contribution of neutrino decay-produced photons to the photon background. Based upon this, the following lifetime limit

---

[13]Here, and throughout the following discussion, "light" will refer to neutrinos of mass less than an MeV, and "heavy" will refer to neutrinos of mass greater than an MeV (but less than M_Z).

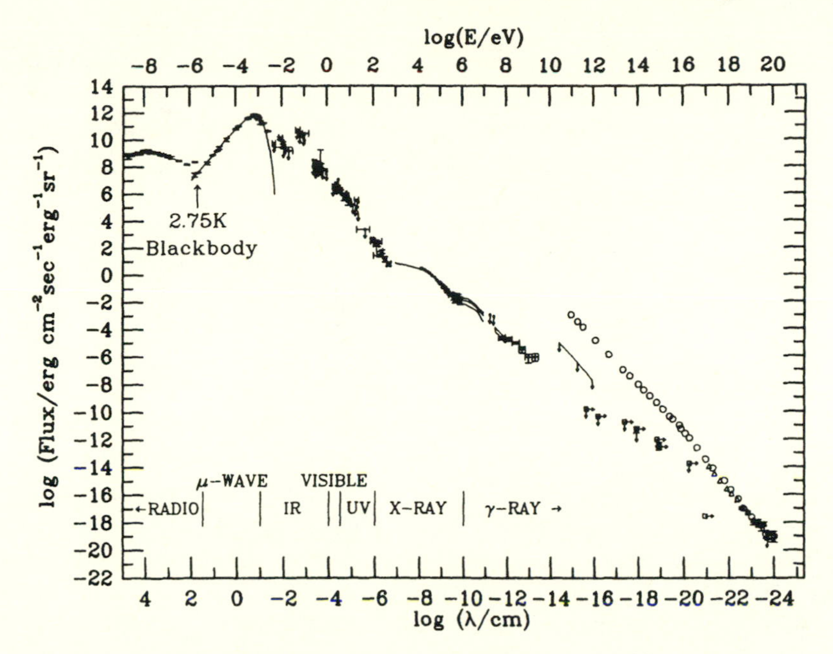

Fig. 5.5: The diffuse photon background for $10^5 \geq \lambda \geq 10^{-24}$ cm. Vertical arrows indicate upper limits, horizontal arrows indicate integrated flux ($> E$). Open circles and triangles indicate the total cosmic ray flux (photons plus hadrons) which places an upper limit to the photon flux (from [17]).

results:

$$\tau_{\rm sec} \geq \begin{cases} 10^{23} m_{\rm eV} & \text{light neutrinos} \\ 10^{25} m_{\rm GeV}^{-2} & \text{heavy neutrinos,} \end{cases} \tag{5.107}$$

applicable for neutrino lifetimes $\tau \gtrsim 3 \times 10^{17}$sec. The forbidden region of the mass–lifetime plane is shown in Fig. 5.6.

• $t_{rec} \simeq 6 \times 10^{12} (\Omega_0 h^2)^{-1/2}$sec $\leq \tau \leq t_U$: If neutrinos decay after recombination, but before the present epoch, then the decay-produced photons will not interact and should appear today in the diffuse photon background. Again, for simplicity assume that each neutrino decay produces one photon of energy $m/2$. Then the present flux of such photons is

$$\begin{aligned} \frac{d\mathcal{F}_\gamma}{d\Omega} &= \frac{n_\nu c}{4\pi} \\ &\simeq 3 \times 10^{11}\ \text{cm}^{-2}\,\text{sr}^{-1}\,\text{sec}^{-1} \qquad \text{light neutrinos} \\ &\simeq 4 \times 10^{4}\ m_{\rm GeV}^{-3}\ \text{cm}^{-2}\,\text{sr}^{-1}\,\text{sec}^{-1} \quad \text{heavy neutrinos} \end{aligned} \tag{5.108}$$

where we have assumed that when the neutrino species decays it is non relativistic, so that each decay-produced photon has energy $E \simeq m/2(1+z_D)$ today, where $(1+z_D) \simeq 3.5 \times 10^{11}\,(\Omega_0 h^2)^{-1/3}\tau_{\rm sec}^{-2/3}$. Comparing these flux estimates to our rough estimate of the diffuse background flux we obtain the constraints,

$$\begin{aligned} m &\lesssim 2 \times 10^6 (\Omega_0 h^2)^{-1/3}\tau_{\rm sec}^{-2/3}\,\text{eV} \qquad \text{light neutrinos} \\ m &\gtrsim 8 \times 10^{-3} (\Omega_0 h^2)^{1/6}\tau_{\rm sec}^{1/3}\,\text{GeV} \qquad \text{heavy neutrinos,} \end{aligned} \tag{5.109}$$

applicable for neutrino lifetimes in the range $3.5 \times 10^{11}(\Omega_0 h^2)^{-1/3}\text{sec} \lesssim \tau \lesssim 3 \times 10^{17}\text{sec}$. For very light neutrino species the assumption that the species decays when it is non relativistic breaks down. If the species decays after $t = t_{therm} \simeq 10^6$ sec and before the present epoch, and is relativisic when it decays, the decay-prduced photons will be comparable in energy and number to the CMBR photons, and will cause significant distortions to the CMBR [17]. Thus a neutrino species that decays while relativistic in the time interval $10^6 \lesssim t \lesssim 3 \times 10^{17}$ sec is forbidden. The excluded region is $200 \lesssim (\tau_{\rm sec}/m_{\rm eV}) \lesssim 4 \times 10^{20}(\Omega_0 h^2)^{1/3}$, for

$$m_{\rm eV} \lesssim \begin{cases} 3.5 \times 10^8 (\Omega_0 h^2)^{-1/3}\tau_{\rm sec}^{-2/3} & \tau_{\rm sec} \gtrsim 4.4 \times 10^{10}(\Omega_0 h^2)^{-2} \\ 4.6 \times 10^6 \tau_{\rm sec}^{-1/2} & \tau_{\rm sec} \lesssim 4.4 \times 10^{10}(\Omega_0 h^2)^{-2}. \end{cases} \tag{5.110}$$

The forbidden regions of the mass–lifetime plane are shown in Fig. 5.6.

• $t_{therm} \simeq 10^6\text{sec} \leq \tau \leq t_{rec}$: For neutrino decays that occur during this epoch, the decay-produced photons can scatter with electrons, which can in turn scatter with CMBR photons, thereby changing the spectral shape of the CMBR. However, during this epoch processes that alter the number of photons in the CMBR, e.g., the double Compton process, $\gamma + e \rightarrow \gamma + \gamma + e$, are not effective (i.e., $\Gamma < H$). Therefore, the result of dumping significant amounts of electromagnetic energy density from neutrino decays is a Bose-Einstein spectrum (with $\mu_\gamma \neq 0$) for the CMBR. As discussed in Chapter 1, the CMBR is to a very good precision a black body. Thus, any electromagnetic energy density resulting from neutrino decays during this epoch must be much less than that in the CMBR itself. Recalling that

$$\begin{aligned} \frac{\rho_\nu}{\rho_\gamma} &= \frac{m Y_\infty s}{\rho_\gamma} \\ &\simeq 0.1 m/T \qquad \text{light neutrinos} \end{aligned}$$

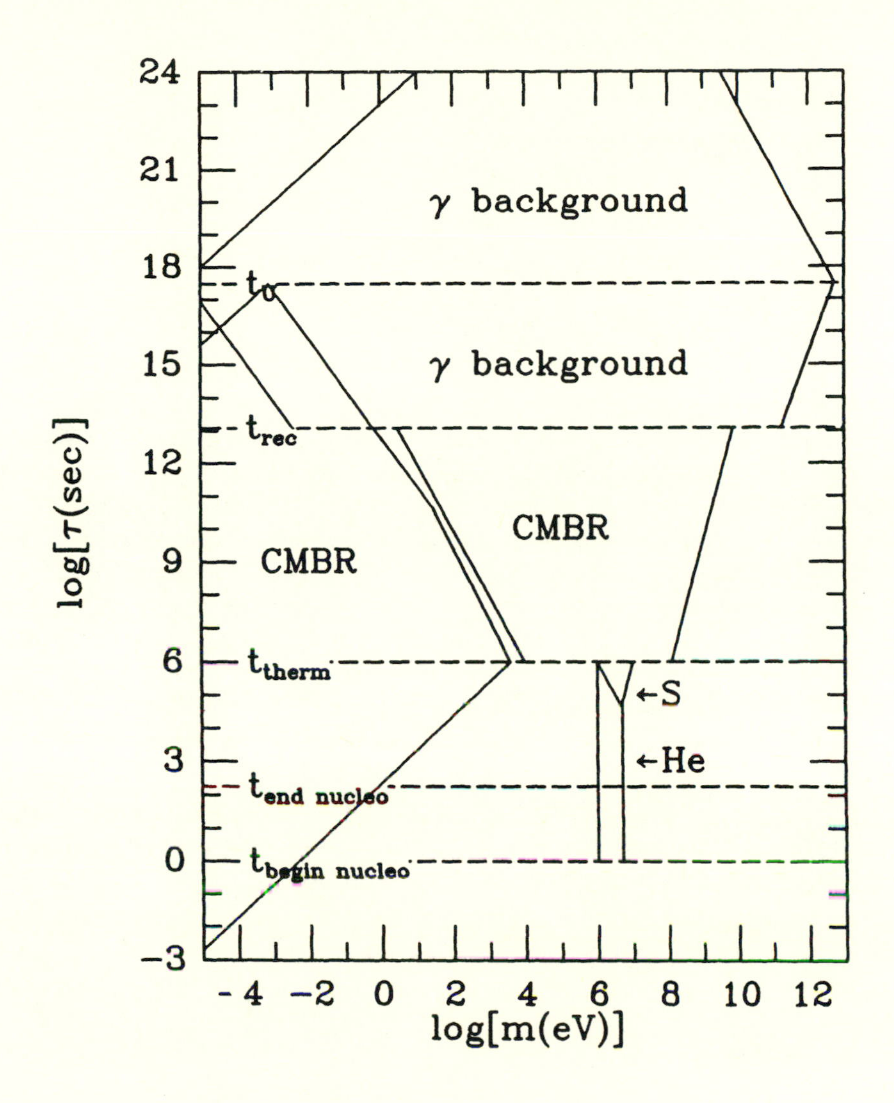

Fig. 5.6: Cosmological limits to the mass and lifetime of an unstable neutrino species that decays radiatively.

$$\frac{\rho_\nu}{\rho_\gamma} \simeq 2 \times 10^{-8} m_{\rm GeV}^{-3} m/T \qquad \text{heavy neutrinos,} \qquad (5.111)$$

and requiring that $\rho_\nu/\rho_\gamma \lesssim 1$, we obtain the following limits for a neutrino species that decays during this epoch:

$$m \lesssim 10^7 \tau_{\rm sec}^{-1/2}\,{\rm eV} \qquad \text{light neutrinos}$$
$$m \gtrsim 4 \times 10^{-3}\,\tau_{\rm sec}^{1/4}\,{\rm GeV} \qquad \text{heavy neutrinos,} \qquad (5.112)$$

where we have taken $t_{\rm sec} \simeq T_{\rm MeV}^{-1/2}$. These limits are applicable for neutrino lifetimes in the range $10^6{\rm sec} \lesssim \tau \lesssim 10^{13}{\rm sec}$. The forbidden region of the mass–lifetime plane is shown in Fig. 5.6.[14]

• $t_{end\ nucleo} \simeq 3$ min $\leq \tau \leq t_{therm}$: For neutrino decays that occur during this epoch, the decay-produced photons can be thermalized into the CMBR because both Compton and double Compton scattering are effective ($\Gamma > H$). However, in so doing the entropy per comoving volume is increased (as we discussed in Section 5.3). This has the effect of decreasing further the present value of η relative its value during primordial nucleosynthesis: In the standard scenario, $\eta = 4\eta_{BBN}/11$. As discussed in Chapter 1, luminous matter (necessarily baryons) provides $\Omega_{LUM} \sim 0.01$, and thus provides direct evidence that today $\eta \gtrsim 4 \times 10^{-11}$. On the other hand, as discussed in Chapter 4, primordial nucleosynthesis indicates that at the time of nucleosynthesis η corresponded to a present value of (3 to 10) $\times 10^{-10}$. Thus any entropy production after the epoch of nucleosynthesis must be less than a factor of $\sim 10^{-9}/4 \times 10^{-11} \sim 30$. Recalling our formula for entropy production by a decaying species we obtain the bound

$$30 \gtrsim S_f/S_i \simeq 1.83 \langle g_*^{1/3} \rangle^{3/4} \frac{m Y_\infty \tau^{1/2}}{m_{Pl}^{1/2}}, \qquad (5.113)$$

which leads to the limits

$$10^9 \gtrsim m_{\rm eV}\,\tau_{\rm sec}^{1/2} \qquad \text{light neutrinos}$$
$$10^7 \gtrsim m_{\rm GeV}^{-2}\,\tau_{\rm sec}^{1/2} \qquad \text{heavy neutrinos,} \qquad (5.114)$$

---

[14] A neutrino species that decays after nucleosynthesis and produces photons of energy greater than 30 MeV can lead to photofission of the light elements produced during nucleosynthesis. Constraints that follow from this are discussed in the paper by Lindley which is preprinted in Chapter 4 of *Early Universe: Reprints*.

applicable for neutrino lifetimes in the range 200 sec $\lesssim \tau \lesssim 10^6$ sec. This bound too is shown in Fig. 5.6.

• $t_{begin\ nucleo} \simeq 1$ sec $\leq \tau \leq t_{end\ nucleo}$: If the neutrino lifetime is longer than about a sec, then massive neutrinos can contribute significantly to the mass density of the Universe during nucleosynthesis, potentially leading to an increase in ^{4}He production. Recall, only the equivalent of 1 additional neutrino species can be tolerated without overproducing ^{4}He. One additional neutrino species is about equivalent to the energy density contributed by photons. Since the crucial epoch is when the neutron-to-proton ratio freezes out ($t \sim 1$ sec, $T \sim 1$ MeV), the constraint that follows is $(\rho_\nu/\rho_\gamma)_{T\simeq \mathrm{MeV}} \lesssim 1$. This results in the mass limit

$$m \gtrsim 5 \times 10^{-3}\mathrm{GeV} \quad \text{heavy neutrinos.} \tag{5.115}$$

Note there is no corresponding limit for a light species because a light species is just one additional relativistic neutrino species. This limit, which is applicable to a heavy neutrino species with lifetime greater than about 1 sec, is shown in Fig. 5.6.

• $\tau \ll 1$ sec: A neutrino species that decays earlier than about 1 sec after the bang disappears without leaving much of a cosmological trace. Its decay products thermalize before primordial nucleosynthesis, and its only effect is to increase the entropy per comoving volume. If we understood the origin of the baryon-to-entropy ratio in great detail, and could predict its "pre-nucleosynthesis" value, then we could use entropy production by the decaying neutrino species to obtain constraints for very short lifetimes.

• *Astrophysical Implications:* Neutrino decay into visible modes can have "astrophysical" effects too. As the detection of neutrinos from SN 1987A dramatically demonstrated, type II supernovae are a copious source of neutrinos. The integrated flux of neutrino-decay-produced photons from type II supernovae that have occurred throughout the history of the Universe can be used to obtain a very stringent bound to acceptable neutrino masses and lifetimes.

Each type II supernova releases about 3×10^{53} ergs of energy in thermal neutrinos with average energy about 12 MeV—or about $N_{\nu\bar{\nu}} \simeq 5\times 10^{57}$ neutrinos and antineutrinos of each species. The historical (last 1000 yr) type II rate in our own galaxy is about 1 per 30 yr (give or take a factor of 3), and the observed extragalactic rate is roughly $1.1h^2$per 100 yr per $10^{10} L_{B\odot}$ [15]. Using the measured mean blue luminosity density of the Universe, $L_{B\odot} \sim 2.4h \times 10^8 L_{B\odot}\,\mathrm{Mpc}^{-3}$, this translates into a present type II rate (per volume) of $\Gamma_{SN} \simeq 2.5h^3 \times 10^{-85}\,\mathrm{cm}^{-3}\,\mathrm{sec}^{-1}$. Assuming that the type

II rate has been constant over the history of the Universe, the differential photon number flux from the decay of supernova-produced neutrinos is

$$\frac{d\mathcal{F}_\gamma}{d\Omega dE} = \frac{9}{5\sqrt{2}} \frac{\Gamma_{SN} t_U^2 N_{\nu\bar{\nu}}}{4\pi \langle E_\nu \rangle \tau / m} \frac{1}{\langle E_\nu \rangle^{1/2} E^{1/2}}. \tag{5.116}$$

For simplicity we have assumed that the supernovae neutrinos are mono-energetic: $E_\nu = \langle E_\nu \rangle \simeq 12$ MeV, that each decay-produced photon carries half the energy of the parent neutrino, and a flat Universe. Comparing the expected photon number flux at energy $\langle E_\gamma \rangle = \langle E_\nu \rangle / 6 \simeq 2$ MeV,

$$\langle E_\gamma \rangle \frac{d\mathcal{F}_\gamma}{d\Omega dE} \simeq \frac{1}{2} \frac{\Gamma_{SN} t_U^2 N_{\nu\bar{\nu}} m}{4\pi \langle E_\nu \rangle \tau}, \tag{5.117}$$

with the measured diffuse γ-ray flux at a few MeV, $3 \times 10^{-3} \mathrm{cm}^{-2}\, \mathrm{sr}^{-1}\, \mathrm{s}^{-1}$, we obtain the following constraint:

$$\tau_{\rm sec} \gtrsim 5 \times 10^{12} (\Gamma_{SN} / 3 \times 10^{-85} \mathrm{cm}^{-3} \mathrm{sec}^{-1}) m_{\rm eV}. \tag{5.118}$$

This bound applies to neutrino species light enough to be produced in supernovae ($m \lesssim 10$ MeV) and which decay outside the envelope of the exploding star ($\tau_{\rm sec} \gtrsim 10^{-5} m_{\rm eV}$).[15] This constraint is shown in Fig. 5.7.[16]

For a neutrino species that decays within the envelope of the exploding star, and thereby deposits energy in the envelope a different bound can be derived. Any energy deposited by neutrino decays in the envelope will be thermalized and radiated in the visible part of the spectrum. The energy radiated by SN 1987A in the visible was only about 10^{47} ergs, while each neutrino species carries off about 10^{53} ergs! The energy which is deposited in the envelope by a hypothetical unstable neutrino species is

$$E_{DEP} \simeq N_{\nu\bar{\nu}} \langle E_\nu \rangle \min[R_{BSG} / \tau_{LAB}, 1] \tag{5.119}$$

$$\simeq \min[10^{48} m_{\rm eV} / \tau_{\rm sec}\, \mathrm{ergs},\ 10^{53}\, \mathrm{ergs}] \tag{5.120}$$

where $R_{BSG} \sim 3 \times 10^{12}$ cm is the radius of the envelope of the progenitor blue super giant (Sanduleak -69 202, by name), and $\tau_{LAB} = \langle E_\nu \rangle \tau / m$ is

[15] In deriving (5.116) we have assumed the lab lifetime, $\tau_{LAB} \simeq \langle E_\nu \rangle \tau / m$, is larger than the age of the Universe ($\tau \gtrsim 10^{11} m_{\rm eV}$ sec). If not, $d\mathcal{F}_\gamma / d\Omega \simeq \Gamma_{SN} t_U N_{\nu\bar{\nu}} / 8\pi \simeq 1$ $\mathrm{cm}^{-2} \mathrm{sr}^{-1} \mathrm{s}^{-1}$—which is ruled out, independent of m and τ.

[16] Based upon γ-ray observations of SN 1987A made by the SMM spacecraft a similar, more direct, and slightly more restrictive bound obtains [15].

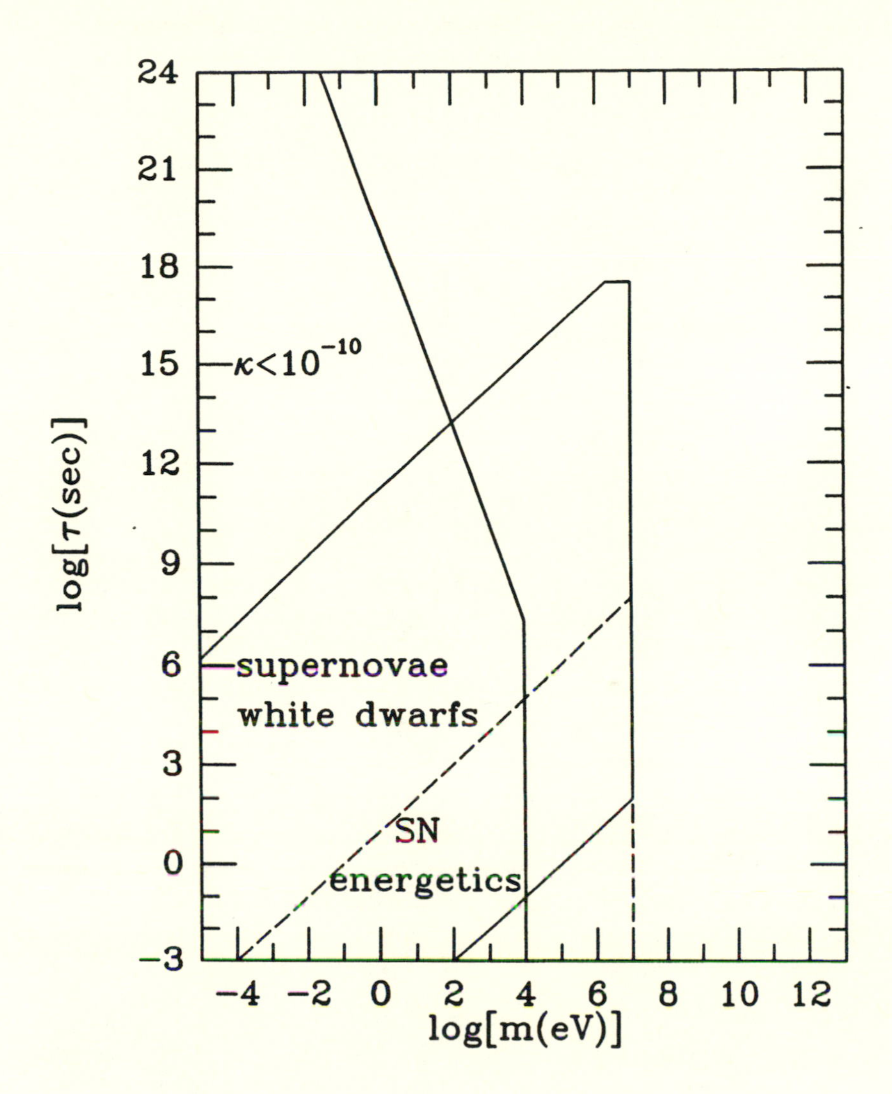

Fig. 5.7: Astrophysical limits to the mass and lifetime of an unstable neutrino that decays radiatively.

the neutrino lifetime in the rest frame of the supernova. Comparing this to the observed energy of 10^{47} ergs, we obtain the bounds

$$\begin{array}{ll} m_{\rm eV}/\tau_{\rm sec} \lesssim 0.1 & (\tau_{\rm sec} \gtrsim 10^{-5} m_{\rm eV}) \\ m_{\rm eV} \gtrsim 10^{7} & (\tau_{\rm sec} \lesssim 10^{-5} m_{\rm eV}) \end{array} \quad (5.121)$$

This constraint too is shown in Fig. 5.7.

A neutrino species that can decay radiatively, $\nu_j \longrightarrow \nu_i + \gamma$, necessarily has an electromagnetic coupling that can be quantified as a transition magnetic moment, $\mu_{ij} = \kappa_{ij}(e/2m_e)$. The transition magnetic moment and neutrino mass and lifetime are related by

$$\begin{aligned} \tau^{-1} &= \alpha_{EM}\kappa^2 m^3/8m_e^2, \\ \kappa &= 0.44\tau_{\rm sec}^{-1/2} m_{\rm eV}^{-3/2}, \end{aligned} \quad (5.122)$$

where we have assumed $m_j \gg m_i$. The transition moment leads to an electromagnetic correction to ν–e scattering. Laboratory limits to ν–e scattering through the transition moment leads to the bound $\kappa_{e\mu} \lesssim 10^{-8}$, or

$$\tau_{\rm sec} \gtrsim 2 \times 10^{15} m_{\rm eV}^{-3} \qquad (\nu_\mu \to \nu_e + \gamma). \quad (5.123)$$

Further, such a transition moment leads to neutrino pair emission from white dwarfs and red giants through the process *plasmon* $\to \nu_i\nu_j$. For $\kappa \sim 10^{-10}$ to 10^{-11} plasmon $\nu\bar{\nu}$ emission can be a very significant cooling mechanism for these objects, and can effect their evolution. Based upon this, a limit of $\kappa_{ij} \lesssim 10^{-10}$ or so has been derived for $m \lesssim 10$ keV (see, e.g., the paper of Beg, Marciano, and Ruderman, in [15]). This translates to the limit

$$\tau_{\rm sec} \gtrsim 2 \times 10^{19} m_{\rm eV}^{-3} \qquad (m \lesssim 10 \text{ keV}). \quad (5.124)$$

All of the astrophysical and cosmological constraints just discussed are summarized in Figs. 5.6 and 5.7. These constraints serve to illustrate how a large variety of cosmological and astrophysical observations can be used to probe particle properties in regimes beyond the reach of the terrestrial laboratory. We return in Chapter 10 to discuss another such very interesting example: the axion.

5.6 Concluding Remarks

In the first part of this Chapter we have focused on the treatment of non-equilibrium processes in the expanding Universe. As we have stressed, and will continue to stress, departures from equilibrium are responsible for most of the cosmological relics—without such departures the past history of the Universe would be irrelevant. The Boltzmann equation is the primary theoretical tool for dealing with non-equilibrium in the expanding Universe, and we have used it to consider recombination, the freeze out of the relic abundance of a massive particle species, and entropy production by the out-of-equilibrium decay of a massive relic.

Of particular interest was the relic abundance of a massive particle species. As we discussed in Chapters 1 and 4, most of the mass density in the Universe is dark, and if $\Omega_0 \gtrsim 0.15$, the dark matter is unlikely to be baryonic. The prime suspects in that case are relic particles from the early history of the Universe—e.g., a light neutrino species of mass $92h^2$ eV, or a heavy neutrino species of mass a few GeV would provide $\Omega_0 \simeq 1$. As we will discuss in Chapter 9, there are almost countless other candidates—axions, photinos, higgsinos, superheavy monopoles, etc., and most of the candidates arise as thermal relics. In Chapters 7 and 10, we will discuss the other class of relics, non-thermal relics; they include the axion, the monopole, and cosmic string.

"Cold relics" have the interesting property that their present abundance is inversely proportional to their mass and annihilation cross section. Consider a hypothethical species X, with mass m, 2 degrees of freedom, and an annihilation cross section, $\langle\sigma_A|v|\rangle = a \times 10^{-37}$ cm^2 (i.e., $n = 0$). It follows from our discussion of freeze out that

$$\begin{aligned}
x_f &= 18 + \ln[a(m/\mathrm{GeV})] \\
Y_\infty &= \frac{2.7 \times 10^{-9}}{a(m/\mathrm{GeV})} \\
\Omega_X h^2 &= \frac{0.043 x_f}{a} \simeq \frac{0.77}{a}
\end{aligned} \tag{5.125}$$

where we have taken $g_* = 60$. Note that up to logarithmic factors, i.e., $x_f \sim 18 + \ln\cdots$, $\Omega_X h^2$ is determined by the annihilation cross section, which is parameterized by a. It is also apparent that for $\Omega_X h^2$ to be of order unity, the annihilation cross section must be characteristic of a weak process. It is now clear why a heavy neutrino, and many of the supersymmetric partners, are prime dark matter candidates—all interact (with ordinary matter) with roughly weak strength.[16]

Because all interactions of relic dark matter particles with ordinary matter—annihilation into ordinary matter, pair production, elastic scattering, etc.—are set by the annihilation cross section (and its scaling with energy or temperature), the fact that $\Omega_X h^2 \sim 1$ fixes the annihilation cross section to be order 10^{-36} to 10^{-37} cm^2 has a multitude of implications. The production of dark matter particles at accelerators, or their direct detection necessarily involves "weak processes." There have been a variety of interesting proposals for the detection of relic dark matter particles—observing γ-rays from their annihilations in the galactic halo, observing high energy neutrinos from the annihilations of relic dark matter particles that are captured by and accumulate in the sun, and detecting the tiny (order keV) energy they deposit when they scatter with ordinary matter in a cryogenic, bolometric detector. All of the relevant cross sections are set by $\Omega_X h^2$—and fortunately for us, a weak cross section appears large enough to make the aforementioned detection schemes feasible.

5.7 References

Rules and conventions for calculating matrix elements, as well as matrix elements for some of the processes considered in this Chapter can be found in many references, e.g.,

1. C. Quigg, *Gauge Theories of the Strong, Weak, and Electromagnetic Interactions* (Benjamin/Cummings, Menlo Park, Calif. 1983); J. D. Bjorken and S. D. Drell, *Relativistic Quantum Mechanics* (McGraw-Hill, New York, 1964).

Statistical mechanics in the expanding Universe is also discussed in

2. R. V. Wagoner, in *Physical Cosmology*, eds. J. Audouze, R. Balian, and D. N. Schramm (North-Holland, Amsterdam, 1980); J. Bernstein, *Kinetic Theory in the Expanding Universe* (Cambridge Univ. Press, Cambridge, 1988).

The first discussions of the freeze out of massive particle annihilation were

[16]This presumes that the scale of the masses of the supersymmetric partners is, as is currently popular, of order the weak scale.

3. Ya. B. Zel'dovich, *Zh. Eksp. Teor. Fiz.* **48**, 986 (1965); Ya. B. Zel'dovich, L. B. Okun, and S. B. Pikelner, *Usp. Fiz. Nauk*, **84**, 113 (1965); H.-Y. Chiu, *Phys.Rev. Lett.* **17**, 712 (1966).

An early numerical treatment of freeze-out was given by

4. S. Wolfram, *Phys. Lett.* **82B**, 65 (1979).

Convenient approximate analytic formulae for the decoupling of a massive species and its relic abundance are given in

5. J. Bernstein, L. Brown, and G. Feinberg, *Phys. Rev. D* **32**, 3261 (1985); R. Scherrer and M. S. Turner, *Phys. Rev. D* **33**, 1585 (1986); **34**, 3263 (E) (1986).

The neutrino mass bound usually referred to as the Cowsik-McClelland bound is given in

6. R. Cowsik and J. McClelland, *Phys. Rev. Lett.* **29**, 669 (1972). It was first derived by G. Gerstein and Ya. B. Zel'dovich, *Zh. Eksp. Teor. Fiz. Pis'ma Red.* **4**, 174 (1966); and it was also discussed by G. Marx and A. Szalay, in *Neutrino '72*, eds. A. Frenkel and G. Marx (OMKDT-Technoinform, Budapest, 1972), p. 123.

Post freeze out annihilation is discussed in

7. M. H. Reno and D. Seckel, *Phys. Rev. D* **38**, 3441 (1988); J. Hagelin, R. Parker, and A. Honberg, *Phys. Lett.* **215B**, 397 (1988); J. Hagelin and R. Parker, *Nucl. Phys.* **B**, in press (1989); J. A. Frieman, E. W. Kolb, and M. S. Turner, *Phys. Rev. D*, in press (1989).

The Lee-Weinberg bound is discussed in

8. B. W. Lee and S. Weinberg, *Phys. Rev. Lett.* **39**, 165 (1977), although it was discovered independently by many people: P. Hut, *Phys. Lett.* **69B**, 85 (1977); K. Sato and H. Kobayashi, *Prog. Theor. Phys.* **58**, 1775 (1977); M. I. Vysotskii, A. D. Dolgov, and Ya. B. Zel'dovich, *JETP Lett.* **26**, 188 (1977).

The fact that the annihilation of massive Majorana particles into light fermions is suppressed at low energies because Fermi statistics requires p-wave annihilation was emphasized by

9. H. Goldberg, *Phys. Rev. Lett.* **50**, 1419 (1983). The effect on the Lee-Weinberg bound was first noted by L. L. Krauss, *Phys. Lett.* **128B**, 37 (1983). Detailed numerical calculations of the Lee-Weinberg bound for both Dirac and Majorana neutrinos are given in E. W. Kolb and K. A. Olive, *Phys. Rev. D* **33**, 1202 (1986).

Evasion of the Lee-Weinberg bound by decay of the massive neutrino was pointed out by

10. D. A. Dicus, E. W. Kolb, and V. L. Teplitz, *Phys. Rev. Lett.* **39**, 168 (1977).

Cosmological and astrophysical limits to the lifetime of unstable massive neutrino species have been discussed by many authors:

11. The total energy density of the Universe: M. I. Vysotsky, Ya. B. Zel'dovich, M. Yu. Khlopov, and V. M. Chechetkin, *Zh. Eksp. Teor. Fiz. Pis'ma* **26**, 200 (1977); *ibid* **27**, 533 (1978); K. Sato and H. Kobayashi, *Prog. Theor. Phys.* **58**, 1775 (1977); D. A. Dicus, E. W. Kolb, and V. L. Teplitz, *Phys. Rev. Lett.* **39**, 168 (1977); *Ap. J.* **221**, 327 (1978); T. Goldman and G. J. Stephenson, *Phys. Rev. D* **16**, 2256 (1977).

12. Galaxy formation: K. Freese, E. W. Kolb, and M. S. Turner, *Phys. Rev. D* **27**, 1689 (1983); G. Steigman and M. S. Turner, *Nucl. Phys.* **253B**, 375 (1985).

13. The spectrum of the CMBR: K. Sato and H. Kobayashi, *Prog. Theor. Phys.* **58**, 1775 (1977); J. E. Gunn, B. W. Lee, I. Lerche, D. N. Schramm, and G. Steigman, *Ap. J.* **223**, 1015 (1978); D. A. Dicus, E. W. Kolb, and V. L. Teplitz, *Ap. J.* **221**, 327 (1978); R. Cowsik, *Phys. Rev. Lett.* **39**, 784 (1977).

14. Primordial nucleosynthesis: K. Sato and H. Kobayashi, *Prog. Theor. Phys.* **58**, 1775 (1977); D. A. Dicus, E. W. Kolb, V. L. Teplitz, and R. V. Wagoner, *Phys. Rev. D* **17**, 1529 (1978); S. Miyama and K. Sato, *Prog. Theor. Phys.* **60**, 1703 (1977).

15. Stellar evolution and supernova: R. Cowsik, *Phys. Rev. Lett.* **39**, 784 (1977); S. W. Falk and D. N. Schramm, *Phys. Lett.* **79B**, 511 (1978); M. A. B. Beg, W. J. Marciano, and M. Ruderman, *Phys. Rev. D*

17, 1395 (1978); M. Fukugita and S. Yazaki, *Phys. Rev. D* **36**, 3817 (1987). The type II supernova rate is discussed in G. A. Tammann, *Ann. NY Acad. Sci.* **302**, 61 (1977) and S. van den Bergh, R. D. McClure, and R. Evans, *Ap. J.* **323**, 44 (1987). The limit based upon SN 1987A is discussed in E. W. Kolb and M. S. Turner, *Phys. Rev. Lett.* **62**, 509 (1989).

16. The mass–lifetime limits appear in the reviews of A. Dolgov and Ya. B. Zel'dovich, *Rev. Mod. Phys.* **53**, 1 (1981); M. S. Turner, in *Neutrino 81*, eds. R. J. Cence, E. Ma, and A. Roberts (Univ. Hawaii Press, Honolulu, 1981), p. 95; L. L. Krauss *Phys. Rep.*, in press (1990).

A summary of the diffuse photon background can be found in

17. M. T. Ressell and M. S. Turner, *Comments on Astrophysics*, in press (1989).

Limits to the properties of supersymmetric particles can be found in

18. H. Pagels and J. Primack, *Phys. Rev. Lett.* **48**, 223 (1983); S. Weinberg, *Phys. Rev. Lett.* **48**, 1303 (1983); J. Ellis, J. S. Hagelin, D. V. Nanopoulos, K. A. Olive, and M. Srednicki, *Nucl. Phys.* **B238**, 453 (1984).

For magnetic monopoles see the discussion and references in Chapter 7. For particles from higher-dimensional theories, see

19. E. W. Kolb and R. Slansky, *Phys. Lett.* **135B**, 378 (1984).

For mirror fermions, see

20. G. Senjanović, F.Wilczek, and A. Zee, *Phys. Lett.* **141B** 389 (1984); J. Bagger and S. Dimopoulos, *Nucl. Phys.* **B244**, 242 (1984).

It is likely that for any known, suspected, postulated, hoped-for, or dreaded particle there exist cosmological limits to its properties.

The idea of detection of elementary particle dark matter by the energy deposited in a cryogenic bolometric detector was proposed by

21. M. Goodman and E. Witten, *Phys. Rev. D* **31**, 3059 (1985). Reprinted in *Early Universe: Reprints.*

For references to low-background cryogenic detectors, see

22. A. Drukier and L. Stodolsky, *Phys. Rev. D* **30**, 2295 (1984); A. Drukier, K. Freese, and D. Spergel, *Phys. Rev. D* **33**, 3495 (1986); B. Cabrera, L. L. Krauss, and F. Wilczek, *Phys. Rev. Lett.* **55**, 25 (1985); S. P. Ahlen, et al., *Phys. Lett.* **195B**, 603 (1987); D. O. Caldwell, et al., *Phys. Rev. Lett.* **61**, 510 (1988). For an up to date discussion of dark matter detection techniques and the expected detection rates, see J. Primack, D. Seckel, and B. Saudolet, *Ann. Rev. Nucl. Part. Sci.* **38**, 751 (1988); K. Griest, *Phys. Rev. D* **38**, 2357 (1988); *Phys. Rev. Lett.* **61**, 666 (1988).

For detection of elementary particle dark matter via its annihilation products, see

23. M. Srednicki, S. Theisen, and J. Silk, *Phys. Rev. Lett.* **56**, 263 (1985); S. Rudaz, *Phys. Rev. Lett.* **56**, 2128 (1986); S. Rudaz and F. W. Stecker, *Ap. J.*, in press (1988); T. K. Gaisser, G. Steigman, and S. Tilav, *Phys. Rev. D* **34**, 2206 (1986); M. Srednicki, K. A. Olive, and J. Silk, *Nucl. Phys.* **B279**, 804 (1987).

6

BARYOGENESIS

6.1 Overview

The goal of grand unification (or, grand unified theories—GUTs) is to unify the strong, weak, and electromagnetic interactions and the quarks and leptons within the framework of a gauge field theory based upon a simple or semi-simple, non-Abelian symmetry group, e.g., $SU(5)$, $SO(10)$, E_6. Such theories make two startling predictions: new interactions that violate baryon number (B) and lepton number (L) and thereby lead to the instability of the proton, and the existence of stable, superheavy magnetic monopoles (in the form of topologically-stable configurations of gauge and Higgs fields). Both predictions have very significant cosmological consequences.

The B-violating interactions are today apparently very, very weak as evidenced by the longevity of the proton.[1] Roughly speaking, they can be characterized by a coupling constant analogous to the Fermi coupling constant, but at least 25 orders of magnitude smaller:

$$G_{\Delta B} \sim M^{-2} \lesssim 10^{-30}\,\mathrm{GeV}^{-2} \tag{6.1}$$

where $M \sim$ the energy scale of unification, which is likely to be 10^{14} GeV or greater. However, as discussed in Chapter 3, at temperatures comparable to, or greater than, M, B-violating forces (if they exist) should have strength comparable to all the other interactions of Nature. As we shall see, these interactions can allow a baryon-symmetric Universe to evolve

[1]Current limits to the proton lifetime are in excess of 10^{31} to 10^{32} yr, depending upon the particlular mode of decay [2]. This fact rules out the simplest unified theory, $SU(5)$, as the $SU(5)$ prediction for the proton lifetime is $10^{29\pm1}$ yr. Nevertheless, because of its simplicity we will often use $SU(5)$ for purposes of illustration.

a baryon asymmetry of the magnitude required to explain the present baryon-to-photon ratio. The baryogenesis scenario is one of the great triumphs of particle cosmology, and in the absence of direct evidence for proton decay, baryogenesis may provide the strongest, albeit indirect, evidence for some kind of unification of the quarks and leptons.

As we shall discuss in Chapter 7, the other important prediction of grand unification—superheavy magnetic monopoles—leads to a serious cosmological problem. In the standard cosmology, superheavy magnetic monopoles are grossly overproduced; so much so that it seems safe to say that the standard cosmology and the (simplest) unified theories are incompatible. This dark cloud is not without a silver lining; the search for a solution to the monopole problem led to the inflationary Universe scenario.

6.2 Evidence for a Baryon Asymmetry

Antimatter is rare on earth. It exists in "large" quantities ($10^{11}\bar{p}$'s $\sim 10^{-13}$ g) only in the antiproton accumulators at Fermilab and CERN. Antimatter is also rare in the solar system. The fact that Neil Armstrong survived his "one small step" is evidence that the moon is made of matter. Solar cosmic rays indicate that the sun too is composed of matter. Planetary probes have visited eight of the nine planets, and their survival dramatically demonstrates that the solar system is made of matter.

Cosmic rays provide samples of material from throughout the entire galaxy, and probably from distant galaxies as well. Antiprotons are seen in cosmic rays at about the 10^{-4} level compared to protons [3]. The magnitude of the antiproton flux is consistent with the hypothesis that the antiprotons are secondaries produced by cosmic ray collisions with the ISM, and does not seem to indicate the presence of antimatter in the galaxy, even at the 10^{-4} level. The flux of anti-^{4}He is less than 10^{-5} that of ^{4}He, and there has yet to be an indisputable detection of an antinucleus. Cosmic rays are solid evidence that there is a galactic asymmetry between baryons and antibaryons, and that this asymmetry is maximal, i.e., all baryons and essentially no antibaryons.

The evidence on larger scales is somewhat more tenuous. Many clusters of galaxies contain intracluster gas, as evidenced by x-ray emission. If both matter galaxies and antimatter galaxies existed in the same cluster there should also be strong γ-ray emission from nucleon-antinucleon annihilations. The absence of such a γ-ray flux is evidence that nearby clusters of galaxies (like Virgo) which typically contain 10^{13} to $10^{14} M_\odot$ of material are either all baryons or all antibaryons. There is little or no information on scales larger than clusters of galaxies. The evidence against the existence of large amounts of antimatter in the Universe is reviewed in [4].

If there is a significant amount of antimatter in the Universe, it must be segregated from matter on scales at least as large as $10^{12}\ M_\odot$, and probably larger than $10^{14}\ M_\odot$. On the face of it, this does not preclude a baryon symmetric Universe. However, as we discussed in Chapter 5, in a locally-baryon-symmetric Universe nucleons and antinucleons remain in chemical equilibrium down to a temperature of $\sim$ 22 MeV, when $n_b/s = n_{\bar{b}}/s \simeq 7\times 10^{-20}$, a number that is 9 orders of magnitude smaller than the observed value of n_b/s. In order to avoid the "annihilation catastrophe" an unknown physical mechanism would have to operate at a temperature greater than 38 MeV, the temperature when $n_b/s = n_{\bar{b}}/s \simeq 8\times 10^{-11}$, and separate nucleons and antinucleons. However, the horizon at that time only contained $\sim 10^{-7}\ M_\odot$, and so causality precludes separating out chunks even approaching a solar mass, let alone $10^{14}\ M_\odot$.

The most reasonable conclusion then is that the Universe at early times ($T \gtrsim 38$ MeV) possessed an asymmetry between the number of baryons and the number of antibaryons which prevented the annihilation catastrophe. As discussed in Chapter 3, this asymmetry is characterized by the baryon-to-entropy ratio $B \equiv n_B/s = 3.81\times 10^{-9}\Omega_B h^2$, which in the absence of B-nonconserving interactions and/or entropy production remains constant throughout the course of the expansion. From our knowledge of primordial nucleosynthesis we know that $B \simeq \eta/7 \simeq$ (4 to 7) $\times 10^{-10}/7 \simeq$ (6 to 10) $\times 10^{-11}$. Although the baryon asymmetry is maximal today, i.e., no antimatter, at high temperatures ($T \gtrsim 1$ GeV) thermal quark-antiquark pairs were present in great numbers ($n_q \sim n_{\bar{q}} \sim n_\gamma$), so that the baryon asymmetry observed today corresponds to a tiny quark-antiquark asymmetry at early times ($t \lesssim 10^{-6}$ sec):

$$\frac{n_q - n_{\bar{q}}}{n_q} \simeq 3\times 10^{-8}. \tag{6.2}$$

That is, for every 30 million antiquarks, there were 30 million and 1 quarks present! A very tiny asymmetry indeed. The baryogenesis scenario provides a very attractive means by which this curious (but crucial) number could arise dynamically in a Universe that is initially baryon symmetric, or possibly even irrespective of any initial baryon asymmetry present.[2]

[2]While it now seems most fruitful to try to understand the origin of the small

6.3 The Basic Picture

There are three basic ingredients necessary to generate a non-zero baryon number from an initially baryon symmetric state. (i) *Baryon Number Violation:* There must obviously be a violation of baryon number. If baryon number is conserved in all interactions, the present baryon asymmetry can only reflect asymmetric initial conditions. (ii) *C and CP Violation:* Even in the presence of B-nonconserving interactions a baryon asymmetry will not develop unless both C (charge conjugation) and CP (charge conjugation combined with parity) are violated: In the absence of a preference for matter or antimatter, B-nonconserving reactions will produce baryon and antibaryon excesses at the same rate, thereby maintaining zero net baryon number. Both C and CP violation are necessary to supply such an arrow. Put concisely, baryon number is odd under both C and CP. (iii) *Non-Equilibrium Conditions:* In chemical equilibrium the entropy is maximal when the chemical potentials associated with all non-conserved quantum numbers vanish. Further, particle and antiparticle masses are guaranteed to be equal by CPT invariance. Thus, in thermal equilibrium the phase space density of baryons and antibaryons, given by $[1+\exp((p^2+m^2)/T)]^{-1}$, are necessarily identical, implying that $n_b = n_{\bar{b}}$.

Within the context of unified gauge theories and the expanding Universe all three ingredients are at hand. GUTs predict the existence of B-nonconserving interactions; both C and CP are observed to be violated microscopically in Nature—in the interactions of K^0 and $\bar{K}^0$ mesons; and as we have emphasized the Universe has often undergone departures from thermal equilibrium. In a prescient paper in 1967, almost a decade before the advent of GUTs, Sakharov [5] identified these ingredients that are necessary for the Universe to dynamically evolve a baryon asymmetry.

To illustrate the mechanics of baryogenesis, consider a particle X that decays to quark/lepton final states qq ($B = 2/3$) and $\bar{q}\bar{l}$ ($B = -1/3$) (see Table 6.1). Since the two final states have different baryon number, the decays of X, $\bar{X}$ violate B. Note that CPT invariance requires the equality of the decay rates of the X and $\bar{X}$ bosons; C and CP are violated if the branching ratio of the X to the qq final state ($= r$) is unequal to the branching ratio of the $\bar{X}$ to the $\bar{q}\bar{q}$ final state ($= \bar{r}$); that is, $r \neq \bar{r}$.

---

initial baryon-to-entropy ratio, previously much effort was devoted to trying to explain the reciprocal of this ratio: the very large entropy per baryon. In the context of a cold or tepid bang, many researchers sought to invoke dissipative processes to create the observed entropy per baryon, e.g., by the dissipation of initial anisotropy, and/or inhomogeneity.

particle		final state	branching ratio	B
X	$\longrightarrow$	qq	r	2/3
X	$\longrightarrow$	$\bar{q}\bar{l}$	$1-r$	$-1/3$
$\bar{X}$	$\longrightarrow$	$\bar{q}\bar{q}$	$\bar{r}$	$-2/3$
$\bar{X}$	$\longrightarrow$	$q\bar{l}$	$1-\bar{r}$	1/3

Table 6.1: Final states and branching ratios for X, $\bar{X}$ decay.

Imagine a box containing equal numbers of X and $\bar{X}$ bosons, i.e., symmetric initial conditions. The mean net baryon number produced by the decay of an X is equal to $B_X = r(2/3) + (1-r)(-1/3)$, and that produced by the decay of an $\bar{X}$ is equal to $B_{\bar{X}} = \bar{r}(-2/3)+(1-\bar{r})(1/3)$. The mean net baryon number produced by the decay of an X, $\bar{X}$ pair is just $\varepsilon \equiv B_X + B_{\bar{X}} = r - \bar{r}$. The baryon number produced vanishes, of course, if C or CP is conserved ($r = \bar{r}$). If there are no further baryon number violating reactions, then a net baryon asymmetry will persist after all the X, $\bar{X}$ bosons decay.

As we shall see in more detail, this very simple picture illustrates well the so-called out-of-equilibrium decay scenario. In the expanding Universe the initial abundance of X, $\bar{X}$ bosons is thermal: $n_X = n_{\bar{X}} \sim n_\gamma$. Provided that the X is massive enough, the interactions that decrease the number of X, $\bar{X}$'s are ineffective for $T < m_X$, and X, $\bar{X}$ bosons become overabundant as T falls below the X-boson mass. Eventually, when $T \ll m_X$, X, $\bar{X}$ bosons freely decay without "back" reactions—inverse decays and other B-violating reactions are suppressed by the low temperature. The resulting baryon number-to-entropy ratio is $n_B/s \sim \varepsilon n_X/g_* n_\gamma \sim \varepsilon/g_*$. The remainder of this section is devoted to various refinements of the out-of-equilibrium scenario.

First, let's consider the necessary ingredients—B, C, and CP violations and non-equilibrium conditions—in more detail.

• *B Violation*: The existence of baryon number violation seems to be a generic feature of GUTs. When the strong and electroweak interactions are unified, quarks and leptons typically appear as members of a common irreducible representation of the gauge group. Thus gauge bosons mediate interactions that transform quarks into leptons or antiquarks, and thereby violate B. On dimensional grounds the lifetime of the proton should be $\tau_p \sim (G_{\Delta B}^2 m_p^5)^{-1} \sim \alpha_{GUT}^{-2} M^4 m_p^{-5}$. The stability of the proton, $\tau \gtrsim 10^{31}$ to 10^{32} yr, implies that such additional gauge bosons must be very massive: $M \gtrsim 10^{14}$GeV or so. In addition, there are also likely to be Higgs bosons with B-nonconserving interactions. The typically weaker couplings of Higgs bosons that mediate baryon-number violation allow them to have somewhat smaller masses, perhaps as low as 10^{10} GeV. In both cases the large mass of the intermediate boson is responsible for the feebleness of baryon-number violation today. This suppression (relative to the familiar interactions) is of course overcome at the extremely high temperatures that should have existed shortly after the big bang, and interactions that violate baryon number should have been just as potent as all other interactions. The requirement of B violation arises naturally in GUTs. A gauge or Higgs boson that violates baryon number in its decays will be generically denoted as X.

- *C and CP Violation*: C is maximally violated in the weak interactions, so C violation in the decay of the X boson should not be a fundamental problem. CP violation is observed in the neutral kaon system, with dimensionless strength of 10^{-3}. Since its origin is not well understood, it is easy to imagine that C and CP violation manifest themselves in all sectors of the theory, including the superheavy boson sector—and of course at some level C and CP violation must occur in the superheavy sector due to loop corrections involving the light quarks. As we shall see, a C, CP violation of only $\varepsilon \sim 10^{-8}$ or so is required to produce the observed value of n_B/s.

To explicitly see how C, CP violation enters, consider a system with two superheavy bosons, X and Y, with baryon number violating decays. The generalization of ε defined above is

$$
\begin{aligned}
\varepsilon_X &= \sum_f B_f \frac{\Gamma(X \to f) - \Gamma(\bar{X} \to \bar{f})}{\Gamma_X} \\
\varepsilon_Y &= \sum_f B_f \frac{\Gamma(Y \to f) - \Gamma(\bar{Y} \to \bar{f})}{\Gamma_Y},
\end{aligned}
\tag{6.3}
$$

where the sum runs over all final states f, state f has baryon number B_f, and Γ_X (Γ_Y) is the total X (Y) decay width. For simplicity, assume there are but two final states for X and Y decay, and that the interaction Lagrangian is given by

$$
\mathcal{L} = g_1 X i_2^\dagger i_1 + g_2 X i_4^\dagger i_3 + g_3 Y i_1^\dagger i_3 + g_4 Y i_2^\dagger i_4 + \text{h.c.} \tag{6.4}
$$

where i_1, i_2, i_3, i_4 are fermion states (quarks and leptons), and the g_i are coupling strengths (which can be complex). This Lagrangian leads to the

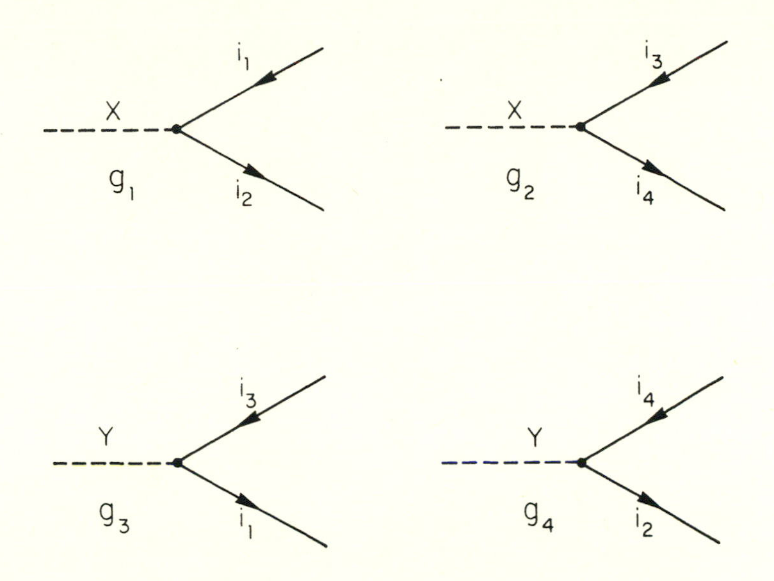

Fig. 6.1: Lowest-order Feynman graphs for X and Y decay.

decay processes: $X \to \bar{i}_1 + i_2$, $\bar{i}_3 + i_4$, and $Y \to i_1 + \bar{i}_3$, $i_2 + \bar{i}_4$. The lowest order diagrams for X and Y decay are shown in Fig. 6.1. The lowest order processes cannot contribute to ε, as $\Gamma(X \to \bar{i}_1 i_2) = |g_1|^2 I_X = \Gamma(\bar{X} \to i_1 \bar{i}_2) = |g_1^*|^2 I_{\bar{X}}$ where the kinematic factors $I_X = I_{\bar{X}}$ arise from the phase space integrals. The first non-zero contribution to ε comes from the interference of the lowest order graphs in Fig. 6.1 with the one-loop corrections shown in Fig. 6.2. These interference terms are given by

$$\begin{aligned} \Gamma(X \to \bar{i}_1 i_2) &= g_1 g_2^* g_3 g_4^* I_{XY} + (g_1 g_2^* g_3 g_4^* I_{XY})^* \\ \Gamma(\bar{X} \to i_1 \bar{i}_2) &= g_1^* g_2 g_3^* g_4 I_{XY} + (g_1^* g_2 g_3^* g_4 I_{XY})^*, \end{aligned} \tag{6.5}$$

where the phase-space factors I_{IJ} (I, $J = X, Y$) now also include the kinematic factors arising from integrating over the internal momentum loop due to J exchange in I decay. If the intermediate particles (i_1, i_2, i_3, i_4) in the loop are kinematically allowed to propagate on shell, as will be the case if X, Y are superheavy bosons and i_{1-4} are light quarks and leptons, the quantity I_{IJ} will be complex. As we will see shortly, the complexity of I_{IJ} is crucial to baryogenesis.

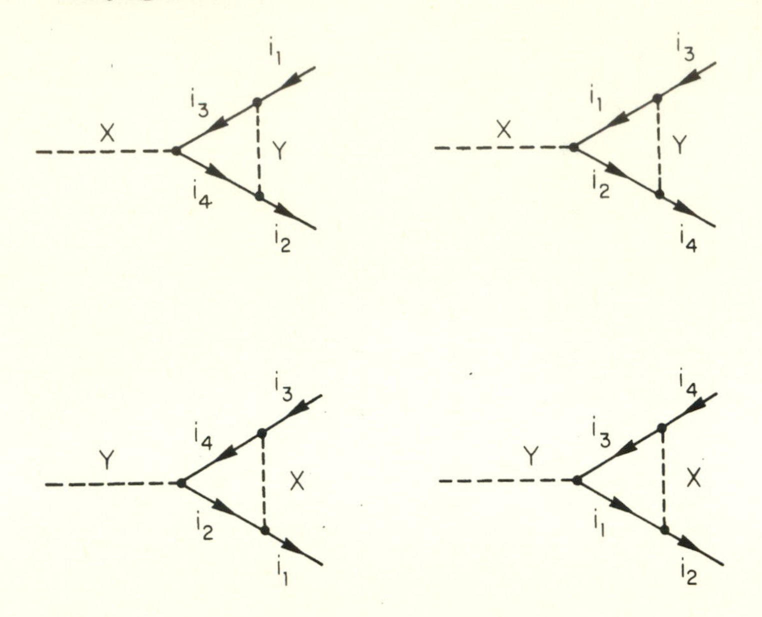

Fig. 6.2: One-loop corrections to X and Y decay.

The difference between $X \to \bar{i}_1 i_2$ and $\bar{X} \to i_1 \bar{i}_2$ is given by

$$\begin{aligned}\Gamma(X \to \bar{i}_1 i_2) - \Gamma(\bar{X} \to i_1 \bar{i}_2) &= 2iI_{XY}\text{Im}(g_1 g_2^* g_3 g_4^*) \\ &\quad + 2iI_{XY}^*\text{Im}(g_1^* g_2 g_3^* g_4) \\ &= 4\text{Im}I_{XY}\text{Im}(g_1^* g_2 g_3^* g_4). \end{aligned} \tag{6.6}$$

After a similar calculation for the other decay mode, we find that

$$\varepsilon_X = \frac{4}{\Gamma_X}\text{Im}I_{XY}\text{Im}(g_1^* g_2 g_3^* g_4)\left[(B_{i_4} - B_{i_3}) - (B_{i_2} - B_{i_1})\right]. \tag{6.7}$$

Repeating the same calculations for the Y decay modes, we find $\varepsilon_Y = -\varepsilon_X$.

From this exercise we see that 3 things are required in order to have $\varepsilon \neq 0$. First, there must be *two* baryon number violating bosons, each with mass greater than the sum of the (fermion) masses in the internal loops; otherwise, $\text{Im}I_{XY} = 0$. Second, C, CP violation arises from the interference of loop diagrams with the tree graph, and manifests itself in complex coupling constants. Thus, in general we expect ε to be of the order of α^N, where α characterizes the coupling constant of the loop particle(s) and N is the number of loops in the lowest-order diagram that interferes with the tree graph to give rise to $\varepsilon \neq 0$. Note that for this reason, $\varepsilon \sim \alpha^N$ is expected to be small. Third, the X and Y particles in the above example must not be degenerate in mass, or the baryon number produced by X decays will precisely cancel that produced by Y decays.

- *Departure from Thermal Equilibrium*: The necessary non-equilibrium condition is provided by the expansion of the Universe. Recall that if the expansion rate is faster than key particle interaction rates, departures from equilibrium can result. Here the departure from equilibrium will be the overabundance of X, $\bar{X}$ bosons. Assume that at some very early time when $T \gg m_X$ (e.g., the Planck time), X, $\bar{X}$ bosons are present in equilibrium numbers, $n_X = n_{\bar{X}} \simeq n_\gamma$. In LTE $n_X = n_{\bar{X}} \simeq n_\gamma$ for $T \gtrsim m_X$, and $n_X = n_{\bar{X}} \simeq (m_X T)^{3/2} \exp(-m_X/T) \ll n_\gamma$ for $T \lesssim m_X$ (see Fig. 6.3). Equilibrium numbers of X, $\bar{X}$ bosons will only be present provided that the interactions which create and destroy X, $\bar{X}$ bosons (decay, annihilation, and their inverse processes) are occurring rapidly on the expansion time scale: $\Gamma \gtrsim H$. The annihilation process is "self-quenching" since $\Gamma_{ANN} \propto n_X$, and the decay process is most important for maintaining equilibrium numbers of X, $\bar{X}$ bosons. For simplicity then, we will ignore the annihilation process.

The conditions necessary for a departure from equilibrium can be quantified. The decay rate of the X, denoted by Γ_D, the inverse decay rate (i.e., X production), denoted as Γ_{ID}, $2 \leftrightarrow 2$ B-nonconserving scattering processes mediated by X, $\bar{X}$ boson exchange (e.g., $\bar{i}_1 + i_2 \leftrightarrow \bar{i}_3 + i_4$, $\Delta B = (B_4 - B_3) - (B_2 - B_1) \neq 0$), denoted as Γ_S, and the expansion rate, denoted as H, are given by

$$\Gamma_D \simeq \alpha m_X \begin{cases} m_X/T & T \gtrsim m_X \\ 1 & T \lesssim m_X \end{cases}$$

$$\Gamma_{ID} \simeq \Gamma_D \begin{cases} 1 & T \gtrsim m_X \\ (m_X/T)^{3/2} \exp(-m_X/T) & T \lesssim m_X \end{cases}$$

$$\Gamma_S \simeq n\sigma \simeq T^3 \alpha^2 \frac{T^2}{(T^2 + m_X^2)^2}$$

$$H \simeq g_*^{1/2} T^2 / m_{Pl}, \tag{6.8}$$

where $\alpha \sim g^2/4\pi$ measures the coupling strength of the X boson. Note that decays are inhibited at high temperatures by the usual time dilation factor, and that inverse decays are suppressed at low temperatures as the

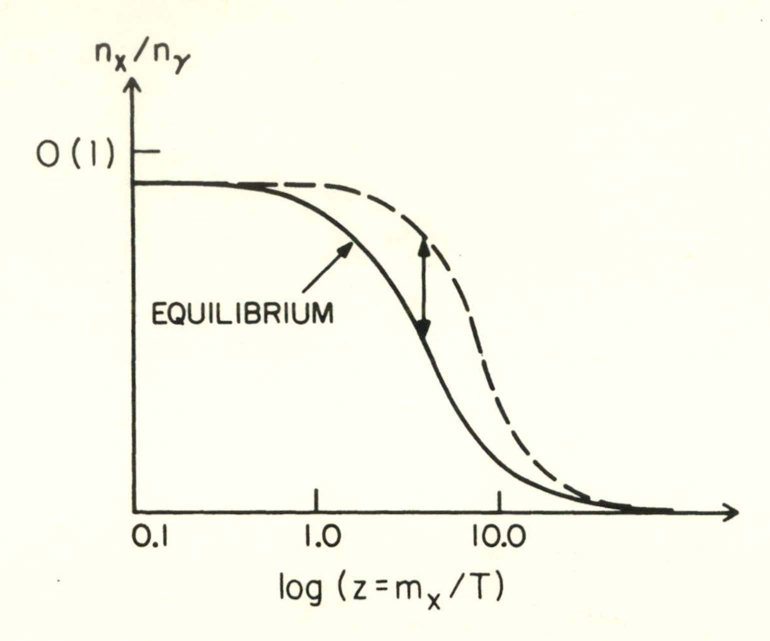

Fig. 6.3: The equilibrium number of X, $\bar{X}$ bosons per comoving volume, n_X^{EQ}/n_γ (solid), and the actual number per comoving volume, n_X/n_γ (dashed). The departure from equilibrium is the overabundance of X, $\bar{X}$ bosons for $T \lesssim m_X$ (arrow).

fraction of decay products with sufficient energy to produce an X or $\bar{X}$ is suppressed by a Boltzmann factor. At high temperatures, the $2 \leftrightarrow 2$ scattering cross section is $\sigma \simeq \alpha^2/T^2$, while at low temperatures $\sigma \simeq \alpha_{GUT}^2 T^2/m_X^4 \simeq G_{\Delta B}^2 T^2$. A schematic comparison of the rates is shown in Fig. 6.4. Note that all the reaction rates depend upon m_X, while H is independent of m_X. The scale of the rates for processes involving X, $\bar{X}$ bosons relative to the expansion rate is set by m_X: For large values of m_X the rates become smaller relative to H, i.e., less effective (see Fig. 6.4).

For purposes of baryogenesis the most important rate is the decay rate, as decays are the mechanism that regulates the number of X, $\bar{X}$ bosons. It is therefore useful to define a quantity

$$K \equiv (\Gamma_D/2H)_{T=m_X} = \frac{\alpha m_{Pl}}{3.3 g_*^{1/2} m_X}. \tag{6.9}$$

which measures the effectiveness of decays at the crucial epoch ($T \sim m_X$) when X, $\bar{X}$ bosons must decrease in number if they are to stay in equilibrium. Note also that for $T \lesssim m_X$, K determines the effectiveness of inverse

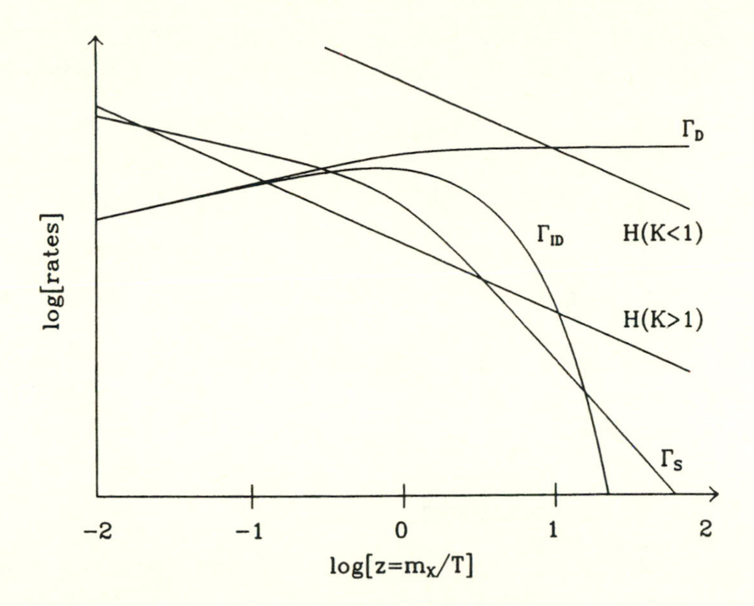

Fig. 6.4: Rates for X, $\bar{X}$ boson interactions and the expansion rate H (schematic). The upper line for H corresponds to a large X-boson mass.

decays and $2 \leftrightarrow 2$ scatterings: $\Gamma_{ID}/H \sim (m_X/T)^{3/2}\exp(-m_X/T)K$ and $\Gamma_S/H \sim \alpha(T/m_X)^5 K$. If $K \ll 1$, then for $T \sim m_X$ the decay rate Γ_D is less than the expansion rate H (equivalently $\tau_X \gtrsim t$), and X, $\bar{X}$ bosons do not decrease in number. Equilibrium will not be maintained and X, $\bar{X}$ bosons will become overabundant (see Fig. 6.3). This overabundance is the requisite departure from thermal equilibrium. In this limit X, $\bar{X}$ bosons just drift along and then eventually decay. In the limit of pure drift and decay (the British limit) it is simple to compute the baryon asymmetry that evolves. When the X, $\bar{X}$ bosons finally decay, $t \sim \Gamma_D^{-1}$ (or $T \sim K^{1/2}m_X$), they are very much overabundant: $n_X = n_{\bar{X}} \sim n_\gamma$, i.e., no exponential suppression. Since each decay produces a mean net baryon number ε, the resulting baryon number density is $n_B \sim \varepsilon n_X \sim \varepsilon n_\gamma$. The entropy density is, $s \sim g_* n_\gamma$, and so the baryon asymmetry produced is

$$B \equiv n_B/s \sim \varepsilon n_\gamma / g_* n_\gamma \sim \varepsilon / g_*. \tag{6.10}$$

Because $K \ll 1$, both inverse decays and $2 \leftrightarrow 2$ B-nonconserving scattering processes are impotent when the decays finally occur and can be safely ignored. Taking $g_* = 10^2$ to 10^3, we see that only a tiny C, CP violation is required: $\varepsilon \sim 10^{-8}$ to 10^{-7} can account for $B \sim 10^{-10}$. The condition that K be much less than 1 translates into a condition on the boson mass:

$$m_X \gg \alpha g_*^{-1/2} m_{Pl} \sim (\alpha/10^{-2})10^{16}\,\text{GeV}. \tag{6.11}$$

For a gauge boson $\alpha = g_{gauge}^2/4\pi \equiv \alpha_{GUT} \simeq 1/45$, while for a Higgs boson α can be much smaller.

In the other extreme, $K \gg 1$, one expects the abundance of X, $\bar{X}$ bosons to track its equilibrium value as $\Gamma_D \gg H$ for $T \sim m_X$. If the equilibrium abundance is tracked precisely enough, there will be no departure from equilibrium, and no baryon asymmetry will evolve. The intermediate regime, K not too different from 1, is more interesting and to address it we will have to develop a more quantitative model.

6.4 Simple Boltzmann Equations

The tool for studying the precise evolution of the baryon asymmetry is of course the Boltzmann equations. Given a definite GUT one can derive Boltzmann equations to follow the development of the baryon asymmetry of the Universe in the same detail that one does the light element abundances during primordial nucleosynthesis. However, lacking strong motivation for any particular GUT, we will write down Boltzmann equations for a highly simplified "toy model," one which serves to illustrate most of the salient features of baryogenesis, and which can be used to mimic realistic GUTs.

The simple model consists of a massive boson that is self conjugate ($X \equiv \bar{X}$) and has interactions that violate B, C, and CP, "light" (i.e., highly relativistic) particles b and $\bar{b}$ that carry baryon numbers $+1/2$ and $-1/2$, and light particles that carry no baryon number and represent the thermal bath of radiation, and collectively have g_* degrees of freedom. The amplitudes for X decay are taken to be[3]

$$\begin{aligned} |\mathcal{M}(X \to bb)|^2 = |\mathcal{M}(\bar{b}\bar{b} \to X)|^2 &= \frac{1}{2}(1+\varepsilon)|\mathcal{M}_0|^2 \\ |\mathcal{M}(X \to \bar{b}\bar{b})|^2 = |\mathcal{M}(bb \to X)|^2 &= \frac{1}{2}(1-\varepsilon)|\mathcal{M}_0|^2. \end{aligned} \tag{6.12}$$

Note that $\varepsilon_X = [(1/2)(1+\varepsilon)|\mathcal{M}_0|^2 - (1/2)(1-\varepsilon)|\mathcal{M}_0|^2]/[(1/2)(1+\varepsilon)|\mathcal{M}_0|^2 + (1/2)(1-\varepsilon)|\mathcal{M}_0|^2] = \varepsilon$. We will assume that the interactions of the X

[3]CPT invariance relates $M(i \to j)$ and $M(\bar{j} \to \bar{i})$.

bosons and baryons with the thermal bath are rapid enough to maintain kinetic equilibrium,[4] and thus their phase space densities are given by

$$\begin{aligned} f_b(E) &= \exp\left[-(E-\mu)/T\right] \\ f_{\bar{b}}(E) &= \exp\left[-(E+\mu)/T\right] \\ f_X(E) &= \exp\left[-(E-\mu_X)/T\right], \end{aligned} \qquad (6.13)$$

where μ is the baryon chemical potential, μ_X is the X chemical potential, and for simplicity we use Maxwell-Boltzmann statistics. *Kinetic* equilibrium also implies that $\mu \equiv \mu_b = -\mu_{\bar{b}}$, and is maintained by B-conserving interactions with the thermal plasma like $b + \bar{b} \leftrightarrow \gamma + \gamma$. Whether or not *chemical* equilibrium is maintained for baryon number depends upon the rate for baryon number violating reactions: If they are sufficiently rapid, then chemical equilibrium obtains for baryon number and X bosons, so that $\mu = \mu_X = 0$. The baryon number-to-entropy ratio is given by

$$\begin{aligned} B \equiv \frac{n_B}{g_* n_\gamma} = \frac{1}{2}\left(\frac{n_b - n_{\bar{b}}}{g_* n_\gamma}\right) &= g_*^{-1} \sinh\left(\frac{\mu}{T}\right) \\ &= g_*^{-1}\frac{\mu}{T} + \mathcal{O}((\mu/T)^3) \end{aligned} \qquad (6.14)$$

where we have taken the entropy density s to be $g_* n_\gamma$, and $n_\gamma \simeq 2T^3/\pi^2$.

The Boltzmann equation for the number density of X's is easy to obtain. In the spirit of the simple model we will only include decays and inverse decays as they are the most important process for regulating the number density of X's. The equation governing n_X is[5]

$$\begin{aligned} \dot{n}_X + 3Hn_X &= \int d\Pi_X d\Pi_1 d\Pi_2 (2\pi)^4 \delta^4(p_X - p_1 - p_2) \\ &\quad \times \left[-f_X(p_X)|\mathcal{M}_0|^2 + \frac{1}{2}(1+\varepsilon)|\mathcal{M}_0|^2 f_{\bar{b}}(p_1) f_{\bar{b}}(p_2)\right. \end{aligned}$$

[4]If instead one assumes that the X bosons are totally decoupled, the evolution of the baryon asymmetry is affected very little. This is because all the action occurs for $T \lesssim m_X$, when the X bosons are very non-relativistic, so that the temperature dependence of their decay and inverse decay rates is not significant.

[5]Note that $n_X = \exp(\mu_X/T)\int d^3p \exp(-E/T)/2\pi^2 = \exp(\mu_X/T)n_X^{\rm EQ}$, so that either μ_X or n_X can be used to determine the X abundance. Henceforth we will use n_X rather than μ_X.

$$
\begin{aligned}
&\left.+\frac{1}{2}(1-\epsilon)|\mathcal{M}_0|^2 f_b(p_1)f_b(p_2)\right] \\
=& \int d\Pi_X d\Pi_1 d\Pi_2 (2\pi)^4\delta^4(p_X - p_1 - p_2) \\
&\times\left[-f_X(p_X) + f_X^{EQ}(p_X)\right]|\mathcal{M}_0|^2 + \mathcal{O}(\epsilon,\ \mu/T) \\
\equiv& -\Gamma_D(n_X - n_X^{EQ}),
\end{aligned}
\tag{6.15}
$$

where n_X^{EQ} is the equilibrium number density of X's and Γ_D is the thermally-averaged decay width of the X. In expressing the rhs of (6.15) in terms of $(n_X - n_X^{EQ})$ we have used the same "trick" that we did in Chapter 5 when deriving the Boltzmann equation for the relic abundance of a massive particle species. It is convenient to transform to the dimensionless evolution variable $z \equiv m_X/T$ and to follow the number of X's per comoving volume, $X \equiv n_X/s \sim g_*^{-1} n_X/n_\gamma$. After doing so we find

$$
\begin{aligned}
X' &= -z\gamma_D K(X - X_{EQ}) \\
\Delta' &= -X'_{EQ} - z\gamma_D K\Delta,
\end{aligned}
\tag{6.16}
$$

where

$$
\begin{aligned}
\Delta &\equiv X - X_{EQ} \\
K &\equiv \Gamma_D(z=1)/2H(m_X) \simeq \alpha m_{Pl}/3g_*^{1/2} m_X
\end{aligned}
\tag{6.17}
$$

$$
\gamma_D = \Gamma_D(z)/\Gamma_D(z=1) \simeq \begin{cases} z/2 & z \ll 1 \\ 1 & z \gg 1 \end{cases}
\tag{6.18}
$$

$$
X_{EQ} \equiv n_X^{EQ}/s \simeq \begin{cases} g_*^{-1} & z \ll 1 \\ g_*^{-1}(\pi/2)^{1/2} z^{3/2} e^{-z} & z \gg 1. \end{cases}
\tag{6.19}
$$

In a similar manner, one obtains the Boltzmann equation governing the number density of b's:

$$
\begin{aligned}
\dot{n}_b + 3Hn_b =& \int d\Pi_X d\Pi_1 d\Pi_2 (2\pi)^4\delta^4(p_X - p_1 - p_2) \\
&\times[-(1-\epsilon)f_b(p_1)f_b(p_2) + (1+\epsilon)f_X(p_X)]\,|\mathcal{M}_0|^2 \\
&+2\int d\Pi_1\cdots d\Pi_4 (2\pi)^4\delta^4(p_1 + p_2 - p_3 - p_4)
\end{aligned}
$$

$$\times \left[-f_b(p_1)f_b(p_2)|\mathcal{M}'(bb \to \bar{b}\bar{b})|^2\right.$$

$$\left. + f_{\bar{b}}(p_3)f_{\bar{b}}(p_4)|\mathcal{M}'(\bar{b}\bar{b} \to bb)|^2\right]$$

$$+\mathcal{O}\left(\varepsilon n_B + n_B^2\right). \tag{6.20}$$

Note that we have neglected terms that are $\mathcal{O}(\varepsilon n_B,\, n_B^2)$; because n_B/s, $\varepsilon \ll 1$, they will be very small. The corresponding Boltzmann equation for $n_{\bar{b}}$ is obtained from (6.20) by the interchange $b \leftrightarrow \bar{b}$; $\varepsilon \leftrightarrow -\varepsilon$.

The equation for the evolution of the baryon number density is obtained by subtracting the equation for $n_{\bar{b}}$ from that for n_b and multiplying by a factor of 1/2:

$$\dot{n}_B + 3Hn_B = \varepsilon\Gamma_D(n_X - n_X^{EQ}) - n_B(n_X^{EQ}/n_\gamma)\Gamma_D - 2n_Bn_b\langle\sigma|v|\rangle. \tag{6.21}$$

Here and above $|\mathcal{M}'(bb \to \bar{b}\bar{b})|^2$ and $|\mathcal{M}'(\bar{b}\bar{b} \to bb)|^2$ are the squares of the matrix elements for $2 \leftrightarrow 2$ B-nonconserving scatterings, with the part due to real intermediate-state X's removed (this part has already been taken into account in the decay and inverse decay terms). The quantity

$$\langle\sigma|v|\rangle \equiv \int d\Pi_1 \cdots d\Pi_4 (2\pi)^4\delta^4(p_1 + p_2 - p_3 - p_4)$$

$$\times f_b(p_1)f_b(p_2)|\mathcal{M}'(bb \to \bar{b}\bar{b})|^2/n_\gamma^2 \tag{6.22}$$

$$\simeq \frac{A\alpha^2T^2}{(T^2 + m_X^2)^2} \tag{6.23}$$

is the thermally-averaged $2 \leftrightarrow 2$ B-nonconserving scattering cross section (A is a numerical factor which accounts for the number of scattering channels, etc.). In deriving (6.21) for the evolution of n_B, one must take into account the CP-violating part of the $2 \leftrightarrow 2$ scattering amplitude. It is obtained from the decay and inverse decay amplitudes by the use of CPT and unitarity.[6] The CP-violating part of $|\mathcal{M}'(bb \to \bar{b}\bar{b})|^2 - |\mathcal{M}'(\bar{b}\bar{b} \to bb)|^2$ accounts for the crucial $-\varepsilon\Gamma_D n_X^{EQ}$ term.

Note that the source term for net baryon number reflects the three necessary ingredients for baryogenesis: $\varepsilon\Gamma_D(n_X - n_X^{EQ})$ is only non-zero if B, C, and CP are violated, and if non-equilibrium conditions pertain, i.e.,

[6]One might have thought that the CP-violating part of the $2 \leftrightarrow 2$ process would be higher order and could be neglected; this is not the case. For a detailed discussion, see [6].

$n_X \neq n_X^{EQ}$. Note too that the last two terms on the rhs of (6.21) account for the damping of baryon asymmetry by inverse decays and $2 \leftrightarrow 2$ B-nonconserving processes respectively.

Transforming the equation for n_B to one for the baryon number per comoving volume ($\equiv B$) as a function of the evolution variable z we obtain:

$$B' = \varepsilon z K \gamma_D \Delta - z K \gamma_B B \tag{6.24}$$

where

$$\gamma_B \equiv [g_* X_{EQ} \gamma_D + 2n_\gamma \langle \sigma |v| \rangle] / \Gamma_D(z=1) \tag{6.25}$$

$$\simeq \begin{cases} z/2 + A\alpha/z & z \ll 1 \\ z^{3/2} e^{-z} + A\alpha z^{-5} & z \gg 1. \end{cases} \tag{6.26}$$

This equation together with (6.16) forms a coupled system of equations that can be solved for the evolution of $B(z)$. The physics of these equations is quite simple. The departure from equilibrium Δ drives the production of a baryon asymmetry. Inverse decays and $2 \leftrightarrow 2$ scatterings, embodied in γ_B, tend to damp any baryon asymmetry. The changing value of X_{EQ} drives Δ; in the absence of the expansion of the Universe, $T' = X'_{EQ} = 0$, and any departure from equilibrium would decay away exponentially, $\Delta \propto \exp(-Kz^2/2)$. However, because of the expansion, T', $X'_{EQ} \neq 0$, and Δ does not decrease exponentially quickly.

It is straightforward to formally integrate these equations to obtain $\Delta(z)$ and $B(z)$,

$$\begin{aligned} \Delta(z) &= \Delta_i \exp\left[-\int_0^z z' K \gamma_D(z') dz'\right] \\ &\quad - \int_0^z X'_{EQ}(z') \exp\left[\int_{z'}^z z'' K \gamma_D(z'') dz''\right] dz' \qquad (6.27) \\ B(z) &= B_i \exp\left[-\int_0^z z' K \gamma_B(z') dz'\right] \\ &\quad + \varepsilon K \int_0^z z' \Delta(z') \gamma_D(z') \exp\left[-\int_{z'}^z z'' K \gamma_B(z'') dz''\right] dz' \end{aligned}$$

where Δ_i and B_i are the departure from equilibrium and baryon asymmetry at time zero respectively. For the moment we will take both to be zero.

The solution for $B(z)$ can easily be obtained in two limiting regimes: $K \ll 1$ and $K \gg 1$. First consider $K \ll 1$, where we expect out-of-equilibrium decay and a final baryon asymmetry of ε/g_*. In this case the integrating factors are approximately equal to unity, since the arguments of the exponentials are proportional to K. In this limit we have:

$$\begin{aligned} X(z) &\simeq X(0)\exp(-Kz^2/2) \\ B(z) &\simeq \varepsilon[X(0) - X(z)] \\ &\Longrightarrow B_f \equiv B(\infty) = \varepsilon X(0) \sim \frac{\varepsilon}{g_*} \end{aligned} \tag{6.28}$$

That is, the X's decay when $z \sim K^{-1/2}$ ($t \sim \Gamma_D^{-1}$), producing a baryon asymmetry $B \sim \varepsilon/g_*$, in agreement with the qualitative picture discussed earlier.

In the other limiting regime, $K \gg 1$, the solution for $\Delta(z)$ is also easily found:

$$\begin{aligned} \Delta'(z) &\simeq 0 \\ \Delta(z) &\simeq -X'_{EQ}/zK\gamma_D(z) \simeq X_{EQ}/zK, \end{aligned} \tag{6.29}$$

valid for $z \gtrsim 1$. That is, because the rates for both decays and inverse decays are rapid ($\Gamma \gtrsim H$), $\Delta(z)$ tracks the value that keeps Δ' nearly zero. The departure from equilibrium is of order K^{-1}. As noted earlier, $\Delta(z)$ does not decrease exponentially because of the X'_{EQ} source term. Any initial departure from equilibrium (if one exists) does, however, decrease exponentially.

Using the method of steepest descent, the integral for $B_f \equiv B(z=\infty)$ can be evaluated,

$$B_f \simeq \frac{\varepsilon}{g_*}\frac{\pi}{2\sqrt{a}} z_f^{3/2} \exp\left[-z_f - \int_{z_f}^{\infty} zK\gamma_B(z)dz\right], \tag{6.30}$$

where z_f is determined by:

$$z_f K\gamma_B(z_f) = 1 \tag{6.31}$$

and $a \equiv -(zK\gamma_B)'|_{z_f}$. The epoch $z = z_f$ corresponds to the time when the processes that damp the baryon asymmetry, inverse decays and $2 \leftrightarrow 2$ scatterings, freeze out, i.e., $\Gamma_B(z_f) \simeq H(z_f)$. Damping processes are important until $z \sim z_f$ and thereafter become ineffective, allowing the baryon asymmetry present then to freeze in.

If K is greater than 1, but not too large (to be quantified shortly), then during the epoch when B-nonconserving processes are effective ($1 \lesssim z \lesssim z_f$), inverse decays are more important in damping the baryon asymmetry than are $2 \leftrightarrow 2$ scatterings, and z_f is determined by when inverse decays freeze out: $K z_f^{5/2} \exp(-z_f) \simeq 1$. Asymptotically the solution is $z_f \simeq \ln K$; however for $K \lesssim 10^6$, $z_f \simeq 4.2(\ln K)^{0.6}$ is a better fit. Bringing everything together we have,

$$B_f \simeq \frac{\varepsilon}{g_* K z_f} \simeq \frac{0.3(\varepsilon/g_*)}{K(\ln K)^{0.6}}. \tag{6.32}$$

Note that the final baryon asymmetry produced only slowly decreases with increasing K, a fact which is somewhat surprising. Recall that one expected that when $K \gtrsim 1$, corresponding to $\Gamma_D \gtrsim H(m_X)$, an equilibrium abundance of X bosons would be maintained and no baryon asymmetry would evolve. Thus one would have expected B_f to fall precipitously for $K \gtrsim 1$. As we will now discuss, such a rapid decrease in B_f with K only takes place when $2 \leftrightarrow 2$ scatterings are important.

When $K \gg 1$, $2 \leftrightarrow 2$ scatterings are the dominant B damping processes at freeze out ($z \sim z_f$). In this case z_f is determined by $AK\alpha z_f^{-4} \simeq 1$, which implies that $z_f \simeq (AK\alpha)^{1/4}$, from which it follows that

$$B_f \simeq \frac{\varepsilon}{g_*}(AK\alpha)^{1/2} \exp\left[-4(AK\alpha)^{1/4}/3\right]. \tag{6.33}$$

For very large K, when $2 \leftrightarrow 2$ scatterings are the most important B damping process when $z \simeq z_f$, the final baryon asymmetry does decrease exponentially (with $K^{1/4}$).

To determine approximately how large K must be before the exponential fall off sets in, we can compare freeze out for inverse decay processes, $z_f \simeq 4.2(\ln K)^{0.6}$, with freeze out for $2 \leftrightarrow 2$ scatterings, $z_f \simeq (AK\alpha)^{1/4}$. The "crossover" value of K ($\equiv K_C$) is determined by solving

$$K_C(\ln K_C)^{-2.4} \simeq 300/A\alpha. \tag{6.34}$$

For $1 \lesssim K \lesssim K_C$, B_f is given by the power law expression, (6.32), and for $K \gtrsim K_C$, B_f falls off exponentially and is given by (6.33).

Figures 6.5 and 6.6 show the results of numerically integrating the equations for the evolution of the baryon asymmetry and the final asymmetry which results respectively. For a GUT like $SU(5)$ or $SO(10)$, g_*, the total

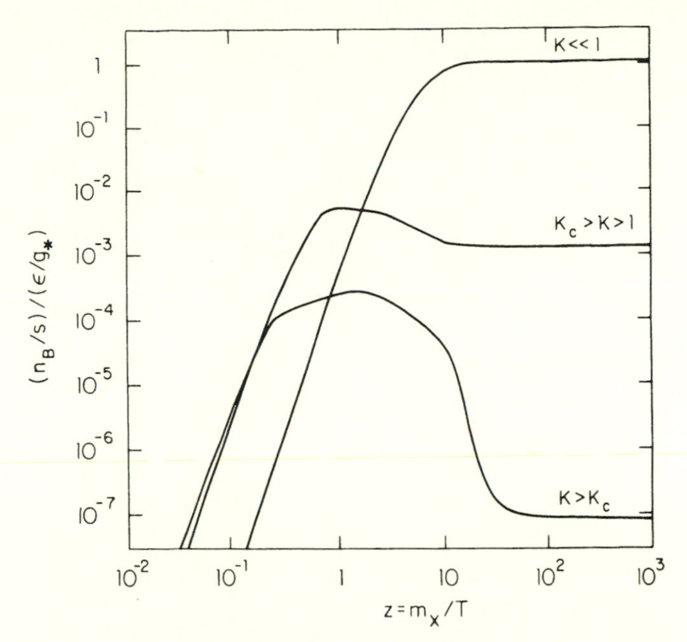

Fig. 6.5: The development of the baryon asymmetry B as a function of $z = m_X/T$ for $K \ll 1$, $K_C \gtrsim K \gtrsim 1$, and $K \gtrsim K_C$.

number degrees of freedom at temperatures of order 10^{14} to 10^{15} GeV is of order 200, and the gauge coupling constant is about 1/45, so that for a gauge boson $K \simeq 7 \times 10^{15}\,\text{GeV}/m_X$. For the XY gauge bosons of $SU(5)$, $m_X \simeq$ few $\times\, 10^{14}$ GeV and $A \simeq$ few $\times\, 10^3$, which implies that $K_{XY} \simeq 30$ and $K_C \simeq 100$. For such a boson

$$B_f \simeq \frac{0.3\varepsilon}{g_* K_{XY} (\ln K_{XY})^{0.6}} \simeq 10^{-5}\varepsilon. \tag{6.35}$$

The mass and couplings of the superheavy Higgs bosons that mediate B nonconservation are far less certain. For the triplet member of the 5-dimensional Higgs multiplet of $SU(5)$, the Higgs coupling to fermions is determined by the fermion masses and $h \simeq (m_{\rm fermion}/m_W)g$ (where $\alpha_H \simeq h^2/4\pi$ and $\alpha_{GUT} = g^2/4\pi$). The heaviest quark mass determines h.[7] For a top quark mass of order 50 GeV, $\alpha_H \simeq$ few $\times\, 10^{-3}$. For $\alpha_H \simeq 10^{-3}$, $K_H \simeq 3 \times 10^{14}\,\text{GeV}/m_X$ and $A \simeq$ few $\times\, 10^3$, so that $K_C \gtrsim 10^3$. Except

[7]To properly estimate the effective value of α_H one should average α_H over all quark species; in the limit of one very heavy quark (like the top quark), that is equivalent to dividing its coupling by the total number of quark species.

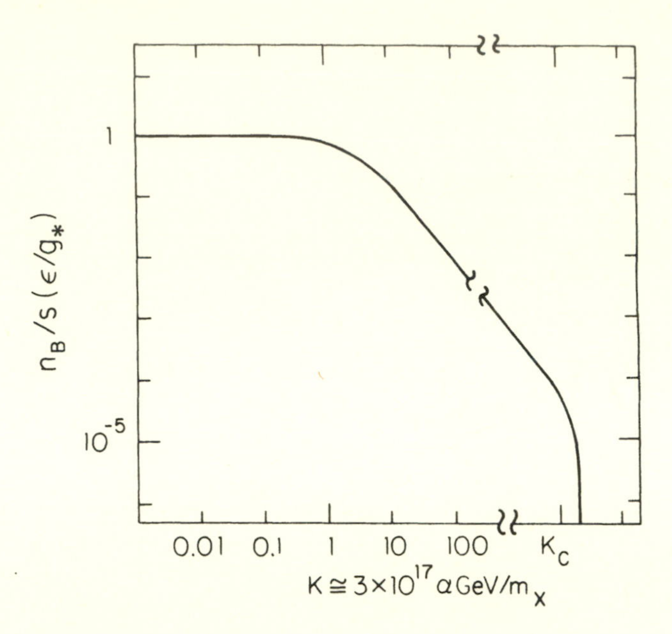

Fig. 6.6: Schematic summary of the final baryon asymmetry B_f, as a function of $K = \alpha m_{Pl}/3g_*^{1/2}m_X$.

for light Higgs masses $K_H \lesssim 1$, so that the asymmetry produced is the maximal value:

$$B_f \simeq \varepsilon/g_*. \tag{6.36}$$

This simple toy model serves very well to illustrate the pertinent features of baryogenesis. More sophisticated systems of Boltzmann equations, based upon realistic GUTs, have been derived to evolve the baryon asymmetry [7]. The equations become quite complicated when proper account is taken of all the particles and interactions, and the final baryon asymmetry is sensitive to unknown Higgs masses and Yukawa couplings. Nevertheless, in principle it is possible to calculate the evolution of the baryon asymmetry in much the same way as one does the production of the light nuclei.

6.5 Damping of Pre-Existing Asymmetries

Thus far we have set to zero any initial baryon asymmetry. We will now consider the damping of initial or pre-existing baryon asymmetries. Since there may be several superheavy bosons whose interactions violate baryon number, with different masses, there may be several epochs during which the baryon asymmetry evolves. Here we consider the damping effect of a single superheavy boson species. Recall that essentially all the evolution of n_B/s takes place when $T \sim m_X$, cf. Fig. 6.5. Suppose that at temperatures $T \gg m_X$ the baryon asymmetry has the value $(n_B/s)_i = B_i$. In this case the solution to the equation governing the evolution of B is the one previously obtained plus a solution to the homogeneous equation

$$B' = -zK\gamma_B B \tag{6.37}$$

that satisfies the initial condition: $B(z \ll 1) = B_i$. The solution to the homogeneous equation is

$$\begin{aligned} B(z) &= B_i \exp\left[-\int_0^z z' K\gamma_B(z')dz'\right] && (6.38)\\ &\simeq B_i \exp[-4(K + A\alpha)] && (6.39) \end{aligned}$$

where the first term in the exponential accounts for the damping by inverse decays and the second by $2 \leftrightarrow 2$ scatterings. Recall for the XY gauge bosons of $SU(5)$, $K_{XY} \simeq 30$: They should strongly damp any pre-existing baryon asymmetry.

Combining this solution with the particular soution found previously we find that

$$B_f = B_i \exp\left[-\int_0^\infty zK\gamma_B dz\right] + \varepsilon f(K) \tag{6.40}$$

where $f(K)$ follows from (6.27), with its limiting forms given by (6.32) for $K_C \gtrsim K \gtrsim 1$ and by (6.33) for $K \gtrsim K_C$. In general the effect of a single superheavy boson species is to damp any pre-existing asymmetry, to an extent determined by K, and to produce an asymmetry proportional to ε. If the masses of the various superheavy bosons are sufficiently different, so that the various epochs of baryon asymmetry evolution are dynamically separate, then the equation above allows one to track the evolution of (n_B/s) through successive epochs of evolution. If the masses are not sufficiently different, then the dynamics is more complicated and the above equation is not applicable.

With this simple model we have neglected a potentially important feature of the evolution of the baryon asymmetry—the existence of so-called non-thermalizing modes. This traces to the fact that we have treated B (and L) as if they are carried by but one species. In fact, baryon number is carried by a variety of species—doublet and singlet quarks (under

$SU(2)$). Because of this, there exist different eigenmodes associated with the damping of baryon number, and this fact can play an important role in the evolution of the baryon number of the Universe. We will now illustrate nonthermalizing modes with a simple example.

Baryon minus lepton number $(B-L)$ is an accidental global symmetry of the $SU(5)$ gauge interactions. For the moment, suppose that it is also effectively conserved by any Higgs interactions, i.e., either conserved by these interactions, or these interactions are impotent ($\Gamma \lesssim H$). We can deduce the structure of the equations governing the damping of both B and L by noting that in the absence of C, CP violation (i.e., $\varepsilon = 0$) B and L non-conserving interactions can only damp B and L (or any combinations of B and L), and thus the equations for B' and L' must be linear in B and L and homogeneous:

$$\begin{aligned} B' &= -(1/2)zK\gamma(z)(B + c_1 L) \\ L' &= -(1/2)zK\gamma(z)(c_2 B + c_3 L), \end{aligned} \tag{6.41}$$

where we have inserted the "kinematic factor" $(1/2)zK\gamma(z)$ based upon our experience above, cf. (6.37). Subtracting the equation for L' from the equation for B' we must obtain zero, as $B-L$ is conserved. This implies $c_2 = 1$ and $c_1 = c_3$. Adding the two equations results in an equation for $(B+L)'$:

$$(B+L)' = -zK\gamma(z)\left[\frac{1+c_1}{2}(B+L) + \frac{1-c_1}{2}(B-L)\right]. \tag{6.42}$$

Since $B-L$ is constant, the second term in the equation would generate a $B+L$ (and hence B and L) asymmetry with $\varepsilon = 0$ unless $c_1 = 1$—in contradiction to the rule that a C, CP violation is necessary. Thus c_1 must be equal to 1. Setting $c_1 = 1$, the equations become

$$\begin{aligned} (B-L)' &= 0 \\ (B+L)' &= -zK\gamma(z)(B+L). \end{aligned} \tag{6.43}$$

That is, $B-L$ and $B+L$ are the two eigenmodes, where $B-L$ does not damp, and $B+L$ damps at a rate given by

$$B + L = (B+L)_i \exp\left[-\int_0^z zK\gamma(z)dz\right]. \tag{6.44}$$

Baryon and lepton number are related to $B+L$ and $B-L$ by

$$B = \frac{1}{2}(B+L) + \frac{1}{2}(B-L), \tag{6.45}$$

$$L = \frac{1}{2}(B+L) - \frac{1}{2}(B-L). \tag{6.46}$$

Now consider the evolution of pre-existing B and L asymmetries B_i and L_i. Asymptotically we have

$$(B-L)_f = (B-L)_i, \tag{6.47}$$

$$(B+L)_f = (B+L)_i \exp\left[-\int_0^\infty zK\gamma dz\right]. \tag{6.48}$$

Transforming back to the B, L basis, we find that

$$B_f = \frac{1}{2}(B-L)_i + \frac{1}{2}(B+L)_i \exp\left[-\int_0^\infty zK\gamma dz\right], \tag{6.49}$$

$$L_f = -\frac{1}{2}(B-L)_i + \frac{1}{2}(B+L)_i \exp\left[-\int_0^\infty zK\gamma dz\right]. \tag{6.50}$$

The projection of B (or L) onto the $B-L$ mode undergoes no damping, while the projection onto the $B+L$ mode does undergo damping. The $(B-L)$ projection of B (or L) is referred to as the non-thermalizing mode. This simple example illustrates a very important point: A pre-existing asymmetry can survive an epoch of strong damping provided that it has a projection onto a non-thermalizing mode. We will shortly make use of this fact.

$B-L$ is a simple example of a non-thermalizing mode; many unified theories have additional non-thermalizing modes, modes that may prove useful in the protection of a baryon asymmetry against subsequent damping effects. With this fact in mind, we should modify our prescription for the effect of a single superheavy species on the evolution of B:

$$B_f = B_{NTi} + B_{Ti} \exp\left[-\int_0^\infty zK\gamma_B(z)dz\right] + \epsilon f(K), \tag{6.51}$$

where B_{Ti} and B_{NTi} represent the "thermalizing" and "non-thermalizing" parts of the pre-existing asymmetry.

6.6 Lepton Numbers of the Universe

Given that the Universe possesses a baryon asymmetry, we have every reason to suspect that it is also endowed with a lepton asymmetry, and common sense (as well as baryogenesis) suggests that the lepton asymmetry should be comparable to the baryon asymmetry. The lepton asymmetry L must reside in electrons and the undetected neutrino seas:

$$L \equiv \frac{n_L}{s} = \frac{n_e - n_{\bar{e}}}{s} + \sum_{i=e,\mu,\tau} \frac{n_\nu - n_{\bar{\nu}}}{s}. \tag{6.52}$$

Therein lies the difficulty: While the apparent charge neutrality of the Universe[8] implies that $(n_e - n_{\bar{e}})/s \simeq 0.95(n_B/s) \sim 10^{-10}$, we have no direct knowledge of the lepton number that resides in the three undetected neutrino seas. The successful predictions of primordial nucleosynthesis provide strong indirect evidence that for each neutrino flavor $(n_\nu - n_{\bar{\nu}})/s$ is much less than unity. However, direct evidence is far less restrictive. The strongest direct limit comes from the mass density contributed by the relic neutrino seas. In the degenerate limit ($|\mu_\nu| \gg T$), a massless neutrino species contributes an amount

$$\rho_{\nu 0} = \mu_\nu^4 / 8\pi^2, \tag{6.53}$$

to the present energy density of the Universe. Requiring that $\Omega_\nu h^2 \lesssim 1$ sets an upper limit to the individual neutrino chemical potentials of

$$\mu_\nu \lesssim 8.9 \times 10^{-3} \text{ eV}, \tag{6.54}$$

or equivalently,

$$\mu_\nu / T_\nu \lesssim 53. \tag{6.55}$$

Since the net lepton number density is given by

$$n_\nu - n_{\bar{\nu}} = \frac{T^3}{6\pi^2}\left[\pi^2\left(\frac{\mu}{T}\right) + \left(\frac{\mu}{T}\right)^3\right] \tag{6.56}$$

this implies the rather weak limit of

$$\frac{n_\nu - n_{\bar{\nu}}}{n_\gamma} \lesssim 3.7 \times 10^3. \tag{6.57}$$

---

[8]The best astrophysical limit to a possible charge asymmetry of the Universe is: $n_Q/s \lesssim 10^{-27}$[8].

Within the framework of baryogenesis one would expect that the magnitude of the lepton asymmetry which evolves is comparable to that of the baryon asymmetry. To be more specific, consider $SU(5)$. The quantity $B-L$ is an accidental global symmetry of the gauge interactions of the theory. Again supposing that $B-L$ is at least effectively conserved by the Higgs interactions we can make a definite statement about the lepton asymmetry that evolves. Assuming an initially B and L symmetric Universe ($B = L = B - L = 0$), any B and L asymmetries that evolve must preserve $B - L = 0$ (as we discussed in the previous section $B - L$, is an non-thermalizing mode in $SU(5)$). Thus the B and L asymmetries that evolve must be identical:

$$\frac{n_B}{s} - \frac{n_L}{s} = 0 \Rightarrow L = B \simeq 10^{-10}. \tag{6.58}$$

On the other hand, suppose that an initial $B-L$ asymmetry did exist:

$$(B-L)_i \neq 0. \tag{6.59}$$

That $B-L$ asymmetry must still persist even after baryogenesis which implies that

$$L_{\text{today}} = B_{\text{today}} - (B-L)_i \tag{6.60}$$

Thus the present lepton asymmetry depends upon the initial B − L asymmetry, and could, if that asymmetry were large, be very different from the baryon asymmetry.

While this example was somewhat contrived, it illustrates a general principle: Due to the possible existence of non-thermalizing modes associated with global symmetries involving both baryon and lepton number, it is possible that with the help of an initial asymmetry in some baryon and/or lepton species, the lepton asymmetry which exists today could be quite different from the baryon asymmetry [9]. While in the standard cosmology one has the possibility of such initial asymmetries, in inflationary models where any initial asymmetries are exponentially diluted away by inflation, one does not have this freedom.

6.7 Way-Out-of-Equilibrium Decay

When computing the baryon asymmetry produced in the out-of-equilibrium decay scenario we have neglected the entropy produced by X's decays, tacitly assuming that it is small. In the standard scenario, where the entropy of the 100's of relativistic degrees of freedom is very large this is probably a reasonable assumption. Let us now consider the opposite extreme, namely where *only* X, $\bar{X}$ bosons are present, and all the entropy in the Universe results from their out-of-equilibrium decays. This case will be of great interest in inflationary models, wherein the X is the scalar field responsible for inflation and its decays account for the reheating of the Universe.

Denote the number density of X's and $\bar{X}$'s by n_X, and that of the net baryon number by n_B. Further assume that each decay of an X, $\bar{X}$ pair produces a mean net baryon number[9] of ε—and, for the moment neglect any damping processes which destroy baryon number, i.e., inverse X, $\bar{X}$ decays and $2 \leftrightarrow 2$ B-nonconserving processes. Suppose also that the decay products of the X, $\bar{X}$ decays are relativistic and quickly thermalize. The equations governing the evolution of n_X and n_B are

$$\begin{aligned}
\dot{n}_X + 3Hn_X &= -n_X/\tau \\
\Rightarrow n_X &= n_{Xi}(R/R_i)^{-3}\exp(-t/\tau) \\
\dot{n}_B + 3Hn_B &= \varepsilon n_X/\tau \\
S^{1/3}\dot{S} &= \left(\frac{2\pi^2}{45}g_*\right)^{1/3} R^4\rho_X/\tau \\
H^2 &= \frac{8\pi}{3m_{Pl}^2}(m_X n_X + \rho_R),
\end{aligned} \tag{6.61}$$

where m_X and τ are the mass and lifetime of the X, subscript i refers to some initial epoch $t_i \ll \tau$, ρ_R is the energy density of the relativistic decay products of X, $\bar{X}$'s, and the entropy in a comoving volume is just

$$S = \frac{2\pi^2}{45}g_* R^3T^3 = \left(\frac{4}{3}\right)^{3/4}\left(\frac{2\pi^2 g_*}{45}\right)^{1/4} R^3\rho_R^{3/4}. \tag{6.62}$$

These equations are just those we derived in Chapter 5 for a decaying particle, supplemented by the equation for the evolution of n_B. The

[9] X, $\bar{X}$ decays might directly produce an excess of baryons over antibaryons, e.g., $X, \bar{X} \rightarrow q$'s + l's, or might produce the excess in several steps: $X, \bar{X} \rightarrow \phi$'s $\rightarrow q$'s + l's, i.e., X, $\bar{X}$ decay to ϕ particles, which subsequently decay to quarks and leptons. In either case it is the net excess baryon number per X, $\bar{X}$ decay that is relevant and is denoted by ε.

solution for n_B is easily obtained:

$$n_B = \varepsilon \frac{R_i^3 n_{Xi}}{R^3}[1 - \exp(-t/\tau)]. \tag{6.63}$$

The physical content of this solution is clear: At late times, $t \gg \tau$, the net baryon number per comoving volume ($\equiv R^3 n_B$) is just ε times the initial number of X's per comoving volume ($\equiv R_i^3 n_{Xi}$). In Chapter 5 we wrote down the solution for the asymptotic value of S (denoted by S_f):

$$S_f^{4/3} = \frac{4}{3} m_X n_{Xi} R_i^4 \int_0^\infty \left(\frac{2\pi^2 g_*}{45}\right)^{1/3} (R(t)/R_i) e^{-t/\tau} d(t/\tau) \tag{6.64}$$

where here the initial entropy in the comoving volume is zero: $S_i = 0$. The total energy density of the Universe is that of the X and its relativistic decay products, and in this circumstance we have previously evaluated the integral in the above equation:

$$S_f^{4/3} = 1.09 \left(\frac{4}{3} m_X n_{Xi} R^4\right) \left(\frac{16\pi^3 m_X n_{Xi}}{135 m_{Pl}^2} g_*\right)^{1/3} \tau^{2/3}. \tag{6.65}$$

Taking the ratio of the final value of the net baryon number per comoving volume to the entropy per comoving volume (S_f), we find

$$B_f \equiv \left(\frac{R^3 n_B}{S}\right)_f = 0.54\varepsilon \frac{(m_{Pl}/\tau)^{1/2}}{m_X} g_*^{-1/4}. \tag{6.66}$$

Recalling that the reheat temperature is defined to be the radiation temperature at $t \simeq \tau$, and that its value is given by

$$T_{RH} \simeq 0.55 g_*^{-1/4} (m_{Pl}/\tau)^{1/2}, \tag{6.67}$$

the baryon asymmetry produced can also be written as

$$\frac{n_B}{s} \simeq \varepsilon \frac{T_{RH}}{m_X}. \tag{6.68}$$

We note that this same result, up to numerical factors, could have been obtained using the simultaneous decay approximation, i.e., by assuming that all the X, $\bar{X}$'s decay at time $t \sim \tau$, reheat the Universe to a temperature $T_{RH}^4 \sim \tau^{-2} m_{Pl}^2$, and produce a net baryon number ε per X, $\bar{X}$ decay.

In the way-out-of-equilibrium decay scenario the baryon asymmetry produced only depends upon the C, CP violation ε and the ratio of the reheat temperature to the mass of the decaying particle. (Since we have ignored "damping processes," one must also verify that they are ineffective during and after the decay epoch.) As we will see in Chapter 8, this mechanism provides an important means for producing the baryon asymmetry in inflationary Universe models, because it bypasses the need for a high reheat temperature—only the ratio of T_{RH} to m_X is relevant.

6.8 Sphalerons

As a result of non-trivial vacuum gauge configurations (i.e., $A_\mu^a \neq 0$), the vacuum structure—so-called θ-vacuum—of non-Abelian gauge theories is very rich.[10] The θ-vacuum structure in the electroweak theory leads to the anomalous non-conservation of baryon number. Transitions between the different vacua are accompanied by a change in the baryon number. This process is usually associated with the instanton solutions that describe the tunnelling between different θ-vacua (see Fig. 6.7). Since the process is inherently non-perturbative, the rate for baryon number non-conservation is proportional to $\exp(-4\pi/\alpha_W)$, where $\alpha_W = g^2/4\pi$. Because $4\pi/\alpha_W \gg 1$, such quantum tunnelling is unimportant today, and was certainly unimportant in the early Universe.

At finite temperature the transitions between different vacua can be driven by thermal effects. Kuzmin, Rubakov, and Shaposhnikov (KRS) have argued that the rate to surmount (rather than tunnel through) the barrier may be appreciable at temperatures of order 100 GeV or so [10]. Since there is a change in baryon number associated with these transitions, the rate for B violation should also be appreciable. In their analysis KRS considered an $SU(2)$ gauge theory, which is equivalent to the Weinberg-Salam model in the limit $\sin^2\theta_W \to 0$, α_W fixed. The Lagrangian for their $SU(2)$ model with gauge coupling g is given by

$$\mathcal{L} = D_\mu \phi^a D^\mu \phi^a - \frac{1}{4} F_{\mu\nu}^a F^{a\mu\nu} - \frac{1}{4}\lambda(\phi^a\phi^a - \sigma^2/2)^2 + \mathcal{L}_{\rm fermions} \qquad (6.69)$$

[10]The θ-vacuum structure of the electroweak theory is analogous to that of QCD, discussed in Chapter 10.

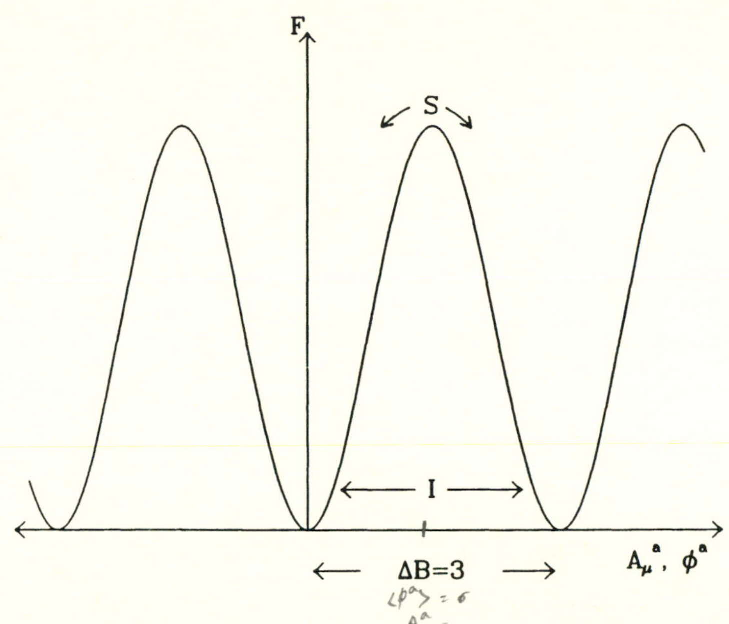

Fig. 6.7: Schematic illustration of the barriers separating different θ-vacua, the instanton tunnelling path through the barrier, and the sphaleron path over the barrier. The vertical axis is the free energy, and the horizontal axis represents gauge and Higgs field configurations. Between adjacent θ-vacua $B + L$ differs by $2N_g$, and B differs by N_g, where N_g is the number of generations (here taken to be 3).

where $D_\mu \equiv \partial_\mu + i(g/2)\tau^a A^a_\mu$, and the gauge field strength tensor is $F^a_{\mu\nu} \equiv \partial_\mu A^a_\nu - \partial_\nu A^a_\mu - g\epsilon^{abc} A^b_\mu A^c_\nu$ (τ^a are the Pauli spin matrices). The basic idea is that thermal fluctuations in the $SU(2)$ gauge field A^a_μ and symmetry-breaking Higgs field ϕ^a can cause transitions that proceed "over the potential barrier" between different θ-vacua, thereby resulting in baryon number violation (see Fig. 6.7). The calculation of the transition rate is related to the calculation of the decay of the false vacuum at finite temperature (to be discussed in the next Chapter). However, in this case the calculation is much more challenging and several crucial approximations must be made. The first approximation is to ignore the effect of the fermion fields, and to consider A^a_μ and ϕ^a as the only dynamical fields. The validity of this approximation will be discussed later. The second approximation is to replace the finite temperature "bounce" action (S_3) by the free energy at the maximum of the barrier between adjacent θ-vacua.

The maximum free energy for this transition is that of a static configuration called the *sphaleron*, given in the $A_0^a = 0$ gauge by

$$
\begin{aligned}
A_{\rm cl}^i &= \frac{i}{g}\frac{\varepsilon_{ijk}x_j\tau_k}{r^2} f(\xi) \\
\phi_{\rm cl}^a &= \frac{\sigma}{\sqrt{2}}\frac{i\vec{\tau}\cdot\vec{\mathbf{x}}}{r}\begin{bmatrix} 0 \\ 1 \end{bmatrix} h(\xi),
\end{aligned}
\tag{6.70}
$$

where $r^2 = |\vec{\mathbf{x}}|^2$ and .

$$
f(0) = h(0) = 0, \qquad f(\infty) = h(\infty) = 1 \tag{6.71}
$$

are functions of the dimensionless parameter $\xi = r/r_0 = rg\sigma$. The free energy of this configuration is given by

$$
\begin{aligned}
F &= \int d^3x \left[\frac{1}{4}F_{\mu\nu}^a F^{a\mu\nu} - D_\mu\phi^a D^\mu\phi^a + \frac{1}{4}\lambda\left(\phi^a\phi^a - \sigma^2/2\right)^2\right] \\
&\simeq \frac{2m_W}{\alpha_W},
\end{aligned}
\tag{6.72}
$$

where $m_W = g\sigma/2$. One can see that the sphaleron solution interpolates between a vacuum state ($|\phi|^2 = \sigma^2/2$) at $r \gg \sigma^{-1}$ and the top of the Higgs potential ($\phi^a = 0$) at $r \ll \sigma^{-1}$. The sphaleron solution is a saddle point in field configuration space: the lowest barrier between two θ-vacua. Moreover, as its name implies (*sphaleron*, Greek for ready to fall) the sphaleron configuration is classically unstable.

With the KRS assumption that the action for the tunnelling rate at finite temperature is given by F, the rate of vacuum transitions should be proportional to $\exp(-F/T)$. Heuristically, this corresponds to the probability of having a sufficiently large thermal fluctuation to take the fields over a barrier of height F. The rate at which the baryon number per comoving volume is violated should be proportional to this rate times the net baryon number per comoving volume, i.e.,

$$
\dot{B} = -cBT\exp(-F/T), \tag{6.73}
$$

where c is a dimensionless constant expected to be of order unity, and the overall factor of T is assumed on dimensional grounds. Since CP is conserved by the gauge and Higgs interactions in this model, a baryon asymmetry cannot be generated by the vacuum transitions. In the early

Universe g, λ, and σ (and hence m_W) are all functions of temperature. The temperature dependence of g and λ is only proportional to $\ln T$ and can be ignored. The temperature dependence of σ around the critical temperature is much more important. At $T \geq T_C$, $\sigma = 0$, and as the temperature passes below the critical temperature, $\sigma^2 \simeq \sigma_0^2(1 - T^2/T_C^2)$ ($\sigma_0 = 246$ GeV).[11]

At lower temperatures, $T \ll T_C$, $F/T \gg 1$ and the rate for sphaleron-induced B violation is exponentially suppressed. However as $T \rightarrow T_C$, $F/T \propto \sigma(T)/T \rightarrow 0$, and the rate for B violation is no longer exponentially suppressed. For a range of temperatures just below T_C, $\Gamma_{\Delta B}/H = (\dot{B}/B)/H \sim 10^{17}c$, corresponding to very rapid B violation.

The sphaleron solution has also been interpreted as a massive, unstable particle whose interactions violate B. With this interpretation we can immediately use the formalism developed earlier to write the Boltzmann equation for the evolution of B,

$$B' = -zK_{\text{sphaleron}}\gamma_B B + \varepsilon z K \gamma_D (X - X_{EQ}), \tag{6.74}$$

where $\varepsilon = 0$ (CP is conserved), and

$$\begin{aligned} z &= m_W(0)/T \\ K_{\text{sphaleron}} &\sim c m_{Pl}/m_W \sim 10^{17}c \\ \gamma_B &\sim \begin{cases} z/2 & (z \lesssim m_W(0)/T_C) \\ (\gamma z)^{3/2}\exp(-\gamma z) & (z \gtrsim m_W(0)/T_C). \end{cases} \end{aligned} \tag{6.75}$$

Here, $\gamma = F/m_W(0) = 4\pi\sigma(T)/g m_W(0)$, and $m_W(0)$ is the zero temperature W mass. Since $K_{\text{sphaleron}} \gg 1$, damping should be very significant, cf. (6.40), although a precise estimate awaits a careful calculation of c.

The potential impact of the KRS calculation certainly warrants more detailed calculations of the finite-temperature transition rate between θ-vacua. In particular, one approximation made by KRS that might be questioned is the neglect of the plasma effects on vacuum transitions. The sphaleron configurations (6.70) are spatially large. Their characteristic size is several σ^{-1}. Within such a volume there are 10^4 to 10^6 particles with weak charge. Since the gauge configuration is purely magnetic

[11] $F/T \propto \sigma(T)/T$ is formally equal to zero for $T \geq T_C$. However the sphaleron solution is not physically meaningful above the critical temperature since the size of the sphaleron (proportional to σ^{-1}) diverges as $\sigma \rightarrow 0$ for $T \geq T_C$. Thus, for $T \geq T_C$, the above formalism is no longer valid.

($A_0^a = 0$) and the plasma is not a magnetic conductor, there are no plasma screening effects for the *static* configuration. However, in the evolution to and from the static configuration (which must occur during transitions between different θ-vacua) there must be time-dependent gauge fields and A_μ^a must have an electric component which *will* be affected by electric screening. The problem of electric screening can be addressed by considering the analogous problem in electrodynamics; i.e., by calculating how long it takes to establish an ordinary magnetic field configuration of size σ^{-1} in a conducting plasma. By time-reversal invariance, the time it takes to establish such a field configuration is the same as it takes the field configuration to decay. In the limit that the conductivity of the Universe is infinite, the time is infinite: In a perfect conductor currents in the plasma do not decay. Of course, the Universe is not a perfect conductor, and the time for the establishment of the coherent field will be proportional to the conductivity. Plasma screening effects lead to a suppression of $\Gamma_{\Delta B}$ by powers of (m_W/T).

If sphaleron-mediated B violation is indeed very effective, then there are three possibilities for the origin of n_B/s: (i) avoidance of sphaleron effects due to a first-order electroweak phase transition in which the Universe becomes trapped in the $\phi^a = 0$ phase until a temperature $T \ll T_C$ where sphaleron effects are unimportant; (ii) production of the baryon asymmetry by sphaleron processes; or (iii) the survival of a baryon asymmetry produced at an earlier epoch as a non-thermalizing mode. A first-order phase transition is a viable possibility so long as the associated entropy production is not so great as to dilute the baryon asymmetry below $B \sim 10^{-10}$. Because sphaleron processes proceed so rapidly, $K_{\rm sphaleron} \gg 1$, possibility (ii) seems unlikely as the factor $f(K)$ will be very small. In addition, sphaleron processes naïvely conserve CP. As we will now describe, possibility (iii) seems the most plausible.

While $B+L$ is anomalous under $SU(2)_L \otimes U(1)_Y$, $B-L$ is not and is conserved. Specifically, sphalerons mediate processes like

$$\begin{aligned} \text{VACUUM} \Rightarrow \varepsilon_{ij}\varepsilon_{kl}\varepsilon_{\alpha\beta\gamma} \Big[& (u_L)_{\alpha i}(d_L)_{\beta j}(u_L)_{\gamma k}(e_L^-)_l \\ & +(c_L)_{\alpha i}(s_L)_{\beta j}(c_L)_{\gamma k}(\mu_L^-)_l + (t_L)_{\alpha i}(b_L)_{\beta j}(t_L)_{\gamma k}(\tau_L^-)_l \Big], \qquad (6.76) \end{aligned}$$

where i, j, k, l are $SU(2)_L$ indices and α, β, γ are $SU(3)_C$ indices. In a sphaleron transition a color-singlet, $SU(2)_L$-singlet neutral "hydrogen-like" object from each generation forms out of the vacuum. Thus, $B+L$ is violated by 6 units and B is violated by 3 units, while $B_i - L_i$ ($i =$

e, μ, τ, or 1, 2, 3) is conserved. Here B_1 is the baryon asymmetry in u and d quarks, etc., L_e is the lepton asymmetry in ν_e's and e^-'s, etc., and $B = B_1 + B_2 + B_3$, $L = L_e + L_\mu + L_\tau$. There are clearly three non-thermalizing modes: $B_i - L_i$. Denote the pre-existing baryon and lepton asymmetries by $(B_i)_{\rm initial}$ and $(L_i)_{\rm initial}$. Weak interaction processes ensure that $B_1 = B_2 = B_3 = B/3$; in the absence of neutrino masses, there are no off-diagonal lepton interactions.[12] Even with neutrino masses these processes may be too slow to ensure that $L_e = L_\mu = L_\tau$, so we will not assume that the lepton asymmetries are equal. In terms of the eigenmodes $B + L$ and $B/3 - L_i$, the baryon and lepton asymmetries are given by:

$$B = \frac{1}{2}(B+L) + \frac{1}{2}\sum_i (B/3 - L_i) \tag{6.77}$$

$$L_i = \frac{1}{6}(B+L) + \sum_{j\neq i}(B/3 - L_j) - \frac{5}{6}\sum_k (B/3 - L_k). \tag{6.78}$$

If there is strong $B + L$ damping by sphalerons, any pre-existing $B + L$ asymmetry will be erased, while the $(B/3 - L_i)$ asymmetries will survive. Thus the post-sphaleron B and L_i asymmetries will be:

$$B_{\rm final} = \frac{1}{2}\sum_i (B/3 - L_i)_{\rm initial} \tag{6.79}$$

$$(L_i)_{\rm final} = \sum_{j\neq i}(B/3 - L_j)_{\rm initial} - \frac{5}{6}\sum_k (B/3 - L_k)_{\rm initial}. \tag{6.80}$$

In the simple case where $L_i = L/3$, this becomes

$$B_{\rm final} = \frac{1}{2}(B - L)_{\rm initial}, \qquad (L_i)_{\rm final} = -\frac{1}{6}(B - L)_{\rm initial}. \tag{6.81}$$

It is clear that the potentially devastating effect of sphaleron-induced B-violation can be circumvented provided that a $B - L$ asymmetry is produced during an earlier epoch of baryogenesis. While $B - L$ is conserved in $SU(5)$ and a $B - L$ asymmetry cannot dynamically evolve, $B - L$ is violated in $SO(10)$, E_6, and many other candidate GUTs, and so it is very likely that any baryon asymmetry that evolves will have a non-zero projection on to $B - L$ which can survive sphaleron damping effects.

[12]That is, interactions that change lepton flavor; e.g., neutrino masses result in interactions like $\nu_e \leftrightarrow \nu_\mu$, etc., and whose strength is proportional to the neutrino mass.

6.9 Spontaneous Baryogenesis

Numerous alternatives to the standard out-of-equilibrium decay scenario have been suggested [11]; although many are interesting, all but one are just variations on the theme of out-of-equilibrium decay, as they all incorporate the basic ingredients, B, C, CP violation and non-equilibrium conditions. The exception, spontaneous baryogenesis [12], is quite novel; CP is not violated and the baryon asymmetry is produced *in equilibrium*! The key to this interesting scenario is a temporary, dynamical violation of CPT invariance.[13]

The crucial feature of this model is the following term in the Lagrangian,

$$\mathcal{L} = \cdots + \frac{1}{f}\partial_\mu \phi\, j_B^\mu \tag{6.82}$$

where j_B^μ is the baryon number current, ϕ is a scalar field, and f is an energy scale of the order of 10^{13} GeV or greater.[14] This term, when non-zero, violates CPT invariance. For simplicity, take ϕ to be spatially constant; then we have

$$\frac{1}{f}\partial_\mu \phi j_B^\mu = \frac{1}{f}\dot{\phi}(n_b - n_{\bar{b}}). \tag{6.83}$$

It is easy to see that this new term shifts the energy of a baryon relative to that of an antibaryon by an amount $2\dot{\phi}/f$; that is, it has the effect of giving baryons a chemical potential $\mu_b = -\dot{\phi}/f$ and antibaryons a chemical potential $\mu_{\bar{b}} = -\mu_b = \dot{\phi}/f$. Because of this fact, in thermal equilibrium there will be a non-zero baryon number density:

$$\begin{aligned} n_B &= n_b - n_{\bar{b}} = \frac{g_b T^3}{6\pi^2}\left[\pi^2 \frac{\mu_b}{T} + \left(\frac{\mu_b}{T}\right)^3\right] \\ &\simeq \frac{g_b \mu_b T^2}{6} \sim -\frac{\dot{\phi} T^2}{f} \end{aligned}$$

[13]In all previous discussions CPT invariance has been assumed. While it is generally believed that CPT invariance is a necessity to formulate a self-consistent quantum field theory, the violation invoked in this scenario is only dynamical.

[14]This is actually the effective Lagrangian, valid at energies and temperatures $T \lesssim f$. The ϕ field, called the *ilion* by the authors [12], is axion-like: the axion couples to the axial vector current, and the ilion couples to the baryon number current. The axion is discussed in Chapter 10.

$$n_B/s \sim -\frac{\dot\phi}{g_* fT}. \tag{6.84}$$

In this scenario, ϕ is initially displaced from its equilibrium point, i.e., where the first derivative of its potential vanishes; as it evolves toward that equilibrium point $\dot\phi \neq 0$, which leads to a chemical potential for baryons.[15] As it does, the equilibrium value of the net baryon number per comoving volume, $B = n_B/s \sim -\dot\phi/g_* fT$, changes. Eventually, when ϕ settles into the minimum of its potential and $\dot\phi = 0$, the equilibrium value of B becomes zero. However, B can only track its equilibrium value so long as B-violating interactions are occurring rapidly (rate $\Gamma \gtrsim H$). If B-violating interactions become ineffective before $\dot\phi = 0$, a non-zero value of B will freeze in, leaving the Universe with a permanent baryon asymmetry. If we denote the epoch when B-violating interactions freeze out by the subscript f, $\Gamma(T_f) \simeq H(T_f)$, then the residual baryon asymmetry is

$$B(T \lesssim T_f) \sim (\dot\phi/g_* fT)|_{T_f} \tag{6.85}$$

Through a temporary dynamical violation of CPT invariance, this novel scenario actually produces the baryon asymmetry in thermal equilibrium and without explicit CP violation.

6.10 Epilogue

The fact the Universe possesses a baryon asymmetry is one of its most fundamental and important features. Its existence prevents the annihilation catastrophe associated with locally-baryon-symmetric cosmologies; its value determines η and hence the outcome of primordial nucleosynthesis, and if baryons dominate the present mass density, it also determines the epoch of matter–radiation equality. In short, $n_B/s \simeq 10^{-10}$ is a number of tremendous cosmological significance.

Baryogenesis provides a very plausible and attractive scenario for the origin of this number based upon unified gauge theories and involving non-equilibrium interactions that violate B, C, CP, and took place around the GUT epoch ($t \sim 10^{-34}$ sec). In the baryogenesis scenario the baryon

[15] In Chapter 8 we will discuss in great detail the evolution of a scalar field in the expanding Universe; for now it suffices to say that the equation of motion is just $\ddot\phi + 3H\dot\phi + V'(\phi) = 0$, where $V(\phi)$ is the scalar potential for ϕ and $V' = \partial V/\partial\phi$. If $V' \neq 0$, ϕ will evolve.

asymmetry that evolves depends upon a statistical mechanics factor, $f(K)$, which is of order unity, times the magnitude of the C, CP violation (ε) divided by g_*. While baryon number violation appears to be a generic prediction of unified gauge theories, our understanding of C, CP violation is far less adequate (even in low-energy physics). Given the existence of C, CP violation in the K^0–$\bar{K}^0$ system, C, CP violation must be incorporated into any viable unified gauge theory—and as we saw this is easily accomplished with complex Higgs couplings. Attempts to relate the C, CP violation necessary for the baryon asymmetry to that observed in the neutral kaon system have been unsuccessful: Within the context of models where CP violation arises through a phase in the Kobayshi-Maskawa matrix,[16] not enough C, CP violation arises in X boson decays to explain the observed value of n_B/s, missing by 8 or so orders of magnitude [13].

If the baryon asymmetry of the Universe did indeed evolve dynamically, then there are a host of cosmological implications. For example, since its value only depends upon K and ε, n_B/s cannot, in the standard cosmology, be spatially-varying. This means that in the context of the standard cosmology baryogenesis precludes isocurvature baryon number fluctuations, and therefore favors curvature fluctuations.[17] In Chapter 4 we mentioned that primordial nucleosynthesis could be used to constrain the amount of entropy generation after nucleosynthesis since it provides a probe of the baryon-to-entropy ratio at that epoch. Likewise, baryogenesis can be used to constrain entropy production after the epoch when the baryon asymmetry is generated. Since the baryon asymmetry produced is at most of the order ε/g_*, and ε is of order α^N ($N \geq 1$), the baryon asymmetry produced is very likely to be much less than about 10^{-4}. Given that the observed value today is about 10^{-10}, we can conclude that the entropy in the Universe cannot have increased by more than a factor of order 10^6 since baryogenesis.

We have described the "standard scenario," out-of-equilibrium decay, in detail and mentioned some alternative mechanisms. Other, even more exotic possibilities have been suggested—production by primordial black holes that evaporate by the Hawking process, by the out-of-equilibrium decay of some species whose decays violate B, C, and CP, but at very low

[16]The Kobayshi-Maskawa, or K-M, matrix parameterizes the mixing of the quark mass eigenstates with the quark weak-interaction eigenstates.

[17]In Chapter 9 we will discuss density perturbations in great detail. If we allow ourselves to deviate from the standard cosmology, either by allowing large-amplitude, large-scale shear [14], or with an inflationary epoch and spontaneous baryogenesis [15], it is possible to produce isocurvature baryon number fluctuations.

temperature ($T \ll 10^{14}$ GeV), and by the temporary existence of a quark or lepton condensate [16].

With the baryogenesis scenario we have traded one piece of "initial data," $(n_B/s)_i$, for a very attractive, dynamical framework, which at present is incomplete. Moreover, if the Universe ever underwent inflation, or if rapid B-violating interactions were present without C, CP violation, one would not have the luxury of relying upon an initial baryon asymmetry to explain n_B/s: Baryogenesis would be mandatory! Once the theory that unifies quarks and leptons is at hand, the baryon asymmetry of the Universe can be calculated in the same way that the light element abundances have been. Until that time, baryogenesis provides a compelling argument that there is some kind of quark–lepton unification.

6.11 References

1. For a review of baryogenesis, see E. W. Kolb and M. S. Turner, *Ann. Rev. Nuc. Part. Sci.* **33**, 645 (1983).

2. For a recent review of proton-decay experiments, see D. Perkins, *Ann. Rev. Nucl. Part. Sci.* **34**, 1 (1984).

3. E. A. Bogomolov, et al., *Proc. of the 16th Intl. Cosmic Ray Conference* (Kyoto, 1979), Vol. 2, p. 330; R. L. Golden, et al., *Phys. Rev. Lett.* **43**, 1196 (1979); M. V. K. Apparao, *Can. J. Phys.* **46**, S654 (1968); A. Buffington, et al., *Ap. J.* **248**, 1179 (1981); S. P. Ahlen, et al., *Phys. Rev. Lett.* **61**, 145 (1988); R. Streitmatter, et al., (LEAP collaboration), in preparation (1989).

4. G. Steigman, *Ann. Rev. Astron. Astrophys.* **14**, 339 (1976).

5. A. D. Sakharov, *JETP Letters* **5**, 24 (1967).

6. E. W. Kolb and S. Wolfram, *Nucl. Phys.* **B172**, 224 (1980); J. N. Fry, K. A. Olive, and M. S. Turner, *Phys. Rev. D* **22**, 2953 (1980).

7. Boltzmann equations for $SU(5)$ have been derived by J. N. Fry, K. A. Olive, and M. S. Turner, *Phys. Rev. D* **22**, 2953 (1980); **22**, 2977 (1980), and for both $SU(5)$ and $SO(10)$ by J. A. Harvey, E. W. Kolb, D. B. Reiss, and S. Wolfram, *Nucl. Phys.* **B201**, 16 (1982).

8. R. A. Lyttleton and H. Bondi, *Proc. roy. Soc. (London)* **A252**, 313 (1959).

9. J. A. Harvey and E. W. Kolb. *Phys. Rev. D*, **24**, 2090 (1981).

10. V. A. Kuzmin, V. A. Rubakov, and M. E. Shaposhnikov, *Phys. Lett.* **155B**, 36 (1985).

11. Alternative baryogenesis scenarios involving primordial black holes have been considered by M. S. Turner, *Phys. Lett.* **B89**, 155 (1979); D. Lindley, *Mon. Mot. R. astron. Soc.* **196**, 317 (1981); A. D. Dolgov, *Zh. Eksp. Teor. Fiz.* **79**, 337 (1980); A. F. Grillo, *Phys. Lett.* **94B**, 364 (1980); and L. L. Krauss, *Phys. Rev. Lett.* **49**, 1459 (1982). Baryogenesis in supersymmetric unified models has been studied by E. W. Kolb and S. Raby, *Phys. Rev. D* **27**, 2990 (1983); J. N. Fry and M. S. Turner, *Phys. Lett.* **125B**, 379 (1983); H. E. Haber, *Phys. Rev. D* **26**, 1317 (1982). Models in which the baryon asymmetry evolves at very low temperature ($T \ll 10^{14}$ GeV) have been discussed by S. Dimopoulos and L. Hall, *Phys. Lett.* **196B**, 135 (1987); M. Claudson, L. Hall, and I. Hinchliffe, *Nucl. Phys.* **B241**, 309 (1984).

12. A. Cohen and D. Kaplan, *Phys. Lett.* **199B**, 251 (1987).

13. G. Segrè and M. S. Turner, *Phys. Lett.* **B99**, 339 (1981); J. A. Harvey, E. W. Kolb, D. B. Reiss, and S. Wolfram, *Phys. Rev. Lett.* **47**, 391 (1981).

14. J. D. Barrow and M. S. Turner, *Nature* **291**, 469 (1981); J. R. Bond, E. W. Kolb, and J. Silk, *Ap. J.* **255**, 341 (1982).

15. M. S. Turner, A. Cohen, and D. Kaplan, *Phys. Lett.* **216B**, 20 (1989).

16. I. Affleck and M. Dine, *Nucl. Phys.* **B249**, 361 (1985).

7

PHASE TRANSITIONS

7.1 High-Temperature Symmetry Restoration

One of the most important concepts in modern particle theory is that of spontaneous symmetry breaking (SSB). The idea that there are underlying symmetries of Nature that are not manifest in the structure of the vacuum appears to play a crucial role in the unification of the forces. In all unified gauge theories—including the standard model of particle physics—the underlying gauge symmetry is larger than that of our vacuum, whose symmetry is that of $SU(3)_C \otimes U(1)_{EM}$. Of particular interest for cosmology is the theoretical expectation that at high temperatures, symmetries that are spontaneously broken today were restored [1], and that during the evolution of the Universe there were phase transitions, perhaps many, associated with the spontaneous breakdown of gauge (and perhaps global) symmetries. In particular, we can be reasonably confident that there was such a phase transistion at a temperature of order 300 GeV and a time of order 10^{-11} sec, associated with the breakdown of $SU(2)_L \otimes U(1)_Y \rightarrow U(1)_{EM}$. Moreover, the vacuum structure in many spontaneously broken gauge theories is very rich: Topologically stable configurations of gauge and Higgs fields exist as domain walls, cosmic strings, and monopoles. In addition, classical configurations that are not topologically stable, so-called nontopological solitons, may exist and be stable for dynamical reasons. Interesting examples include soliton stars, Q-balls, nontopological cosmic strings, and so on [2].

The cosmological production, and subsequent implications, of such topological defects will occupy much of this Chapter. The possibility that the Universe undergoes inflation during a phase transition will be the subject of the next Chapter. Before discussing topological defects and their production in cosmological phase transitions, we will review some general

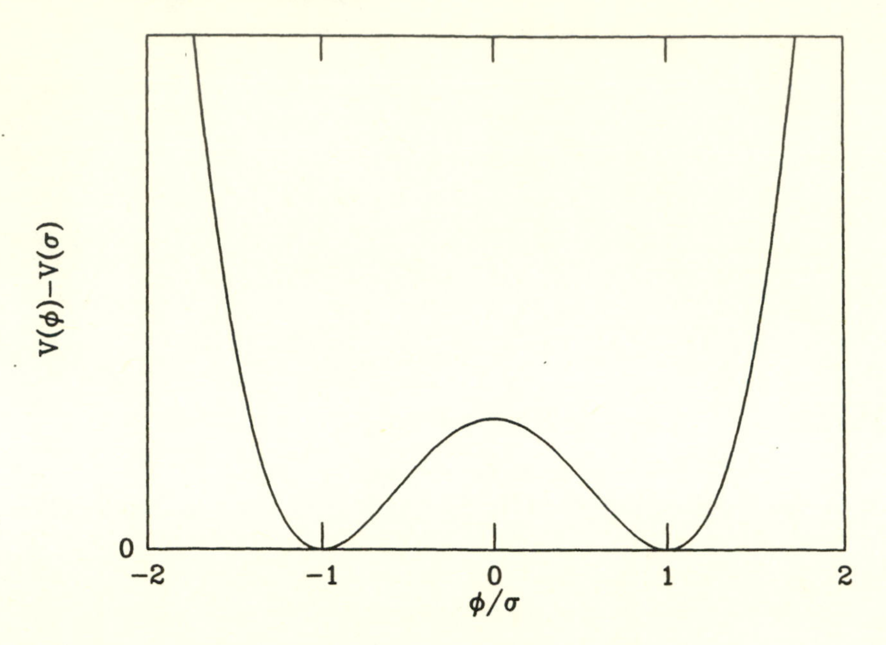

Fig. 7.1: An example of a potential that implements SSB.

aspects of high-temperature symmetry restoration and phase transitions.

7.1.1 A simple model

To illustrate some of the features of finite temperature effects, consider a real scalar field described by the Lagrangian density

$$\begin{aligned}\mathcal{L} &= \frac{1}{2}\partial_\mu\phi\partial^\mu\phi - V(\phi),\\ V(\phi) &= -\frac{1}{2}m^2\phi^2 + \frac{1}{4}\lambda\phi^4.\end{aligned} \tag{7.1}$$

The potential $V(\phi)$ is shown in Fig. 7.1. Note that the Lagrangian is invariant under the discrete symmetry transformation $\phi \leftrightarrow -\phi$. The minima of the potential (determined by the conditions $V' \equiv \partial V/\partial\phi = 0$ and $V'' \equiv \partial^2 V/\partial\phi^2 > 0$), denoted by $\sigma_\pm$, and the value of the potential and its second derivative at the minima, are given by

$$\sigma_\pm = \pm\sqrt{\frac{m^2}{\lambda}},$$

$$V(\sigma_\pm) = -\frac{m^4}{4\lambda},$$
$$V''(\sigma_\pm) = 2m^2. \qquad (7.2)$$

Since $V(+(m^2/\lambda)^{1/2}) = V(-(m^2/\lambda)^{1/2})$, both $\sigma_\pm = \pm(m^2/\lambda)^{1/2}$ are equivalent minima of the potential. On the other hand, although $V'(0) = 0$, $\langle\phi\rangle = 0$ is an unstable extremum of the potential because $V''(0) < 0$. Since the quantum theory must be constructed about a stable extremum of the classical potential, the ground state of the system is either $\langle\phi\rangle = +(m^2/\lambda)^{1/2}$ or $\langle\phi\rangle = -(m^2/\lambda)^{1/2}$, and the reflection symmetry $\phi \leftrightarrow -\phi$ present in the Lagrangian is broken by the choice of a vacuum state.[1] A symmetry of the Lagrangian not respected by the vacuum is said to be spontaneously broken. The mass of the physical boson of the theory is determined by the curvature of the potential about the true ground state: $M^2 = V''(\sigma_\pm) = 2m^2 = 2\lambda\sigma_\pm^2$.

The stress tensor for a scalar field ϕ is given by

$$T_{\mu\nu} = \partial_\mu\phi\partial_\nu\phi - \mathcal{L}g_{\mu\nu}, \qquad (7.3)$$

Taking ϕ to be constant, $\phi = \langle\phi\rangle$, we find that $T^{\mu\nu} = V(\langle\phi\rangle)g^{\mu\nu}$, so that the energy density of the vacuum is

$$\langle T^0{}_0\rangle \equiv \rho_V = -\frac{m^4}{4\lambda}. \qquad (7.4)$$

The contribution of the vacuum energy to the energy density of the Universe today can at most be comparable to the critical density $\rho_C = 1.88 \times 10^{-29}h^2$ g cm$^{-3} \simeq 10^{-46}$ GeV^4. A larger vacuum energy would lead to a present expansion rate greater than that observed. Since the vacuum energy must be so incredibly small relative to any other fundamental energy scale, it is tempting to require $\rho_V = 0$. This can be accomplished by adding the constant $+m^4/4\lambda = \lambda\langle\phi\rangle^4/4$ to the Lagrangian. This constant term will not affect the equations of motion or the quantum theory; its sole effect is to make the present vacuum energy vanish. By adding this constant, we can write the potential in the form

$$V(\phi) = \frac{\lambda}{4}(\phi^2 - \sigma^2)^2, \qquad \sigma^2 = \sigma_\pm^2. \qquad (7.5)$$

[1]Here and throughout $\langle\cdots\rangle$ denotes the vacuum expectation value of the quantity, e.g., $\langle\phi\rangle$ = vacuum expectation value of the field ϕ.

We should note that at present there is no physical principle that requires ρ_V to be so small. The apparent minuteness of the present value of the vacuum energy remains an unexplained puzzle.

The phenomenon of high-temperature symmetry restoration can be understood in several ways. A physical approach is to consider the effect of finite temperature (or density) upon the propagation of a particle. If one naïvely attempts to construct a theory about $\langle\phi\rangle = 0$ using the potential of (7.5), one finds that the mass squared of the scalar field is negative: $V''(0) = -\lambda\sigma^2$. The imaginary mass results in exponentially growing solutions, with ϕ growing until it finds the true ground state. However, if the ϕ field is in contact with a heat bath, the interaction of ϕ particles with particles in the thermal bath will, in general, damp this exponential growth. A way to quantify this damping is to assign to the ϕ a temperature-dependent "plasma mass," which on dimensional grounds must be of the form $m^2_{\text{plasma}} = a\lambda T^2$, where a is a numerical constant of order unity. At finite temperature the effective mass of the scalar field about the classical solution $\langle\phi\rangle = 0$ is $m_T^2 = -\lambda\sigma^2 + m^2_{\text{plasma}}$. At temperatures where $m_T^2 < 0$, $\langle\phi\rangle = 0$ will be an unstable point, signalling SSB; while at temperatures where $m_T^2 \geq 0$, the effective mass will be real, and $\langle\phi\rangle = 0$ becomes a stable, classical minimum of the potential. Clearly, there is some critical temperature, $T_C \simeq \sigma/a^{1/2}$, where $m_T^2 = 0$; above this critical temperature $\langle\phi\rangle = 0$ is a stable minimum and the symmetry is restored.

The rigorous approach to symmetry restoration is to account for the effect of the ambient background gas in the calculation of higher-order quantum corrections to the classical potential [3].[2] The *finite-temperature, effective potential* is the free-energy density associated with the ϕ field, $V_T(\phi_c) = \rho_\phi - Ts_\phi$. From $V_T(\phi_c)$, one can compute the energy density: $\rho_\phi = V_T(\phi_c) + Ts_\phi$, where $s_\phi = -\partial V_T(\phi_c)/\partial T$ and the pressure $p_\phi = -V_T(\phi_c)$. To one loop in these quantum corrections, the full potential is given by [3]

$$V_T(\phi_c) = V(\phi_c) + \frac{T^4}{2\pi^2}\int_0^\infty dx\, x^2 \ln\left[1 - \exp[-(x^2 + M^2/T^2)^{1/2}]\right], \quad (7.6)$$

where $V(\phi_c)$ is the zero-temperature, one-loop potential which is a function of the classical field ϕ_c, and $M^2(\phi_c) = -m^2 + 3\lambda\phi_c^2$. The zero-temperature

[2] To be completely accurate and technically correct, we use ϕ_c to denote the classical value of the field ϕ.

one-loop effective potential is[3]

$$V(\phi_c) = -\frac{1}{2}m^2\phi_c^2 + \frac{1}{4}\lambda\phi_c^4 + \frac{1}{64\pi^2}M^4 \ln\left(\frac{M^2}{\mu^2}\right), \tag{7.7}$$

where μ is an arbitrary mass scale which can be related to the renormalized coupling constants, and $M(\phi_c)$ has been defined above.

At high temperatures ($T \gg T_C$), $V_T(\phi_c)$ can be expanded in T:

$$V_T(\phi_c) = V(\phi_c) + \frac{\lambda}{8}T^2\phi_c^2 - \frac{\pi^2}{90}T^4 + \cdots, \tag{7.8}$$

$$\rho_\phi = V(\phi_c) - \frac{\lambda}{8}T^2\phi_c^2 + \frac{\pi^2}{30}T^4 + \cdots. \tag{7.9}$$

We recognize the $-\pi^2T^4/90$ term in (7.8) as the free-energy (or minus the pressure) of a massless spin-0 boson, and the $\pi^2T^4/30$ term in (7.9) as the energy density of a spin-0 boson; both terms arise due to the thermal bath of ϕ particles. Although not usually done, one might refer to the ϕ_c-dependent terms in (7.9) as the "temperature-corrected" potential [not to be confused with the finite-temperature effective potential $V_T(\phi_c)$]. We see that $\partial^2V/\partial\phi_c^2$, evaluated at $\phi_c = 0$, changes sign at a temperature $T_C \simeq 2m/\lambda^{1/2} = 2\sigma$. Below T_C the extremum at $\phi_c = 0$ is a local maximum, and above T_C it is the global minimum, so that $\phi_c = 0$ is the ground state. At temperatures above T_C the reflection symmetry is restored since $\phi_c = 0$ is the ground state.

The phase transition from the symmetric phase to the broken phase in this model is second order. In general, a symmetry-breaking phase transition can be first or second order. The temperature dependence of $V_T(\phi_c)$ for a first-order phase transition is shown in Fig. 7.2. For $T \gg T_C$ the potential is quadratic, with only one minimum at $\phi_c = 0$. When $T = T_1$, a local miminum develops at $\phi_c \neq 0$. For $T = T_C$, the two minima become degenerate, and below T_C, the $\phi_c \neq 0$ minimum becomes the global minimum. If for $T \leq T_C$ the extremum at $\phi_c = 0$ remains a local minimum, there must be a barrier between the minima at $\phi_c = 0$ and $\phi_c \neq 0$. Therefore, the change in ϕ_c in going from one phase to the other must be discontinuous, indicating a first-order phase transition.

[3]The only interactions in the theory we are considering are the self interactions of the ϕ particles that arise from the $\lambda\phi^4$ term. If the ϕ field interacts with other fields there will be additional one-loop corrections to the potential.

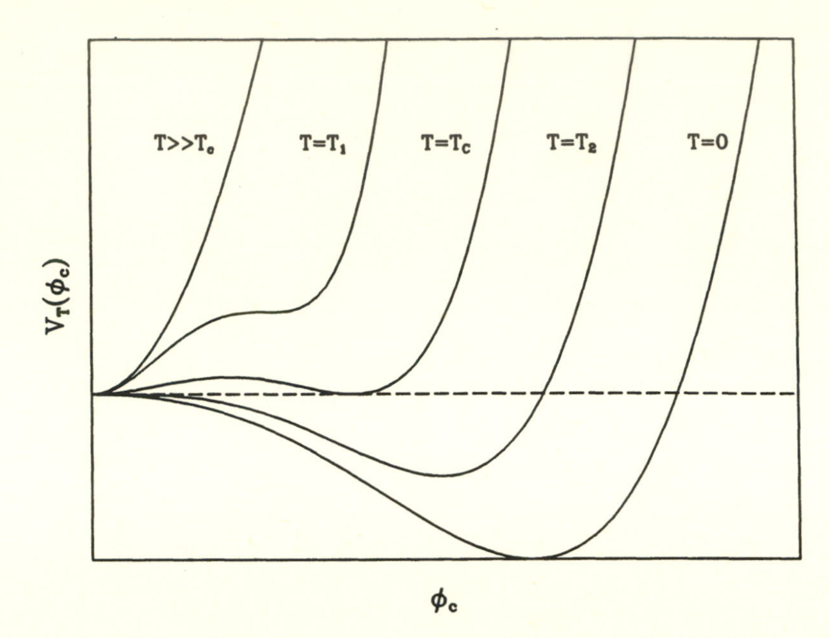

Fig. 7.2: The temperature dependence of $V_T(\phi_c)$ for a first-order phase transition. Only the ϕ_c-dependent terms in $V_T(\phi_c)$ are shown.

Moreover, the transition cannot take place classically, but must proceed either through quantum or thermal tunnelling.[4] Finally, when $T = T_2$ the barrier disappears and the transition may proceed classically. For a second-order transition there is no barrier at the critical temperature, and the transition occurs smoothly.

7.1.2 Tunnelling

In a cosmological phase transition ϕ evolves from the high-temperature, symmetric minimum $\phi = 0$, to the low-temperature, symmetry-breaking minimum $\phi = \sigma$. In the case of a first-order transition there is a potential barrier separating the two minima. In this Subsection we will discuss how the barrier is overcome through tunnelling.

Let us first review the dynamics of quantum tunnelling in the semiclassical theory as developed by Coleman [4]. In so doing, we will illustrate tunnelling with an example of relevance for cosmological phase transitions,

[4]We use the term *thermal tunnelling* to refer to tunnelling at finite temperature, and the term *quantum tunnelling* to refer to tunnelling at zero temperature. Both involve the nucleation of bubbles of the true vacuum which then expand at the speed of light.

namely a potential $V(\phi)$ with a local "metastable" minimum at $\phi = 0$ separated from the global "true" minimum at $\phi = \sigma$ by a barrier. A simple potential with the desired properties is[5]

$$
\begin{aligned}
V(\phi) &= \frac{\lambda}{4}\phi^2(\phi - \phi_0)^2 - \frac{\lambda}{2}\epsilon\phi_0\phi^3 \\
\sigma &= \phi_0\left[\frac{3(1+\epsilon) + \sqrt{9(1+\epsilon)^2 - 8}}{4}\right],
\end{aligned}
\tag{7.10}
$$

shown in Fig. 7.3. Note that as $\epsilon \to 0$, the global minimum σ approaches ϕ_0, and the minima at $\phi = 0$ and $\phi = \sigma$ become degenerate. While the methods we discuss here are of general applicability, for the purpose of pedagogy, we will illustrate them with the specific example of potential (7.10).

Transition to the true vacuum state by quantum tunnelling occurs through the nucleation of bubbles of the energetically favored asymmetric phase ($\phi = \sigma$), which then expand outward at the speed of light [4]. The first step in the calculation of the probability for bubble nucleation is solving the *Euclidean* equation of motion for $\phi(t_E, \vec{x})$[6]

$$
\Box_E\phi - V'(\phi) = \frac{d^2\phi}{dt_E^2} + \nabla^2\phi - V'(\phi) = 0 \tag{7.11}
$$

where as usual, $V'(\phi) \equiv dV/d\phi$. The boundary conditions to be imposed upon the solution are: $d\phi(0,\vec{x})/dt_E = 0$; and $\phi(\pm\infty,\vec{x})$ equal to its metastable minimum ($\phi = 0$ for our example). This is referred to as the "bounce" solution, because ϕ starts from the metastable minimum at $t_E = -\infty$, comes to rest at some value of ϕ near the global minimum at $t_E = 0$, and bounces back to the starting point at $t_E = +\infty$.

The probability of bubble nucleation per unit volume per unit time is

---

[5]Note that for simplicity we have not adjusted the potential to give $V = 0$ at the global minimum. In this Subsection we only discuss tunnelling in Minkowski space where the zero of the potential is irrelevant. When gravitational effects are taken into account, the tunnelling rate depends upon the absolute value of the metastable vacuum energy. The corrections to the Minkowski space results are of order V/m_{Pl}^4—and therefore generally small; for details, see [5].

[6]That is, $t_E = it$. One can think of the tunnelling rate as being associated with a classical motion in imaginary time because the decay rate is related to the imaginary part of the energy.

then given by

$$\Gamma = A\exp(-S_E) \tag{7.12}$$

where S_E is the Euclidean action for the solution of (7.11):

$$S_E(\phi) = \int d^3x dt_E \left[\frac{1}{2}\left(\frac{d\phi}{dt_E}\right)^2 + \frac{1}{2}(\vec{\nabla}\phi)^2 + V(\phi)\right]. \tag{7.13}$$

The calculation of the pre-exponential factor A is in general quite complicated:

$$A = \left(\frac{S_E(\phi)}{2\pi}\right)^2 \left(\frac{\det'[-\Box_E + V''(\phi)]}{\det[-\Box_E + V''(0)]}\right)^{-1/2}, \tag{7.14}$$

where prime indicates that zero eigenvalues of the operator are to be omitted when computing the determinant.[7] Luckily, for most applications an estimate for A based on dimensional grounds will suffice—all the "action" is in the exponential.

All possible solutions to (7.11) contribute to the tunnelling rate; and the one with least action makes the largest contribution to the tunnelling rate. In flat space, at zero temperature, the least-action Euclidean solution has $O(4)$ symmetry, in which case ϕ is only a function of r, where $r^2 = t_E^2 + |\vec{x}|^2$, and the $O(4)$-Euclidean equation of motion for $\phi(r)$ is:

$$\frac{d^2\phi}{dr^2} + \frac{3}{r}\frac{d\phi}{dr} - V'(\phi) = 0. \tag{7.15}$$

The boundary conditions for the $O(4)$ solution are: $d\phi/dr = 0$ at $r = 0$; and $\phi = 0$ at $r = +\infty$. The value of ϕ at $r = 0$, denoted as ϕ_e, is implicitly determined by the other two boundary conditions, and is called the "escape point." Starting with $\phi = \phi_e$ and $d\phi/dr = 0$ at $t_E = 0$, ϕ will evolve to the value $\phi = 0$ at $t_E = \infty$. In general, ϕ_e is somewhere between $\phi = 0$ and $\phi = \sigma$.

The operational method used to find the escape point can be illustrated with the potential of (7.10). If one imagines ϕ to be the "position" of a particle and r to be the "time" variable, (7.15) describes the classical motion of a particle under the influence of the inverted potential $-V(\phi)$ with a velocity-dependent "friction" force. If ϕ starts at rest from the

[7]In general there are 4 zero modes associated with the translational invariance of Euclidean space. The factor $(S_E(\phi)/2\pi)^2$ is the result of a product of four $(S_E(\phi)/2\pi)^{1/2}$ factors, one each from the four zero modes.

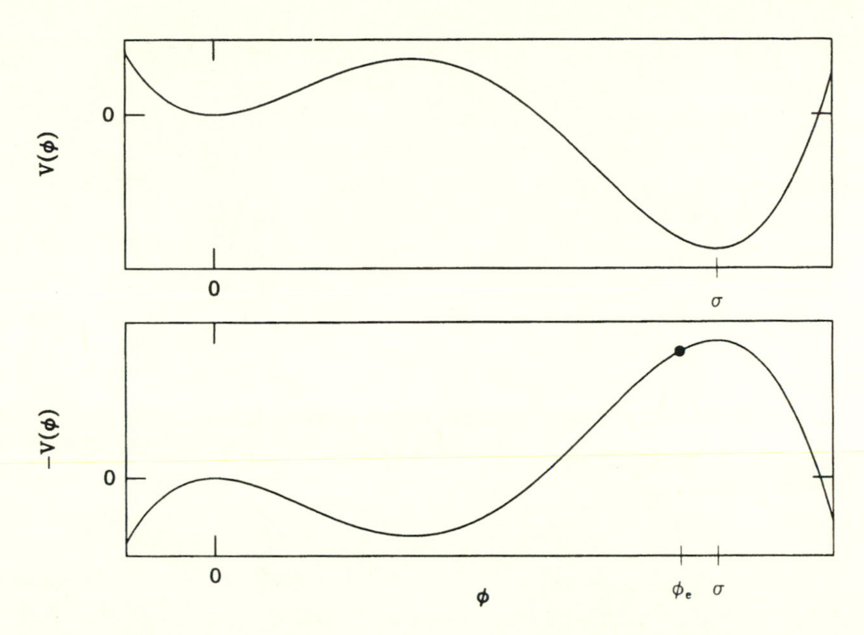

Fig. 7.3: Graph of potential (7.10) used to illustrate quantum tunnelling. The $O(4)$-Euclidean equation of motion (7.15) describes the evolution of a "particle" moving under the influence of the "inverted" potential $-V$ subject to a time-dependent friction force. A particle released from the escape point $\phi = \phi_e$ will overcome the friction and roll to rest at $\phi = 0$ at infinite Euclidean time.

escape point, it will have just enough energy to overcome the friction force and come to rest at $\phi = 0$ at infinite r. If ϕ is released at rest to the left of the escape point, it will "undershoot" the desired final configuration—after some finite r, ϕ will come to rest and then reverse direction before reaching $\phi = 0$. If ϕ is released at rest to the right of the escape point, it will "overshoot" the desired final configuration—after some finite r it will pass $\phi = 0$ and continue toward $\phi = -\infty$. Once the escape point is identified by trial, and mostly error, (7.15) can be solved for $\phi(r)$. Once the solution $\phi(r)$ is known, the Euclidean action is obtained from (7.13) by a simple change of integration variables:

$$S_E = 2\pi^2 \int_0^\infty r^3 dr \left[\frac{1}{2} \left(\frac{d\phi}{dr} \right)^2 + V(\phi) \right] . \tag{7.16}$$

How is one to interpret the Euclidean solution $\phi(r)$? For simplicity, suppose that the transition of $\phi(r)$ from $\phi(r = 0) = \phi_e$ to $\phi(r = \infty) = 0$

occurs suddenly at $r = R$, so that very crudely: $\phi(r) \simeq \phi_e$ for $r \lesssim R$, and $\phi(r) \simeq 0$ for $r \gtrsim R$. (We will shortly give a specific example of just such a solution.) When $\phi(r)$ is expressed in terms of Lorentzian time, $\phi(r) = \phi[(|\vec{x}|^2 - t^2)^{1/2}]$, one can immediately glean the Minkowski-space interpretation of the $O(4)$-Euclidean solution: A spherical bubble of radius R, within which $\phi = \phi_e$, is nucleated at time $t_E = t = 0$, and expands outward at the speed of light.[8] The equation of motion for the surface of the bubble is $(|\vec{x}|^2 - t^2) = R^2$; or $|\vec{x}_{\rm surface}| = (t^2 + R^2)^{1/2}$; as $t \to \infty$, $d|\vec{x}_{\rm surface}|/dt \to 1$. Outside the bubble, i.e., $|\vec{x}| \gtrsim |\vec{x}_{\rm surface}|$, space is still in the false vacuum: $\phi = 0$.[9]

In general, a closed-form analytic solution to equation (7.15) cannot be found. However in the "thin-wall" approximation, where the difference in energy between the metastable and true vacua are small compared to the height of the barrier, it is possible to find a simple, approximate analytic expression for S_E. In the thin wall limit ($\epsilon \to 0$ in our example), the minima are nearly degenerate. In order to overcome the friction term, ϕ_e must be very close to σ. The form of the solution can be understood by again using the particle analogy: For $\epsilon \ll 1$, the inverted potential is very flat near the escape point, and the particle sits near $\phi = \sigma$ for a very long time until it builds up "speed." Then, at some large value of r, say $r \sim R$, the particle rapidly makes the transition across the "dip" in $-V(\phi)$, and finally slowly comes to rest at $\phi = 0$. Recall that $\phi(r)$ corresponds to the bubble profile, and R to the bubble radius.

Because the transition from $\phi \sim \sigma$ to $\phi \sim 0$ occurs over a small interval of r, the thickness of the bubble is small—hence the name "thin wall." To estimate the thickness of the wall we can solve (7.15) for $r \gtrsim R$. Since it takes a long "time" (large r) before things start rolling, the friction term, which is proportional to r^{-1}, may be neglected in this limit, and the equation of motion for $r \sim R$ is quite simple:

$$d^2\phi/dr^2 = V'(\phi) \simeq V_0'(\phi), \tag{7.17}$$

where V_0 is the potential in the limit of exact degeneracy of the minima

[8] The evolution of ϕ beyond the escape point $\phi = \phi_e$ is classical. The configuration $\phi = \phi_e$ is classically unstable, and ϕ simply "rolls down" the potential $V(\phi)$ to $\phi = \sigma$ in a time of order σ^{-1}. We will return to the classical evolution of a scalar field displaced from its minimum in Chapter 8.

[9] Had we identified the nucleation event with some time $t < 0$, then the bubble would first contract to size R, and then expand—an unphysical solution. Identification of the nucleation event with some time $t > 0$, corresponds to the nucleation of a bubble of radius greater than R; R, then, is the radius of the smallest possible bubble.

($\epsilon = 0$ in our example). The solution to the equation of motion ignoring the damping term is

$$\frac{d\phi}{dr} = -\sqrt{2V_0(\phi)} \qquad r = \int^{\phi} \frac{d\phi'}{\sqrt{2V_0(\phi')}} . \tag{7.18}$$

For potential (7.10), the solution to (7.18) is

$$\phi^{TW}(r) = \frac{1}{2}\phi_0 \left[1 - \tanh \frac{r - R}{\Delta}\right], \tag{7.19}$$

where the thin-wall (TW) solution is only valid for $r \sim R$ (where $0 \ll \phi \ll \phi_0$), and the bubble-wall thickness is $\Delta = \phi_0^{-1}\sqrt{8/\lambda}$. As we shall see shortly, the bubble radius is $R = (\epsilon\phi_0\sqrt{2\lambda}\,)^{-1}$, so that $\Delta/R = 2\epsilon/\lambda$. Hence the term "thin wall" is well justified provided that $\epsilon \ll 1$.

We now have a complete, albeit approximate, form for the thin-wall bounce solution:

$$\phi(r) = \begin{cases} \phi_0 & r \ll R \\ \phi^{TW}(r) & r \simeq R \\ 0 & r \gg R\,. \end{cases} \tag{7.20}$$

Examples of thin-wall and thick-wall solutions are shown in Fig. 7.4.

The Euclidean action is now straightforward to compute in terms of R:

$$\begin{aligned} S_E &= 2\pi^2 \int_0^\infty r^3 dr \left[\frac{1}{2}\left(\frac{d\phi}{dr}\right)^2 + V(\phi)\right] \\ &= -2\pi^2 \frac{R^4}{4}\Delta V + 2\pi^2 R^3 \int_0^{\phi_0} d\phi\sqrt{2V_0(\phi)} \\ &= -\frac{1}{4}\pi^2 R^4 \lambda\epsilon\phi_0^4 + \frac{1}{6}\pi^2 R^3 \sqrt{\lambda}\,\phi_0^3, \end{aligned} \tag{7.21}$$

where ΔV is the *magnitude* of the difference in energy density between the two minima ($\Delta V > 0$); in our example $\Delta V = \epsilon\lambda\phi_0^4/2$. The first line in (7.21) is true in general; the second line is true in the thin-wall approximation; and the third-line is specific to our example here. The first term in the second or third lines is "vacuum energy" associated with the period where $r < R$, $\phi \sim \sigma$, and $V(\phi) = -\Delta V$; while the second term is the "gradient energy" associated with the bubble wall.

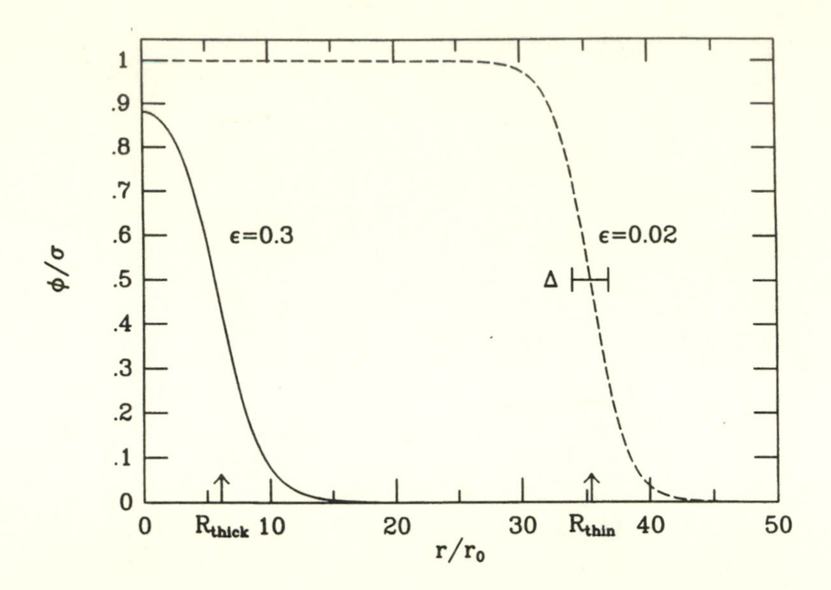

Fig. 7.4: The $O(4)$-Euclidean solution $\phi(r)$ for a thin-wall (dashed line) and thick-wall (solid line) bubble; here $r_0 = \phi_0^{-1}/\sqrt{\lambda}$. The solution corresponds to the bubble profile: $\phi \sim \phi_e$ (true vacuum) inside the bubble, and $\phi \sim 0$ (false vacuum) outside the bubble. Note that for the thin-wall bubble, the wall thickness (Δ) is much less than the bubble radius.

The final step in the evaluation of the Euclidean action is the determination of R by the minimization of S_E ($dS_E/dR = 0$):

$$R = 3(\Delta V)^{-1} \int_0^{\phi_0} d\phi \sqrt{2V(\phi)}. \qquad (7.22)$$

For our example, $R = (\sqrt{2\lambda}\,\epsilon\phi_0)^{-1}$. The final expression for S_E in the thin-wall limit is

$$S_E = \frac{27\pi^2 \left[\int_0^{\phi_0} d\phi \sqrt{2V_0(\phi)}\right]^4}{2(\Delta V)^3}. \qquad (7.23)$$

For our example, $S_E = \pi^2/48\lambda\epsilon^3$. Note that for our example S_E only depends upon the dimensionless parameters λ and ϵ, and that in the thin-wall limit, i.e., $\epsilon \ll 1$, the bubble action S_E is much larger than one—as one would expect since the energy difference between the two minima is very small.

The tunnelling rate at finite temperature is computed by following the same procedure above, remembering that field theory at finite temperature is equivalent to Euclidean field theory, periodic in imaginary time with period T^{-1} [6]. Thus, the finite-temperature tunnelling rate is found by solving (7.11) subject to the additional condition that $\phi(t_E, \vec{\mathbf{x}}) = \phi(t_E + T^{-1}, \vec{\mathbf{x}})$, and computing S_E according to (7.13). The periodicity in the Euclidean time variable t_E reduces the integral for S_E to

$$\begin{aligned} S_E &= S_3/T \\ S_3 &= \int d^3x \left[\frac{1}{2}(\nabla^2\phi)^2 + V_T(\phi)\right], \end{aligned} \tag{7.24}$$

where $V_T(\phi)$ is the temperature-dependent potential discussed in the previous Subsection.

At sufficiently high temperatures, the least-action solution has $O(3)$ symmetry [6], and (7.11) becomes

$$\frac{d^2\phi}{ds^2} + \frac{2}{s}\frac{d\phi}{ds} - V_T'(\phi) = 0, \tag{7.25}$$

where $s^2 = \vec{\mathbf{x}}^2$. The solution must satisfy the boundary conditions: $\phi(s = \infty)$ is the metastable minimum, and $d\phi/ds$ evaluated at $s = 0$ vanishes.

After a simple change of integration variables, the Euclidean action for the $O(3)$-invariant solution is

$$S_3 = 4\pi \int s^2 ds \left[\frac{1}{2}\left(\frac{d\phi}{ds}\right)^2 + V_T(\phi)\right]. \tag{7.26}$$

As in the zero-temperature case, the "bounce" solution is straightforward to obtain numerically. As before, the prefactor A is of less importance, and more difficult to compute.

The connection between zero-temperature and finite-temperature tunnelling is manifest: At temperatures much less than the inverse radius of the bubble—a scale that is set by the potential, $R^{-1} \sim \phi_0$ in our example—the T^{-1} periodicity of t_E becomes irrelevant as $T^{-1} \to \infty$, and the $O(4)$ solution has the least action. In the opposite limit, at very high temperatures, the periodicity of t_E becomes very relevant, the $O(3)$ solution has the least action, and $S_E \to S_3/T$.

7.1.3 Electroweak symmetry breaking

Let us illustrate the dynamics of a SSB phase transition with a worked example—the phase transition associated with the breaking of the electroweak symmetry. In this case the symmetry of the theory is an $SU(2)_L \otimes U(1)_Y$ gauge symmetry. In the electroweak theory the full gauge symmetry is broken down to $U(1)_{EM}$ by a complex $SU(2)_L$ doublet of scalar fields Φ, with potential $V(\Phi) = -m^2\Phi^\dagger\Phi + \lambda(\Phi^\dagger\Phi)^2$. The complex doublet can be expressed in terms of four real scalar fields:

$$\Phi = \frac{1}{\sqrt{2}} \begin{pmatrix} \phi_1 + i\phi_2 \\ \phi + i\phi_3 \end{pmatrix}. \tag{7.27}$$

Note that $\Phi^\dagger\Phi = (\phi_1^2 + \phi_2^2 + \phi^2 + \phi_3^2)/2$. Since the potential depends only upon $\Phi^\dagger\Phi$, the "direction" of the mimimum in Φ-space is undetermined, and its choice arbitrary. Following standard conventions, we pick the direction of $\langle\Phi\rangle$ to lie along the real part of the neutral component of Φ, i.e., along the ϕ direction. The potential for ϕ is now exactly the same as the potential of (7.1), and the vacuum expectation value for ϕ is $\langle\phi\rangle^2 = m^2/\lambda = \sigma^2$. Electroweak SSB gives masses to the W and Z gauge bosons as well as to fermions:

$$\begin{aligned} M_W^2 &= \frac{1}{4} g^2 \sigma^2, \\ M_Z^2 &= \frac{1}{4}(g^2 + g'^2)\sigma^2 = \frac{M_W^2}{\cos^2\theta_W}, \\ M_f^2 &= \frac{1}{2} h_f^2 \sigma^2, \end{aligned} \tag{7.28}$$

where g and g' are the gauge coupling constants of $SU(2)_L$ and $U(1)_Y$, and h_f is the Yukawa coupling of fermion f to the scalar doublet Φ, $\mathcal{L}_f = -h_f \bar{\Psi}_f \Phi \Psi_f$. The weak-mixing, or Weinberg, angle is defined by $g'/g = \tan\theta_W$ ($\sin^2\theta_W = 0.229 \pm 0.004$; $e = g\sin\theta_W$). As before, the mass of the physical scalar, or Higgs, particle is (at tree level) $\sqrt{2}m$. The value of the vacuum expectation value σ is determined from the effective four-Fermi interaction: $G_F/\sqrt{2} = g^2/8M_W^2 = 1/2\sigma^2$, which gives $\sigma = 246$ GeV. Since the top-quark mass and the Higgs mass are still undetermined, h_t and λ are as yet unknown.

Because the Higgs couples to both gauge bosons and fermions, there are additional contributions to the one-loop effective potential. At zero temperature, the one-loop effective potential is

$$
\begin{aligned}
V(\phi_c) &= -\frac{1}{2}m^2\phi_c^2 + \frac{1}{4}\lambda\phi_c^4 \\
&\quad + \frac{1}{64\pi^2}(-m^2 + 3\lambda\phi_c^2)^2 \ln\left(\frac{-m^2 + 3\lambda\phi_c^2}{\mu^2}\right) \\
&\quad + \frac{3}{1024\pi^2}[2g^4 + (g^2 + g'^2)^2]\phi_c^4 \ln\left(\frac{\phi_c^2}{\mu^2}\right) \\
&\quad - \frac{3}{64\pi^2}h_t^4\phi_c^4 \ln\left(\frac{\phi_c^2}{\mu^2}\right),
\end{aligned}
\tag{7.29}
$$

where, for simplicity, we have included the fermion loop corrections only for the fermion with the largest Yukawa coupling (assumed to be the top quark). The origin of the terms in (7.29) is manifest: The first line is the tree-level potential; the second line is the one-loop potential due to scalar loops [cf. (7.7)]; the penultimate line is from gauge boson loops; and the last line is the one-loop contribution from the top quark. Note that the sign of the fermion contribution to the one-loop potential is opposite to that of the boson contributions.

The gauge-boson contribution to the $\phi_c^4 \ln(\phi_c^2)$ term is 1.75×10^{-4}; the top-quark contribution is $-5.19 \times 10^{-4}(M_t/100 \text{ GeV})^4$; and the Higgs boson contribution is $9.73 \times 10^{-5}(M/100 \text{ GeV})^4$. A priori, all three contributions could be comparable. Let us first consider the case where the Higgs mass is small ($M \lesssim 100$ GeV) and the Higgs contribution to the one-loop potential can be ignored. We can then write the potential (7.29) as

$$
\begin{aligned}
V(\phi_c) &= -\frac{1}{2}m^2\phi_c^2 + \frac{1}{4}\lambda\phi_c^4 + B\phi_c^4 \ln\left(\frac{\phi_c^2}{\mu^2}\right) \\
&= -\frac{1}{2}(2B + \lambda)\sigma^2\phi_c^2 + \frac{1}{4}\lambda\phi_c^4 + B\phi_c^4 \ln\left(\frac{\phi_c^2}{\sigma^2}\right).
\end{aligned}
\tag{7.30}
$$

Here we have used the fact that $V'(\sigma) = 0$ implies $m^2 = (\lambda + 2B)\sigma^2$, and $B = 1.75 \times 10^{-4} - 5.19 \times 10^{-4}(M_t/100 \text{ GeV})^4$. The Higgs mass is $M^2 = V''(\sigma) = 2(\lambda + 6B)\sigma^2$. The existence of a minimum at $\phi_c = \sigma$ requires that $V''(\sigma) \propto \lambda + 6B > 0$. The potential (7.30) is shown in Fig. 7.5 for several values of the Higgs mass.

Let us now turn our attention to the case where $B \geq 0$, i.e., $M_t \lesssim 76$

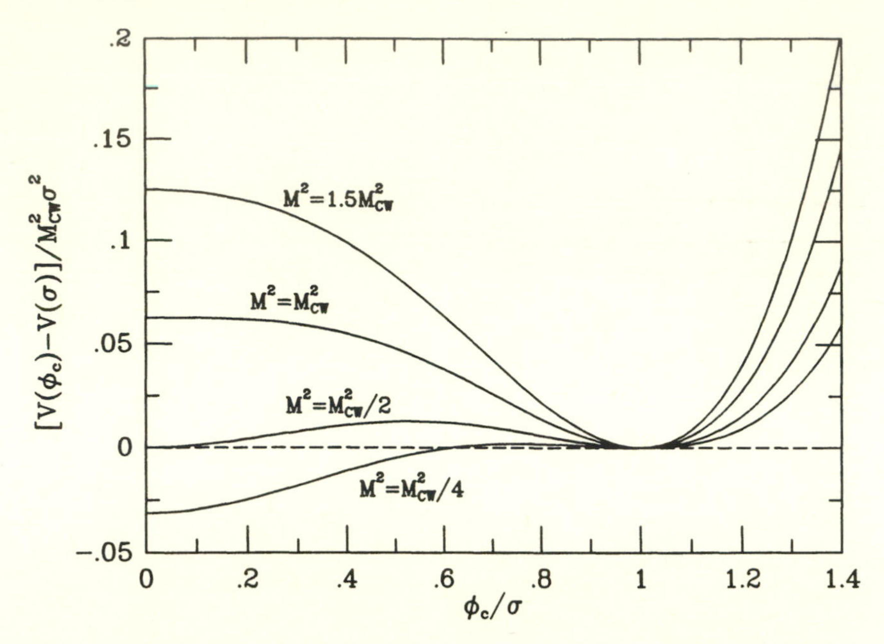

Fig. 7.5: The one-loop Higgs potential in the electroweak model for different values of the Higgs mass. Here, one-loop scalar contributions have been ignored, and it is assumed that $B \geq 0$. M_{CW} is the Coleman-Weinberg mass, $M_{CW} = \sqrt{8B}\sigma$. At one loop, the Coleman-Weinberg mass is about 9 GeV for $M_t = 0$, and it vanishes for $M_t \gtrsim 76$ GeV. Note that if $M \leq M_{CW}$ there is a barrier between $\phi_c = 0$ and $\phi_c = \sigma$, and that if $M \leq M_{CW}/\sqrt{2}$, $\phi_c = \sigma$ is no longer the global minimum of the theory.

GeV.[10] The extrema of the potential are at $\phi_c = 0$ and $\phi_c = \sigma$. Of particular interest for the development of the phase transition is the question of whether, at zero temperature, $\phi_c = 0$ is also a minimum of the potential. The condition for $\phi_c = 0$ to be a minimum is $V''(0) \geq 0$, which is satisfied for $2B + \lambda \leq 0$, or equivalently, $M^2 \leq M^2_{CW}$ where M_{CW} is the Coleman-Weinberg mass, $M^2_{CW} = 8B\sigma^2 \simeq (9\text{ GeV})^2$.[11] Therefore, if $M \leq M_{CW}$,

[10]The lower bound to the top quark mass is already in excess of 70 GeV; by the time that this book appears in print the bound may well be in excess of 76 GeV—or the top quark may have been discovered! Once M_t is known to exceed 76 GeV, the one-loop correction to $V(\phi_c)$ changes sign (B becomes negative). For all intents and purposes, if $M_t \gg 76$ GeV, the electroweak phase transition is second order irrespective of the mass of the Higgs. Because the range of possibilities for electroweak SSB is so rich if $M_t \leq 76$ GeV, we have assumed such—risking that our assumption will be falsfied! In any case, our discussion here serves as an exposition for a generic SSB phase transition.

[11]For $M \leq M_{CW}$, the tree-level potential has no SSB minimum, and symmetry breaking is driven by loop corrections. $M = M_{CW}$ is known as the Coleman-Weinberg

$\phi_c = 0$ is also a minimum, separated from $\phi_c = \sigma$ by a barrier, and the transition is first order. We will shortly see that this presents a problem because the tunnelling rate from $\phi_c = 0$ to $\phi_c = \sigma$ is very small.

In the event that $M \leq M_{CW}$ and $\phi_c = 0$ is a local minimum, we must require that the SSB minimum $\phi_c = \sigma$ have a lower free energy than the minimum at $\phi_c = 0$; otherwise the Universe would have remained in the symmetric minimum, and electroweak symmetry breaking would not have occurred. $V(0) \geq V(\sigma)$ is satisfied for $\lambda + 4B \geq 0$, or equivalently, $M^2 \geq M_{CW}^2/2$. This bound is referred to as the Linde-Weinberg bound [8].

Now consider the potential at finite temperature. As in the previous example, the finite-temperature potential will have a temperature-dependent piece in addition to the zero-temperature part. The temperature-dependent part receives a contribution from all particles that couple to the scalar field, including the scalar field itself. The one-loop potential at finite temperature can be written as a sum of integrals similar to the one in (7.6), of the form

$$F_{\pm}[X(\phi_c)] \equiv \pm \int_0^\infty dx\; x^2 \ln\left[1 \mp \exp[-(x^2 + X(\phi_c)/T^2)^{1/2}]\right] \qquad (7.31)$$

(F_+ applies to boson loops and F_- to fermion loops). For the electroweak model, $V_T(\phi_c)$ is given by [cf. (7.6)]

$$\begin{aligned} V_T(\phi_c) \;&=\; V(\phi_c) + \frac{T^4}{2\pi^2}\Big\{6F_+[g^2\phi_c^2/4] + 3F_+[(g^2+g'^2)\phi_c^2/4] \\ &\qquad + F_+[M^2(\phi_c)] + 12F_-[h_t^2\phi_c^2/2]\Big\}, \qquad (7.32) \end{aligned}$$

where, as before, $V(\phi_c)$ is the one-loop potential at zero temperature, and for simplicity we have included only the ϕ_c-dependent terms. For $T \gg \sigma$, the terms proportional to T^4 are just given by minus the pressure of a gas of the massless fermions and bosons that couple to ϕ.

It is convenient to study the phase transition in different regimes, determined by the Higgs mass. First, consider the cruiserweight regime for the Higgs mass,[12] 1 TeV $\gtrsim M \gtrsim$ 100 GeV, where scalar loops dominate (7.32). In this limit the transition is second order and proceeds like

---

limit [7].

[12]If the Higgs mass is in the heavyweight division, M greater than a few TeV., $\lambda \gtrsim 1$, and the Higgs is strongly coupled. The perturbative, high-temperature analysis given here is not strictly applicable (although it may be qualitatively correct).

the transition in the simple toy model discussed in Section 7.1.1, with $T_C \simeq 2m/\lambda^{1/2} = 2\sigma \simeq 500$ GeV.

Now consider the electroweak phase transition for Higgs masses in the welterweight range: 100 GeV $\gtrsim M > M_{CW}$. Here the vector and fermion loops dominate (7.32) and the transition is (weakly) first order. Determination of T_C is difficult; however, T_C is not the temperature of greatest interest. Thermal and quantum tunnelling rates are slow, and the transition does not proceed until $T = T_2 < T_C$, when the barrier vanishes and the transition can proceed classically. The transition temperature T_2 is determined by the condition that $V_T''(0) = 0$ and is easily found:

$$T_2^2 = \frac{16m^2}{3g^2 + g'^2 + 8h_t^2} = \frac{8M^2}{1.32 + 2.65(M_t/100 \text{ GeV})^2}. \tag{7.33}$$

The effective temperature for the phase transition is T_2. From a temperature of T_C until a temperature of about T_2, the Universe is trapped in the metastable, false-vacuum state $\phi_c = 0$, during which time it is said to "supercool" (just as supercooling liquid water below 0°C). When the barrier disappears ($T \simeq T_2$), the phase transition to the SSB vacuum takes place, and the vacuum energy is released, heating the Universe to a temperature of order T_C, and thereby increasing the entropy density by a factor of about $(T_C/T_2)^3$. Since for Higgs in this weight class, T_C is less than twice T_2, there is only a small amount of supercooling and associated entropy production; hence the transition is weakly first order.

The transition will be strongly first order for Higgs masses in the bantamweight division, $M_{CW} \geq M \geq M_{CW}/\sqrt{2}$. In this range there will *always* be a barrier between the metastable and true vacuum ($T_2 \to 0$ as $M \to M_{CW}$), and the transition must proceed via tunnelling. From Fig. 7.5 it is clear that, as is usually the case, the thin-wall approximation is not valid and a numerical calculation of the bounce action is necessary. Numerical calculations [9] give a quantum tunnelling action $S_E \geq 10^4$ for $M \lesssim M_{CW}$, and of course, $S_E \to \infty$ for $M \to M_{CW}/\sqrt{2}$ where the two minima are exactly degenerate. The minimum value of the action relevant for thermal tunnelling, S_3/T, is about 10^3 for $M \lesssim M_{CW}$, and it too diverges as $M \to M_{CW}/\sqrt{2}$. Since the tunnelling rate is proportional to the exponential of the action, the transition is never completed for Higgs masses in the bantamweight division.

Witten [12], however, pointed out that there is an additional physical phenomenon that must be considered. The extreme supercooling will eventually be terminated when QCD interactions become strong, at a temperature $T \sim 200$ MeV. Below this temperature, chiral symmetry is dynamically broken and a quark condensate forms, which is signalled by $\langle\bar{q}q\rangle$ developing a non-zero vacuum expectation value, $\langle\bar{q}q\rangle \simeq \Lambda_{QCD}^3$ (q is the quark field). Since the Higgs Yukawa coupling makes a contribution to the potential $\Delta V = -h_q\phi\bar{q}q$, the $\langle\bar{q}q\rangle$ condensate will result in an effective linear term in the Higgs potential which destabilizes the symmetric minimum, thereby driving the transition. However, the large amount of supercooling will result in a big entropy release, $S_{\rm after}/S_{\rm before} \sim (T_C/200\ {\rm MeV})^3 \sim 10^9$, which dilutes the baryon number present before the phase transition. Such a large reduction in the baryon-to-entropy ratio is probably unacceptable, and thus cosmological considerations constrain the Higgs mass to be greater than about 1.1 or so times M_{CW} [12].

Finally, for Higgs masses in the flyweight division, $M \leq M_{CW}/\sqrt{2}$, the true ground state at zero temperature is $\phi_c = 0$. Since this is also the ground state at high temperature, the Universe will remain in the symmetric state, and the electroweak symmetry will never be broken—in strong contradiction to reality! In sum, cosmological considerations of the electroweak phase transition result in the limit $M \gtrsim 1.1 M_{CW}$.

While the preceeding discussion applies specifically to electroweak symmetry breaking, it provides an excellent paradigm for a generic SSB phase transition, e.g., GUT symmetry breaking. Even if accelerator experiments rule out the possibility that one-loop corrections play an important role in electroweak symmetry breaking, the lessons learned by this worked example are still applicable to other phase transitions. Our discussion in this Section will be of great utility when we turn to inflation in the next Chapter. In this Section we only touched upon the interesting phenomena of SSB and finite-temperature symmetry restoration; we refer the interested reader to the more detailed reviews that exist, e.g. [10].

7.2 Domain Walls

We have seen that SSB is an intergal part of modern particle physics, and provided that temperatures in the early Universe exceeded the energy scale of a broken symmetry, that symmetry should have been restored. How can we tell if the Universe underwent a series of SSB phase transitions? One possibility is that symmetry-breaking transitions were not "perfect," and that false vacuum remnants were left behind, frozen in the form of topological defects: domain walls, strings, and monopoles [11].

As the first example of a topological defect associated with spontaneous symmetry breaking, consider the domain wall. The simple scalar model of Section 7.1.1 can be used to illustrate domain walls. Recall that the Lagrangian for a real scalar field that undergoes SSB can be written as

$$\mathcal{L} = \frac{1}{2}(\partial_\mu \phi)^2 - \frac{1}{4}\lambda(\phi^2 - \sigma^2)^2. \tag{7.34}$$

The $\mathcal{Z}_2$ reflection symmetry of the Lagrangian, i.e., invariance under $\phi \rightarrow -\phi$, is spontaneously broken when ϕ takes on the vacuum expectation value $\langle\phi\rangle = +\sigma$ or $\langle\phi\rangle = -\sigma$. So far we have assumed that all of space is in the same ground state, but this need not be the case! Imagine that space is divided into two regions. In one region of space $\langle\phi\rangle = +\sigma$, and in the other region of space $\langle\phi\rangle = -\sigma$. Since the scalar field must make the transition from $\phi = -\sigma$ to $\phi = +\sigma$ smoothly, there must be a region where $\phi = 0$, i.e., a region of false vacuum. This transition region between the two vacua is called a domain wall. As we shall see, domain walls can arise whenever a $\mathcal{Z}_2$ (or any discrete) symmetry is broken.

Consider an infinite wall in the x–y plane at $z = 0$. At $z = -\infty$, $\phi = -\sigma$, and at $z = +\infty$, $\phi = +\sigma$. The equation of motion for ϕ is

$$-\frac{\partial^2 \phi}{\partial z^2} + \lambda\phi(\phi^2 - \sigma^2) = 0. \tag{7.35}$$

The solution to the equation of motion, subject to the boundary conditions above, is

$$\phi_W(z) = \sigma \tanh(z/\Delta), \tag{7.36}$$

where the "thickness" of the wall is characterized by $\Delta = (\lambda/2)^{-1/2}\sigma^{-1}$. This solution is illustrated in Fig. 7.6. While we will not prove it here, it should be clear that the domain wall is topologically stable; the "kink" at $z = 0$ can move around or wiggle, but it can't disappear (except by meeting up with an antikink and annihilating).

The finite, but non-zero, thickness of the wall is easy to understand. The terms contributing to the surface energy density include a gradient term, proportional to $\Delta \times (\nabla\phi)^2 \sim \sigma^2/\Delta$, and a potential energy term, proportional to $\Delta \times V(\phi) \sim \Delta\lambda\sigma^4$. The gradient term is minimized by making the wall as thick as possible, while the potential term is minimized by making the wall as thin as possible. The balance between these terms results in a wall of thickness $\Delta \sim \lambda^{-1/2}\sigma^{-1}$.

The stress tensor for the domain wall is obtained by substitution of the wall solution (7.35) into the expression for the stress-energy tensor for a

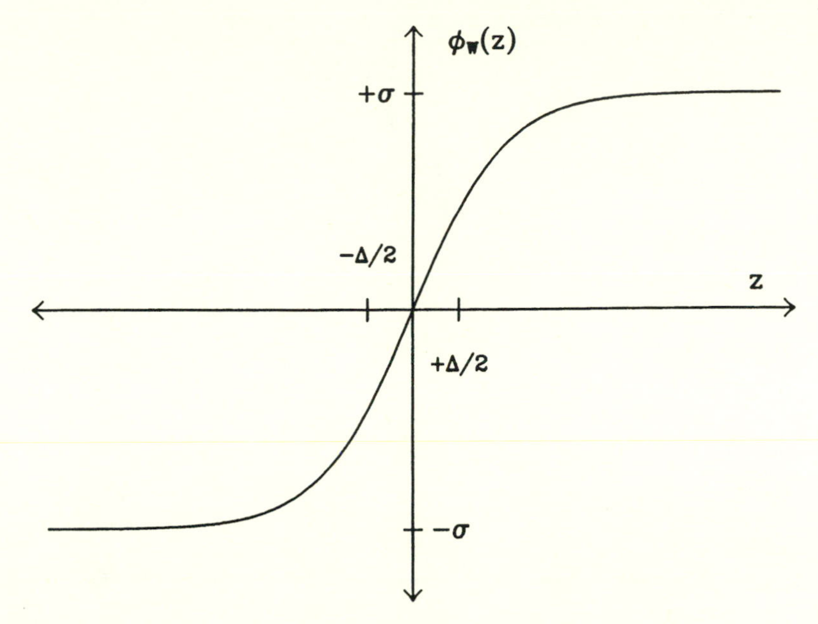

Fig. 7.6: The solution $\phi_W(z)$ for an infinite wall in the x–y plane.

scalar field, cf. (7.3):

$$T^{\mu}_{\ \nu} = \frac{\lambda}{2}\sigma^4 \cosh^{-4}(z/\Delta)\mathrm{diag}(1,1,1,0). \tag{7.37}$$

Note that the z-component of the pressure vanishes, and that the x- and y-components of the pressure are equal to minus the energy density. The energy density associated with the domain wall, i.e., $T^0_{\ 0}$, as a function of z is shown in Fig. 7.7. The surface energy density associated with the wall, given by

$$\eta \equiv \int T^0_{\ 0} dz = \frac{2\sqrt{2}}{3}\lambda^{1/2}\sigma^3, \tag{7.38}$$

is identical to the integrated, transverse components of the stress, $\int T^i_{\ i} dz$. That is, the surface tension in the wall is precisely equal to the surface energy density. Because of this fact walls are inherently relativistic, and their gravitational effects are inherently non-Newtonian (and very interesting).

For the stress tensor $T^{\mu}_{\ \nu} = \mathrm{diag}(\rho,\ -p_1,\ -p_2,\ -p_3)$, the Newtonian limit of Poisson's equation is $\nabla^2\phi = 4\pi G(\rho + p_1 + p_2 + p_3)$ (here, of course, ϕ is the Newtonian gravitational potential). For the planar domain wall $p_3 = 0$ and $p_1 = p_2 = -\rho$, so $\nabla^2\phi = -4\pi G\rho$. The "wrong" sign on the

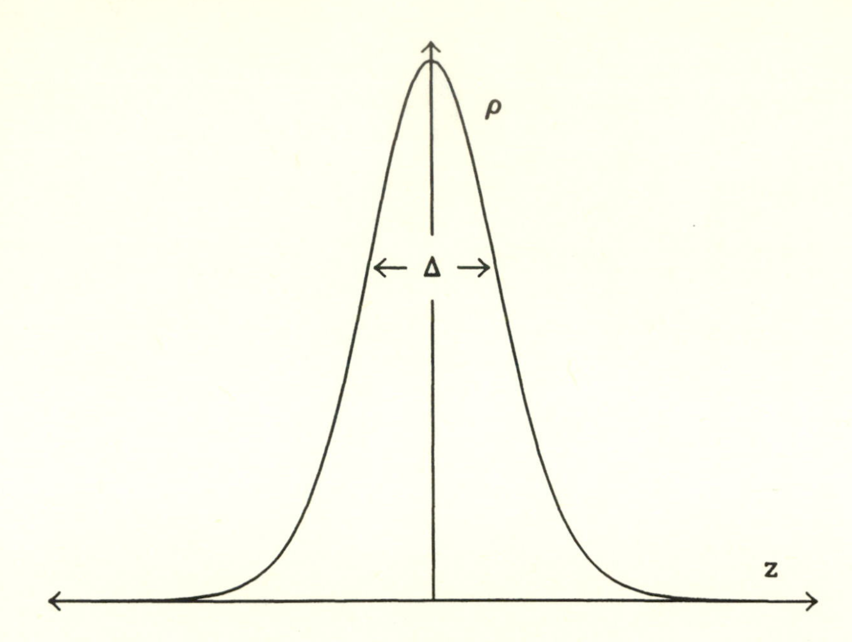

Fig. 7.7: The energy density in the wall as a function of z.

rhs of this equation implies that (gravitational) test particles are repelled by an infinite domain wall, and that two infinite domain walls repel one another [13]. This strange gravitational behavior pertains only for infinite, planar walls. At large distances the gravitational field of a spherical wall of radius R is that of a particle of mass $m \sim 4\pi R^2 \eta$.

Whenever the Universe undergoes a phase transition associated with the spontaneous symmetry breaking of a discrete symmetry, domain walls will inevitably form. The production mechanism will be discussed in more detail in Section 7.5; for the moment, simply assume that at the time of the phase transition, the correlation length ξ of the scalar field vacuum expectation value (VEV) is finite. For any two points separated by a distance $D \gtrsim \xi$ the VEV is uncorrelated, and there is a 50% chance that the two points will be in different vacua. If they are, they must be separated by a domain wall. Therefore, a phase transition associated with the SSB of a $\mathcal{Z}_2$ symmetry should lead to a network of domain walls, with walls separated by a typical distance ξ (see Fig. 7.8). In addition, a typical wall in the network will be curved, with curvature radius characterized by ξ. Within the network there will be both infinite walls and finite walls (i.e., a region of space surrounded by a domain wall, a vacuum bag if you will).

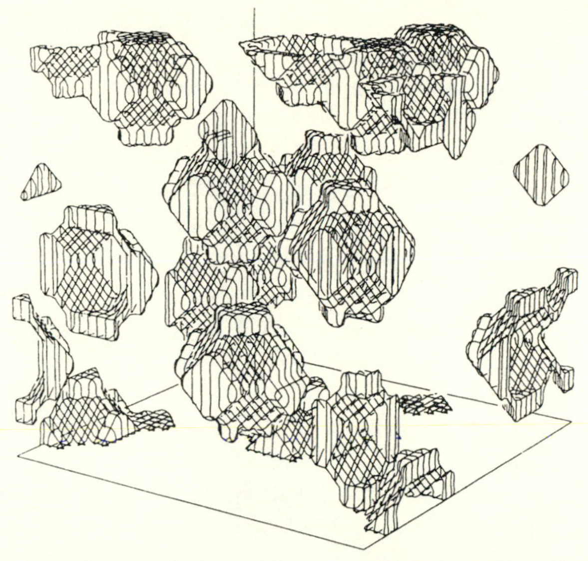

Fig. 7.8: A domain wall network formed in a numerical simulation of a cosmological phase transition where a Z_2 symmetry is spontaneously broken. The contour lines indicate the domain walls that separate the regions of different vacua (from [14]).

Just as a membrane with a surface tension evolves to minimize its surface area, a curved wall will evolve so as to minimize its surface area. The evolution of a wall can, in principle, be followed by solving the equations of motion for the scalar field itself. Often it is easier to use the equations of motion for an idealized thin wall.[13] The equations that describe the minimization of the surface area of the wall can be derived from an action principle. As the two-dimensional wall moves in space, it sweeps out a three-dimensional (two space, one time) surface. The action whose minimization yields the equation of motion for the wall is given by

$$S_{\text{Wall}} = -\eta \int dV = -\eta \int \sqrt{-\gamma}\, d^3\sigma, \tag{7.39}$$

where dV is the invariant space-time volume of the "world tube" swept out by the wall, and $\sigma^i = (\tau,\ \sigma^1,\ \sigma^2)$ are coordinates on the wall (τ is timelike and $\sigma^{1,2}$ are spacelike). The space-time coordinates of the wall's

[13]There are two corrections to the idealized "thin" wall approximation: (i) finite thickness effects due to the curvature of the wall, of order Δ/R ($R =$ curvature radius of the wall); and (ii) gravitational effects, of order $\Delta/d_{EH} \simeq (\sigma/m_{Pl})^2$ where $d_{EH} \simeq (G\eta)^{-1}$ is the distance to the wall's event horizon.

world tube are $x^\mu(\tau,\ \sigma^1,\ \sigma^2)$, and $\gamma = \det(\gamma_{\mu\nu})$, where

$$\gamma_{\mu\nu} = \partial_\mu x^\alpha(\tau,\ \sigma^1,\ \sigma^2)\partial_\nu x^\beta(\tau,\ \sigma^1,\ \sigma^2)g_{\alpha\beta}. \tag{7.40}$$

The physical content of the wall action is manifest: A wall evolves so as to minimize its world-tube volume. Explicit equations of motion for $x^\mu(\tau,\ \sigma^1,\ \sigma^2)$, of course, depend upon the choice of coordinates.

Solving for the cosmic evolution of a domain-wall network is quite involved, but some gross features can be studied by considering the dynamics of a perfect gas of walls, moving with velocity v, inside a box of volume $V \gg \xi$. The perfect gas assumption neglects any dissipative effects due to the motions or interactions of the walls.

For simplicity, let us first assume that there are N infinite, planar walls in the x–y plane within the volume V. Because of the symmetry of this configuration, the stress tensor will be a function only of z. The stress tensor at the point z will receive contributions from all the walls. If walls reside at points z^i $(i = 1,\ \cdots,\ N)$, the stress tensor will be $T_{\mu\nu}(z) = \sum_{i=1}^N T^W_{\mu\nu}(z^i - z)$, where $T^W_{\mu\nu}(z^i - z)$ is the energy-momentum tensor at point z for a wall located at $z = z^i$, cf. (7.37) with $z \to z - z^i$. $T_{\mu\nu}(z)$ can also be expressed as an integral of a sum of δ-functions: $T_{\mu\nu}(z) = \int dz' T^W_{\mu\nu}(z - z')\sum_{i=1}^N \delta(z' - z^i)$. In the limit of many walls, it is a good approximation to replace the sum over δ-functions by $f(z')$, the average number of walls per unit length between z' and $z' + dz'$, normalized such that $\int f(z')dz' = N$. The *average* energy-momentum tensor of the wall gas will be

$$\langle T_{\mu\nu}\rangle \equiv \frac{\int dz\, T_{\mu\nu}(z)}{\int dz} = \frac{\int dz \int dz'\, f(z')T^W_{\mu\nu}(z - z')}{\int dz}. \tag{7.41}$$

If the average wall separation is $\langle L\rangle$, we may approximate $f(z)$ by $f(z) \sim \langle L\rangle^{-1}$, and $\langle T_{\mu\nu}\rangle$ is given by

$$\langle T_{\mu\nu}\rangle \simeq \frac{1}{\langle L\rangle}\int T^W_{\mu\nu}(z)dz \equiv \frac{1}{\langle L\rangle}W_{\mu\nu}. \tag{7.42}$$

The tensor $W_{\mu\nu}$ is the stress-energy "surface density" (e.g., $W^0{}_0 = \eta$). Thus, the energy density of the above wall configuration is $\rho_W \equiv \langle T^0{}_0\rangle = \eta/\langle L\rangle$, cf. (7.38).

If the walls are moving with average velocity v in the $+\hat{\mathbf{z}}$ direction with respect to an observer at rest in the box, it is straightforward to compute $W_{\mu\nu}$ by the appropriate Lorentz transformation. The tangential components of $W_{\mu\nu}$ will not be affected, and $W_{\mu\nu}$ becomes

$$W^{\mu}_{\ \nu} = \begin{pmatrix} \gamma^2\eta & 0 & 0 & \gamma^2\eta v \\ 0 & -\eta & 0 & 0 \\ 0 & 0 & -\eta & 0 \\ \gamma^2\eta v & 0 & 0 & \gamma^2\eta v^2 \end{pmatrix}, \tag{7.43}$$

where $\gamma \equiv (1 - v^2)^{-1/2}$. The form of $W_{\mu\nu}$ for walls moving in the $-\hat{\mathbf{z}}$ direction may be obtained from (7.43) with the substitution $v \to -v$. The off-diagonal terms disappear upon averaging over walls moving in the $\pm\hat{\mathbf{z}}$ directions. Repeating the procedure for walls moving in the $\pm\hat{\mathbf{x}}$ and $\pm\hat{\mathbf{y}}$ directions, the average stress tensor for a perfect gas of domain walls is

$$\langle T^{\mu}_{\ \nu}\rangle = \frac{\eta}{3\langle L\rangle} \begin{pmatrix} 3\gamma^2 & 0 & 0 & 0 \\ 0 & (v^2\gamma^2 - 2) & 0 & 0 \\ 0 & 0 & (v^2\gamma^2 - 2) & 0 \\ 0 & 0 & 0 & (v^2\gamma^2 - 2) \end{pmatrix}. \tag{7.44}$$

With the usual definitions of the energy density and the pressure, $\rho_W \equiv \langle T^0_{\ 0}\rangle$ and $p_W \equiv -\langle T^i_{\ i}\rangle$, the equation of state for the wall gas is

$$p_W = (v^2 - 2/3)\rho_W. \tag{7.45}$$

Recall from Section 3.1 that if the pressure can be expressed in terms of the energy density as $p = w\rho$, the energy density evolves as $\rho \propto R^{-3(1+w)}$, and the scale factor evolves as $R(t) \propto t^{2/3(1+w)}$. From (7.45) we see that the dynamics of the expansion and evolution of the wall system depend upon the average velocity of the wall gas. The evolution of ρ_W and $R(t)$ for several values of v are given in Table 7.1.

The result for non-relativistic walls is $\rho_W \propto R^{-1}$. In this limit the wall gas stretches conformally with the expansion, with the distance between walls increasing as R, the surface area of the walls growing as R^2, and the number density of walls decreasing as R^{-3}. Since the mass of a wall is proportional to its area, and the wall energy density is proportional to the product of the wall mass and the wall number density, the energy density of the walls decreases as R^{-1}. One expects the velocities associated with any domain walls created in the early Universe to be rapidly red shifted away, so cosmological walls should quickly come to dominate both the

v	w	ρ_W	$R(t)$
0	$-2/3$	R^{-1}	t^2
$1/\sqrt{3}$	$-1/3$	R^{-2}	t^1
$\sqrt{2}/\sqrt{3}$	0	R^{-3}	$t^{2/3}$
1	$1/3$	R^{-4}	$t^{1/2}$

Table 7.1: The evolution of a wall-dominated Universe for several values of v.

radiation ($\rho_R \propto R^{-4}$) and the matter ($\rho_M \propto R^{-3}$) energy densities, and thereby drastically alter the standard cosmology.

The existence of large-scale domain walls in the Universe today can be precluded simply based upon their contribution to the total mass density. A domain wall of size $H_0^{-1} \simeq 10^{28}h^{-1}$cm would have a mass of order $M_{\rm wall} \sim \eta H_0^{-2} \sim 4 \times 10^{65}\lambda^{1/2}(\sigma/100\text{GeV})^3$ grams, or about a factor of $10^{10}\lambda^{1/2}(\sigma/100\ \text{GeV})^3$ times that of the total mass within the present Hubble volume. Walls would also lead to large fluctuations in temperature of the CMBR unless σ is very small: $\delta T/T \simeq G\eta H_0^{-1} \simeq 10^{10}\lambda^{1/2}(\sigma/100\ \text{GeV})^3$. Apparently, domain walls are cosmological bad news unless the energy scale and/or coupling constant associated with them are very small [15].

The existence of domain wall solutions for this simple model traces to the existence of the disconnected vacuum states: $\langle\phi\rangle = \pm\sigma$. The general mathematical criterion for the existence of topologically stable domain walls for the symmetry breaking pattern $\mathcal{G} \rightarrow \mathcal{H}$ is that $\mathbf{\Pi}_0(\mathcal{M}) \neq \mathcal{I}$, where $\mathcal{M}$ is the manifold of equivalent vacuum states $\mathcal{M} \equiv \mathcal{G}/\mathcal{H}$, and $\mathbf{\Pi}_0$ is the homotopy group that counts disconnected components. In the above example, $\mathcal{G} = \mathcal{Z}_2$, $\mathcal{H} = \mathcal{I}$, $\mathcal{M} = \mathcal{Z}_2$, and $\mathbf{\Pi}_0(\mathcal{M}) = \mathcal{Z}_2 \neq \mathcal{I}$.

7.3 Cosmic Strings

The next example of a topological defect is the cosmic string, a one-dimensional structure. As we shall see, cosmic strings are much more palatable to a cosmologist than domain walls. A simple model that illustrates the cosmic string is the Abelian Higgs model, a spontaneously broken $U(1)$ gauge theory.[14] The Lagrangian of the model contains a $U(1)$

[14]Note, the $U(1)$ gauge symmetry discussed here is not that of electromagnetism; $U(1)_{EM}$ is of course unbroken.

gauge field, A_μ, and a *complex* Higgs field, Φ, which carries $U(1)$ charge e,

$$\mathcal{L} = D_\mu \Phi D^\mu \Phi^\dagger - \frac{1}{4} F_{\mu\nu} F^{\mu\nu} - \lambda(\Phi^\dagger \Phi - \sigma^2/2)^2, \tag{7.46}$$

where

$$\begin{aligned} F_{\mu\nu} &= \partial_\mu A_\nu - \partial_\nu A_\mu, \\ D_\mu \Phi &= \partial_\mu \Phi - ieA_\mu \Phi. \end{aligned} \tag{7.47}$$

We immediately recognize that the theory is spontaneously broken as $V(\Phi)$ is minimized for $\langle|\Phi|\rangle^2 = \sigma^2/2$. The physical states after SSB are a scalar boson of mass $M_S^2 = 2\lambda\sigma^2$ and a massive vector boson of mass $M_V^2 = e^2\sigma^2$.

The complex field Φ can be written in terms of two real fields: $\Phi = (\phi + i\phi_1)/\sqrt{2}$. *If* the VEV is chosen to lie in the real direction, then the potential becomes $V(\phi) = (\lambda/4)(\phi^2 - \sigma^2)^2$, where $\langle|\Phi|\rangle = \langle\phi\rangle/\sqrt{2}$. However, energetics do not determine the phase of $\langle\Phi\rangle$ since the vacuum energy depends *only* upon $|\Phi|$; this fact follows from the $U(1)$ gauge symmetry. Defining the phase of the VEV by $\langle\Phi\rangle = (\sigma/\sqrt{2})\exp(i\theta)$, we see that $\theta = \theta(\vec{x})$ can be position dependent. However, Φ must be single valued; i.e., the total change in θ, $\Delta\theta$, around any closed path must be an integer multiple of 2π. Imagine a closed path with $\Delta\theta = 2\pi$. As the path is shrunk to a point (assuming no singularity is encountered), $\Delta\theta$ cannot change continuously from $\Delta\theta = 2\pi$ to $\Delta\theta = 0$. There must, therefore, be one point contained within the path where the phase θ is undefined, i.e., $\langle\Phi\rangle = 0$. The region of false vacuum within the path is part of a tube of false vacuum. Such tubes of false vacuum must either be closed or infinite in length, otherwise it would be possible to deform the path around the tube and contract it to a point without encountering the tube of false vacuum. In most instances, these tubes of false vacuum have a characteristic transverse dimension far smaller than their length, so they can be treated as one-dimensional objects and are called "strings."[15]

The string solution to the equations of motion for the Lagrangian in (7.46) was first found by Nielsen and Olesen [16]. At large distances from an infinite string in the z-direction, their solution is

$$\Phi \longrightarrow (\sigma/\sqrt{2})\exp(iN\theta),$$

[15]There should be no confusion between superstrings and the (cosmic) strings considered here.

$$A_\mu \longrightarrow -ie^{-1}\partial_\mu \left[\ln(\sqrt{2}\Phi/\sigma)\right], \tag{7.48}$$

where θ is the polar angle in the x–y plane, and N is the "winding number" of the string.[16] Note this choice for A_μ and Φ results in a finite-energy solution, since at large distances from the string, $F_{\mu\nu} \to 0$, and $D_\mu\Phi \to 0$ (the $\partial_\mu\Phi$ and A_μ contributions to $D_\mu\Phi$ exactly cancel). There is no general solution to the coupled equations of motion for Φ and A_μ. However, the Higgs field Φ and the θ-component of the gauge field for a string solution can be well approximated by

$$\begin{aligned}\Phi &= \frac{\sigma}{\sqrt{2}}[1-\exp(-r/r_1)]\exp(-i\theta),\\ A_\theta &= \frac{[1-\exp(-r/r_2)]^2}{er},\end{aligned} \tag{7.49}$$

where r_1 and r_2 are proportional to σ^{-1} and depend upon the coupling constants e and λ [17]. The results of a variational calculation for Φ and A_θ are shown in Fig. 7.9.

The Hamiltonian (or energy) per unit length for an infinite string in the z-direction is given by

$$\begin{aligned}\mu &\equiv \frac{dH}{dz} = \int r dr d\theta\, \mathcal{H} = -\int r dr d\theta\, \mathcal{L}\\ &= \int_0^\infty \int_0^{2\pi} r dr d\theta \left[\left|\frac{\partial\Phi}{\partial r}\right|^2 + \left|\frac{1}{r}\frac{\partial\Phi}{\partial\theta} - ieA_\theta\Phi\right|^2 + V(\Phi) + \frac{B^2}{2}\right]\end{aligned} \tag{7.50}$$

where $\vec{\mathbf{B}} = \vec{\nabla} \times \vec{\mathbf{A}}$ is the "magnetic" field associated with the $U(1)$ gauge field. The total magnetic flux within the string (or vortex, as it is often referred to) is $N(2\pi/e)$.[17] Because there are no closed-form solutions for A_μ and Φ, μ cannot be calculated in closed form, except for the case $M_V = M_S$

[16] In general we will be interested only in $N = 1$; for $M_V/M_S < 1$, it is energetically favorable for a string with $N > 1$ to decay into N strings with $N = 1$. Even if $N > 1$ strings are stable, the number of $N > 1$ strings formed in the phase transition is much less than the number of $N = 1$ strings [18].

[17] The strings discussed here are called gauge strings. There are also strings associated with the SSB of a global $U(1)$ symmetry (i.e., $e = 0$). Such strings are called global (or axionic) strings. Other than the fact that their energy per length is logarithimically divergent (cut off by the distance to the next string), their cosmological properties are very similar to gauge strings.

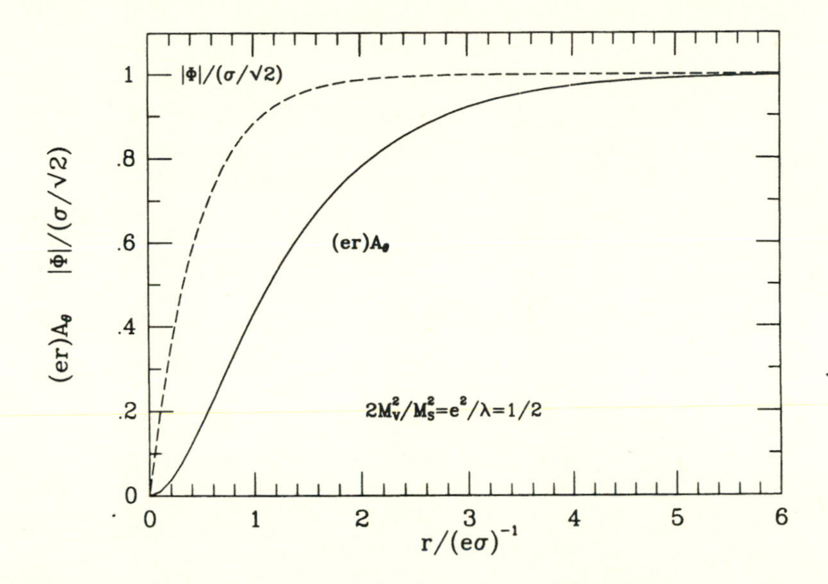

Fig. 7.9: Solutions for $A_\theta(r)$ and $|\Phi(r)|$ for the case $2M_V^2/M_S^2 = e^2/\lambda = 1/2$. Note that for $r \gtrsim 3(e\sigma)^{-1}$ the solutions approach the Nielsen-Olesen solution (7.48). Thus, the "thickness" of the string is approximately σ^{-1}.

where the equations of motion for A_μ and Φ decouple. In this case $\mu = \pi\sigma^2$. In general, μ is a slowly varying function of $M_V^2/M_S^2 = e^2/2\lambda$. The results of a variational calculation for μ as a function of e^2/λ are shown in Fig. 7.10.

It can be shown that the stress-energy tensor associated with a long, thin, straight string is given by [19]

$$T^\mu_{\ \nu} = \mu\delta(x)\delta(y)\mathrm{diag}(1,0,0,1), \tag{7.51}$$

where, as above, μ is the mass per unit length of the string. Note that the pressure is negative—i.e., it is a string tension—and equal to $-\mu$. Like domain walls, strings are intrinsically relativistic.

Far from a string loop of radius R, the gravitational field of the string is that of a particle of mass $M_{\rm string} = 2\pi R\mu$. For a loop of size about that of the present horizon, $M_{\rm string} \simeq 10^{18}(\sigma/\mathrm{GeV})^2$ grams. As with domain walls, there are non-Newtonian gravitational effects associated with strings. Recall that for a stress tensor of the form $T^\mu_{\ \nu} = \mathrm{diag}(\rho,\ -p_1,\ -p_2,\ -p_3)$, the Newtonian limit of Poisson's equation is $\nabla^2\phi = 4\pi G(\rho + p_1 + p_2 + p_3)$. For

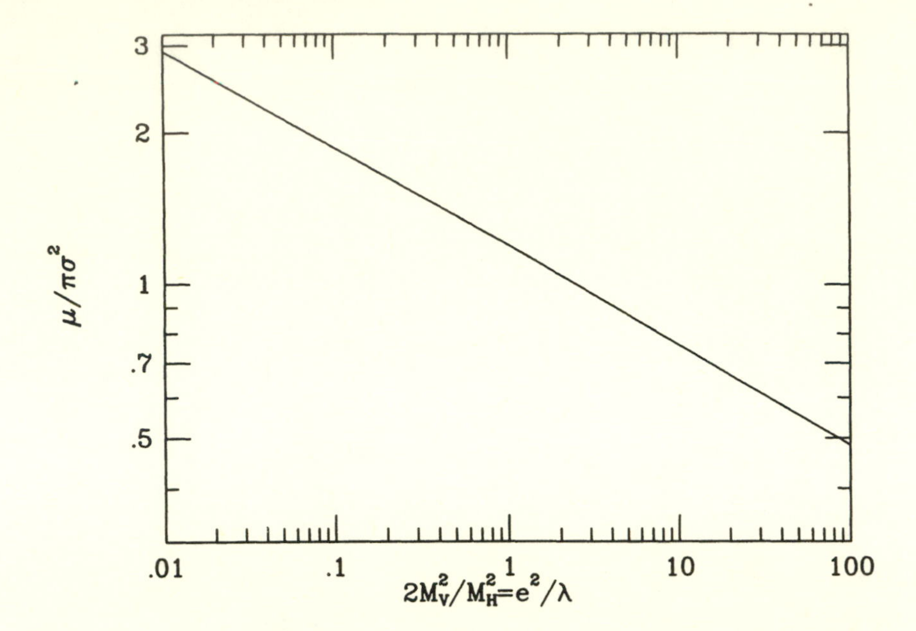

Fig. 7.10: The mass per unit length, μ, of an infinite string in units of $\pi\sigma^2$ as a function of $M_V^2/M_S^2 = e^2/2\lambda$. Note that μ is a very slowly varying function of e^2/λ.

an infinite string in the z direction $p_3 = -\rho$ and $p_1 = p_2 = 0$, and Poisson's equation becomes $\nabla^2\phi = 0$, which suggests that space is flat outside of an infinite straight string.

Vilenkin has solved Einstein's equations for the metric outside an infinite, straight cosmic string in the limit that $G\mu \ll 1$ [20]. In terms of the cylindrical coordinates $(r,\ \theta,\ z)$ the metric is

$$ds^2 = dt^2 - dz^2 - dr^2 - (1 - 4G\mu)^2 r^2 d\theta^2. \tag{7.52}$$

By a transformation of the polar angle, $\theta \to (1-4G\mu)\theta$, the metric becomes the flat-space Minkowski metric: As expected, space-time around a cosmic string is that of empty space. However, the range of the flat-space polar angle θ is only $0 \leq \theta \leq 2\pi(1 - 4G\mu)$ rather than $0 \leq \theta \leq 2\pi$. This is referred to as a conical singularity.[18] The price paid in writing the metric for a cosmic string in Minkowski form is a deficit angle of $\Delta\theta = 8\pi G\mu$. Note that the gravitational effects of the string depend upon the dimensionless

[18] The space is called conical because two-dimensional hypersurfaces in the plane normal to the string are cone shaped.

quantity

$$G\mu \simeq \pi(\sigma/m_{Pl})^2 \simeq 10^{-6}(\sigma/10^{16}\mathrm{GeV})^2. \tag{7.53}$$

As we shall see, all cosmological effects of cosmic strings are characterized by this same parameter.

The conical nature of space around a string leads to several striking effects: double images of objects located behind the string, fluctuations in the microwave background, and the formation of wakes. To understand the formation of double images, consider the simplified situation of an infinite string normal to the plane containing the source and the observer. The conical space is flat space with a wedge of angular size $\Delta\theta$ removed and points along the cuts identified (see Fig. 7.11).[19] Due to this, the observer will see two images of the source, with the angular separation, $\delta\alpha$, between the two images determined by

$$\begin{aligned}\sin(\delta\alpha/2) &= \sin(\Delta\theta/2)\frac{l}{d+l},\\ \delta\alpha &\simeq \Delta\theta\frac{l}{d+l} = 8\pi G\mu\frac{l}{d+l}.\end{aligned} \tag{7.54}$$

Here l (d) is the distance from the string to the source (observer), and the second equation is a small-angle approximation.

The conical metric also leads to discontinuities in the temperature of the microwave background. Imagine as the source, a point on the last scattering surface for the microwave background radiation. An observer at rest with respect to the string will see two images of the same point on the last scattering surface, separated by an angle $\delta\alpha \simeq \Delta\theta$ (for $d \ll l$). Now if the string and observer are not at rest with respect to each other, but instead have a relative velocity v which is perpendicular to the line of sight, the momentum vector of one image will have a small component (order $\Delta\theta$) parallel to the direction of $\vec{v}$, and the other, a small component antiparallel to the direction of $\vec{v}$. The net effect is a small Doppler shift of the radiation temperature [21],

$$\frac{\delta T}{T} \simeq 8\pi G\mu v \tag{7.55}$$

across the string. Based upon this effect and the observed isotropy of

[19] The reader is encouraged to cut along the dotted lines of Fig. 7.11 and to tape the identified edges together. If you ruin the book, you can always purchase another copy.

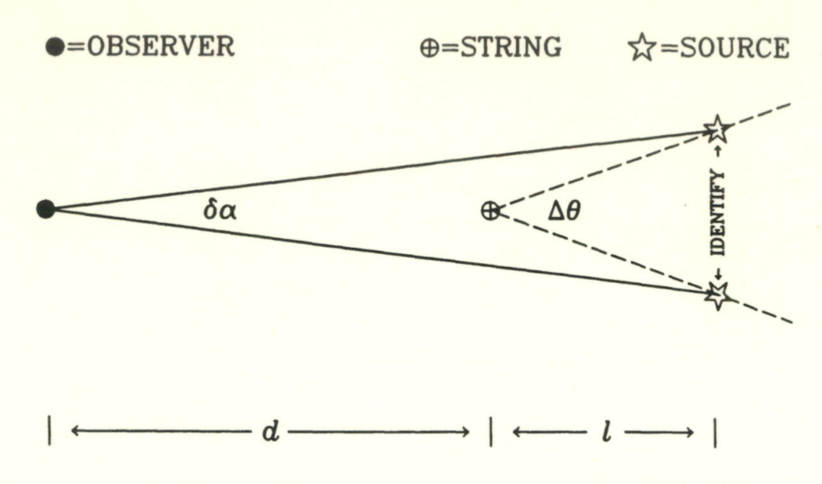

Fig. 7.11: The conical nature of space around an infinite straight string leads to double images of distant sources.

the CMBR, we can conclude any strings that exist at present must be characterized by $G\mu \lesssim 10^{-5}$.

A third interesting, and potentially important, effect of cosmic strings which arises from the conical structure of the space-time around a string are string wakes. Consider a long, straight string moving through the Universe with velocity v. As the string moves past particles in the Universe the particles will be deflected and will acquire a "wake" velocity $v_W \simeq 4\pi G\mu v$, transverse to the direction of motion of the string. If the particles have a very small internal velocity dispersion, e.g., cold-dark matter particles, or baryons after decoupling, then matter on both sides of the passing string will move toward the plane defined by the motion of the string. In a Hubble time, a wedge-shaped sheet of matter, with overdensity of order unity, opening angle $\simeq 8\pi G\mu$, and width vH^{-1}, will form in the wake of the string. The mass of the material within the wake-produced sheet can be considerable, about $8\pi G\mu v^3$ of that in the horizon; likewise, the scale of the thin (thickness/width $\sim 8\pi G\mu$) sheets that are formed is comparable to that of the horizon scale. It has been suggested that the sheets that form in the wakes of long, straight cosmic strings play an important role in structure formation [22].

Just as a wall network is formed when a discrete symmetry is broken, a string network is formed when a $U(1)$ symmetry is broken. The initial network consists of both infinite strings (about 80% by mass) and closed string loops (about 20% by mass). Once formed, the string network will

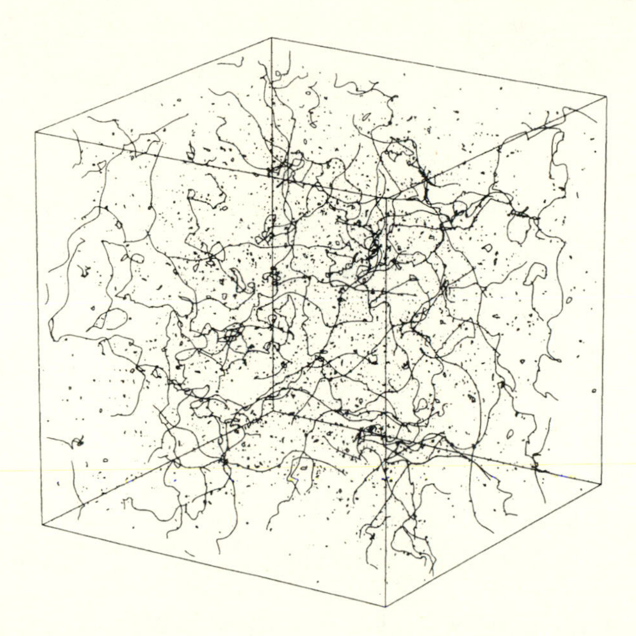

Fig. 7.12: A numerical simulation of a cosmic string network that forms when a $U(1)$ symmetry is spontaneously broken. It consists of infinite (i.e., stretching across the box) strings and finite string loops [23].

be conformally stretched by the expansion of the Universe. In addition, due to the string tension, strings will evolve to straighten themselves out. On scales larger than H^{-1} the conformal stretching dominates, while on scales smaller than H^{-1} the string tension dominates. The evolution of cosmological string networks has been studied by numerical simulation [23, 24], and the results of one such simulation are shown in Fig. 7.12.

In order to get a rough idea of the types of behavior possible for the string network, we can treat cosmic strings as a perfect gas, just as we did in the previous Section with walls. Performing the appropriate averages for cosmic strings of average velocity v, the stress tensor becomes

$$\langle T^{\mu}{}_{\nu}\rangle = \frac{\mu}{3\langle L\rangle^2}\begin{pmatrix} 3\gamma^2 & 0 & 0 & 0 \\ 0 & (v^2\gamma^2-1) & 0 & 0 \\ 0 & 0 & (v^2\gamma^2-1) & 0 \\ 0 & 0 & 0 & (v^2\gamma^2-1) \end{pmatrix}, \tag{7.56}$$

where $\langle L\rangle$ is the average separation between strings, cf. (7.44). The equa-

v	w	ρ_S	$R(t)$
0	$-1/3$	R^{-2}	t^1
$1/\sqrt{2}$	0	R^{-3}	$t^{2/3}$
1	$1/3$	R^{-4}	$t^{1/2}$

Table 7.2: The evolution of a string-dominated Universe for several values of v.

tion of state for the string gas is

$$p_S = (2v_S^2/3 - 1/3)\rho_S. \tag{7.57}$$

As in the domain wall case, the evolution of the Universe and of the string density depends upon the average string velocity. The dependence of $w \equiv p_S/\rho_S$, $\rho_S \propto R^{-3(1+w)}$, and $R(t) \propto t^{2/3(1+w)}$ upon the average string velocity is shown in Table 7.2.

Just as with domain walls, low-velocity strings should rapidly come to dominate the mass density of the Universe since $\rho_S \propto R^{-2}$ (for $v_S = 0$). Thus one might expect that cosmic strings are also a cosmological disaster. However, the evolution of a string network is more complicated than can be shown with the simple string-gas model. Two additional important effects must be included: (i) The intercommutation of intersecting string segments, which leads to the continual chopping up of long strings into smaller loops; (ii) the decay of small loops through the emission of gravitational radiation. These two effects are discussed below.

In order to follow the evolution of a string network one needs the equation of motion for a string. In principle, one should use the classical field equations for ϕ and A_μ. However, for a thin string (thickness much less than radius of curvature), one can use the equation of motion for an infinitely thin string. These equations are derived from the Nambu action,

$$S_{\text{string}} = -\mu \int dA = -\mu \int \sqrt{-\gamma}\, d^2\sigma, \tag{7.58}$$

where dA is the invariant area of the world "sheet" swept out by the motion of the string, and $\sigma^i = (\tau,\ \sigma)$ are the coordinates on the string's worldsheet (τ = timelike, σ = spacelike). The space-time coordinates of the string's world sheet are $x^\mu(\sigma,\ \tau)$, and $\gamma = \det(\gamma_{\mu\nu})$, where $\gamma_{\mu\nu} = \partial_\mu x^\alpha(\sigma,\ \tau)\partial_\nu x^\beta(\sigma,\ \tau)g_{\alpha\beta}$. The physical content of the Nambu action is clear: A string evolves so as to minimize the area of its world sheet. Just as a domain wall tries to minimize its surface area, a string evolves to minimize its length—of course, in the process it will oscillate.

The actual equation of motion for the trajectory of the string depends upon the choice of string coordinates. With an appropriate choice the equations of motion are very simple:

$$\ddot{\vec{\mathbf{x}}} - \vec{\mathbf{x}}'' = 0 \tag{7.59}$$

with $\dot{\vec{\mathbf{x}}}^2 + \vec{\mathbf{x}}'^2 = 1$ and $\dot{\vec{\mathbf{x}}} \cdot \vec{\mathbf{x}}' = 0$. Here $\vec{\mathbf{x}}(\sigma,\, t)$ is the spatial trajectory of the string, t is clock time, σ is proportional to position along the string (starting at some arbitrary point), with prime denoting $\partial/\partial\sigma$ and overdot, $\partial/\partial t$. The oscillatory nature of the solutions is manifest from (7.59).

The motion of a small, closed loop is particularly simple: Owing to its string tension a loop of characteristic radius R oscillates relativistically, with a period $\tau \sim R$. As it oscillates it will radiate gravitational waves due to its time-varying quadrupole moment, $Q \sim \mu R^3$. The power radiated in gravitational waves is given by

$$P_{GW} \simeq G(\dddot{Q})^2 \simeq \gamma_{GW} G\mu^2 \tag{7.60}$$

where γ_{GW} is a numerical constant of order 100 [25]. In a characteristic time τ_{GW} the loop will radiate away its mass-energy, shrink to a point, and vanish:

$$\tau_{GW} \sim \mu R/P_{GW} \sim (\gamma_{GW} G\mu)^{-1} R. \tag{7.61}$$

That is, a loop will undergo about $10^{-2}(G\mu)^{-1}$ oscillations before it disappears.

The other key bit of physics crucial to the evolution of a string network is intercommutation. Two string segments that intersect and in the process of intersection swap partners (rather than passing through one another), are said to have intercommuted. Based upon numerical experiments it appears that the probability for this to occur is nearly unity [26]. Through the process of intercommutation, long string can be cut up into smaller segments and/or loops, thereby regulating the energy of the string network. As we saw earlier, the energy density associated with a long, slow string only decreases as R^{-2}, whereas the energy density associated with a loop decreases as R^{-3} (until it evaporates into gravitational waves).

Numerical experiments [23, 24] with string networks show that after the string network is produced, it rapidly approaches a so-called "scaling solution," where, in a radiation-dominated Universe,[20]

$$\rho_S \simeq 30\mu/t^2, \tag{7.62}$$

so that $\rho_S \propto R^{-4}$, and the ratio of string energy density to radiation remains fixed:

$$\frac{\rho_S}{\rho_r} \simeq 30\frac{32\pi}{3}G\mu. \tag{7.63}$$

Most of this energy density resides in a few infinite strands per horizon volume, with the rest in a "scale invariant" distribution of loops,

$$n_{\rm loop}(E,t)dE \simeq 0.3\left(\frac{\mu t}{E}\right)^{3/2}\frac{dE}{Et^3}. \tag{7.64}$$

Here $n_{\rm loop}(E,t)dE$ is the number density of loops at time t with energy E to $E + dE$ (the *rms* radius of a loop of energy E is just $R = E/2\pi\mu$). These loops extend from nearly horizon size downward and are produced by the chopping up of long strings (and even loops themselves) through self-intersection and intercommutation.

Naïvely, the total energy density in loops is given by

$$\rho_{\rm loop} = \int_{E_{\rm min}}^{\mu t} E n_{\rm loop} dE \simeq 0.6\mu^{3/2}t^{-3/2}E_{\rm min}^{-1/2}. \tag{7.65}$$

Based upon this, one would conclude that $\rho_{\rm loop} \propto t^{-3/2} \propto R^{-3}$, and that $\rho_{\rm loop}$ diverges as $E_{\rm min}^{-1/2}$. Were either *apparent* feature of $\rho_{\rm loop}$ actually true, cosmic strings would be a cosmic catastrophe, as they would quickly come to dominate the mass density of the Universe. Fortunately, neither is true! Recall that a loop undergoes of order $(\gamma_{GW}G\mu)^{-1}$ oscillations before it radiates itself away into gravitational waves. This means that at time t the smallest loop has a characteristic size $R_{\rm min} \sim \gamma_{GW}G\mu t$ and energy $E_{\rm min} \sim \gamma_{GW}G\mu^2 t$. When this fact is taken into account, the actual energy density in loops is

$$\rho_{\rm loop} \sim (\gamma_{GW}G\mu)^{-1/2}\mu/t^2, \tag{7.66}$$

which scales in the same way as the energy density in long strings.

We can easily estimate the energy density of the gravitational wave background produced by the death throes of strings. In the time interval

[20]A similar scaling solution obtains in a matter-dominated Universe, with $\rho_S \simeq 10\mu/t^2$.

$t \to t + dt$, loops of size $R \sim \gamma_{GW} G\mu t \to R + dR \sim \gamma_{GW} G\mu(t + dt)$ decay and convert their energy density into gravity waves. Using (7.65) and the fact that once produced, the energy density of gravity waves is red shifted as R^{-4}, it follows that

$$\dot{\rho}_{GW} + 4H\rho_{GW} \simeq (G\mu/\gamma_{GW})^{1/2}(Gt^3)^{-1}$$
$$\Longrightarrow \quad R^{-4} d(R^4 \rho_{GW})/dt = (G\mu/\gamma_{GW})^{1/2}(Gt^3)^{-1}. \tag{7.67}$$

If we assume that the Universe is radiation dominated and g_* is constant, then $\rho_R = 3/(32\pi Gt^2) \propto R^{-4}$, and (7.67) becomes

$$\frac{d}{dt}\left(\frac{\rho_{GW}}{\rho_R}\right) \simeq \frac{32\pi}{3}\left(\frac{G\mu}{\gamma_{GW}}\right)^{1/2} t^{-1}. \tag{7.68}$$

Integrating from some very early initial time t_i when loops first begin decaying ($t_i \sim (\gamma_{GW} G\mu)^{-1} t_{\rm form}$ where $t_{\rm form} \sim m_{Pl}/\mu$ is the formation time of the string network) until time t, we find

$$\rho_{GW}/\rho_R \simeq \frac{32\pi}{3}(G\mu/\gamma_{GW})^{1/2} \ln(t/t_i). \tag{7.69}$$

The ratio of string-produced gravity waves relative to radiation is proportional to $(G\mu/\gamma_{GW})^{1/2}$ and grows logarithmically with time.

We see then, that the existence of the scaling solution depends in a crucial way upon both string intercommutation and loop evaporation into gravitational waves. In the absence of either effect, strings would have disastrous cosmological consequences. As cosmic strings stand, they are cosmologically safe and have several potentially interesting consequences:

- they leave behind a background of relic gravitational waves;
- relic string present today can lead to temperature fluctuations in the CMBR;
- relic string present today can act as gravitational lenses; and
- string loops, or flattened structures formed in the wakes of strings, can possibly serve as seeds to initiate structure formation in the Universe.

The string-produced gravitational waves affect the yields of primordial nucleosynthesis just as additional relativistic species. The limit "$N_\nu \leq 4$" discussed in Chapter 4 implies that $\rho_{GW}/\rho_R \lesssim 0.16$ during nucleosynthesis. Using (7.69) for ρ_{GW}/ρ_R results in a limit of about $G\mu \lesssim 10^{-5}$ (for a careful treatment and details, see [27]). In addition, if $G\mu \gtrsim 10^{-6}$, the relic gravitational waves present today would have significantly affected timing measurements of the millisecond pulsar—an effect not seen—and therefore such values of $G\mu$ are precluded [28]. As we have previously mentioned the present isotropy limits of the CMBR are consistent with $G\mu \lesssim 10^{-5}$. As discussed, the characteristic separation between string-produced double images is $\delta\alpha \simeq 4\pi G\mu \simeq 3''(G\mu/10^{-6})$. The occurrence of a line, or a "loop," of double images would be striking evidence for a relic cosmic string.[21] Finally, cosmic string characterized by $G\mu \sim 10^{-6}$ or so ($\sigma \sim 10^{16}$ GeV) provides a potentially viable means of seeding structure formation in the Universe. Whether loops, sheets formed by wakes, or a combination of the two, are most important in this regard, depends upon the relative amount of string in loops and infinite strands. The possibility that cosmic string may play an important role in structure formation has spurred a very active area of research in recent years [30].

If the cosmological consequences of cosmic string with $G\mu \sim 10^{-6}$ weren't interesting enough, Witten [31] has shown that in relatively simple extensions of the non-Abelian Higgs model discussed here cosmic string can be superconducting![22] The charge carriers on such strings can be either fermions or bosons, and the critical currents can be as large as 10^{20}A. The superconducting variety of cosmic string has a host of additional, interesting astrophysical and cosmological consequences, including structure formation initiated by the explosive electromagnetic decay of string loops [32] and the production of ultra-high energy cosmic rays [33].

In our discussion of cosmic strings, we have used the simplest example of a spontaneously broken gauge theory for which string solutions exist. In general, there will be string solutions associated with the symmetry breaking $\mathcal{G} \to \mathcal{H}$, if the manifold of degenerate vacuum states, $\mathcal{M} = \mathcal{G}/\mathcal{H}$, contains unshrinkable loops, i.e., if the mapping of $\mathcal{M}$ onto the circle is

---

[21] Interestingly enough, Cowie and Hu recently reported evidence for four "twin galaxies" in a small region of the sky. With a little imagination one can trace out a string loop! While twin galaxies are not an unusual occurrence, these twins have almost identical red shifts [29].

[22] The simplest example is a $U(1)' \otimes U(1)_{EM}$ gauge theory where $U(1)_{EM}$ is the $U(1)$ of electromagnetism, and the coupling between the two $U(1)$'s is such that $U(1)_{EM}$ is spontaneously broken within the string.

non-trivial. This is formally expressed by the statement that topologically stable string solutions exist if $\mathbf{\Pi}_1(\mathcal{M}) \neq \mathcal{I}$. In the above example $\mathcal{G} = U(1)$, $\mathcal{H} = \mathcal{I}$, and $\mathcal{M} = U(1)$. The group $U(1)$ can be represented by the points on a circle, and so $\mathbf{\Pi}_1[U(1)]$ is the mapping of the circle onto itself. Such a mapping is characterized by the winding number of the mapping, i.e., $\theta \to N\theta$ ($N = 0,\ 1,\ \cdots$), so that $\Pi_1(\mathcal{M}) = \mathcal{Z}$, the set of integers.

7.4 Magnetic Monopoles

Domain walls are two-dimensional topological defects, and strings are one-dimensional defects. Point-like defects also arise in some theories which undergo SSB, and remarkably, they appear as magnetic monopoles. A simple model that illustrates the magnetic monopole solution is an $SO(3)$ gauge theory,[23] in which $SO(3)$ is spontaneously broken to $U(1)$ by a Higgs triplet Φ^a, where a is the group space index. The Lagrangian density for this theory is

$$\begin{aligned}
\mathcal{L} &= \frac{1}{2}D_\mu\Phi^a D^\mu\Phi^a - \frac{1}{4}F^a_{\mu\nu}F^{a\mu\nu} - \frac{1}{8}\lambda(\Phi^a\Phi^a - \sigma^2)^2, \\
F^a_{\mu\nu} &= \partial_\mu A^a_\nu - \partial_\nu A^a_\mu - e\epsilon_{abc}A^b_\mu A^c_\nu, \\
D_\mu\Phi^a &= \partial_\mu\Phi^a - e\epsilon_{abc}A^b_\mu\Phi^c. \qquad (7.70)
\end{aligned}$$

Once again, we encounter a theory that undergoes SSB. In this model, two of the three gauge bosons in the theory acquire a mass through the Higgs mechanism. There is also a physical Higgs particle. The masses of the vector and Higgs bosons are

$$\begin{aligned}
M_V^2 &= e^2\sigma^2, \\
M_S^2 &= \lambda\sigma^2. \qquad (7.71)
\end{aligned}$$

The magnitude of $\langle\Phi^a\rangle$ is fixed by the minimization of the potential: $|\Phi| = \sigma$. However, the direction of $\langle\Phi^a\rangle$ in group space is not. This is just a manifestation of the $SO(3)$ gauge symmetry. It should be clear that the lowest energy solution is the one where $\Phi^a(\vec{\mathbf{x}})$ = const ($\vec{\mathbf{x}}$ = spatial coordinate) since this also minimizes the kinetic energy (spatial gradient term). Even if $\Phi^a(\vec{\mathbf{x}}) \neq$ const, the spatial dependence of the direction

[23] $SO(3)$ refers to the group of length-preserving rotations in a three-dimensional "internal" space.

of Φ^a can often be gauged away, i.e., $D_\mu\Phi^a$ made equal to zero by an appropriate gauge configuration $A^a_\mu(\vec{x})$, with finite energy. However, there are Higgs field configurations that cannot be deformed into a configuration of constant Φ^a by a finite-energy gauge transformation.

An example of a configuration that cannot be gauged away is the so-called "hedgehog" configuration, in which the direction of Φ^a in group space is proportional to $\hat{\mathbf{r}}$, where $\hat{\mathbf{r}}$ is the unit vector in ordinary space. This solution is spherically symmetric, and as $r \to \infty$,

$$\begin{aligned} \Phi^a(r,t) &\rightarrow \sigma\hat{\mathbf{r}}, \\ A^a_\mu(r,t) &\rightarrow \varepsilon_{\mu ab}\hat{\mathbf{r}}_b/er. \end{aligned} \tag{7.72}$$

Like the domain wall and the cosmic string solutions, continuity requires that the Higgs field vanish as $r \to 0$. The vanishing of the Higgs field at the origin accounts for the topological stability of the hedgehog: There is no way to smoothly deform the hedgehog into a configuration where $\langle|\Phi^a|\rangle = \sigma$ everywhere. The size of the monopole, i.e., the region over which $\langle|\Phi^a|\rangle \neq \sigma$, is of order σ^{-1} (see Fig. 7.13). The energy of the hedgehog configuration receives contributions from both the vacuum energy associated with $\langle|\Phi^a|\rangle \neq \sigma$ and spatial gradient energy associated with the variation of $\langle\Phi^a\rangle$.

Next consider the long range ($r \to \infty$) magnetic field associated with the hedgehog solution:

$$B^a_i = \frac{1}{2}\varepsilon_{ijk}F^a_{jk} = \frac{\hat{\mathbf{r}}_i\hat{\mathbf{r}}^a}{er^2}. \tag{7.73}$$

Remarkably, it corresponds to the magnetic field of a magnetic monopole with charge of $h = 1/e$, or twice the Dirac charge.[24] The hedgehog is a magnetic monopole! Note that we started with a theory without fundamental magnetic monopoles, and due to the topology of the Higgs VEV, a classical solution with the properties of a magnetic monopole arose. The magnetic monopole solution was first found by 't Hooft and Polyakov [34].

The energy of the magnetic monopole solution is obtained by computing the Hamiltonian for this configuration

$$E = \int \mathcal{H}\ d^3x = -\int \mathcal{L}\ d^3x$$

[24]In CGS units (where $e^2 = \alpha_{EM}$) the charge of a Dirac monopole is $h_D = 1/2e = e/2\alpha \simeq 69e$ (e is the electric charge of the proton). In Heaviside-Lorentz units (where $e^2/4\pi = \alpha_{EM}$), the charge of the monopole is $h_D = 2\pi/e \simeq 69e$.

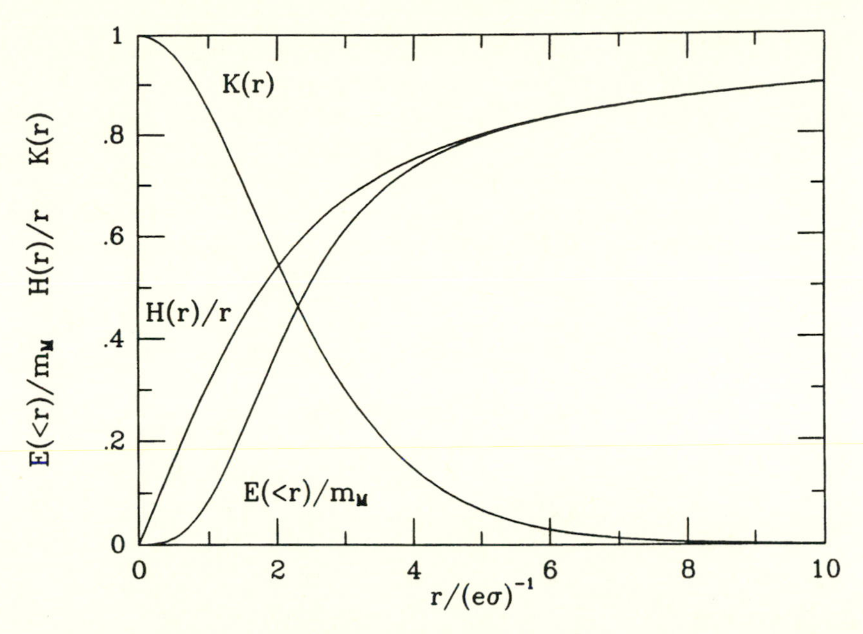

Fig. 7.13: The Higgs field, $\Phi^a = H(r)\hat{\mathbf{r}}^a/er$, the gauge field, $A_i^a = \varepsilon^{aij}\hat{\mathbf{r}}_j(1 - K(r))/er$, and the energy contained within a sphere of radius r, $E(< r)$, for the hedgehog solution in the Prasad-Sommerfield limit ($\lambda/e^2 \to 0$). The "size" of the monopole is roughly σ^{-1}, and its mass is $m_M = 4\pi\sigma/e$.

$$= \frac{4\pi}{e^2}\int_0^\infty dr\left[K'^2 + \frac{(K^2-1)^2}{2r^2} + \frac{H^2K^2}{r^2} + \frac{(rH'-H)^2}{2r^2} + \frac{\lambda r^2}{8e^2}\left(\frac{H^2}{r^2} - \sigma^2 e^2\right)^2\right], \tag{7.74}$$

where $\Phi^a = H(r)\hat{\mathbf{r}}^a/er$, $A_0^a = 0$, $A_i^a = \varepsilon^{aij}\hat{\mathbf{r}}_j(1 - K(r))/er$, and prime denotes $\partial/\partial r$. The functions $H(r)$ and $K(r)$, as well as the monopole mass m_M, depend upon $\sqrt{\lambda}/e = M_S/M_V$ (the ratio of the Higgs to vector boson mass), and in general there are no closed-form solutions. However, in the special case $M_S/M_V \to 0$ (the so-called Prasad-Sommerfield limit [35]), there is a closed-form solution,

$$K(r) = \frac{r/r_0}{\sinh(r/r_0)},$$
$$H(r) = (r/r_0)\coth(r/r_0) - 1; \tag{7.75}$$

where $r_0 = (e\sigma)^{-1}$ (see Fig. 7.13). The monopole mass in the Prasad-Sommerfield limit is given by

$$m_M = 4\pi\frac{\sigma}{e} = \frac{M_V}{\alpha}. \tag{7.76}$$

In general, the mass of the monopole is a monotonically increasing function of $\lambda/e^2 = M_S^2/M_V^2$. The Prasad-Sommerfield solution bounds the monopole mass from below; the upper bound to the monopole mass, the mass in the $M_V/M_S \to 0$ limit, is a factor of 1.787 larger than this value [36].

In general, gauge and Higgs field configurations corresponding to a magnetic monopole exist if the vacuum manifold ($\mathcal{M} = \mathcal{G}/\mathcal{H}$) associated with the symmetry breaking pattern $\mathcal{G} \to \mathcal{H}$ contains non-shrinkable surfaces, i.e., if the mapping of $\mathcal{M}$ onto the two-sphere is non-trivial. Mathematically, this is expressed by the statement that monopoles solutions arise in the theory if $\mathbf{\Pi_2}(\mathcal{M}) \neq \mathcal{I}$. If $\mathcal{G}$ is simply connected, then $\mathbf{\Pi_2}(\mathcal{G}/\mathcal{H}) = \mathbf{\Pi_1}(\mathcal{H})$. If $\mathcal{G}$ is not simply connected, the generalization of the above expression is $\mathbf{\Pi_2}(\mathcal{G}/\mathcal{H}) = \mathbf{\Pi_1}(\mathcal{H})/\mathbf{\Pi_1}(\mathcal{G})$. In the example above, $\mathcal{G} = SO(3)$, $\mathcal{H} = U(1)$,[25] and $\mathbf{\Pi_2}[SO(3)/U(1)] = \mathbf{\Pi_1}[U(1)]/\mathbf{\Pi_1}[SO(3)] = \mathcal{Z}/\mathcal{Z}_2$, the integers mod 2.

Since $\mathbf{\Pi_1}[U(1)] = \mathcal{Z}$, it follows that when any semi-simple group is broken down to a subgroup $\mathcal{H}$ that contains an explicit $U(1)$ factor $\mathbf{\Pi_1}(\mathcal{H}) \neq \mathcal{I}$, and monopole solutions necessarily exist. Since the goal of grand unification is to unify the strong, weak, and electromagnetic interactions within a single gauge group that is ultimately broken down to $SU(3) \otimes SU(2) \otimes U(1)$, it is apparent that the existence of magnetic monopole solutions is a generic prediction of grand unification. Moreover, since the scale of grand unification is expected to be around 10^{14}GeV, the magnetic monopoles associated with GUTs are "superheavy," $m_M \sim 10^{16}$GeV.

Magnetic monopoles have been discussed by theorists and sought after by experimentalists for decades. The existence of the 't Hooft-Polyakov monopole solution has spurred new interest, because unlike the Dirac monopole, which could or could not be put into the theory at the theorist's whim, these topological monopoles must necessarily exist if $\mathbf{\Pi_2}(\mathcal{M}) \neq \mathcal{I}$, and they arise not as fundamental entities in the theory but as classical solutions to the field equations. As we shall discuss shortly, if magnetic

---

[25] $SO(3)$ is not simply connected. It is equivalent to the three-sphere with antipodal points identified. (Warning: Identifying antipodal points can be hazardous and should be attempted only by professionals.)

monopoles are indeed present in the Universe, they have a myriad of interesting astrophysical and cosmological consequences.

7.5 The Kibble Mechanism

We have discussed the three kinds of topological defects associated with spontaneously broken symmetries: the point defect, or monopole; the line defect, or cosmic string; and the sheet defect, or domain wall. The existence and stability of these objects is dictated by topological considerations.

Many spontaneously broken gauge theories predict the existence of one or more of the above topological defects. These objects are inherently non-perturbative and probably cannot be produced in high energy collisions at terrestrial accelerators. It is very likely that the only place they can be produced is in phase transitions in the early Universe. Although monopoles, strings, and domain walls are topologically stable, they are not the minimum energy configurations. However, their production in cosmological phase transitions seems unavoidable. The "unavoidable" cosmological production mechanism is known as the Kibble mechanism [37] and is very much akin to the mechanism for production of various defects in ordinary, laboratory phase transitions.

The Kibble mechanism hinges upon the fact that during a cosmological phase transition any correlation length is always limited by the particle horizon. The particle horizon, first discussed in Section 2.2, is the maximum distance over which a massless particle could have propagated since the time of the bang. It is given by

$$d_H = R(t) \int_0^t \frac{dt'}{R(t')}. \tag{7.77}$$

If $R \propto t^n$ $(n > 1)$, then $d_H = t/(1-n)$.

The correlation length associated with the phase transition sets the maximum distance over which the Higgs field can be correlated. The correlation length depends upon the details of the phase transition and is temperature-dependent. It is related to the temperature-dependent Higgs mass: $\xi \sim M_H^{-1}(T) \sim T^{-1}$. In any case, the fact that the horizon distance is finite in the standard cosmology implies that at the time of the phase transition $(t = t_C,\ T = T_C)$, the Higgs field must be uncorrelated on scales greater than d_H, and thus the horizon distance sets an absolute maximum for the correlation length.

During a SSB phase transition, some Higgs field acquires a VEV. Because of the existence of the particle horizon in the standard cosmology, when this occurs $\langle\phi\rangle$ cannot be correlated on scales larger than $d_H \sim H^{-1} \sim m_{Pl}/T^2$. Therefore, it should be clear that the non-trivial vacuum configurations discussed in the previous Sections will necessarily be produced, with an abundance of order one per horizon volume. While these topological creatures are not the minimum-energy configurations of the Higgs field, they arise as "topological defects" because of the finite particle horizon. Since they are stable, they are "frozen in" as permanent defects when they form.

Consider monopoles as an example of the "freezing in" of topological defects. The direction of the VEV of the Higgs field Φ^a must be random on scales greater than d_H. That means that in different Hubble-sized volumes, $\langle\Phi^a\rangle$ will point in different directions (in group space). The hedgehog corresponds to $\langle\Phi^a\rangle$ changing its direction as $\hat{\mathbf{r}}$ (about some point). Thus one expects that the probability that a monopole configuration will result from a Higgs VEV whose direction is uncorrelated on scales greater than H^{-1} should be of order unity. Therefore, about one monopole (or antimonopole) per horizon volume should arise, $n_M \sim d_H^{-3} \sim T_C^6/m_{Pl}^3$. The entropy density at temperature T_C is $s \sim T_C^3$, so that the resulting monopole-to-entropy ratio is $n_M/s \sim (T_C/m_{Pl})^3$.

Barring significant monopole-antimonopole annihilation, or entropy production, this ratio remains constant and determines the present monopole abundance. For the canonical values associated with GUT SSB, $T_C \sim 10^{14}$GeV and $m_M \sim 10^{16}$GeV, the relic monopole abundance, $n_M/s \sim 10^{-13}$, results in a present monopole mass density of about $10^{11}\rho_C$, a value that is clearly unacceptable. As we shall see in the next Chapter, the best solution to this very serious problem is inflation, whereby the problem is solved by the massive entropy production associated with a cosmological phase transition. The infusion of entropy dilutes the initial value of n_M/s exponentially.

The topological defects associated with SSB are very interesting beasts, whose only plausible production site is a cosmological phase transition wherein they are produced by the Kibble mechanism. We have not discussed the even more bizarre animals in the topological zoo, such as walls bounded by strings, strings terminated by monopoles, or monopoles strung on a string. These hybrid objects can arise in more complicated symmetry breaking patterns [38].

As we have seen, both domain walls and magnetic monopoles are inherently cosmologically dangerous, while cosmic strings may provide the seeds for structure formation if $G\mu \sim 10^{-6}$ or so. If anything can be said with certainty about walls, strings, and monopoles, it is that, if discovered, they would both provide valuable information about particle physics beyond the standard model and have profound cosmological consequences.

7.6 Monopoles, Cosmology, and Astrophysics

In this Section, we will discuss the intriguing astrophysical and cosmological effects of magnetic monopoles in some detail. The effects are so conspicuous—and thus far not seen—that they result in severe constraints to the possible flux of relic monopoles. The astrophysical and cosmological effects of monopoles involve the average monopole flux in the Universe, $\langle F_M \rangle = n_M v_M / 4\pi$, where v_M is the monopole velocity, and the fraction of critical density contributed by monopoles, Ω_M:

$$\begin{aligned} \langle F_M \rangle &\simeq 10^{10} \left(\frac{n_M}{s}\right) \left(\frac{v_M}{10^{-3}\mathrm{c}}\right) \mathrm{cm}^{-2}\mathrm{sr}^{-1}\mathrm{sec}^{-1}, \\ \Omega_M h^2 &\simeq 10^{24} \left(\frac{n_M}{s}\right) \left(\frac{m_M}{10^{16}\mathrm{GeV}}\right). \end{aligned} \quad (7.78)$$

(The use of 10^{-3}c for the typical monopole velocity will be discussed in detail later.)

7.6.1 Magnetic monopoles at birth

Let us begin with a more careful consideration of the monopole abundance produced via the Kibble mechanism. Consider first the case where the transition is either second order or weakly first order. The age of the Universe when $t \simeq t_C$ is given in the standard cosmology by $t_C \simeq 0.3 g_*^{-1/2} m_{Pl}/T_C^2$. For an $SU(5)$ GUT, $T_C \simeq 10^{14}$GeV, $m_M \simeq 10^{16}$GeV, and $t_C \simeq 10^{-34}$s. Using the $SU(5)$ example and taking the correlation length to be t_C^{-1}, we find

$$\frac{n_M}{s} \simeq 10^2 \left(\frac{T_C}{m_{Pl}}\right)^3 \simeq 10^{-13}. \quad (7.79)$$

Preskill [39] has shown that unless n_M/s is greater than about 10^{-10}, annihilation of monopoles and antimonopoles does not significantly reduce the initial monopole abundance. If $n_M/s > 10^{-10}$, he finds that n_M/s is reduced to about 10^{-10} by annihilations.[26] Since the estimate for the initial value of n_M/s arising due to the Kibble mechanism is less than 10^{-10}, we can ignore annihilations. *Assuming* that the expansion has been adiabatic since $T \simeq T_C$, our estimate for n_M/s in (7.79) leads to

$$\begin{aligned}\langle F_M \rangle &\simeq 10^{-3}\left(\frac{T_C}{10^{14}\mathrm{GeV}}\right)^3\left(\frac{v_M}{10^{-3}\mathrm{c}}\right)\ \mathrm{cm^{-2}sr^{-1}sec^{-1}},\\ \Omega_M h^2 &\simeq 10^{11}\left(\frac{T_C}{10^{14}\mathrm{GeV}}\right)^3\left(\frac{m_M}{10^{16}\mathrm{GeV}}\right),\end{aligned}\tag{7.80}$$

a flux that is easy to detect, and a mass density that is unacceptably large (unless $T_C \ll 10^{14}$GeV). As we discussed in Chapter 3, the age of the Universe restricts $\Omega_0 h^2$ to be less than one. It is very clear that the wedding of the simplest GUTs (wherein $T_C \sim 10^{14}$GeV) and the standard cosmology results in a cosmic catastrophe, the so-called "Monopole Problem."

The limit to Ω_M based upon the age of the Universe translates into the limit $T_C \lesssim 10^{11}$GeV for $m_M \simeq T_C/\alpha$. Given the generous estimate taken for the correlation length, even this is probably *not* safe; if one constructed an attractive unified model wherein $T_C \simeq 10^{11}$ GeV, one would certainly want to make a more careful estimate for the correlation length.

If the GUT transition is strongly first order (excluding inflationary Universe scenarios), then the transition will proceed by bubble nucleation at a temperature $T_N \ll T_C$, when the nucleation rate becomes comparable to the expansion rate H. Within each bubble the Higgs field is correlated; however, the Higgs field in different bubbles should be uncorrelated. Thus, one would expect about one monopole per bubble to be produced. To be specific, when the Universe supercools to a temperature T_N, bubbles nucleate, expand, and rapidly fill all of space; if r_b is the typical size of a bubble when this occurs, one expects $n_M \simeq r_b^{-3}$. After the bubbles coalesce, and the Universe reheats, the entropy density is once again $s \simeq g_* T_C^3$, so that the resulting monopole to entropy ratio is $n_M/s \simeq (g_* r_b^3 T_C^3)^{-1}$. Guth and Weinberg [40] have calculated r_b and find that $r_b \simeq (m_{Pl}/T_C^2)/\ln(m_{Pl}^4/T_C^4)$, which leads to a relic monopole abundance of

$$\frac{n_M}{s} \simeq \left[\left(\frac{T_C}{m_{Pl}}\right)\ln\left(\frac{m_{Pl}^4}{T_C^4}\right)\right]^3 . \tag{7.81}$$

---

[26] This can be easily checked by using the method for calculating the decoupling of a massive particle species detailed in Chapter 5.

The logarithmic factor makes the monopole glut even worse than in the case of a second-order phase transition.

There is no simple way around it; the standard cosmology extrapolated back to a temperature of order 10^{14}GeV and the simplest GUTs are incompatible. One (or both) must be modified. Although this result is discouraging, especially when viewed in the light of the great success of baryogenesis, it does provide a valuable piece of information about physics at very high energies and/or the earliest moments of the Universe. In that regard it is a valuable "window" to energies of 10^{14}GeV and times of 10^{-34}sec.

A number of possible solutions to the monopole problem have been suggested. To date, the most attractive is the inflationary Universe scenario, which will be the subject of the next Chapter. The exponential expansion associated with inflation allows a small, sub-horizon sized region of space, within which the Higgs field is correlated, to encompass all of the presently observed Universe. The end result is less than one monopole in the entire observable Universe due to the Kibble mechanism.

Before proceeding to the astrophysical implications of monopoles, let us mention two other solutions to the monopole problem. It has been pointed out that if there is not a complete unification of the forces, e.g., if $\mathcal{G} = \mathcal{H} \otimes U(1)$, or if the full symmetry of the GUT is not restored in the very early Universe (e.g., if the maximum temperature the Universe reached was less than T_C, or if a large lepton number, $n_L/n_\gamma > 1$, prevented symmetry restoration at high temperature), then there would be no monopole problem.

Langacker and Pi [41] have suggested a more interesting solution. It is based upon an unusual (perhaps contrived) symmetry breaking pattern: $SU(5) \to SU(3) \otimes SU(2) \otimes U(1) \to SU(3) \to SU(3) \otimes U(1)$. The first phase transition occurs at $T_C \simeq 10^{14}$GeV, the second transition at T_1, and the final transition at T_2 (note T_1 could be equal to T_C). The key to their solution is the existence of an epoch ($T_1 \geq T \geq T_2$) when the $U(1)$ of electromagnetism is spontaneously broken. During this period the Universe is superconducting and magnetic flux must be confined to flux tubes, leading to the efficient annihilation of the monopoles and antimonopoles that were produced earlier at the GUT transition. The resulting monopole abundance in this scheme is about one per horizon volume *when the superconducting phase ends*, or

$$n_M/s \simeq 10^2(T_2/m_{Pl})^3 \simeq 10^{-46}(T_2/10^3\text{GeV})^3, \tag{7.82}$$

a number safely below the cosmological density limit if T_2 is much less than 10^{14}GeV.[27]

If the glut of monopoles produced as topological defects via the Kibble mechanism can be avoided, then the only cosmic production mechanism is pair production in very energetic particle collisions, e.g., particle(s) + antiparticle(s) $\rightarrow$ monopole + antimonopole. The numbers produced in this way are intrinsically small, because monopole configurations do not exist in the theory until SSB occurs ($T_C \simeq M$ = scale of SSB) and have a mass $m_M \simeq M/\alpha \simeq 100M \simeq 100T_C$. Thermal pair production results in a relic monopole abundance of

$$\frac{n_M}{s} \simeq 10^2 \left(\frac{m_M}{T_{\rm max}}\right)^3 \exp\left(-\frac{2m_M}{T_{\rm max}}\right), \tag{7.83}$$

where $T_{\rm max}$ is the highest temperature reached after SSB [43].

In general, one expects that $m_M/T_{\rm max}$ is at least 100, so that $\Omega_M \lesssim 10^{-40}$ and $\langle F_M \rangle \lesssim 10^{-32}\text{cm}^{-2}\text{sr}^{-1}\text{sec}^{-1}$—a negligible number of monopoles. However, the number produced is *exponentially* sensitive to $m_M/T_{\rm max}$, so that a factor of 3 to 5 change in $m_M/T_{\rm max}$ results in an enormous change in the predicted production. It is not inconceivable that thermal production of monopoles could lead to an interesting relic abundance.

Cosmology seems to make two firm predictions about the relic monopole abundance: That there should be equal numbers of north and south magnetic poles, and that either far too few to detect, or far too many to be consistent with the standard cosmology, should have been produced. The detection of a single superheavy monopole would necessarily send theorists back to their chalkboards!

7.6.2 Magnetic monopoles at adolescence

Leaving aside the question of the relic abundance, what is the fate of any relic monopoles that were produced? As we have discussed, monopoles and antimonopoles do not annihilate in significant numbers; however, they do interact with the ambient charged particles (e.g., monopole + e^- $\leftrightarrow$ monopole + e^-) and thereby stay in *kinetic equilibrium* (KE $\simeq 3T/2$)

---

[27] E. Weinberg has carried out an analysis of very efficient annihilation schemes like this, and argues that initial fluctuations in the local monopole–antimonopole abundances result in a *minimum* relic monopole abundance of $n_M/s \simeq 10^2 T_C T_2^2/m_{Pl}^3$ [42]. If this is correct, it is not clear whether the superconducting scenario could sufficiently reduce n_M/s.

until the epoch of $e^{\pm}$ annihilations ($T \simeq 0.3$ MeV, $t \simeq 10$ s). At the time of $e^{\pm}$ annihilations, monopoles and antimonopoles should have internal velocity dispersions of

$$\langle v_M^2 \rangle^{1/2} \simeq \sqrt{T/m_M} \simeq 30 \text{ cm s}^{-1}(10^{16}\text{GeV}/m_M)^{1/2}. \tag{7.84}$$

After this, monopoles are effectively collisionless, and their velocity dispersion decays as $R(t)^{-1}$. *If* one neglects gravitational and magnetic effects, today monopoles should have a tiny internal velocity dispersion,

$$\langle v_M^2 \rangle^{1/2} \simeq 10^{-8}\text{cm s}^{-1}(10^{16}\text{GeV}/m_M)^{1/2}. \tag{7.85}$$

Since they are collisionless, their velocity dispersion provides their only support against gravitational collapse. With such a small velocity dispersion, monopoles are gravitationally unstable on all scales of astrophysical interest.

After the Universe becomes matter dominated, matter can begin to clump, and structure starts to form. Monopoles also should clump and participate in the formation of structure [44]. However, because they cannot dissipate their gravitational energy, they cannot collapse into the more condensed objects found in the Universe, such as stars, planets, cheesecake, the disk of the galaxy, etc., because the formation of these objects certainly must have involved the dissipation of gravitational energy. Thus, one would expect to find monopoles only in structures whose formation did not require dissipation, such as clusters of galaxies and galactic halos. As we shall see, galaxies with magnetic fields will not provide safe havens for monopoles less massive than 10^{20}GeV.

Although monopoles have a very small internal velocity dispersion initially, there are several astrophysical mechanisms that will inevitably increase their velocities. Like any other form of matter, monopoles will be accelerated by the gravitational fields that arise due to the inhomogeneous distribution of matter in the Universe. The gravitational field in the neighborhood of our galaxy is such that our local group is moving at about 2×10^{-3}c with respect to the CMBR (our peculiar velocity). Monopoles that encounter a galaxy or a cluster of galaxies will be accelerated by their gravitational fields to about 10^{-3}c for a galaxy and to 0.3 to 1×10^{-2}c for a cluster. A typical monopole, however, will never encounter a galaxy or a cluster of galaxies, as the mean free path of a monopole between encounters is $\Delta L_{\text{galaxy}} \simeq 10^{26}$cm and $\Delta L_{\text{cluster}} \simeq 3 \times 10^{28}$ cm, while the distance a monopole travels in a Hubble time is only about $10^{25}h^{-1}(v_M/10^{-3}\text{c})$ cm.

Magnetic monopoles will also be accelerated by magnetic fields. Over the age of the Universe, an intergalactic magnetic field of strength 10^{-11}G will accelerate a monopole to a velocity

$$v_M \simeq 3 \times 10^{-4} c \left(\frac{B}{10^{-11}\mathrm{G}}\right) \left(\frac{10^{16}\mathrm{GeV}}{m_M}\right). \tag{7.86}$$

The strength (and existence!) of any intergalactic magnetic field is still an open question: Limits and measurements have been reported in the range 10^{-9} to 10^{-11}G [45]. The galactic magnetic field on the other hand, is well measured. It has a strength of about 3×10^{-6}G, and a coherence length of about 300 pc. In traversing 300 pc of galactic magnetic field, a monopole will acquire a velocity of

$$v_M \simeq 3 \times 10^{-3} c \left(\frac{10^{16}\mathrm{GeV}}{m_M}\right)^{1/2}. \tag{7.87}$$

While monopoles should emerge from the early Universe with very small velocities, it is clear that contemporary astrophysical processes insure that they will have a significant velocity today—at least of order 10^{-3}c. As we shall see, the typical monopole velocity is important for monopole detection schemes, and for determining the astrophysical effects of monopoles, and the resulting flux limits.

7.6.3 Magnetic monopoles at middle age—flux limits

In discussing the astrophysical implications of GUT monopoles, the three most important properties of a monopole are: (i) its macroscopic mass, $m_M \simeq 10^{16}$ GeV $\simeq 10^{-8}$g—comparable to an amoeba; (ii) its sizeable magnetic charge, $h = n69e$ ($n = \pm 1, \pm 2, \cdots$); and (iii) its ability to catalyze nucleon decay with a cross section of order $10^{-28}\mathrm{cm}^2$. We have discussed the first two properties; shortly we will elaborate on the third and most intriguing property. Based upon these three properties, a variety of very stringent astrophysical bounds can be placed on the flux of relic monopoles.

• *Mass Density:* Theoretical prejudice favors the flat cosmological model (i.e., $\Omega_0 = 1$). As discussed in Chapter 4, big bang nucleosynthesis strongly indicates that baryons contribute much less than critical density, $\Omega_B \lesssim 0.15$. As we have discussed previously, flat rotation curves of galaxies provide strong evidence that most of the mass associated with a galaxy is dark and exists in an extended structure (most likely a spherical halo). Monopoles are certainly a candidate for the dark matter in galaxies and for providing closure density.

The age of the Universe implies that $\Omega_0 h^2 \lesssim 1$; if monopoles are uniformly distributed in the cosmos and have typical velocity $v_M \simeq 10^{-3}$c, this constrains their average flux to be

$$\langle F_M \rangle \lesssim 10^{-14}\mathrm{cm}^{-2}\mathrm{sr}^{-1}\mathrm{sec}^{-1}\left(\frac{m_M}{10^{16}\mathrm{GeV}}\right)^{-1}. \tag{7.88}$$

If monopoles cluster in galaxies, ours in particular, the local galactic flux can be significantly higher. The mass density in the neighborhood of the sun is about 10^{-23}g cm^{-3}; of this, about half is accounted for (stars, gas, dust, etc.). Monopoles can at most provide the other half, resulting in the flux bound

$$\langle F_M \rangle \lesssim 5\times 10^{-10}\mathrm{cm}^{-2}\mathrm{sr}^{-1}\mathrm{sec}^{-1}\left(\frac{m_M}{10^{16}\mathrm{GeV}}\right)^{-1}. \tag{7.89}$$

The actual situation is less favorable; the local unseen matter most certainly must be associated with the disk, whose scale height is a few hundred pc. If it were associated with the halo, the halo mass interior to the solar system would result in an orbital velocity for the solar system of about 1100 km s^{-1}, or five times that observed. The best estimate for the local halo density is $\rho_{\mathrm{halo}} \simeq 0.5\times 10^{-24}$g cm^{-3}. Taking the local monopole mass density to be less than 10^{-24}g cm^{-3}, we obtain the more realistic bound

$$\langle F_M \rangle \lesssim 10^{-10}\,\mathrm{cm}^{-2}\mathrm{sr}^{-1}\mathrm{sec}^{-1}\left(\frac{m_M}{10^{16}\mathrm{GeV}}\right)^{-1}. \tag{7.90}$$

• *Parker Limit:* A monopole by virtue of its magnetic charge will be accelerated by any magnetic field it encounters, and in the process it can gain kinetic energy (KE). Of course, any KE gained must be compensated for by a loss in field energy: $\Delta\mathrm{KE} = -\Delta[(B^2/2)\times\mathrm{Volume}]$. Consider a monopole with charge $h = 1/2e$, which is initially at rest in a region of uniform magnetic field. It will be accelerated by the field, and after moving a distance ℓ, the monopole will have kinetic energy and velocity

$$\begin{aligned}\mathrm{KE} &\simeq hB\ell \simeq 10^{11}\mathrm{GeV}(B/3\times 10^{-6}G)(\ell/300\ \mathrm{pc}),\\ v_{\mathrm{mag}} &\simeq \left(\frac{2hB\ell}{m_M}\right)^{1/2}\end{aligned}$$

$$\simeq 3 \times 10^{-3}c\left(\frac{B}{3\times 10^{-6}\text{G}}\frac{\ell}{300\text{ pc}}\frac{10^{16}\text{GeV}}{m_M}\right)^{1/2}. \quad (7.91)$$

If the monopole is not initially at rest, then the story is a bit different. There are two limiting situations, and they are delineated by the relative sizes of the initial velocity of the monopole, v_0, and the velocity just calculated above, $v_{\rm mag}$. First, if the monopole is moving slowly compared to $v_{\rm mag}$, $v_0 \ll v_{\rm mag} \simeq (2hB\ell/m_M)^{1/2}$, then it will undergo a large velocity change due to the magnetic field, and its change in KE will be well-approximated by (7.91). On the other hand, if $v_0 \gg v_{\rm mag}$, the monopole will be only slightly affected by the magnetic field, and its change in KE will depend upon the direction of its motion relative to the orientation of the magnetic field. In this situation the energy gained by a spatially isotropic flux of monopoles, or a flux of equal numbers of north and south poles, will vanish at first order in the magnetic field strength B, as some poles will lose KE and others will gain KE. However, at second order in B, the distribution of monopoles as a whole will increase its KE, by an amount

$$\langle \Delta \text{KE}\rangle \simeq \frac{1}{4}(hB\ell)\left(\frac{v_{\rm mag}}{v_0}\right)^2 \qquad \text{(per monopole).} \qquad (7.92)$$

Our galactic magnetic field is characterized by $B \simeq 3\times 10^{-6}$G, $\ell \simeq 300$ pc, and $v_{\rm mag} \simeq 3\times 10^{-3}c(10^{16}\text{GeV}/m_M)^{1/2}$. For any monopole within our galaxy, v_0 is set by the virial velocity: $v_0 \simeq 10^{-3}$c. Thus, monopoles less massive than about 10^{17}GeV will undergo large deflections when moving through the galactic field, and their gain in KE is given by (7.91). Because of this energy gain, monopoles less massive than 10^{17}GeV will be ejected from our galaxy in a very short time and will not cluster in our galactic halo. In fact, even the second-order gain in KE will "evaporate" monopoles as massive as 10^{20}GeV in a time comparable to the age of the galaxy [46]. Although gravitational considerations would suggest that monopoles should cluster in galactic halos, galactic magnetic fields should prevent monopoles less massive than 10^{20}GeV from residing there for the age of the Universe.[28]

The "no free lunch" principle, ΔKE $= -\Delta$(Magnetic Field Energy), and the formulae for $v_{\rm mag}$ and $\langle\Delta\text{KE}\rangle$ can be used to derive a limit to the average flux of monopoles in the galaxy. If, as it is commonly believed,

[28]These conclusions are not valid if the magnetic field of the galaxy is, in part, produced by monopoles. This point will be discussed below.

the origin of the galactic magnetic field is due to dynamo action, then the time required to generate/regenerate the field is of the order of a galactic rotation time of $\tau \simeq 10^8$ yr. Demanding that monopoles not drain the field energy in a time shorter than this, i.e., $[(B^2/2)\cdot \mathrm{Vol}]/[\langle F_M\rangle \cdot \langle \mathrm{KE}\rangle \cdot$ Area $\cdot(\pi\text{-sr})] \gtrsim \tau$, results in the following constraints [46, 47]:

$$\begin{aligned}
\langle F_M\rangle &\lesssim 10^{-15}\mathrm{cm}^{-2}\mathrm{sr}^{-1}\mathrm{sec}^{-1}\left(\frac{B}{3\times 10^{-6}\mathrm{G}}\right)\left(\frac{3\times 10^7\mathrm{yr}}{\tau}\right)\\
&\quad\times\left(\frac{r}{30\ \mathrm{kpc}}\right)^{1/2}\left(\frac{300\ \mathrm{pc}}{\ell}\right)^{1/2} \qquad (m_M \lesssim 10^{17}\mathrm{GeV}),\\
\langle F_M\rangle &\lesssim 10^{-16}\mathrm{cm}^{-2}\mathrm{sr}^{-1}\mathrm{sec}^{-1}\left(\frac{m_M}{10^{16}\mathrm{GeV}}\right)\left(\frac{3\times 10^7\mathrm{yr}}{\tau}\right)\\
&\quad\times\left(\frac{300\ \mathrm{pc}}{\ell}\right) \qquad (m_M \gtrsim 10^{17}\mathrm{GeV}).
\end{aligned} \tag{7.93}$$

Here τ is the regeneration time of the field, ℓ is the coherence length of the field, and r is the size of the magnetic field region in the galaxy. Constraint (7.93), valid for monopoles less massive than 10^{17}GeV, is known as the "Parker bound," and has served as a "benchmark" for the sensitivity of monopole search experiments. For more massive monopoles ($m_M \gtrsim 10^{17}$GeV) the "Parker bound" becomes less restrictive; however, the mass density constraint becomes more restrictive. These two bounds together restrict the flux to be less than about $10^{-13}\mathrm{cm}^{-2}\mathrm{sr}^{-1}\mathrm{sec}^{-1}$ regardless of the monopole mass (see Fig. 7.14).

Parker's argument can be applied to the survival of other astrophysical magnetic fields. The most stringent bound obtained by this method follows from consideration of intracluster (IC) magnetic fields. Many rich clusters are observed to have IC magnetic fields with typical strength $B \sim 10^{-7}$G, and coherence length $\ell \simeq 100$ kpc. The "Parker limit" from IC fields is $\langle F_M\rangle \lesssim 10^{-18}\mathrm{cm}^{-2}\mathrm{sr}^{-1}\mathrm{sec}^{-1}$ [48]. Because our knowledge of the existence and persistence time of IC fields is less secure, this bound is less reliable than the bound from our own galactic field.

One might imagine the Parker bound being circumvented if monopoles themselves participate in the maintenance of the galactic magnetic field. This can occur if a coherent monopole magnetic plasma mode is excited, and monopoles only "borrow" the KE they gain from the magnetic field, returning it a half cycle later [46, 49]. In order for this to work the monopole oscillations must maintain both spatial and temporal coherence; if they

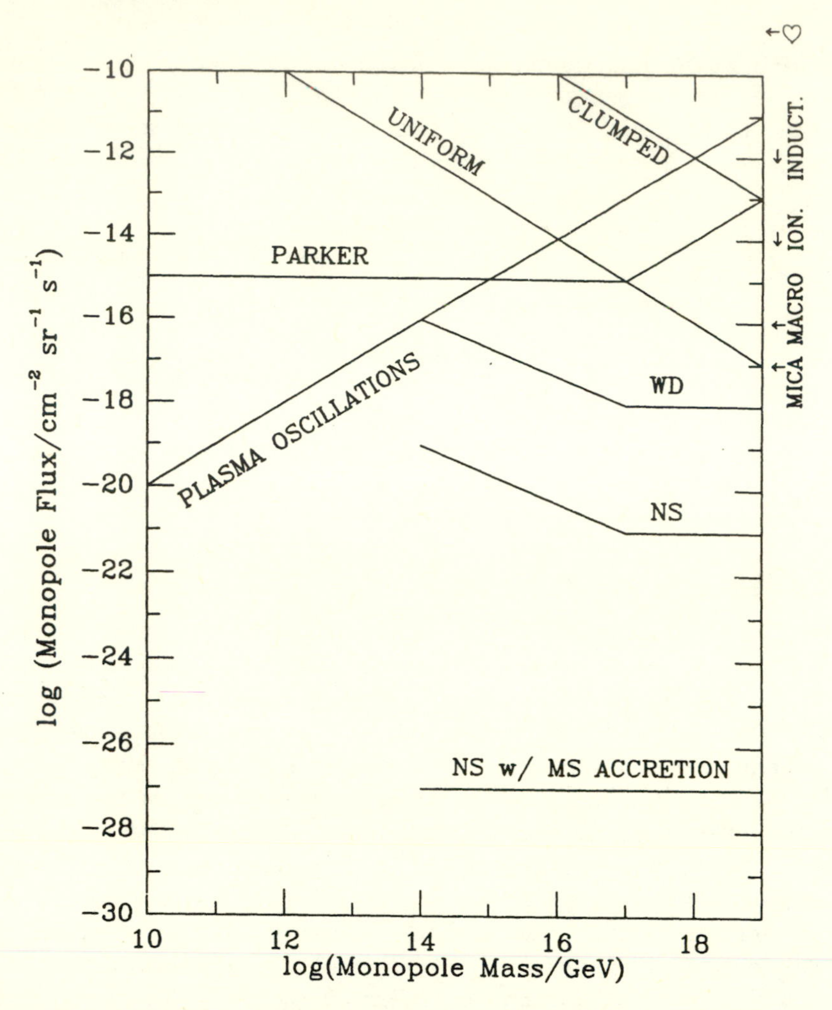

Fig. 7.14: Summary of the astrophysical/cosmological limits to the monopole flux as a function of monopole mass. Whenever necessary, the monopole velocity was assumed to be 10^{-3}c. The mass-density limits are for monopoles uniformly distributed in the cosmos (*UNIFORM*) and clustered within the galaxy (*CLUMPED*). The limits labeled *WD*, *NS*, and *NS w/ MS ACCRETION* are based upon monopole catalysis with $(\sigma v) = 10^{-28}\mathrm{cm}^2$. The limit *w/ MS ACCRETION* includes monopoles accreted during the main-sequence phase of the NS progenitor. The WD and NS limits become less restrictive for $m_M \lesssim 10^{17}$GeV, as such monopoles should have velocities larger than 10^{-3}c, which lowers the capture cross section. The line labeled *Plasma Oscillations* indicates the *lower* limit to the flux predicted in the magnetic plasma oscillation scenario. The flux implied by Cabrera's Valentine's Day event [57] is so indicated. Current flux limits, as well as the expected sensitivity of the MACRO detector [58, 59] are also indicated.

do not, "phase-mixing" (Landau damping) will cause the oscillations to damp rapidly, so that the previous bounds apply. A necessary condition for temporal coherence to be maintained is that the phase velocity of the oscillations $v_{ph} \simeq \omega_{pl}(\ell/2\pi)$ be greater than the gravitational velocity dispersion of the monopoles (about 10^{-3}c); ℓ is the wavelength of the relevant mode. The monopole plasma frequency is given by

$$\omega_{pl} = \left(\frac{4\pi h^2 n_M}{m_M}\right)^{1/2}, \tag{7.94}$$

where n_M is the monopole number density. The condition that $v_{ph} \gtrsim \langle v^2\rangle^{1/2} \simeq 10^{-3}$c implies a *lower* bound to the monopole flux of

$$\begin{aligned} \langle F_M \rangle &\gtrsim \frac{1}{4} m_M \langle v^2\rangle^{3/2}(h\ell)^{-2} \\ &\gtrsim 10^{-14}\left(\frac{m_M}{10^{16}\mathrm{GeV}}\right)\left(\frac{1\ \mathrm{kpc}}{\ell}\right)^2 \mathrm{cm}^{-2}\mathrm{sr}^{-1}\mathrm{sec}^{-1}, \end{aligned} \tag{7.95}$$

and an *upper* bound to the oscillation period: $\tau = 2\pi/\omega_{pl} \lesssim \ell/\langle v^2\rangle^{1/2} \simeq 3\times 10^6\mathrm{yr}(\ell/1\ \mathrm{kpc})$—a very short time compared to other galactic timescales.

While it is intriguing that coherent monopole effects might allow a monopole flux larger than the Parker bound, and that monopoles might even play a role in the generation of the galactic magnetic field, there are several hurdles that must be overcome if this possibility is to be realized. How is spatial coherence maintained in the inhomogeneous environment of our galaxy? What prevents small-scale instabilities from efficiently draining the oscillations through Landau damping? Last, and most important, the *lower* bound to F_M predicted in such a scenario is at variance with the experimental *upper* limit to F_M unless the monopole mass is less than about 10^{15}GeV.

• *Monopole Catalysis:* Perhaps the most intriguing property of the monopole is its ability to catalyze nucleon decay with a strong interaction cross section $(\sigma v) \simeq 10^{-28}\mathrm{cm}^2$. Since the symmetry of the unified gauge group is restored at the monopole core, one would expect, on geometric grounds, that monopoles should catalyze nucleon decay with a cross section $\sigma \simeq M^{-2} \simeq 10^{-56}\mathrm{cm}^2$ (M^{-1} is the size of the monopole core). Such a small cross section would be of little interest. However, Rubakov [50] and Callan [51] have shown that due to the singular nature of the potential between the s-wave of a fermion and a magnetic monopole, the fermion wave function is literally sucked into the core. Due to s-wave sucking, the catalysis cross section saturates the unitarity bound: $(\sigma v) \simeq (\text{fermion energy})^{-2}$. The actual catalysis reaction and value of the cross section depends upon the unified theory.[29] In $SU(5)$ for example, $M + n \to M + \pi^- + e^+$, or $M + p \to M + \pi^0 + e^+$.

Needless to say, monopole catalysis has great potential to produce astrophysical fireworks. While the nuclear reaction that powers the sun, $4p \to {}^4\text{He} + 2e^+ + 2\nu_e$, proceeds at a weak interaction rate (due to the first step: $p + p \to D + e^+ + \nu_e$) and releases only about 0.7% of the rest mass involved, monopole catalysis proceeds at a strong interaction rate and releases 100% of the rest mass of the nucleon. Monopole catalysis of nucleon decay has the potential to liberate energy at a prodigious rate,

$$\dot{\varepsilon}_{\text{cat}} = n_N(\sigma v) \simeq 3 \times 10^3 (\sigma v)_{-28} \left(\rho_N/\text{g cm}^{-3}\right) \quad \text{erg s}^{-1} \qquad \text{(per monopole)}, \tag{7.96}$$

where $(\sigma v)_{-28} = (\sigma v)/10^{-28}\text{cm}^2$ and ρ_N is the density of nucleons. Only about 10^{28} monopoles at the center of the sun are needed to produce the entire solar luminosity of $4 \times 10^{33}\text{erg s}^{-1}$.

Because of their awesome power to release energy via catalysis, there can't be too many monopoles in astrophysical objects like stars, planets, etc.; otherwise the sky would be aglow in all wavebands from the energy released by monopoles.[30] The measured luminosities of neutron stars (some as low as $3 \times 10^{30}\text{erg s}^{-1}$); white dwarfs (some as low as 10^{29}erg s^{-1}); Jupiter (10^{25}erg s^{-1}); and the Earth ($3 \times 10^{20}\text{erg s}^{-1}$) imply upper limits to the number of monopoles in these objects: some neutron stars, $N_M \lesssim 10^{12}(\sigma v)_{-28}^{-1}$ monopoles; some white dwarfs, $N_M \lesssim 10^{18}(\sigma v)_{-28}^{-1}$ monopoles; Jupiter, $N_M \lesssim 10^{20}(\sigma v)_{-28}^{-1}$ monopoles; and the Earth, $N_M \lesssim 3 \times 10^{15}(\sigma v)_{-28}^{-1}$ monopoles. In order to translate these limits into bounds to the monopole flux and abundance we need to know how many monopoles are predicted to reside in each of these objects.

Although planets, stars, etc., should be monopole free at the time of their formation, they will accumulate monopoles during their lifetimes by capturing monopoles that impinge upon them. The number captured by

---

[29] Monopoles are expected to catalyze nucleon decay through GUT effects in their cores and through the weak anomaly. The latter effect is less important, except for monopoles that do not have the full GUT symmetry restored in their cores. For details, see [52].

[30] The energy released by catalysis in these objects would, of course, be thermalized and radiated from the surface of the objects.

an object is

$$N_M = (4\pi R^2)(\pi - \text{sr})(1 + 2GM/Rv_M^2)\langle F_M \rangle \varepsilon \tau, \tag{7.97}$$

where M, R and τ are the mass, radius, and age of the object, v_M is the monopole velocity, and ε is the efficiency with which the object captures monopoles that strike its surface. The quantity $(1 + 2GM/Rv_M^2)$ is just the ratio of the gravitational capture cross section to the geometric cross section. The efficiency of capture depends upon the mass and velocity of the monopole and its rate of energy loss in the object.[31] Main sequence stars of mass $(0.6 \text{ to } 30)M_\odot$ will capture monopoles less massive than about 10^{18}GeV with velocities less than 10^{-3}c with good efficiency ($\varepsilon \simeq 1$), and in its main sequence lifetime a star will capture approximately $10^{40}\langle F_M \rangle$ monopoles, relatively independent of the mass of the star [53]. Neutron stars will capture monopoles less massive than about 10^{20}GeV with velocities less than 10^{-3}c with unit efficiency, capturing about 10^{37} monopoles in 10^{10}yr. Planets like Jupiter can stop monopoles less massive than about 10^{16}GeV with velocities less than 10^{-3}c, accumulating about 10^{38} monopoles in 10^{10}yr. A planet like the Earth can stop only light, or slowly moving, monopoles (for $m_M = 10^{16}$GeV, v_M must be less than about 3×10^{-5}c).

The most stringent limit to F_M follows from considering neutron stars [54]. A variety of techniques have been used to obtain limits to the luminosities of neutron stars. We will discuss but one of the methods and resulting monopole flux limit here; we refer the interested reader to a longer review [55].

PSR 1929+10 is an old (age of about 3×10^6yr) radio pulsar whose distance from the Earth is about 60 pc. The Einstein x-ray observatory was used to measure the luminosity of this pulsar, and it was determined to be $\mathcal{L} \simeq 3 \times 10^{30}$erg s^{-1}, corresponding to a surface temperature of about 30 eV, making it the coolest neutron star yet observed. In its tenure as a neutron star it should have captured 10^{33} monopoles. The measured luminosity sets a limit to the number of monopoles in PSR 1929+10, $N_M \leq 10^{12}(\sigma v)_{-28}^{-1}$, which by using [7.97] can be used to bound $\langle F_M \rangle$ [56]:

$$\langle F_M \rangle \lesssim 10^{-21}(\sigma v)_{-28}^{-1}\,\text{cm}^{-2}\text{sr}^{-1}\text{sec}^{-1}. \tag{7.98}$$

[31] Monopoles lose energy due to electronic interactions, hadronic interactions, and atomic transitions induced between Zeeman-split levels. In a non-degenerate gas, electronic losses (due to eddy currents) dominate, and $dE/dx \simeq (10 \text{ GeV/cm})(\rho/\text{g cm}^{-3})(v_M/\text{c})(10^7 \text{ K}/T)^{1/2}$.

Progenitors of neutron stars are main sequence (MS) stars of mass (1 to 30)$M_\odot$, which were either too massive to become white dwarfs (WDs) or evolved to the WD state and were pushed over the Chandrasekhar limit by accretion from a companion star. MS stars in the mass range (1 to 30)$M_\odot$ during their MS lifetime capture (10^{39} to 10^{41})$\langle F_M \rangle$ monopoles. The progenitor of PSR 1929+10 should have captured at least 10^6 times more monopoles during its MS phase than during its life as a neutron star, and it is likely that a fair fraction of these monopoles should still be in the neutron star. If these monopoles are included, the bound based upon PSR 1929+10 improves significantly, to

$$\langle F_M \rangle \lesssim 10^{-27} (\sigma v)^{-1}_{-28}\, \mathrm{cm}^{-2}\mathrm{sr}^{-1}\mathrm{sec}^{-1} \tag{7.99}$$

—corresponding to a flux of less than one monopole per Earth per year!

It is hoped that by this point the reader appreciates that monopoles are extremely interesting objects, whose relic abundance is a priori completely unknown, and whose relic flux is severely constrained by a host of astrophysical/cosmological arguments. Now, what about their detectability? It is clear that GUT monopoles cannot be created in the lab because of their enormous mass, and so their detection must necessarily involve searching for relic cosmological monopoles [59]. Experimental searches for relic monopoles can be divided into three categories: induction, ionization, and "exotic." The induction experiments rely upon the fact that when a monopole moves through a loop of wire, it induces a current, $I = 4\pi h/L$, independent of all its other properties, i.e., velocity, mass, catalysis of nucleon decay (L = inductance of the loop). This current, of course, arises due to the change in magnetic flux in the loop. Several induction-type searches have been mounted, all utilizing well-shielded, superconducting loops. While such detectors are conceptually very beautiful because their response to a monopole is so clean, they are difficult to build on a scale sufficiently large to make them competitive with the other detectors. Cabrera pioneered this type of detector, and his first detector recorded a current jump consistent with a monopole of Dirac charge on 14 February 1982, corresponding at that time to a flux of $6 \times 10^{-10}\mathrm{cm}^{-2}\mathrm{sr}^{-1}\mathrm{sec}^{-1}$ [57]. Since then, he and others have improved the sensitivity of inductive monopole searches by several orders of magnitude without seeing another event.

Ionization detectors rely on the fact that as a monopole moves through matter it loses energy through its interactions with (bound) electrons. To calculate this energy loss is not a simple matter—a typical monopole should be moving with a velocity of about 10^{-3}c, while the orbital velocity of a bound electron is about $v = \alpha c \simeq 10^{-2}$c. Roughly speaking, an atomic electron feels a Lorentz force due to a monopole, which is comparable to that of a charged particle with "charge" $e \simeq \alpha c \times h \simeq e/2$. Thus, one would naively expect a monopole to lose energy at 1/4 the rate of a slowly moving proton.[32] In addition, for very-slowly-moving monopoles, there are undoubtedly threshold effects: Simple kinematical considerations imply that the maximum recoil energy a monopole can impart to an electron is $\Delta E \sim 2m_e v_e v_M \sim 7$ eV$(v_M/10^{-3}$c). Thus, at very low velocities ($v_M <$ 10^{-3}c), a monopole's energy loss should drop precipitously.

Despite the drawbacks of an ionization detector, its great advantage is the relative ease with which such a detector can be fabricated on large scales. The best experimental limits to the monopole flux are provided by ionization detectors. At present, a very large (football-field class) ionization detector is being built in the underground laboratory housed within the Gran Sasso tunnel—MACRO, for Monopole, Astrophysics, and Cosmic Ray Observatory [58]. It should reach a sensitivity level of about 10^{-16}cm^{-2}sr^{-1}sec^{-1} and will be on the air soon.

Monopole searches based upon the catalysis process have also been carried out. The signature of monopole catalysis is a string of catalyzed nucleon decays. These searches too have been unsuccessful; however, the flux limits are somewhat difficult to quote, as they depend upon both (σv) and v_M.

Exotic detectors is a grab bag, catch-all category, even including some half-baked ideas! Buford Price and his collaborators [61] carried out a very clever and interesting monopole search whose integration time was about 0.45×10^9 yr! Their search relies upon two facts: that monopoles are highly penetrating—a monopole moving at $v \simeq 10^{-3}$c will pass through the Earth—and that a monopole that picks up a nucleus of charge $Z \gtrsim 10$ (e.g., Al) on its journey through the Earth will cause damage to the lattice structure of any mica it passes through. By using track-etch techniques they examined a relatively small sample of mica (area about 13.5 cm^2) which had been buried in Minas Gerais, Brazil, at a depth of about 5 km for 0.45×10^9yr. Sadly, no candidate monopole track was found. Their clever experiment was truly a "monopole search," and it is difficult to translate their negative finding into a flux limit because of the intermediate steps involved (probability of a monopole picking an Al nucleus, etc.). However, making some reasonable assumptions, their flux sensitivity is estimated to be 10^{-17}cm^{-2}sr^{-1}sec^{-1}! Fig. 7.14 summarizes present and future monopole

---

[32] This estimate is borne out by more sophisticated analyses [60].

search experiments.

Finally, the aforementioned half-baked ideas. The astrophysical constraints discussed above suggest that any flux of relic monopoles must be very small—if they catalyze nucleon decay, smaller than one per Earth per year, making an Earth-based detector impractical. Perhaps it is possible to turn these astrophysical arguments around to actually detect monopoles! If monopoles catalyze nucleon decay they should keep old neutron stars warm; with the next generation of x-ray observatories (e.g., AXAF) it might be possible to exploit this fact to use old neutron stars as monopole detectors. Along a slightly different vein, any monopoles that collect in the sun will catalyze nucleon decays within the sun, and some fraction of the time these catalyzed nucleon decays will produce high-energy neutrino. Such neutrinos can be detected in large underground detectors such as Kamiokande II (KII), Frejus, Soudan, and Irvine-Brookhaven-Michigan (IMB), and they can be identified by the fact that they point back to the sun. It is interesting to mention that these detectors were originally built to search for nucleon decay.

After our fantasy excursion through the world of monopoles, astrophysics, and cosmology, there are a few thoughts we would like to leave the reader with. (1) The superheavy magnetic monopoles predicted by GUTs are exceedingly interesting objects, which, if they exist, must be relics of the earliest moments of the Universe. (2) Along with proton decay and neutrino masses, they are one of the few predictions of GUTs that can be studied in our present "low-energy" environment. (3) Because of the glut of monopoles that should have been produced as topological defects in the very early Universe, the simplest GUTs and the standard cosmology extrapolated back to times as early as 10^{-34}sec are not compatible. This is a very important piece of information about physics at very high energies and/or the earliest moments of the Universe. (4) There is no *believable* prediction for the flux of relic, superheavy magnetic monopoles. (5) Based upon astrophysical considerations, we can be reasonably certain that the flux of relic monopoles is small. Since it is not obligatory that monopoles catalyze nucleon decay at a prodigious rate, a *firm* upper limit to the flux is provided by the Parker bound: $\langle F_M \rangle \lesssim 10^{-15}\text{cm}^{-2}\text{sr}^{-1}\text{sec}^{-1}$ ($m_M \lesssim 10^{17}$GeV). It is very likely that the flux must be even smaller, say less than $10^{-18}\text{cm}^{-2}\text{sr}^{-1}\text{sec}^{-1}$ or even $10^{-21}\text{cm}^{-2}\text{sr}^{-1}\text{sec}^{-1}$. (6) Lest we be pessimistic, it is still possible that monopoles are the dark matter and/or will be detected!

7.7 References

1. D. A. Kirzhnits, *JETP Lett.* **15**, 745 (1972); D. A. Kirzhnits and A. D. Linde, *Phys. Lett.* **42B**, 471 (1972) [Reprinted in *Early Universe: Reprints*]; A. D. Linde, *JETP Lett.* **19**, 320 (1974).

2. Nontopological solitons are discussed in R. Friedberg, T. D. Lee, and A. Sirlin, *Phys. Rev. D* **15**, 1964 (1976). A scenario for their formation in cosmological phase transitions can be found in J. A. Frieman, G. B. Gelmini, M. Gleiser, and E. W. Kolb, *Phys. Rev. Lett.* **60**, 2001 (1988). Variations on the theme of nontopological solitons include Q-balls, S. Coleman, *Nucl. Phys.* **B262**, 263 (1985); neutrino balls, B. Holdom, *Phys. Rev. D* **36**, 1000 (1987); quark nuggets, E. Witten, *Phys. Rev. D* **30**, 272 (1984); and soliton stars, R. Ruffini and S. Bonazzola, *Phys. Rev.* **187**, 1767 (1969); T. D. Lee and Y. Pang *Phys. Rev. D* **35**, 3678 (1987).

3. L. Dolan and R. Jackiw, *Phys. Rev. D* **9**, 3320 (1974) [Reprinted in *Early Universe: Reprints*]; S. Weinberg, *Phys. Rev. Lett.* **9**, 3357 (1974).

4. S. Coleman, *Phys. Rev. D* **15**, 2929 (1977) [Reprinted in *Early Universe: Reprints*].

5. S. Coleman and F. De Luccia, *Phys. Rev. D* **21**, 3305 (1980).

6. A. Linde, *Nucl. Phys.* **B216**, 421 (1983).

7. S. Coleman and E. J. Weinberg, *Phys. Rev. D* **7**, 1888 (1973).

8. S. Weinberg, *Phys. Rev. Lett.* **36**, 294 (1976); A. D. Linde, *Phys. Lett.* **B70**, 306 (1977).

9. A. D. Linde, *Phys. Lett.* **B70**, 306 (1977); **B92**, 119 (1980); P. J. Steinhardt, *Nucl. Phys.* **B179**, 492 (1981); A. H. Guth and E. J. Weinberg, *Phys. Rev. Lett.* **45**, 1131 (1980).

10. M. Sher, *Phys. Rep.* **179**, 273 (1989); E. Weinberg, *Phys. Rep.*, in press (1989).

11. Some reviews of the cosmological production of topological defects are: A. Vilenkin, *Phys. Rep.* **121**, 263 (1985) [Reprinted in *Early Universe: Reprints*]; T. W. B. Kibble, *J. Phys.* **A9**, 1387 (1976); J. Preskill, *Ann. Rev. Nucl. Part. Sci.* **34**, 461 (1984).

12. E. Witten, *Nucl. Phys.* **B177**, 477 (1981).

13. A. Vilenkin, *Phys. Lett.* **133B**, 177 (1983); P. Sikivie and J. Ipser, *Phys. Rev. D* **30**, 712 (1984).

14. L. Kawano, Ph. D. thesis, The University of Chicago (1989).

15. Ya. B. Zel'dovich, I. Yu. Kobzarev, and L. B. Okun, *Sov. Phys. JETP* **40**, 1 (1975).

16. H. B. Nielsen and P. Olesen, *Nucl. Phys.* **B61**, 45 (1973).

17. C. T. Hill, H. M. Hodges, and M. S. Turner, *Phys. Rev. D* **37**, 263 (1988).

18. H. M. Hodges, unpublished.

19. D. Förster, *Nucl. Phys.* **B81**, 84 (1974).

20. A. Vilenkin, *Phys. Rev. D* **23**, 852 (1981). The conical structure persists even when one considers the fully coupled gravity-string equations; see, R. Gregory. *Phys. Rev. Lett.* **59**, 740 (1987).

21. A. Stebbins, *Ap. J.* **327**, 584 (1988); F. Bouchet, D. Bennett, and A. Stebbins, *Nature*, **355**, 410 (1988).

22. T. Vachaspati, *Phys. Rev. Lett.* **57**, 1655 (1986); J. Charlton, *Ap. J.* **325**, 521 (1988); A. Stebbins, et al., *Ap. J.* **322**, 1 (1987).

23. A. Albrecht and N. Turok, *Phys. Rev. Lett.* **54**, 1868 (1985); in press (1989); *Phys. Rev. D*, in press (1989).

24. D. B. Bennett and F. R. Bouchet, *Phys. Rev. Lett.* **60**, 257 (1988).

25. T. Vachaspati and A. Vilenkin, *Phys. Rev. D* **31**, 3052 (1985).

26. E. P. S. Shellard, *Nucl. Phys.* **B283**, 624 (1987); R. Matzner, *Computers in Physics* **2**, 51 (1988); K. J. M. Moriaty, E. Myers, and C. Rebbi, in *Cosmic Strings: The Current Status*, eds. F. Accetta, et al. (World Scientific, Singapore, 1988).

27. R. L. Davis, *Phys. Lett.* **161B**, 285 (1985); D. P. Bennett, *Phys. Rev. D* **34**, 3592 (1986).

28. A. Albrecht and N. Turok, *Phys. Rev. Lett.*, in press (1989).

29. L. Cowie and Hu, *Ap. J.* **318**, L33 (1987).

30. N. Turok and R. Brandenberger, *Phys. Rev. D* **33**, 2175 (1986); R. Brandenberger, et al., *Phys. Rev. Lett.* **59**, 2371 (1987); R. J. Scherrer, *Ap. J.* **320**, 1 (1987); A. Melott and R. J. Scherrer, *Nature*, **328**, 691 (1987); R. J. Scherrer, A. Melott, and E. Bertschinger, *Phys. Rev. Lett.* **62**, 379 (1989); E. Bertschinger, *Ap. J.* **316**, 489 (1987).

31. E. Witten, *Nucl. Phys.* **B246**, 557 (1985).

32. J. P. Ostriker, C. Thompson, and E. Witten, *Phys. Lett.* **180B**, 231 (1986).

33. C. T. Hill, D. N. Schramm, and T. P. Walker, *Phys. Rev. D* **36**, 1007 (1987).

34. G. 't Hooft, *Nucl. Phys.* **B79**, 276 (1974); A. Polyakov, *JETP Lett.* **20**, 194 (1974).

35. M. K. Prasad and C. M. Sommerfield, *Phys. Rev. Lett.* **35**, 760 (1975).

36. T. W. Kirkman, et al., *Phys. Rev. D* **24**, 999 (1981); A. N. Schellekens, et al., *Phys. Rev. Lett.* **50**, 1242 (1983).

37. See T. W. B. Kibble, in reference [11].

38. T. W. B. Kibble, in *Cosmology and Particle Physics* (World Scientific, Singapore, 1987), p. 171.

39. J. Preskill, *Phys. Rev. Lett.* **43**, 1365 (1979); Ya. A. Zel'dovich and M. Y. Khlopov, *Phys. Lett.* **79B**, 239 (1978).

40. A. H. Guth and E. Weinberg, *Nucl. Phys.* **B212**, 321 (1983).

41. P. Langacker and S.-Y. Pi, *Phys. Rev. Lett.* **45**, 1 (1980).

42. E. Weinberg, *Phys. Lett.* **126B**, 441 (1983).

43. M. S. Turner, *Phys. Lett.* **115B**, 95 (1982).

44. T. Goldman, E. W. Kolb, and D. Toussaint, *Phys. Rev. D* **23**, 867 (1981); D. A. Dicus, D. Page, and V. L. Teplitz, *Phys. Rev. D*, **26**, 1306 (1982); J. Fry, *Ap. J.* **246**, L93 (1981); J. Fry and G. M. Fuller, *Ap. J.* **286**, 397 (1984).

45. C. Heiles, *Ann. Rev. Astron. Astro.* **14**, 1 (1976); G. L. Verschuur, *Fund. Cosmic Phys.* **5**, 113 (1979); J. P. Vallee, *Astrophys. Lett.* **23**, 85 (1983).

46. M. S. Turner, E. N. Parker, and T. Bogdan, *Phys. Rev. D* **26**, 1296 (1982).

47. E. N. Parker, *Ap. J.* **160**, 383 (1970).

48. Y. Raphaeli and M. S. Turner, *Phys. Lett.* **B121** 115 (1983).

49. E. Salpeter, S. Shapiro, and I. Wasserman, *Phys. Rev. Lett.* **49** 1114 (1982); J. Arons and R. Blandford, *Phys. Rev. Lett.* **50**, 544 (1983).

50. V. A. Rubakov, *JETP Lett.* **33**, 644 (1981); *Nucl. Phys.* **B203**, 311 (1982);

51. C. Callan, *Phys. Rev. D* **25**, 2141 (1982); **26**, 2058 (1982).

52. A. Sen, *Nucl. Phys.* **B251**, 1 (1985).

53. J. A. Frieman, K. Freese, and M. S. Turner, *Ap. J.* **335**, 844 (1988).

54. E. W. Kolb, S. Colgate, and J. A. Harvey, *Phys. Rev. Lett.* **49**, 1373 (1982); S. Dimopoulos, J. Preskill, and F. Wilczek, *Phys Lett.* **119B**, 320 (1982).

55. E. W. Kolb and M. S. Turner, *Ap. J.* **286**, 702 (1984).

56. K. Freese, M. S. Turner, and D. N. Schramm, *Phys. Rev. Lett.* **51**, 1625 (1983).

57. B. Cabrera, *Phys. Rev. Lett.* **48**, 1378 (1982).

58. C. DeMarzo, et al., *Nuovo Cim.*, **9C**, 281 (1986).

59. For a review of experimental techniques and flux limits from induction and ionization monopole searches, see D. Groom, *Phys. Rep.* **140**, 323 (1985).

60. S. Ahlen and G. Tarlé, *Phys. Rev. D* **27**, 688 (1983).

61. P. B. Price, S. Guo, S. Ahlen, and R. Fleisher, *Phys. Rev. Lett.* **52**, 1265 (1984). A discussion of some of the fundamental uncertainties can be found in the paper of D. Seckel and E. W. Kolb, in *Proceedings of the Santa Fe Meeting*, eds. T. Goldman and M. M. Nieto (World Scientific, Singapore, 1984) .

8

INFLATION

As we have discussed at some length in Chapters 1–4, the standard, hot big bang cosmology is remarkably successful: It provides a reliable and tested account of the history of the Universe from at least as early as the time of the synthesis of the light elements ($t \simeq 10^{-2}$ to 10^2 sec and $T \simeq 10$ to 0.1 MeV) until today ($t \simeq 15$ Gyr and $T \simeq 2.75$ K). Moreover, it also provides a sensible framework for discussing the earliest history of the Universe. At present, there are no observational or experimental data that are at variance with this simple, but very elegant, description of our Universe. The standard cosmology is quite an achievement, comparable to that of the standard model of low-energy particle physics, the $SU(3) \otimes SU(2) \otimes U(1)$ gauge theory of the strong, weak, and electromagnetic interactions.

Like its counterpart in particle physics, the standard cosmology is not without its shortcomings—loose ends that point to some grander theory that goes beyond it. These loose ends are not inconsistencies within the hot big bang model itself; rather, they involve questions that the model in the splendor of its success allows one to ask, but for which the model has yet to provide answers. The cosmological puzzles that the standard cosmology raises involve a handful of very fundamental cosmological facts that the model can accommodate, but whose ultimate origin it has yet to elucidate.

8.1 Shortcomings of the Standard Cosmology

• *Large-Scale Smoothness*: The Robertson-Walker metric describes a space that is homogeneous and isotropic. But the question remains—why is space homogeneous and isotropic? There are other possibilities, including homogeneous but anisotropic spaces, and inhomogeneous spaces. Moreover, the most generic cosmological solutions to Einstein's equations are not isotropic or even homogeneous: The Robertson-Walker space-time is a very special one. The most precise indication of the smoothness of the Universe is provided by the CMBR, which is uniform to about a part in 10^4 on angular scales from $10''$ to 180° (see Fig. 1.6). Were the Universe very inhomogeneous, or were the expansion anisotropic, comparable temperature anisotropies in the CMBR would exist.

If the entire observable Universe were in causal contact when the radiation last scattered, it might be imagined that microphysical processes, such as Compton scattering, could have smoothed out any temperature fluctuations, and a single temperature would have resulted. However, within the standard cosmology this could not have happened because of the existence of particle horizons (discussed earlier in Chapter 3).[1] The size of the particle horizon at a given epoch is conveniently expressed in terms of the entropy within a horizon volume:

$$S_{\rm HOR} = s\frac{4\pi}{3}t^3 \simeq \begin{cases} 0.05g_*^{-1/2}(m_{\rm Pl}/T)^3 & (t \lesssim t_{\rm EQ}) \\ 3\times 10^{87}\,(\Omega_0 h^2)^{-3/2}(1+z)^{-3/2} & (t \gtrsim t_{\rm EQ}). \end{cases} \tag{8.1}$$

Note that the entropy contained within the horizon at early times was much less than that today, about 10^{88}. The entropy within the horizon at recombination, when typical photons in the CMBR last scattered, was $S_{\rm HOR}(z \simeq 1100) \simeq 10^{83}$. This means that the present Hubble volume consisted of about 10^5 causally disconnected regions at recombination, so that causal processes could not have effected the smoothness. We can express this fact in another way: The Hubble distance at recombination, i.e., $H^{-1}(t_{rec})$, subtends an angle of only about 0.8° on the sky today, yet the CMBR is uniform across the sky.

Furthermore, at the time of primordial nucleosynthesis, the entropy within the horizon was $S_H(t_{\rm nucleo}) \simeq 10^{63}$, or only about 10^{-25} of that in the present Hubble volume. Remarkably, the synthesis of the light elements seems to have been nearly identical in the 10^{25} causally independent regions that comprise the present horizon volume.[2]

The first untidy fact about the standard cosmology is that there is no

---

[1] Of course, raising this point about the standard model is somewhat paradoxical, as the FRW is exactly isotropic and homogeneous. However, one can imagine a perturbed FRW model, with inhomogeneities and anisotropies that are greater than 10^{-4}. Such a model would still have particle horizons, and the discussion of the *smoothness* problem would not be paradoxical.

[2] If the baryon-to-entropy ratio varied significant in the different causally distinct regions—but had the same average value we observe today—then the light element abundances would differ greatly from those measured [1].

physical explanation for why the Universe on very large scales is so very smooth. This is frequently referred to as the horizon problem.

• *Small-Scale Inhomogeneity*: Although the Universe is apparently very smooth on large scales, there is a plethora of structure on smaller scales: stars, galaxies, clusters, voids, and superclusters, ranging in size from much less than 1 Mpc to 100 Mpc (and perhaps larger). As we shall see in Chapter 9, the standard cosmology provides a very nice framework for understanding the origin of such a rich abundance of small-scale structure. Once the Universe becomes matter-dominated, small, primeval density inhomogeneities grow via the Jeans (or gravitational) instability into the rich array of structure present today. While the relic photons did not take part in the gravitational collapse that gave rise to the structure, they remain as a fossil record of the primeval inhomogeneity, and they indicate that the initial density fluctuations must have been no larger than about a part in 10^4. The existence of abundant structure in the Universe today then poses another puzzle for the standard cosmology—What is the origin of the primeval inhomogeneity?

Density inhomogeneities are usually expressed in a Fourier expansion

$$\frac{\delta\rho(\vec{\mathbf{x}})}{\rho} = (2\pi)^{-3}\int \delta_k \exp(-i\vec{\mathbf{k}}\cdot\vec{\mathbf{x}})d^3k, \tag{8.2}$$

where ρ is the mean density of the Universe, $\vec{\mathbf{k}}$ is the comoving wavenumber associated with a given mode, and δ_k is its amplitude. So long as a density perturbation is of small amplitude ($\delta\rho/\rho \ll 1$), its physical wavenumber and wavelength scale simply with $R(t)$: $k_{\rm phys} = k/R(t)$, $\lambda_{\rm phys} = R(t)(2\pi/k)$. More often than not, $R(t)$ is normalized so that $R(\text{today}) = 1$ and $\lambda_{\rm phys} = \lambda$. Once a perturbation becomes non-linear, it separates from the general expansion and maintains an approximately constant physical size. Often it is also convenient to describe a given mode by the invariant mass contained within a sphere of radius $\lambda/2$:

$$M(\lambda) \equiv \frac{\pi}{6}\lambda^3_{\rm phys}\rho_{\rm NR} \simeq 1.5\times 10^{11}\,(\Omega_0 h^2)(\lambda/{\rm Mpc})^3 M_\odot. \tag{8.3}$$

Thus a galactic-mass perturbation corresponds to $\lambda_{\rm GAL} \sim$ Mpc.[3]

[3]Today, on the scale of a galaxy, $\delta\rho/\rho \sim 10^5$, and so galaxies separated from the expansion long ago. When we say $\lambda_{\rm GAL} \sim$ Mpc, we mean that *had* galactic perturbations *not* gone non-linear by the present, they would have a characteristic size of about 1 Mpc today.

For gaussian fluctuations, all the information about the perturbations is contained in the power spectrum, $|\delta_k|^2$; a commonly used, intuitive measure of the degree of inhomogeneity on a given scale is the *rms* mass fluctuation on that scale, which is related to $|\delta_k|^2$ by,

$$\begin{aligned}
\left\langle(\delta\rho/\rho)^2\right\rangle_\lambda &\equiv V^{-1}\int d^3x\left[\frac{\int d^3r W(r)\rho(\vec{x}+\vec{r})}{\int d^3r W(r)\rho(\vec{r})}-1\right]^2 \\
&= V^{-1}(2\pi)^{-3}\int 4\pi k^2 dk|\delta_k|^2|W(k)|^2/V_W^2 \\
&\simeq V^{-1}\left(k^3|\delta_k|^2/2\pi^2\right)_{k\sim 2\pi/\lambda} \qquad (8.4)
\end{aligned}$$

where $W(r)$ is a *window function* of characteristic scale λ (i.e., $W(r)\sim 1$ for $r\ll\lambda$, and $W(r)\ll 1$ for $r\gg\lambda$), $V_W = 4\pi\int W(r)r^2dr$ is the volume defined by the window function, and V is the volume of the fundamental cube (periodic boundary conditions must be imposed to Fourier expand the density field; for more details, see the next Chapter).

In the absence of a definite model for the origin of density fluctuations, it is usually assumed that the fluctuation spectrum is without a preferred scale, i.e., is a power law, $|\delta_k|^2\propto k^n$, for which it follows that $(\delta\rho/\rho)_\lambda\propto M^{-n/6-1/2}$.

As we shall see, a very convenient way to describe the spectrum of density perturbations is by specifying the amplitude of each mode when it crosses inside the horizon, i.e., when $\lambda_{\rm phys}\sim H^{-1}$.[4] It is convenient because at horizon crossing and earlier, one can identify $(\delta\rho/\rho)_{\rm HOR}$ as the fluctuation in the Newtonian potential. Of some interest is the so-called Harrison-Zel'dovich spectrum of constant-curvature fluctuations ($n=-3$), for which $(\delta\rho/\rho)_{\rm HOR}$ = const. Such a spectrum is not only pleasingly simple but also has the nice feature that it need not be cutoff at long or short wavelengths to avoid $(\delta\rho/\rho)_\lambda\gtrsim 1$, which for long wavelength modes would lead to excessive CMBR temperature fluctuations, and for short wavelength modes excessive black hole production.

Our understanding of structure formation is not good enough to be able to determine the spectrum of primeval density perturbations from present observations. However, based upon the fact that non-linear structures exist today on scales from 1 Mpc to 10 Mpc or so, and the fact that fluctuations in the linear regime grow as $R(t)$ during the matter-dominated epoch, we

[4]Of course, by so doing, one is specifying the amplitudes of different modes at different epochs.

can infer that perturbations of amplitude 10^{-5} or so must have existed on these scales at the epoch of matter–radiation equality. At present, it is fair to say that, depending upon the nature of the dark matter, a range of spectra (including the Harrison-Zel'dovich spectrum) are plausibly consistent with the structure we observe.

During the radiation-dominated epoch, the mass in particles that are non-relativistic today (baryons and whatever else) contained within the horizon is given by

$$M_{\rm HOR}(t) \simeq 0.29 g_*^{-1/2} (\Omega_0 h^2)(T/{\rm MeV})^{-3} M_\odot. \tag{8.5}$$

From this, it is clear that at very early times perturbations on the scales of cosmological interest were well outside the horizon. (This is illustrated in Fig. 8.4.) Herein lies the difficulty: If one imagines causal, microphysical processes acting during the earliest moments of the Universe and giving rise to primeval density perturbations, the existence of particle horizons in the standard cosmology apparently precludes producing inhomogeneities on the scales of interest.[5] The finite horizon size strikes again!

- *Spatial Flatness–Oldness*: For FRW models the quantity Ω_0, along with H_0, specifies our present cosmological model,[6] in that Ω_0 and H_0 determine the radius of curvature, $R^2_{\rm curv} = H_0^{-2}/|\Omega_0 - 1|$, and the density, $\rho_0 = (3H_0^2/8\pi G)\Omega_0$. As we have discussed in Chapter 1, the present observational data probably only restrict Ω_0 to lie in the interval [0.01, few]. The fact that Ω_0 is not significantly different from unity implies that $R_{\rm curv} \sim H_0^{-1}$ and $\rho_0 \sim \rho_C$.

This may not seem so remarkable, but when one takes account of the fact that Ω changes with time, cf. (3.14),

$$\Omega(t) = 1/[1 - x(t)], \tag{8.6}$$

$$x(t) = (k/R^2)/(8\pi G\rho/3) \propto \begin{cases} R(t)^2 & (t \lesssim t_{EQ}) \\ R(t) & (t \gtrsim t_{EQ}), \end{cases} \tag{8.7}$$

the implications are quite astonishing! Because Ω varies as $1/(1 - x)$,

---

[5]It is possible for a phase transition, or similar event, to give rise to *isocurvature* fluctuations at very early times, which later give rise to density perturbations. Cosmic strings provide such an example. The differences between curvature and isocurvature fluctuations are discussed in the next Chapter.

[6]We are ignoring questions of global topology here.

earlier Ω was closer to unity. Specifically,

$$
\begin{aligned}
|\Omega(10^{-43}\,\text{sec}) - 1| &\lesssim \mathcal{O}(10^{-60}), \\
|\Omega(1\,\text{sec}) - 1| &\lesssim \mathcal{O}(10^{-16}).
\end{aligned} \tag{8.8}
$$

Moreover, it also implies that the radius of curvature of the Universe was enormous compared to the Hubble radius:

$$
\begin{aligned}
R_{\text{curv}}(10^{-43}\,\text{sec}) &\gtrsim 10^{30}H^{-1}, \\
R_{\text{curv}}(1\,\text{sec}) &\gtrsim 10^{8}H^{-1}.
\end{aligned} \tag{8.9}
$$

While no law of physics precludes such, viewed in the context of initial data, this suggests that our FRW model was very special indeed, characterized by the following initial data at the Planck epoch

$$
\begin{aligned}
|\Omega - 1| &\lesssim \mathcal{O}(10^{-60}), \\
R_{\text{curv}} &\gtrsim \mathcal{O}(10^{30})H^{-1}, \\
|(k/R^2)|/(8\pi G\rho/3) &\lesssim \mathcal{O}(10^{-60}).
\end{aligned} \tag{8.10}
$$

Were this not the case, we'd be in a very sorry state today! Were all of the above quantities of order unity at the Planck time, the Universe would have either recollapsed in a "jiffy" (*few* $\times 10^{-43}$ sec) for $k > 0$, or the Universe would have reached a temperature of 3 K at the ripe old age of 10^{-11} sec for $k < 0$.[7] Put another way, the natural time scale for cosmology is the Planck time, 10^{-43} sec, and our FRW model has survived some 10^{60} Planck times without recollapsing or becoming curvature dominated.

• *Unwanted Relics*: Within the context of unified gauge theories there are a variety of stable, superheavy particle species that should have been produced in the early Universe, survive annihilation, and contribute too generously to the present energy density, i.e., $\Omega_X \gg 1$. Monopoles provide the most notable example. Further, any particle species with very a large mass typically has a very small annihilation cross section. As we learned in Chapter 5, for thermal relics $\Omega_X \propto \langle\sigma_A|v|\rangle^{-1}$, so that any very massive,

[7]The crucial assumption in the discussion above is adiabaticity, i.e., $T \propto R^{-1}$, which we will see is grossly violated in inflationary models. And in fact, the ***flatness puzzle*** can also be described in terms of entropy: The invariant entropy within a curvature volume, $S_{\text{CV}} \sim sR^3_{\text{curv}}$, for our FRW model is enormous, $S_{\text{CV}} \gtrsim 10^{88}$.

stable particle species is likely to suffer from the "Hitler complex" (the urge to dominate the Universe).

The standard cosmology has no mechanism of ridding the Universe of relics that are overproduced early in the history of the Universe.

• *Cosmological Constant*: The most general form of Einstein's equations (consistent with general covariance) includes Einstein's notorious cosmological constant:

$$R_{\mu\nu} - \frac{1}{2}g_{\mu\nu}\mathcal{R} = 8\pi G T_{\mu\nu} + \Lambda g_{\mu\nu}. \tag{8.11}$$

It is simple to see that a cosmological term is equivalent to an additional form of stress-energy characterized by constant energy density and isotropic pressure,

$$\rho_{\rm VAC} = \Lambda/8\pi G, \qquad p_{\rm VAC} = -\Lambda/8\pi G. \tag{8.12}$$

The equation of state for vacuum energy is $w = p/\rho = -1$. As we discussed in Chapter 7, modulo the factor of $8\pi G$, a cosmological term is identical to the stress-energy associated with the vacuum.[8] In modern cosmology, the terms vacuum energy and cosmological term are used almost synonymously.

As we discussed in Chapter 3, the vacuum energy FRW solution (de Sitter solution) is rather interesting,

$$\begin{aligned} \rho_{\rm VAC} &= \text{const}, \\ H &= (8\pi G \rho_{\rm VAC}/3)^{1/2} = \text{const}, \\ R(t) &\propto \exp(Ht). \end{aligned} \tag{8.13}$$

Rather than slowing, the expansion accelerates with time, and the "total energy within a comoving volume," $E_{\rm TOT} \propto R^3 \rho_{\rm VAC}$, grows exponentially. These facts of course owe to the "elastic" nature of vacuum energy: The pressure of the vacuum is more correctly described as a tension, and the $-pdV$ work the vacuum does increases, rather than decreases, the energy content of a comoving volume. That the expansion accelerates rather than slows can be understood in Newtonian terms, as the effective source of gravity, $\rho + 3p = -2\rho_{\rm VAC}$, is negative.

---

[8]Based upon the simple criterion of covariance, "vacuum stress-energy" must necessarily be of the form $T^{\mu\nu}_{\rm VAC} = \rho_{\rm VAC} g^{\mu\nu}$.

These unusual properties of vacuum energy are crucial to inflationary cosmology and are seemingly in contradiction to the present state of the expanding Universe. Based upon observations, we can probably be confident that $\rho_{\rm VAC}$ (or $\Lambda/8\pi G$) is at most comparable to the critical density, which in convenient units is $8.07 \times 10^{-47}h^2\,{\rm GeV}^4$. This fact leads to the final conundrum on our list, the smallness of the present vacuum energy relative to any fundamental scale in physics.

As mentioned above, no known symmetry forbids a cosmological term, and on general grounds one might have expected $\Lambda/8\pi G \sim m_{Pl}^4$, a possibility that is at variance with reality by a mere 122 orders of magnitude! Moreover, as we discussed in Chapter 7, the energy densities associated with the scalar potentials employed for SSB change by order σ^4 during a phase transition (σ = energy scale of SSB), and no known symmetry principle in quantum field theory requires that the present vacuum energy be zero (or even $\ll \sigma^4$).[9] The absolute value of the energy associated with the scalar potential has no effect on SSB, nor on any routine calculations in quantum field theory, e.g., cross sections, etc. The gravitational effects of vacuum energy are stupendous: Even the relatively small change in vacuum energy density that accompanies the quark/hadron transition, $\Delta V \sim {\rm GeV}^4$, is cosmologically intolerable today.

Since any present vacuum energy must be very small, perhaps even zero, it is a routine practice to adjust all scalar potentials so that $V(\sigma) = 0$ (σ = the minimum of the potential). This simple act raises the possibility that vacuum energy could play an important role in the early Universe. Were a scalar field displaced from the minimum of its potential, an enormous vacuum energy would exist, of order $10^{56}\,{\rm GeV}^4$ for the grand unification scale. As we shall see, it is this energy that powers inflation. While inflation offers solutions to all of the other cosmological puzzles, it sheds no light on the problem of the cosmological constant.

The problems mentioned here do not invalidate the standard cosmology in any way. All can be accommodated by the standard cosmology, even if they are not explained by it. The goal of cosmology is to explain the present state of the Universe on the basis of physical law, and one would hope that physical law would some day more clearly elucidate the above features of our Universe. Inflation is an attractive candidate for doing so.

---

[9]In supersymmetric theories the vacuum energy contributions of fermions exactly cancel those of bosons resulting in zero vacuum energy density. However, supersymmetry is not a symmetry of the low-energy world; at best, supersymmetry might require $\rho_{\rm VAC} \lesssim M_W^4 \simeq 10^8\,{\rm GeV}^4$, where M_W is the effective scale of supersymmetry breaking.

The first three of these shortcomings all involve the issue of initial data—which to some is a religious matter. In any case, the initial data for our FRW model are apparently very special. This point was made specific by Collins and Hawking [2] who showed that the set of initial data that evolve to a Universe qualitatively similar to ours is of measure zero. The other shortcomings in our list, in one way or another, involve the wedding of the standard cosmology with modern quantum field theory, a union that has also provided us with a very attractive explanation for the cosmic asymmetry between matter and antimatter and gives rise to the inflationary Universe.

The dilemma of initial data is by no means a new problem, nor is inflation the first solution that has been proposed. Before proceeding to discuss inflation, we should briefly mention some of the previous attempts. Misner and others [3] have advocated the "chaotic cosmology" approach, the hope being that a more generic, anisotropic, and inhomogeneous initial space-time would smooth itself out through dissipative processes, producing the enormous entropy of our Universe. Penrose has advocated the possibility that only certain kinds of initial singularities are permitted in Nature [4]. Many have suggested that the resolution to these shortcomings lurks within the Planck epoch, where (unknown) quantum gravitational effects provide the answers. A recent offshoot of the last two approaches is that of quantum cosmology and the wavefunction of the Universe, advocated by Hartle and Hawking [5], Vilenkin [6], and others. The idea proposed here is that the initial data for classical cosmology will arise as a consequence of a proper quantum treatment of space-time at very early times (quantum cosmology is discussed in Chapter 11). Perhaps the most unusual "explanation" involves a retreat to the *anthropic principle*: Unless the Universe and the laws of physics were as they are, intelligent life could not have developed to discover and discuss them.[10]

As it will become clear, the inflationary solution is based upon *classical* gravitational theory and well formulated, albeit highly speculative, ideas in modern quantum field theory. As we shall see, in all viable inflationary models, inflation occurs at an energy scale of order 10^{14} GeV or so (times of order 10^{-34} sec). If the inflationary solution proves correct, pre-inflationary history is rendered irrelevant, save for the possibility that it provides the initial state for inflation.

---

[10]It is unclear to one of the authors how a concept as lame as the "anthropic idea" was ever elevated to the status of a principle.

8.2 Inflation—The Basic Picture

The basic idea of inflation is that there was an epoch when vacuum energy was the dominant component of the energy density of the Universe, so that the scale factor grew exponentially. During such an epoch (known as a de Sitter phase), a small, smooth, and causally coherent patch of size less than H^{-1} can grow to such a size that it easily encompasses the comoving volume that becomes the entire observable Universe today. While the concept of inflation is now understood to be much more general than being associated with SSB, we will begin our discussion of the basics of inflation in the context of a symmetry-breaking phase transition. We do so, not only for reasons of definiteness and history, but also because such a connection is an attractive feature of inflation—should it prove to be true.

Consider a symmetry-breaking phase transition characterized by an energy scale σ, e.g., GUT symmetry breaking where $\sigma \simeq 10^{14}$ GeV. As we discussed in Chapter 7, for temperatures greater than order σ, the finite temperature potential $V_T(\phi)$ is minimized for $\langle\phi\rangle = 0$, and the symmetry is restored. In a first order transition, a second minimum develops for $\langle\phi\rangle = \sigma \neq 0$ as the temperature approaches the critical temperature, T_C. At the critical temperature the two minima are exactly degenerate, and at temperatures below T_C the symmetry-breaking minimum is the global minimum of the potential (see Fig. 7.2).

As with any phase transition, energetics are not the entire story. The transition from $\langle\phi\rangle = 0$ to $\langle\phi\rangle = \sigma$ does not take place instantly as soon as it is favored by energetics; the details and time required are determined by the dynamics of the theory. Unlike the typical laboratory setting, in the early Universe things are changing rapidly; for a GUT energy scale of 10^{14} GeV, the expansion time scale is only about 10^{-34} sec, and the length of time required for the phase transition can be significant compared to the age of the Universe.[11] This fact makes for interesting physics!

The first step in the transition to the symmetry-breaking ground state involves the negotiation of any barrier between $\langle\phi\rangle = 0$ and $\langle\phi\rangle = \sigma$. This can proceed either by quantum/thermal tunnelling with the nucleation of bubbles, or by the disappearance of the barrier at some temperature below T_C. For our purposes, the details are not of interest, only the fact that in some region of the Universe the scalar field ϕ is spatially uniform and

[11]Supposing that $\rho \sim M^4$, it follows that $H^{-1} \sim 1/\sqrt{G\rho} \sim M^{-1}(m_{Pl}/M)$. One can see that the possibility for the expansion time scale to be greater than the physics timescale (about M^{-1}) is made easier the closer M is to m_{Pl}.

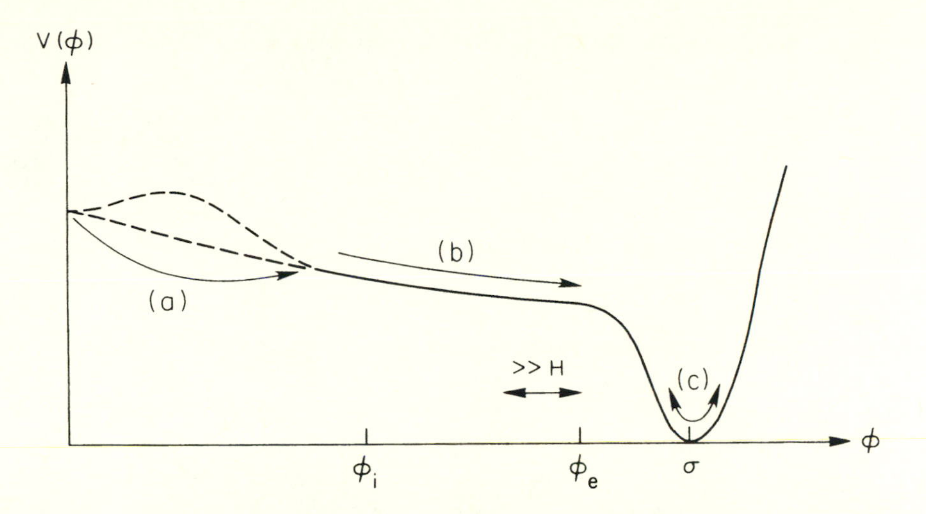

Fig. 8.1: Schematic illustration of an inflationary potential. Notice the "flatness" of the potential—a feature common to all inflationary models. The three qualitative phases of inflation are depicted: (a) barrier penetration (if necessary); (b) slow roll; and (c) coherent oscillations about the minimum of the potential.

is beyond the barrier but still far from the minimum of the potential (see Fig. 8.1).[12]

Once ϕ has made its way around the barrier (if there is one), its evolution to the true vacuum is classical (and downhill at that). As we will discuss in the next Section the classical equation of motion for ϕ is given by

$$\ddot{\phi} + 3H\dot{\phi} + V'(\phi) = 0, \tag{8.14}$$

just the equation for a ball rolling down a hill with friction (here provided by the expansion of the Universe). We will consider the motion of ϕ in great detail in the next Section; for now, the crucial aspect of the evolution of ϕ is the time required for ϕ to roll to its minimum. Provided that the scalar potential is sufficiently flat, this time Δt can be long compared to the expansion time scale: that is, $H\Delta t \gg 1$. During the time it takes ϕ to evolve to σ the Universe is endowed with an enormous vacuum energy, $\rho_{\rm VAC} \simeq V(\phi = 0) \equiv M^4$, and once the temperature of the Universe falls below a temperature $T \sim T_C \simeq M$, this vacuum energy will dominate the

[12]In the language of Section 7.1.2, if tunnelling is involved the escape point (here denoted by ϕ_i) should be very far from the symmetry breaking minimum ($\phi = \sigma$); i.e., $\phi_i \ll \sigma$. This occurs for very flat potentials.

energy density of the Universe. When that occurs the Universe begins a de Sitter phase, expanding exponentially:

$$R(t) \propto \exp(Ht); \qquad H^2 = \frac{8\pi G}{3} V(\phi = 0) \simeq M^4/m_{Pl}^2. \tag{8.15}$$

To be definite, let us assume that the time required for ϕ to evolve to $\phi = \sigma$ is $\Delta t = 100H^{-1}$. During the de Sitter phase of exponential expansion the scale factor grows by the enormous factor of $\exp(100) \simeq 3 \times 10^{43}$! Using our canonical value of $M \simeq 10^{14}$ GeV, the expansion time scale H^{-1} is 10^{-34} sec and the roll time is $\Delta t \simeq 10^{-32}$ sec—a lot happens in a short time. The period during which ϕ slowly rolls to its minimum is referred to as "slow rollover" and is a key feature of all viable models of inflation. While a roll-down time of 10^{-32} sec may not seem that slow, compared to the expansion time scale of the Universe (in this example 10^{-34} sec), it is.

As ϕ nears the minimum of its potential, the potential "steepens" and the evolution of ϕ hastens; as ϕ approaches and inevitiably overshoots the minimum of its potential it begins to oscillate about $\phi = \sigma$ on a time scale short compared to the Hubble time and determined by the curvature of the potential. The enormous vacuum energy of the scalar field then exists in the form of spatially coherent oscillations of the ϕ field, corresponding to a condensate of zero-momentum ϕ particles. Particle creation, or equivalently, the decay of ϕ particles to the other, lighter fields to which it couples, will damp these oscillations. And as the decay products thermalize, the Universe is reheated; in the case that the conversion of vacuum energy is prefectly efficient, $T_{\rm RH}^4 \simeq V(\phi = 0) \simeq M^4$, resulting in a reheat temperature, $T_{\rm RH} \simeq M$. The evolution of ϕ from $\phi = 0$ to the symmetry-breaking minimum ($\phi = \sigma$) is illustrated in Fig. 8.1.

This is the broad-brush picture. To make the illustration more concrete, and to show how the cosmological puzzles are addressed, let us use the following parameters: $M = 10^{14}$ GeV; $H^{-1} = 10^{-34}$ sec; $\Delta t = 100H^{-1} = 10^{-32}$ sec; and $T_C = T_{\rm RH} = 10^{14}$ GeV. Further, let the initial smooth patch have a size $H^{-1} = 10^{-23}$ cm, although it will become clear that its initial size is of little consequence. The crucial feature of inflation is the massive entropy production that occurs during the "reheating" process. The initial entropy contained within the inflating patch is $S_i \simeq T_C^3(H^{-1})^3 \simeq 10^{14}$, far less than that in our present Hubble volume. As the patch expands exponentially, it "supercools," i.e., $T \propto \exp(-Ht)$, with the entropy within the patch remaining fixed. The big event is reheating, during which the temperature rises again to 10^{14} GeV, and the entropy within the patch increases to $S_f \simeq \exp(3H\Delta t)T_{\rm RH}^3(H^{-1})^3 \simeq 10^{144}$. The final size of the patch is $\exp(H\Delta t)H^{-1} \simeq 3 \times 10^{20}$ cm. The highly nonadiabatic reheating process increases the entropy within the inflating patch by a factor of 10^{130}—inflation that is large even by Brazilian standards! The non-adiabaticity of reheating is illustrated in Fig. 8.2.

The enormous entropy increase solves three of the four cosmological problems just like that. First, it solves the problem of *large-scale smoothness*: The smooth patch, which contained only a small fraction of the entropy of the presently observed Universe before inflation, after inflation contains an entropy that is many times greater than that of the observable Universe. The comoving volume that encompasses all that we can see today fits easily within the smooth patch after inflation—10^{56} times over. Seen another way, using the fact that $R \propto T^{-1}$ after inflation, we find that our current Hubble volume had a size of only about 30 cm when the temperature of the Universe was 10^{14} GeV, which is much, much smaller than the post-inflation size of the patch ($\simeq 3 \times 10^{20}$ cm).

Next, it solves the problem of *spatial flatness-oldness*: During inflation the energy density of the Universe remained constant, while the curvature radius grew exponentially, so that the ratio $x = (k/R^2)/(8\pi G\rho/3)$ decreased by a factor of about exp(200). Thus, the radius of curvature of the Universe today should still be much, much greater than the present Hubble radius, thereby "explaining" the flatness of the Universe and "predicting" that Ω_0 should be exponentially close to 1.

Finally, the increase in entropy solves the problem of *unwanted relics*: Consider an unwanted relic that was produced before inflation, with an abundance given by $(n_X/s)_i$. After inflation, its relic abundance is reduced exponentially by the same factor that the entropy increased: $(n_X/s)_f = \exp(-300)(n_X/s)_i$. Moreover, the initial entropy of the Universe is irrelevant; the heat we see today was, to 130 decimal places, all produced during reheating. Needless to say, the baryon asymmetry of the Universe must be produced after inflation; provided that the reheat temperature is sufficiently high this can occur in the usual way (see Chapter 6).

As described to this point, inflation does not solve the *small-scale inhomogeneity* problem. The post-inflation patch is precisely smooth, because at the classical level the homogeneous ϕ field necessarily results in a uniform radiation temperature after reheating. However, density perturbations do arise due to quantum fluctuations in the scalar field ϕ. And of course, inflation sheds no light on the *cosmological constant* problem.

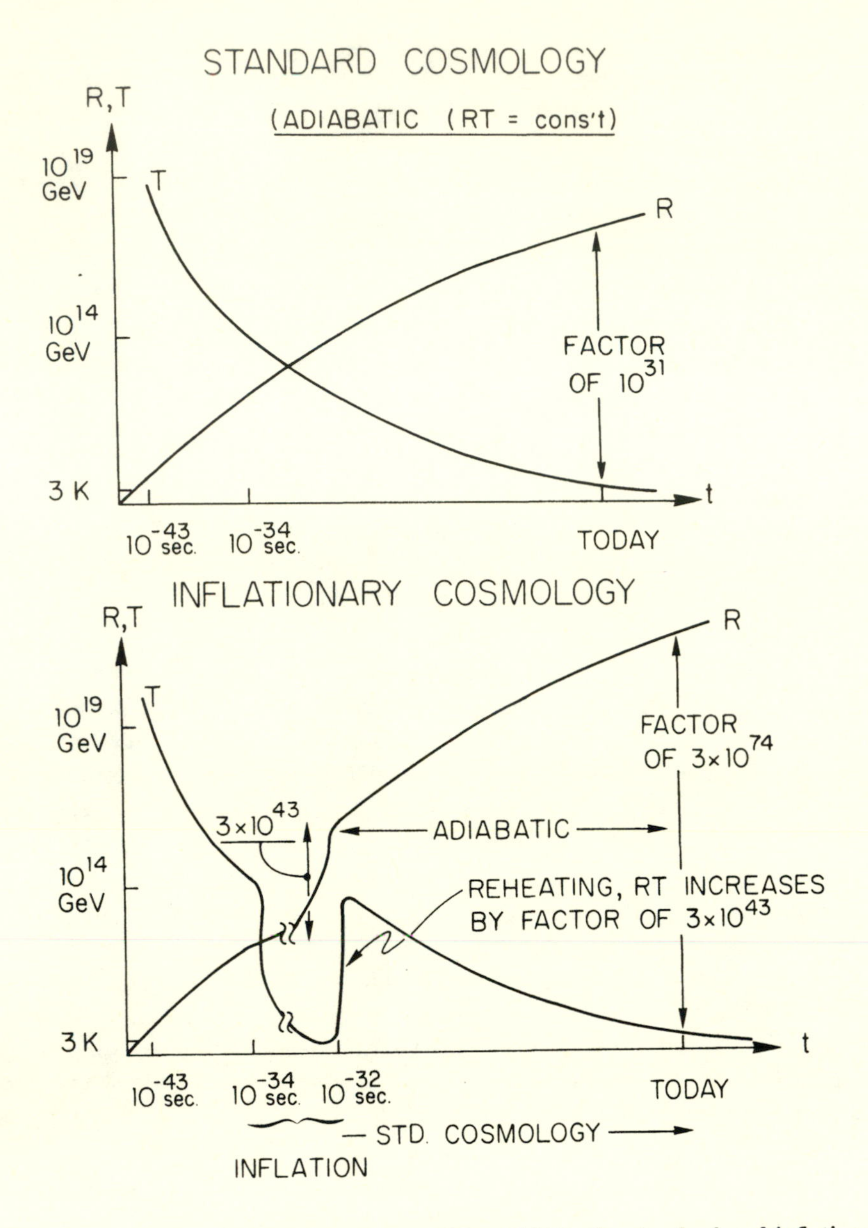

Fig. 8.2: Comparison of the evolution of R and T in the standard and inflationary cosmologies. Note the enormous jump in entropy ($S \propto R^3T^3$) at the end of inflation.

The version of inflation described above is not the original model proposed by Guth in 1981 [7]; it is the variant proposed in 1982 by both Linde [8], and Albrecht and Steinhardt [9], often referred to as "new inflation," or more appropriately, "slow-rollover" inflation. As the latter name suggests, the key feature is the slow evolution of the scalar field to the minimum of the scalar potential. Virtually all models of inflation today are based upon this simple principle.

Guth's original (and unsuccessful) model for inflation, now usually referred to as "old inflation," differed from the present picture of inflation in one crucial respect. In old inflation, the period of exponential expansion occurred while the scalar field was trapped in the $\langle\phi\rangle = 0$ false-vacuum state, and not while ϕ was slowly, but inevitiably, rolling down to $\phi \doteq \sigma$. Therefore, achieving successful inflation required that the tunnelling transition rate be small, implying that the nucleation of true-vacuum bubbles was rare. On the other hand, the latent heat (false-vacuum energy) needed for reheating was stored in the kinetic energy of bubble walls, so that reheating had to proceed via bubble collisions. In the end, the small tunnelling rate—required for sufficient inflation—precluded bubble collisions from reheating the Universe: The phase transition was never completed and most of the Universe continued to inflate forever [10].

8.3 Inflation as Scalar Field Dynamics

As it is presently understood, and stated in the most general terms, inflation involves the dynamical evolution of a weakly coupled scalar field that was at one time displaced from the minimum of its potential. As such, the key to understanding the mechanics of inflation is scalar field dynamics in the expanding Universe.

In order to make the analysis of the evolution of a scalar field manageable, it is necessary to make some simplifying assumptions. All of these assumptions turn out to quite reasonable, and as we shall show later, are quite well justified. They are as follows.

- A background RW space-time with expansion rate given by

$$H^2 = 8\pi G\rho_\phi/3 - k/R^2 \tag{8.16}$$

where ρ_ϕ is the energy density associated with the scalar field. While *assuming* an RW space-time may seem like a tautology, not only does it make the analysis tractable, but as we shall see in Section 8.6, it is well justified, since shortly after inflation begins, space-time rapidly approaches a RW model. Ignoring the other forms of energy density is a good approximation since they rapidly red shift away.

• A homogeneous scalar field (at least on a scale comparable to H^{-1}) with initial value $\phi_i \neq \sigma$, where σ is the global, zero-energy minimum of the potential, i.e., $V(\sigma) = V'(\sigma) = 0$. Provided that any inhomogeneity is sufficiently small (to be quantified later), it can be ignored as spatial gradients in ϕ are red shifted away rapidly. Only the zero-momentum mode of the scalar field will be important.

• Quantum fluctuations in the scalar field will be treated as small corrections to the classical motion of the field, i.e.,

$$\phi(t) = \phi_{\rm cl} + \Delta\phi_{\rm QM}; \qquad \Delta\phi_{\rm QM} \ll \phi_{\rm cl}. \tag{8.17}$$

In the next Section we will discuss the important role of quantum fluctuations, and we will justify this assumption also. For brevity we will always omit the subscript "cl."

Consider a minimally coupled scalar field ϕ with Lagrangian density given by[13]

$$\mathcal{L} = \partial^\mu\phi\partial_\mu\phi/2 - V(\phi) = \dot{\phi}^2/2 - V(\phi). \tag{8.18}$$

For the moment we will ignore the interactions that ϕ must necessarily have with other fields in the theory. As it turns out, they must be very weak for inflation to be successful, so this is a self-consistent assumption. The stress-energy for the ϕ field is

$$T^{\mu\nu} = \partial^\mu\phi\partial^\nu\phi - \mathcal{L}g^{\mu\nu}. \tag{8.19}$$

With the assumption that ϕ is spatially homogeneous, $T^{\mu\nu}$ takes the form of a perfect fluid, with energy density and pressure given by

$$\rho_\phi \;=\; \dot{\phi}^2/2 + V(\phi) \quad [+(\nabla\phi)^2/2R^2] \tag{8.20}$$

$$p_\phi \;=\; \dot{\phi}^2/2 - V(\phi) \quad [-(\nabla\phi)^2/6R^2)] \tag{8.21}$$

For comparison, we have also shown the spatial gradient terms in parentheses (∇ is with respect to comoving coordinates). From the form of the

[13]Minimally coupled refers to the coefficient of the $+(1/2)\xi\mathcal{R}\phi^2$ term in the Lagrangian density; $\xi = 0$ corresponds to minimal coupling, while $\xi = 1/6$ corresponds to conformal coupling. The equation of motion written in terms of the conformal-time variable η is $\psi'' - (\xi - 1/6)R^2\mathcal{R}\psi - \nabla^2\psi + R^4\partial V/\partial\psi = 0$, where prime denotes $d/d\eta$, $\psi = R\phi$, and ∇ is with respect to comoving coordinates. Note that for $\xi = 1/6$, the effects of space-time curvature disappear.

spatial gradient contributions to the stress-energy, it is clear that their importance is damped as $\exp(-2Ht)$, and so they quickly become irrelevant. Moreover, it is also clear that were they the dominant form of stress energy, the scalar field would behave like a fluid with $p = -\rho/3$, which would result in $R(t) \propto t$, and not inflation.

The classical equation of motion for ϕ can be obtained either by varying the action ($S_\phi = \int d^4x\sqrt{-g}\,\mathcal{L}$), or using the conservation of energy-momentum ($T^{\mu\nu}{}_{;\nu} = 0$). Either way it follows that

$$\ddot{\phi} + 3H\dot{\phi} + \Gamma_\phi\dot{\phi} + V'(\phi) = 0. \tag{8.22}$$

The $3H\dot{\phi}$ term arises due to the expansion of the Universe and corresponds to the red shifting of the momentum of the field ($\dot{\phi}$) by the expansion of the Universe. The $\Gamma_\phi\dot{\phi}$ term requires further explanation as it does not follow from the Lagrangian density as written. As we discussed in the previous Section, the spatially coherent oscillations of the ϕ field about $\phi = \sigma$ correspond to a condensate of zero-momentum ϕ particles of mass $m_\phi^2 = V''(\sigma)$, which decay due to quantum particle creation of other fields that couple to ϕ. The damping of these oscillations leads to the all-important reheating of the Universe. The damping of these oscillations by quantum particle creation is precisely equivalent to the decay of ϕ particles into the other, lighter species to which they couple, and Γ_ϕ is none other than the decay width of the ϕ particle [11]. For example, suppose that the ϕ particle decays into two very light fermions and couples to these fermions with strength h; then the decay width $\Gamma_\phi = h^2 m_\phi/8\pi$.

This equation of motion, which is the same as that for a ball rolling down a hill with friction into a valley, has two qualitatively different regimes, each of which has a simple analytical solution.[14] The two regimes are: (1) the slow-roll regime, where friction dominates and ϕ rolls at "terminal velocity;" and (2) the rapid oscillation regime, where the motion of ϕ is that of a damped, simple-harmonic oscillator. We will consider each in turn.

• *Slow Roll:* In this regime the $\ddot{\phi}$ term is negligible, and the $\Gamma_\phi\dot{\phi}$ particle creation term is not operative.[15] During the slow roll the equation of

[14] Of course, it is also very simple to solve for the motion of ϕ numerically; in most instances this is not required as the analytical solutions are very accurate.

[15] Significant particle creation takes place only when the scalar field is oscillating about $\phi = \sigma$; moreover, this form for the particle creation term is correct only when ϕ is oscillating about $\phi = \sigma$ [11].

motion reduces to

$$3H\dot{\phi} = -V'(\phi). \tag{8.23}$$

That is, friction due to the expansion is balanced by the acceleration due to the slope of the potential. The consistency condition for neglecting $\ddot{\phi}$ requires

$$\begin{aligned} |V''(\phi)| &\ll 9H^2 \simeq 24\pi V(\phi)/m_{Pl}^2, \\ |V'(\phi)m_{Pl}/V(\phi)| &\ll (48\pi)^{1/2}. \end{aligned} \tag{8.24}$$

These two conditions determine the duration of slow roll, and they also ensure that during the slow roll the kinetic energy of the scalar field ($\dot{\phi}^2/2$) is much less than that of its potential energy, so that

$$H^2 \simeq \frac{8\pi}{3m_{Pl}^2} V(\phi). \tag{8.25}$$

Using equations (8.23) and (8.25) it is simple to calculate the number of e-folds of growth in the scale factor that occur as ϕ rolls from ϕ_1 to ϕ_2:

$$\ln(R_2/R_1) \equiv N(\phi_1 \to \phi_2) \equiv \int_{t_1}^{t_2} H dt = -\frac{8\pi}{m_{Pl}^2} \int_{\phi_1}^{\phi_2} \frac{V(\phi)}{V'(\phi)} d\phi, \tag{8.26}$$

where we have used the fact that $dt = d\phi/\dot{\phi}$.

For a polynomial potential, $V''(\phi) \sim V'(\phi)/\phi$, and we see that $N(\phi_1 \to \phi_2) \sim 3H^2/V''$, which by the first slow-roll condition must be $\gg \mathcal{O}(1)$. That is, for any potential flat enough to satisfy the slow-roll conditions, many e-folds of growth in the scale factor occur as the field rolls down its potential. The total amount of inflation is given by, $N_{\rm TOT} \equiv N(\phi_i \to \phi_e)$, where ϕ_i is the initial value of ϕ, and ϕ_e is determined by when the slow-rolling condition is no longer satisfied. Usually it suffices to take $\phi_e = \sigma$.

• *Coherent Oscillations:* During this regime $V'' \gg H^2$, and ϕ evolves rapidly on the expansion time scale. Further, once ϕ reaches the bottom of its potential, it will begin to oscillate with frequency $\omega^2 = V''(\sigma)$. In the rapid oscillation regime it is useful to rewrite (8.22) as

$$\dot{\rho}_\phi + 3H\dot{\phi}^2 + \Gamma_\phi \dot{\phi}^2 = 0, \tag{8.27}$$

obtained by multiplying (8.22) by $\dot{\phi}$. As ϕ is rapidly oscillating about σ, $\dot{\phi}^2$ oscillates sinusoidally and can be replaced by its average over an oscillation cycle[16]

$$\langle \dot{\phi}^2 \rangle_{\rm cycle} = \rho_\phi. \tag{8.28}$$

With this substitution for $\langle \dot{\phi}^2 \rangle$, the equation for the evolution of ρ_ϕ becomes

$$\dot{\rho}_\phi + 3H\rho_\phi + \Gamma_\phi \rho_\phi = 0. \tag{8.29}$$

We recognize this equation as an old friend: the equation governing the decay of a massive particle species, which we discussed in Section 5.3. The solution to this equation is given by (5.63):

$$\rho_\phi = M^4 (R/R_{\rm osc})^{-3} \exp[-\Gamma_\phi (t - t_{\rm osc})], \tag{8.30}$$

where "osc" labels the epoch when the coherent oscillations commence, and M^4 is the vacuum energy of the scalar field at that time.

We would like to keep our discussion of the reheating process as general as possible. Let us assume only that the decay products of the ϕ particle are very light relative to the ϕ itself, so that they will be highly relativistic. Then the additional equations required to describe reheating are:

$$\begin{aligned} \dot{\rho}_R + 4H\rho_R &= \Gamma_\phi \rho_\phi, \\ H^2 &= 8\pi G(\rho_\phi + \rho_R)/3, \end{aligned} \tag{8.31}$$

where ρ_R is the energy density in the relativistic decay products. The equation of motion for ρ_R follows from energy conservation, cf. (5.67). Once the decay products interact sufficiently to thermalize, $\rho_R = g_* \pi^2 T^4/30$.

The set of equations describing the reheating process, (8.29) and (8.31), is identical to that governing the out-of-equilibrium decay of a massive, NR particle species, which was discussed in Section 5.3. We will briefly review the salient features of the solutions here.

From $t = t_{\rm osc} \simeq H^{-1} \sim m_{Pl}/M^2$ until $t \simeq \Gamma_\phi^{-1}$, NR ϕ particles (coherent ϕ oscillations) dominate the mass density, and so the Universe is matter-dominated: $R(t) \propto t^{2/3}$. Because the Universe has undergone extreme supercooling during the de Sitter phase, just as the oscillations commence $\rho_R = 0$. During the ϕ-dominated epoch, an approximate solution for ρ_R is given by

$$\rho_R \simeq \frac{m_{Pl}^2 \Gamma_\phi}{10\pi t}[1 - (t/t_{\rm osc})^{-5/3}]$$

[16]Recalling that for a simple harmonic oscillator $\langle V \rangle = \langle \dot{\phi}^2/2 \rangle = \rho_\phi/2$, we see that $\langle p_\phi \rangle \equiv \langle \dot{\phi}^2/2 - V(\phi) \rangle$ vanishes—the coherent ϕ oscillations truly behave as NR matter!

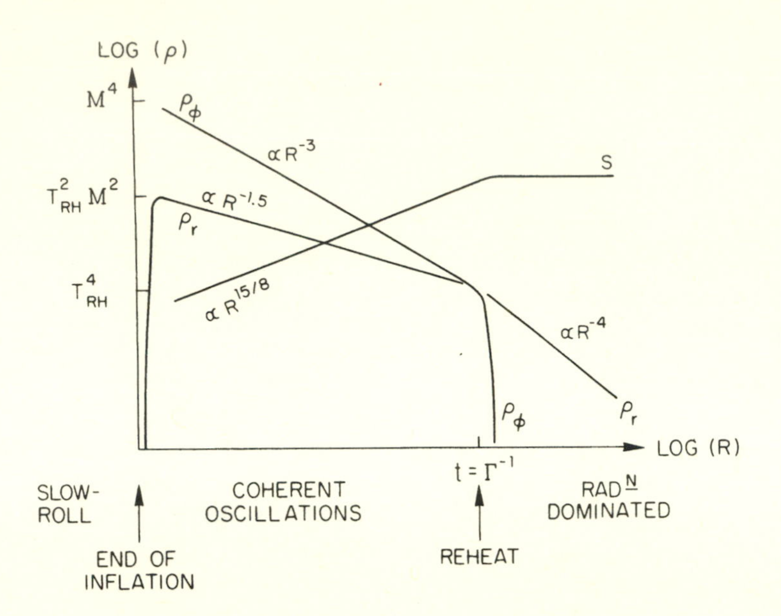

Fig. 8.3: Summary of the evolution of ρ_ϕ, ρ_R, and S during reheating.

$$\simeq \frac{(6/\pi)^{1/2}}{10} m_{Pl}\Gamma_\phi M^2 (R/R_{\rm osc})^{-3/2}[1-(R/R_{\rm osc})^{-5/2}]. \quad (8.32)$$

That is, ρ_R rapidly increases from 0 to a value of about $m_{Pl}\Gamma_\phi M^2$, and thereafter it decreases as $R^{-3/2}$. Yes, just as we discussed in Chapter 5, after the initial transient, the temperature during reheating actually decreases! Thus, the maximum temperature achieved during reheating is

$$T_{\rm max} \simeq 0.8 g_*^{-1/4} M^{1/2} (\Gamma_\phi m_{Pl})^{1/4}. \quad (8.33)$$

However, during this period, the all-important entropy per comoving volume, $S \propto R^3 \rho_R^{3/4}$, increases as $R^{15/8}$.

When $t \simeq \Gamma_\phi^{-1}$, the ϕ particles begin to decay rapidly, the Universe becomes radiation dominated, and the entropy per comoving volume levels off. This is the beginning of the ordinary adiabatic, radiation-dominated phase of the standard cosmology. The temperature at the beginning of the radiation-dominated epoch is

$$T_{\rm RH} \equiv T(t=\Gamma_\phi^{-1}) \simeq 0.55 g_*^{-1/4} (m_{Pl}\Gamma_\phi)^{1/2}. \quad (8.34)$$

Note that as a result of the ϕ-dominated epoch the reheat temperature is determined by Γ_ϕ and not the initial vacuum energy, and that in general $T_{\rm RH}$ is less than M. During the coherent-oscillation phase much of the initial vacuum energy is red shifted away, by a factor of $M^4/T_{\rm RH}^4$.[17]

While $T_{\rm max} \equiv T(t = t_{\rm osc}) \simeq (T_{\rm RH}M)^{1/2}$ is the maximum temperature attained during the reheating process, for all pratical purposes $T_{\rm RH}$ is the "reheat" temperature. This is because during the coherent-oscillation epoch the entropy per comoving volume is increasing, and the abundance of any relic that might be produced—e.g., monopoles, baryon number, axions, etc.—is subsequently diluted by the entropy release. Fig. 8.3 summarizes the reheating process.

Shortly after, or possibly during, reheating one crucial event must take place: baryogenesis, as any pre-inflationary baryon asymmetry has been exponentially diluted away. There are at least two ways baryogenesis can occur: (1) via the usual out-of-equilibrium decay of X-bosons, or (2) directly through the decay of the ϕ particles themselves. The first possibility was discussed at length in Chapter 6; however, there is one important difference in this case: The X-bosons must first be created from the thermal bath. For this to happen, $T_{\rm RH}$ must be greater than, or at least comparable to, the X-boson mass. The second possibility was also discussed in Section 6.7, as way-out-of-equilibrium decay. In this case the baryon asymmetry that results is $n_B/s \simeq \epsilon\, T_{\rm RH}/m_\phi$.

Armed with our knowledge of the reheating process we can calculate the amount of inflation, i.e., number of e-folds of exponential growth, required to solve the *smoothness* and *flatness* problems. To solve the *smoothness* problem our inflating patch must, at the end of reheating, contain an entropy of greater than about 10^{88}. Suppose the initial size of the patch is given by the Hubble radius at the start of inflation, $H^{-1} \sim m_{\rm Pl}/M^2$. During inflation the patch grows by a factor of $\exp(N_{\rm TOT})$ and during reheating by a factor of $R_{\rm RH}/R_{\rm osc} \simeq (M^4/T_{\rm RH}^4)^{1/3}$. Thus the final entropy within the patch is

$$S_{\rm patch} \simeq \exp(3N_{\rm TOT})(M^4/T_{\rm RH}^4)H^{-3}T_{\rm RH}^3 \simeq \exp(3N_{\rm TOT})\frac{m_{\rm Pl}^3}{M^2T_{\rm RH}}. \quad (8.35)$$

[17] If the decay width of the ϕ particle is very large, $\Gamma_\phi \gtrsim H \simeq M^2/m_{Pl}$, which is not usually the case, the coherent ϕ oscillations decay in less than a Hubble time. In this case there is no ϕ-dominated epoch, and $T_{\rm RH} \sim M$—perfect conversion of vacuum energy to radiation.

Requiring that $S_{\rm patch}$ be greater than 10^{88}, we find that

$$N_{\rm TOT} \gtrsim N_{\rm min} = 53 + \frac{2}{3}\ln(M/10^{14}\,{\rm GeV}) + \frac{1}{3}\ln(T_{\rm RH}/10^{10}\,{\rm GeV}). \quad (8.36)$$

Taking both M and $T_{\rm RH}$ to vary between 1 GeV and 10^{19} GeV, $N_{\rm min}$ varies only from 24 to 68. Also notice that the amount of inflation required decreases, albeit logarthmically, with decreasing reheat temperature. While a long ϕ-dominated phase degrades the reheat temperature, it increases the entropy produced.

Now for the *flatness* problem. The curvature radius before inflation is related to H and the value of Ω before inflation ($\equiv \Omega_i$) by

$$(R_{\rm curv})_i = \frac{H^{-1}}{|\Omega_i - 1|^{1/2}}. \quad (8.37)$$

During inflation the curvature radius grows by a factor of $\exp(N_{\rm TOT})$ and during the ϕ-dominated phase by a factor of $(M^4/T_{\rm RH}^4)^{1/3}$. Using these facts we can compute the entropy contained within a volume whose radius is given by the curvature radius:

$$S_{\rm curv} \equiv (R_{\rm curv})_{\rm RH}^3 T_{\rm RH}^3 \simeq \frac{\exp(3N_{\rm TOT}) m_{Pl}^3/(M^2 T_{\rm RH})}{|\Omega_i - 1|^{3/2}}, \quad (8.38)$$

which is, up to the factor involving Ω_i, precisely the same as the entropy within the patch after inflation! Since the entropy within a curvature volume remains constant (so long as the expansion is adiabatic), solving the *flatness* problem requires sufficient inflation so that $S_{\rm curv}$ is greater than that within our current Hubble volume, or $N_{\rm TOT} \geq N'_{\rm min}$, where

$$N'_{\rm min} = 53 + \frac{2}{3}\ln(M/10^{14}\,{\rm GeV}) + \frac{1}{3}\ln(T_{\rm RH}/10^{10}\,{\rm GeV}) + \frac{1}{2}\ln(|\Omega_i - 1|), \quad (8.39)$$

that is, up to the term involving Ω_i, the same amount of inflation as is required to solve the *smoothness* problem. Moreover, we see that the present ratio of the curvature radius to the Hubble radius is given by

$$R_{\rm curv}/H_0^{-1} = \frac{\exp(N_{\rm TOT} - N_{\rm min})}{|\Omega_i - 1|^{1/2}} \quad (8.40)$$

and that the present value of Ω is given by

$$|\Omega_0 - 1| = \exp[2(N_{\rm min} - N_{\rm TOT})]|\Omega_i - 1|. \qquad (8.41)$$

In successful models of inflation, $N_{\rm TOT}$ is typically much larger than $N_{\rm min}$, so that $S_{\rm patch} \gg 10^{88}$, and $|\Omega_0 - 1| \ll 1$.

8.4 Density Perturbations and Relic Gravitons

At the classical level, the fact that ϕ is constant within the inflationary patch ensures that the temperature within the patch is uniform after inflation—reheating is entirely controlled by the value of ϕ within the inflating patch. However, inflation does have the means, both kinematically and dynamically, to produce density perturbations on cosmologically interesting scales. The dynamics of producing density perturbations involves the quantum mechanical fluctuations that must arise in a scalar field in de Sitter space.

Let's begin by discussing the kinematical aspect of inflation that makes the generation of density perturbations possible. In the standard cosmology, all comoving scales λ cross the horizon only once:[18] They start larger than the horizon, i.e., $\lambda_{\rm phys} \gg H^{-1}$, and then cross inside the horizon at some later time. For comoving scales that come inside the horizon during the radiation-dominated epoch, $\lambda \lesssim 13(\Omega_0 h^2)^{-1}$ Mpc, the temperature of the Universe at horizon crossing is:

$$T_{\rm HOR}(\lambda) \simeq 63\,{\rm eV} g_*^{-1/2} g_{*S}^{1/3} (\lambda/{\rm Mpc})^{-1}. \qquad (8.42)$$

For those scales that re-enter the horizon when the Universe is matter dominated, $\lambda \gtrsim 13(\Omega_0 h^2)^{-1}$ Mpc, the temperature is

$$T_{\rm HOR} = 950\,{\rm eV}(\Omega_0 h^2)^{-1}(\lambda/{\rm Mpc})^{-2}. \qquad (8.43)$$

Since the Hubble parameter is constant during inflation, all cosmologically interesting scales begin sub-horizon sized, cross outside the Hubble radius during inflation, and later again cross back inside the horizon (at the usual epoch). This cosmological "good bye" and "hello again" feature

[18]While the Hubble radius H^{-1} and the horizon distance are conceptually different, in the standard cosmology they are essentially equal, and we will use the terms interchangeably. To be precise, it is always the Hubble radius, the distance that provides the "horizon for microphysics," that we mean to use.

of inflation makes it kinematically feasible to imprint density perturbations on interesting scales and is illustrated in Fig. 8.4. Roughly speaking, quantum fluctuations on a given scale arise when that scale is sub-horizon sized; then as the scale crosses outside the horizon these fluctuations "freeze in" as classical, metric perturbations and finally re-enter the horizon as density perturbations.

In order to analyze the spectrum of fluctuations that results, we will need to know when a given scale crosses outside the horizon during the inflationary epoch. As can be clearly seen in Fig. 8.4, in inflationary cosmology it is "first out—last in." A convenient way of specifying when a given scale crossed outside the horizon is by the number of e-folds between horizon-crossing and the end of inflation. We know from the calculation done at the end of the previous Section that the scale corresponding to the present Hubble radius, $\lambda \sim 3000$ Mpc, left the horizon

$$N_{3000\mathrm{Mpc}} = N_{\min} = 53 + \frac{2}{3}\ln(M/10^{14}\,\mathrm{GeV}) + \frac{1}{3}\ln(T_{\mathrm{RH}}/10^{10}\,\mathrm{GeV}) \quad (8.44)$$

e-folds before the end of inflation—which is why inflation has to last at least this long to solve the *smoothness* problem. In general, the scale λ must have crossed outside the horizon $\ln(3000\,\mathrm{Mpc}/\lambda)$ e-folds later, so that

$$N_\lambda = 45 + \ln(\lambda/\mathrm{Mpc}) + \frac{2}{3}\ln(M/10^{14}\,\mathrm{GeV}) + \frac{1}{3}\ln(T_{\mathrm{RH}}/10^{10}\,\mathrm{GeV}); \quad (8.45)$$

see Fig. 8.4.

Since the potentials of interest for inflation are very flat in the region of the potential where inflation occurs, $|V''| \ll H^2$, it is a very good approximation to treat the scalar field ϕ as massless during the slow-roll part of inflation. A massless, minimally coupled scalar field in de Sitter space has a spectrum of quantum mechanical fluctuations characterized by [12]

$$(\Delta\phi)_k^2 \equiv V^{-1}k^3|\delta\phi_k|^2/2\pi^2 = (H/2\pi)^2, \quad (8.46)$$

where the fluctuations in the scalar field have been decomposed into their Fourier components,

$$\delta\phi_k = \int d^3x \exp(i\vec{\mathbf{k}}\cdot\vec{\mathbf{x}})\phi(\vec{\mathbf{x}}). \quad (8.47)$$

Note that the fluctuation power on each scale k, $(\Delta\phi)_k$, is just given by the Gibbons-Hawking temperature associated with the de Sitter space event

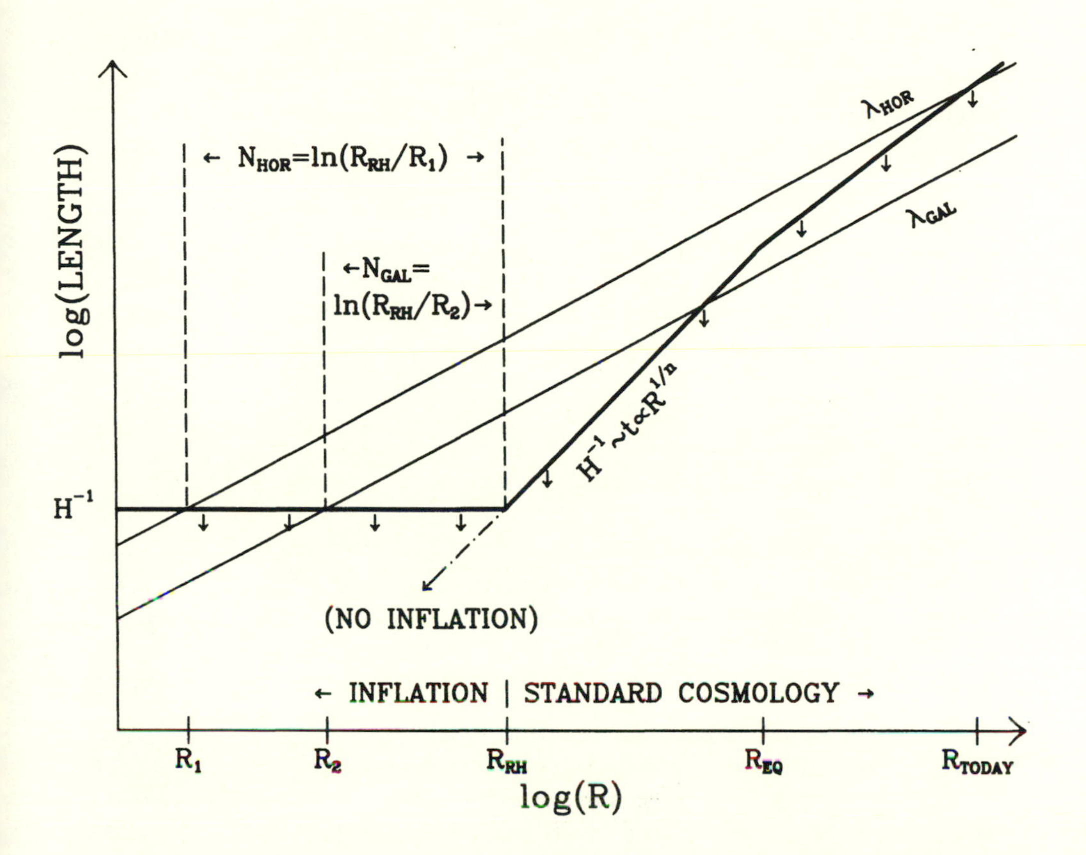

Fig. 8.4: The evolution of the physical size of the comoving scale, λ, and of the Hubble radius, H^{-1}, in the standard and the inflationary cosmologies. In the standard cosmology (i.e., no inflation) a given scale crosses the horizon but once; while in the inflationary cosmology all scales begin sub-horizon sized, cross outside the horizon ("good bye") during inflation, and re-enter again ("hello again") during the post-inflationary epoch. Note that the largest scales cross outside the horizon first and re-enter last. The growth in the scale factor $[N = \ln(R_{\rm RH}/R)]$ between the time a scale crosses outside the horizon during inflation and the end of inflation is also indicated. For a galaxy, $N_{\rm GAL} = \ln(R_{\rm RH}/R_1) \sim 45$, and for the present horizon scale, $N_{\rm HOR} = \ln(R_{\rm RH}/R_2) \sim 53$. Causal microphysics operates only on scales less than H^{-1} (indicated by arrows). During inflation $H^{-1} \equiv H_I^{-1} = \text{const}$, and in the post-inflation era, $H^{-1} \sim t \propto R^{1/n}$ ($n = 1/2$—radiation dominated, $n = 2/3$—matter dominated).

horizon, $T_{\rm GH} = H/2\pi$. The *mean square* fluctuation in ϕ is given by:

$$(\Delta\phi)^2 = \frac{V^{-1}}{2\pi^2}\int k^2 dk|\delta\phi_k|^2 = N(t)(H/2\pi)^2, \tag{8.48}$$

where $N(t) \simeq Ht$ is the number of e-folds of inflation since inflation began.[19] As one would have expected, the size of the quantum fluctuations in ϕ is set by $H/2\pi$.

Causal microphysics operates only on distance scales less than order the Hubble radius, as the Hubble radius represents the distance a light signal can travel in an expansion time. As each mode k crosses outside the horizon, it decouples from microphysics and "freezes in" as a classical fluctuation.[20] The classical evolution of a given mode k is governed by:

$$\delta\ddot{\phi}_k + 3H\delta\dot{\phi}_k + k^2\delta\phi_k/R^2, = 0 \tag{8.49}$$

whose solution for super-horizon-sized modes ($k \ll RH$) is very simple: $\delta\phi_k = \text{const}$.

Of more importance is the energy density perturbation that arises due to the fluctuations in ϕ. Since the potential energy density depends upon the scalar field ϕ, fluctuations in ϕ give rise to perturbations in the energy density:

$$\delta\rho_\phi = \delta\phi(\partial V/\partial\phi). \tag{8.50}$$

In Chapter 9 we will discuss the evolution of classical, scalar metric (or density) perturbations in some detail. The evolution of density perturbations is particularily subtle for modes outside the horizon, as the gauge freedom of general relativity makes it difficult to sort out the physical modes from the gauge artifacts.[21] Since microphysics is impotent on scales greater than the Hubble radius one might expect that evolution of density perturbations would be basically kinematical and quite simple. In fact that is the case.

The evolution of the gauge invariant quantity ζ is particularly simple for super-Hubble-sized perturbations: $\zeta = \text{const}$ [15]. At horizon crossing,

---

[19]The factor of N arises from integrating over all the modes that have crossed outside the horizon up to time t; for further discussion see [13].

[20]The "freezing in" of quantum fluctuations in ϕ as classical fluctuations is discussed in some detail in [14].

[21]The quantity $\delta\rho/\rho$ is not gauge invariant. For sub-horizon-sized modes this fact causes little difficulty, as a Newtonian analysis suffices; however, for super-horizon-sized modes it can be very problematic. We will discuss the subtleties in the next Chapter.

$\lambda_{\rm phys} \simeq H^{-1}$, ζ has a particularily simple form: $\zeta = \delta\rho/(\rho+p)$. In the ordinary matter-dominated (MD) or radiation-dominated (RD) phase, ζ at horizon crossing is, up to a factor of order unity, equal to $\delta\rho/\rho$. Thus, to calculate the amplitude of a density perturbation when it crosses back inside the horizon, $(\delta\rho/\rho)_{\rm HOR}$, one only needs to compute ζ at the time the fluctuation crossed outside the horizon during inflation. During inflation $\rho_\phi + p_\phi = \dot\phi^2$ is very much smaller than $\rho_\phi \simeq V(\phi)$, so that ζ is much larger than $\delta\rho/\rho$. Equating the values of ζ at the two horizon crossings we find:

$$\begin{pmatrix} 3/4 & ({\rm RD}) \\ 1 & ({\rm MD}) \end{pmatrix} \left(\frac{\delta\rho}{\rho}\right)_{\rm HOR} = \zeta_{N_\lambda} = \left(\frac{\delta\phi V'}{\dot\phi^2}\right)_{N_\lambda} \simeq \left(\frac{H^2}{\dot\phi}\right)_{N_\lambda}, \quad (8.51)$$

where we have used the fact that $\delta\phi \simeq H/2\pi$ and the slow-roll equation of motion, $V' = -3H\dot\phi$.

During inflation H and $\dot\phi$ vary rather slowly; moreover, the modes of cosmological interest, say 1 Mpc to 3000 Mpc, crossed outside the horizon during a period spanning only a small fraction of the total inflationary epoch—about 8 e-folds out of the total of 60 or so.[22] As a result, the spectrum of perturbations predicted is very nearly scale invariant—the Harrison-Zel'dovich spectrum [16].[23] This is a very robust and generic prediction of inflation, which traces to the time-translation invariance of de Sitter space: As each mode crosses outside the horizon, it has the same physical size ($\sim H^{-1}$), and the Universe has the same expansion rate; thus each scale has impressed upon it a wrinkle of the same amplitude. Further, because of the nature of quantum fluctuations, the induced density fluctuations are gaussian (provided that the self-coupling of the scalar field is small).[24]

On the other hand, the amplitude of the spectrum of constant curvature fluctuations is model-dependent, depending upon $H^2/\dot\phi$. In order to seed the observed structure in the Universe, perturbations of amplitude 10^{-5} to 10^{-4} or so are probably required. On the basis of the measured

[22]As we shall come to appreciate, the natural "time variable" for inflation is the number of e-folds, and since the horizon-crossing epoch is specified by $N_\lambda \propto \ln\lambda$, all cosmologically interesting scales cross outside the horizon during a narrow interval in this variable.

[23]In most models the amplitude of the perturbations varies by less than a factor of 2 over the scales from 1 Mpc to 3000 Mpc [17].

[24]If the inflaton field is strongly coupled, it is possible that different modes may have non-linear couplings, resulting in non-gaussian fluctuations (non-random phases between different modes) [18].

isotropy of the CMBR, the amplitude can be no larger than 10^{-4}. As we shall see in the next Section, finding a model that achieves this is no mean feat. Arranging for acceptable density perturbations results in the most restrictive constraint on inflationary potentials (and thereby the underlying particle physics models) and necessitates the fact that inflationary potentials must be extremely flat.

Before moving on, let's be very precise about the amplitude of the density perturbations. The *rms* mass fluctuation is related to the *rms* variance in ϕ by

$$(1 \text{ or } 3/4)[k^{3/2}|\delta_k|/\sqrt{2}\pi] = (\Delta\zeta)_{N_\lambda} \simeq [k^{3/2}|\delta\phi_k|/\sqrt{2}\pi]\,(V'/\dot{\phi}^2). \qquad (8.52)$$

Modes k that re-enter the horizon while the Universe is still radiation dominated, i.e., $\lambda \lesssim 13h^{-2}$ Mpc, do so as sound waves in the photon-baryon plasma, with amplitude [15],

$$(\delta\rho/\rho)_{\rm HOR}\ (\equiv k^{3/2}|\delta_k|/\sqrt{2}\pi) \ = \ \left(\frac{2H^2}{\pi\dot{\phi}}\right)_{N_\lambda}. \qquad (8.53)$$

Perturbations in non-interacting, relic particles—such as axions, photinos, etc.— must, by the equivalence principle, have the same amplitude, but of course do not oscillate. Rather they grow logarithmically, by a factor of 2 to 3, by the time of matter–radiation equality.

Modes k that re-enter the horizon when the Universe is dominated by matter, i.e., $\lambda \gtrsim 13h^{-2}$ Mpc, do so as growing mode perturbations with amplitude,

$$(\delta\rho/\rho)_{\rm HOR} \simeq \left(\frac{H^2}{5\pi\dot{\phi}}\right)_{N_\lambda}. \qquad (8.54)$$

As we shall discuss in the next Chapter, it is the spectrum of density perturbations at the epoch of matter–radiation equality that provides the "initial data" for the structure formation problem. In the next Chapter we shall "translate" these horizon-crossing amplitudes into amplitudes specified at the epoch of matter–radiation equality.

All massless, or very light ($m^2 \ll H^2$), fields are excited in de Sitter space in a similar manner, with a spectrum given by (8.46). Once the excited modes re-enter the horizon, they propagate as particles. That is, de Sitter excitation ultimately results in particle production [19]. As a very interesting example, consider the graviton field. Gravitons are the propagating modes associated with transverse, traceless tensor metric perturbations, and they behave as a minimally coupled scalar field with two degrees of freedom. Thus, we can analyze graviton production by considering two minimally coupled scalar fields, $\phi_{+,\times}$, which are related to the dimensionless tensor metric perturbation $h^i{}_j$ by

$$\begin{aligned} h_{+,\times} &= \sqrt{16\pi G}\,\phi_{+,\times}, \\ h^i{}_j &= h_+\,\mathbf{e}_+ + h_\times\,\mathbf{e}_\times, \end{aligned} \tag{8.55}$$

where $\mathbf{e}_+ = \hat{e}_x \otimes \hat{e}_x - \hat{e}_y \otimes \hat{e}_y$ and $\mathbf{e}_\times = \hat{e}_x \otimes \hat{e}_y + \hat{e}_y \otimes \hat{e}_x$ are the polarization tensors for the two graviton modes.

The mean fluctuation in $h_{+,\times}$ is obtained from our previous formula for a minimally coupled, scalar field, cf. (8.46), by multiplying by $\sqrt{16\pi G}$, and is given by

$$(\Delta h_{+,\times})_k \equiv k^{3/2}|h_k|/\sqrt{2}\pi = \frac{2}{\sqrt{\pi}}\frac{H}{m_{Pl}}. \tag{8.56}$$

Just as for a massless, minimally coupled scalar field the amplitude of a given mode k remains constant while that mode is outside the horizon. When a given graviton mode re-enters the horizon, the associated tensor metric fluctuations thereafter propagate as gravitons. The energy density in gravitons associated with a given mode as it crosses back into the horizon is simple to estimate:[25]

$$\begin{aligned} k\frac{d\rho_{\rm graviton}}{dk} &= \frac{k^2[(\Delta h_+)_k^2 + (\Delta h_\times)_k^2]}{16\pi G} \\ &\simeq \frac{k_{\rm phys}^2}{2\pi^2 G}\left(\frac{H}{m_{Pl}}\right)^2 \simeq \frac{H_{\rm HOR}^2 H^2}{2\pi^2}, \\ \frac{k}{\rho_{\rm HOR}}\frac{d\rho_{\rm graviton}}{dk} &\simeq \frac{4}{3\pi}\left(\frac{H}{m_{Pl}}\right)^2, \end{aligned} \tag{8.57}$$

where the subscript "HOR" refers to the epoch of horizon crossing in the post-inflationary Universe. By carrying out this calculation carefully, one can compute the spectrum of relic gravitational radiation produced by

[25]The energy density in gravitational waves is given in terms of the transverse, traceless metric perturbations by $\rho_{\rm graviton} = [\dot{h}_\times^2 + \dot{h}_+^2]/(16\pi G)$, which in turn can be related to $(\Delta h_{+,\times})_k$ by $\rho_{\rm graviton} = (16\pi G)^{-1}\int[(\Delta h_+)_k^2 + (\Delta h_\times)_k^2]k\,dk$.

inflation that should be present today. The results of such a calculation are shown in Fig. 8.5 [20].

From (8.57) we see that the energy density per octave ($kd\rho_{\rm graviton}/dk$) in gravitons at horizon crossing is roughly $(H/m_{Pl})^2$ times that of the total energy of the Universe (H is the expansion rate during inflation). Once a graviton mode crosses inside the horizon, its energy density evolves as R^{-4}, just as a relativistic species. If, for simplicity, we assume that g_* is constant, ρ_R also evolves as R^{-4}, so that $(kd\rho_{\rm graviton}/dk)/\rho_R$ remains constant. For modes that cross inside the horizon while the Universe is radiation dominated, 10^{-7} Mpc(GeV/$T_{\rm RH}$) $\lesssim \lambda \lesssim 13h^{-2}$ Mpc, $\rho_{\rm HOR} \simeq \rho_R$, so it follows that

$$\frac{k}{\rho_R}\frac{d\rho_{\rm graviton}}{dk} \simeq \frac{4}{3\pi}\left(\frac{H}{m_{Pl}}\right)^2 = \frac{32}{9}\left(\frac{M}{m_{Pl}}\right)^4 ,$$

$$\Omega_{\rm graviton}(k)h^2 \simeq 10^{-4}\left(\frac{M}{m_{Pl}}\right)^4 , \tag{8.58}$$

where $\Omega_{\rm graviton}(k) = (kd\rho_{\rm graviton}/dk)/\rho_C$. That $\Omega_{\rm graviton}(k)$ is independent of k for these modes accounts for the plateau in Fig. 8.5.

A graviton mode of particular interest is the one that is just crossing the horizon at the present epoch, i.e., $\lambda \sim 3000$ Mpc. That mode corresponds to a quadrupole metric perturbation on the scale of the present horizon, and as such it gives rise to quadrupole anisotropies in the CMBR of amplitude $(\Delta h)_k \simeq H/m_{Pl}$. Based upon this fact and the present isotropy level of the CMBR, we can conclude that the value of the Hubble parameter during inflation must have been less than about $10^{-4}m_{Pl}$, or that the vacuum energy associated with inflation must have been less than about $10^{-8}m_{Pl}^4$. That is, the energy scale of inflation must be very far removed from that of the Planck scale.[26]

The de Sitter phase excitations of other very light fields may also lead to interesting levels of particle creation, e.g., the axion field or possibly even the photon field; see [19, 21]. In the case of the photon field, it is possible that the de Sitter-space-produced fluctuations ultimately lead to the generation of large-scale primeval magnetic fields. This is a particularily interesting possibility, as the origin of the primeval magnetic fields needed

---

[26] Of course, that does not exclude the possibility that inflation lasted a very long time, i.e., $N_{\rm TOT} \gg 60$, starting with a vacuum energy comparable to m_{Pl}^4 which had fallen to less than $10^{-8}m_{Pl}^4$ by the time the present Hubble scale crossed outside the horizon (about 60 e-folds before the end of inflation).

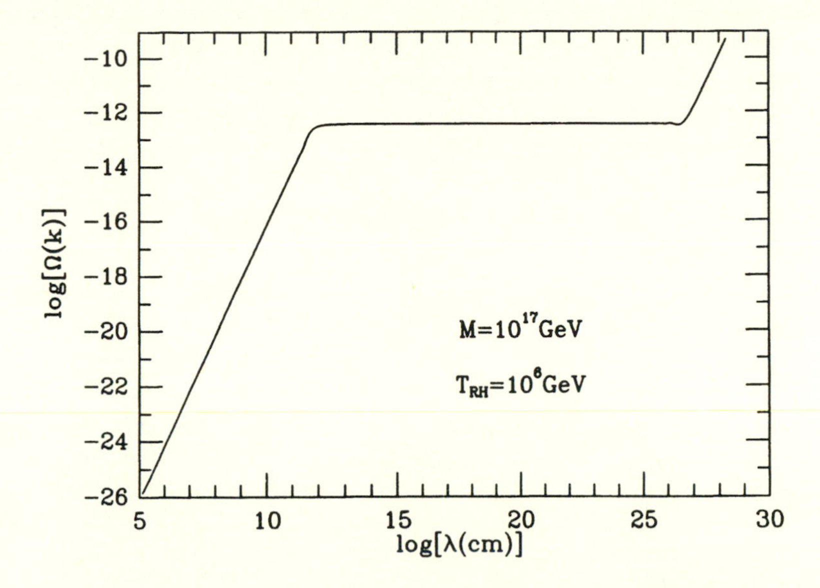

Fig. 8.5: The present spectrum of relic gravitons produced during inflation. $\Omega_{\rm graviton}(k)$ is the fraction of critical density contributed per octave; from [20].

to "seed" the ubiquitous astrophysical magnetic fields observed today is still an unsolved mystery [21].

8.5 Specific Inflationary Models

We have just discussed the mechanics of inflation in general. Now we will turn to specific models, i.e., specific examples of scalar potentials and the particle theories that motivate them. We should warn the reader at the outset that there is no "standard model" of inflation; in fact, there are no truly satisfactory models. There are, however, several very simple models that serve as a framework for discussing inflation, just as the now-defunct $SU(5)$ model serves as a simple example when discussing unified gauge theories.

As will become apparent very quickly, all models of inflation involve very flat scalar potentials, usually characterized by the coefficient of the ϕ^4 term in the potential having a value of order 10^{-14} or so. The first model of slow-rollover (or new) inflation was based upon an $SU(5)$ GUT that was spontaneously broken down to $SU(3)\otimes SU(2)\otimes U(1)$ by a 24-dimensional Higgs field [8, 9]. The scalar potential for this Higgs field was of the

Coleman-Weinberg form (see Chapter 7). As we have done in previous examples of SSB, we can write the scalar potential for this multi-dimensional Higgs field solely in terms of its magnitude in the $SU(3) \otimes SU(2) \otimes U(1)$ direction. The one-loop, zero-temperature Coleman-Weinberg potential written this way is given by (7.21) with $\lambda + 2B = 0$:

$$V(\phi) = B\sigma^4/2 + B\phi^4 \left[\ln(\phi^2/\sigma^2) - 1/2\right], \tag{8.59}$$

where $B = 25\alpha_{\rm GUT}^2/16 \simeq 10^{-3}$, $\alpha_{\rm GUT} \simeq 1/45$, and the minimum of the potential occurs at $\phi = \sigma \simeq 2 \times 10^{15}$ GeV. As discussed in Chapter 7, the $\phi^4 \ln(\phi^2/\sigma^2)$ term responsible for SSB arises due to one-loop quantum corrections from the $SU(5)$ gauge bosons. Because of the absence of a mass term, i.e., no $m^2\phi^2$ term, the potential is very flat near the origin. For $\phi \ll \sigma$, the potential is well approximated by,

$$\begin{aligned} V(\phi) &\simeq B\sigma^4/2 - \lambda\phi^4/4, \\ V'(\phi) &\simeq -\lambda\phi^3, \\ \lambda &\simeq |4B\ln(\phi^2/\sigma^2)| \simeq 80B \simeq 0.1. \end{aligned} \tag{8.60}$$

Very close to $\phi = 0$,

$$\begin{aligned} V(\phi) &\simeq B\sigma^4/2, \\ H^2 &\simeq \frac{4\pi}{3}\frac{B\sigma^4}{m_{Pl}^2} \simeq (10^{10}\,\mathrm{GeV})^2, \end{aligned} \tag{8.61}$$

where we have neglected the energy density associated with the thermal bath of particles, as it rapidly red shifts away.

We discussed Coleman-Weinberg SSB for the electroweak theory in Chapter 7, and much of that discussion is directly applicable here. The SSB phase transition is first order, and at a temperature above the critical temperature (about 10^{14} GeV) a symmetry-breaking minimum develops. For temperatures below the critical temperature, a temperature-dependent barrier separates the true vacuum, $\phi = \sigma$, from the metastable false vacuum, $\phi = 0$. The height of the barrier ($\sim T^4$) and the position of the barrier ($\phi \sim T$) both vary with temperature. Above the critical temperature $\langle\phi\rangle = 0$; at temperatures below the critical temperature the Higgs field ϕ remains trapped in the metastable minimum ($\phi = 0$). Finally, when the temperature of the Universe drops to 10^9 GeV or so, the action for thermal tunnelling becomes order unity, and the metastable $\phi = 0$ false vacuum state becomes unstable. In analogy with similar condensed matter phenomenon, it is expected that at around this temperature the Universe will undergo *spinodal decomposition* [9]. That is, it breaks up into irregularly shaped regions within which ϕ is beyond the barrier and is approximately constant, with a value of order $T \sim 10^9$ GeV. At this point the evolution of ϕ within a "fluctuation region" is downhill.

In the slow-roll, friction-dominated regime we can easily solve for the evolution of ϕ using the approximate form of the potential, (8.60), and the slow-roll equation of motion, $3H\dot{\phi} = -V'(\phi)$,

$$
\begin{aligned}
N(\phi_i \to \phi) &= \int_{t_i}^{t} H dt = \frac{8\pi}{m_{Pl}^2} \int_{\phi_i}^{\phi} \frac{V(\phi)}{-V'(\phi)} d\phi \\
&= \frac{3H^2}{2\lambda} \left(\frac{1}{\phi_i^2} - \frac{1}{\phi^2} \right) \\
&= \frac{\pi}{2} \frac{(\sigma/m_{Pl})^2}{|\ln(\phi^2/\sigma^2)|} \left(\frac{\sigma^2}{\phi_i^2} - \frac{\sigma^2}{\phi^2} \right), \qquad (8.62)
\end{aligned}
$$

where $\phi_i \sim 10^9$ GeV is the initial value of the scalar field in the fluctuation region. The slow-roll approximation is valid until $|V''(\phi_e)| \simeq 9H^2$, or for

$$
\phi^2 \lesssim \phi_e^2 \simeq \frac{3H^2}{\lambda} \simeq 30H^2. \qquad (8.63)
$$

The total number of e-folds during slow roll, $N_{\rm TOT} = N(\phi_i \to \phi_e)$, depends upon ϕ_i. Taking $\phi_i \simeq 10^9$ GeV $\sim H/10$, we see that $N_{\rm TOT}$ is almost 1000—easily enough to solve the *smoothness* problem. However, such a small value for ϕ_i belies our assumption that quantum fluctuations are a small perturbation to the classical trajectory, as $H/2\pi \sim 10^9$ GeV is of the same order as ϕ_i. If we used the mean variance in ϕ to make this comparison, $\Delta\phi = N^{1/2}(H/2\pi)$, the situation would be even more troubling. Moreover, it is clear that in order to obtain sufficient inflation ϕ_i *must* be comparable to $H/2\pi$, so that quantum fluctuations invalidate our semi-classical analysis. Even if ϕ initially had such a small value, we would expect quantum fluctuations to "push" ϕ sufficently far out on the potential to terminate inflation, i.e., beyond $\phi = \phi_e$. Apparently, the very flat Coleman-Weinberg potential is not flat enough. The excessive size of the quantum fluctuations relative to the classical value of ϕ will become even more apparent when we calculate the resulting density perturbations.

Recall that the amplitude of the inflation-produced density perturbations is set by the value of $H^2/\dot{\phi}$, 50 or so, e-folds before the end of inflation, cf. (8.45). Using our slow-roll solutions it is a simple matter to evaluate $(\delta\rho/\rho)_{\rm HOR}$,

$$N(\phi \to \phi_e) = \frac{3H^2}{2\lambda}\left(\frac{1}{\phi^2} - \frac{1}{\phi_e^2}\right) \simeq \frac{3H^2}{2\lambda}\frac{1}{\phi^2},$$

$$(\delta\rho/\rho)_{\rm HOR} \simeq \left(H^2/\dot{\phi}\right)_{N_\lambda} \simeq \left(\frac{3H^3}{-V'}\right)_{N_\lambda} \simeq \left(\frac{8\lambda}{3}\right)^{1/2} N_\lambda^{3/2}, \quad (8.64)$$

which for $N_\lambda \simeq 50$ leads to density perturbations of order 10^2—hardly small enough to be called perturbations, and clearly at variance with the isotropy of the CMBR.[27] The fact that the resulting density fluctuations are so large is directly traceable to the fact the value of ϕ during inflation is comparable to $H/2\pi$.

From the expression for $(\delta\rho/\rho)_{\rm HOR}$, the smallness of the deviation from a scale-invariant spectrum is easy to see: Since $N(\lambda) \simeq 50 + \ln(\lambda/{\rm Mpc})$, the deviation is proportional to $[1 + \ln(\lambda/{\rm Mpc})/50]^{3/2}$. The smallness of the deviation from scale-invariant spectrum is generic to all inflationary models and owes to the fact that $(\delta\rho/\rho)_{\rm HOR}$ depends upon N_λ, which only varies logarithmically with the comoving wavelength.

Reheating in the Coleman-Weinberg model for inflation is not a problem. The 24-dimensional Higgs of $SU(5)$ couples to all the $SU(5)$ gauge and Higgs fields that acquire superheavy masses with strength $g_Y \sim m/\sigma$, where m is the mass the field acquires from SSB. Since the masses that these particles acquire are comparable to the GUT scale, their Yukawa couplings are not too much smaller than unity. The decay width of the ϕ field is order $\Gamma_\phi \sim g_Y^2 m_\phi$, where m_ϕ is the mass of the ϕ. For any reasonable value of g_Y, Γ_ϕ is greater than H, so that reheating is 100% efficient, i.e., $T_{\rm RH} \simeq \sigma$. Reheating proceeds without a hitch!

The Coleman-Weinberg $SU(5)$ model, the first model of slow-rollover inflation, is clearly a failure. In its demise, however, it points to a simple cure for its difficiencies: Reduce the value of B (and thereby λ). It should be clear from (8.62) and (8.64) for $N_{\rm TOT}$ and $(\delta\rho/\rho)_{\rm HOR}$ that both problems are solved by reducing the value of B. The first problem, the validity of the semi-classical approach, requires a modest reduction, a factor of 100 or

[27] Note, the λ in N_λ, refers to the comoving wavelength of a fluctuation, and should not be confused with the λ in the scalar potential.

so. The density perturbation problem is more serious; in order to achieve perturbations of order 10^{-5} or so, B must be reduced to a value of about 10^{-15}! In the context of a unified model where ϕ is the (adjoint) Higgs field which implements GUT symmetry breaking this is all but impossible. The reason is simple: The size of the one-loop quantum corrections due to the gauge bosons is fixed by the gauge coupling constant; in order to have $B \simeq 10^{-15}$, the gauge coupling would have to be very, very small, $\alpha \simeq 3 \times 10^{-8}$, far too small to fit into any framework of unification.[28]

With such a small value for B, the potential is very, very flat and squat; i.e., the height of the potential ($\sim B\sigma^4$) is much, much less than σ^4. Because B is so small, the number of e-folds of inflation is enormous: taking $\phi_i = 1000H$, the number of e-folds is about 10^7! All of these features—tiny quartic coupling, short, squat potential, and an enormous amount of inflation—are generic to models that successfully implement inflation.

It seems that in order to have a sufficiently flat scalar potential, the field responsible for inflation, referred to by some as the *inflaton* field, must be very weakly coupled to all fields (including gauge fields), so that one-loop corrections to the scalar potential do not interfere with the extreme flatness required. This strongly suggests that the inflaton field must be a very weakly coupled, gauge singlet field. This fact has important implications for reheating. Because the inflaton is so weakly coupled, its decay width is necessarily small, and this makes reheating difficult, as $T_{\rm RH} \simeq (\Gamma_\phi m_{Pl})^{1/2}$. To be specific, suppose the inflation couples to a fermion species with strength g_Y. The one-loop corrections to the scalar potential due to this fermion species will be of order, $g_Y^4 \ln(\phi^2/\sigma^2)\phi^4$. In order that they not lead to a ϕ^4 term larger than about $10^{-15}\phi^4$, g_Y must be less than about 3×10^{-4}. On the other hand, the decay width of the inflaton into these fermions is $\Gamma_\phi \sim g_Y^2 m_\phi$—suppressing g_Y also reduces the decay width and reheat temperature. A lower reheat temperature makes it more difficult to generate the baryon asymmetry.

Although the original Coleman-Weinberg $SU(5)$ model for slow-rollover inflation was ultimately a failure, it pointed the way for the construction of a number of somewhat more successful models and to a set of rules for successfully implementing slow-rollover inflation [17]. The first model we will discuss, a rather modest extension of the original new inflation model, was suggested by Pi [22] and Shafi and Vilenkin [23]. In this model, the inflaton is a weakly coupled, gauge-singlet field, whose potential is of the Coleman-

[28]One could of course try to cancel off the one-loop corrections of the gauge bosons against a fermion species; this effectively happens in some supersymmetric models.

Weinberg form. The radiative corrections arise from the inflaton's couplings to other scalar fields in the theory. In addition to implementing inflation, the inflaton also breaks Peccei-Quinn symmetry[29] and indirectly leads to GUT symmetry breaking. While in this model the inflaton is no longer the 24-dimensional Higgs field, it couples to the 24-dimensional Higgs field by a term in the Lagrangian of the form $+\lambda_\Phi \phi^2|\Phi_{24}|^2$. As the inflaton moves away from $\phi = 0$, it induces a negative mass-squared for the 24-dimensional Higgs and thereby triggers GUT symmetry breaking by the 24-dimensional Higgs. Apart from the couplings of the inflaton to other fields in the theory, its scalar potential is identical to that of the original model, cf. (8.59), except $B \simeq 10^{-15}$—which is why we wrote the Coleman-Weinberg inflationary potential (8.59) in the form that we did!

Supersymmetry is an idea that has received a great deal of attention in particle physics in recent years. Not only does it provide a means of stabilizing greatly discrepant mass scales, e.g., the weak and Planck scales, and small couplings, but when it is promoted to a local symmetry (known as supergravity) it provides the means of unifying gravity with the other interactions. Supersymmetric and low-energy supergravity models have been proposed as a means to stabilize the small couplings required in inflationary potentials against radiative corrections. A particularly simple supergravity model of inflation was proposed by Holman, Ramond, and Ross [24]. Their model starts with a superpotential of the form, $W = I + S + G$, where I, S, and G are the inflation, SUSY, and GUT sectors of the theory, respectively. For I they choose the very simple form

$$I = (\Delta^2/M)(\phi - M)^2, \qquad (8.65)$$

where $M \equiv m_{Pl}/\sqrt{8\pi}$ is the reduced Planck mass, and Δ is some undetermined, intermediate mass scale (between the weak scale and the Planck scale). As we shall see, the value of Δ is determined by inflation! The scalar potential is related to the superpotential I, by

$$V_I(\phi) = \exp(|\phi|^2/M^2)\left\{|\partial I/\partial\phi + \phi^* I/M^2|^2 - 3|I|^2/M^2\right\}; \qquad (8.66)$$

[29]The axion and Peccei-Quinn (PQ) symmetry are the subject of Chapter 10. In this model the inflaton is a complex scalar field, and it carries a PQ charge. When it acquires its vacuum expectation value it thereby breaks PQ symmetry. The axion corresponds to the phase degree of freedom of the inflaton field.

from which it follows that

$$V_I(\phi) = \Delta^4 \exp(|\phi|^2/M^2)\left\{\phi^6/M^6 - 4\phi^5/M^5 + 7\phi^4/M^4 - 4\phi^3/M^3 - \phi^2/M^2 + 1\right\}. \quad (8.67)$$

Note that $V(\phi = M) = V'(\phi = M) = 0$. This potential can be expanded in a power series in ϕ/M,

$$V_I(\phi) = \Delta^4\left\{1 - 4\phi^3/M^3 + 6.5\phi^4/M^4 - 8\phi^5/M^5 + ...\right\}. \quad (8.68)$$

Once this is done, it is very easy to analyze the inflationary scenario that results from this potential:

$$\begin{aligned} N(\phi \to M) &\simeq \frac{1}{12}\frac{M}{\phi} - \frac{1}{3}, \\ N_{\rm TOT} &= N(\phi_i \to M), \\ \left(\frac{H^2}{\dot\phi}\right)_{N_\lambda} &\simeq \frac{12}{\sqrt{3}}\frac{\Delta^2}{M^2}N_\lambda^2. \end{aligned} \quad (8.69)$$

Note that the amplitude of the density perturbations produced is proportional to Δ^2; taking $\Delta \simeq 3 \times 10^{-5}M$ results in density perturbations of an acceptable size and corresponds to an intermediate scale of about 10^{14} GeV, which is a very interesting energy scale!

For this model one can readily compute the reheat temperature. The decay width of the inflaton is $\Gamma_\phi \simeq m_\phi^3/M^2 \simeq \Delta^6/M^5$, where the ϕ mass is $m_\phi^2 = 8e\Delta^4/M^2$ (here $e = 2.71828\cdots$). The resulting reheat temperature is about 10^6 GeV, too low to produce the baryon asymmetry in the standard way, i.e., by out-of-equilibrium decay. However, in this model the baryon asymmetry is directed produced by the decays of the inflaton particles themselves, so that $n_B/s \simeq \epsilon T_{\rm RH}/m_\phi \simeq 10^{-4}\epsilon$.

On the face of it this model seems to have achieved successful inflation without resort to a tiny coupling constant. However, the dimensionless coefficient of the ϕ^4 term is $6.5(\Delta/M)^4 \simeq 10^{-17}$. Depending upon how you look at it, the small coupling has either been eliminated, or at least finessed, by expressing it in terms of $(\Delta/M)^4$.

In general, supersymmetric models introduce several additional problems. For such models the high-temperature minimum of the potential is

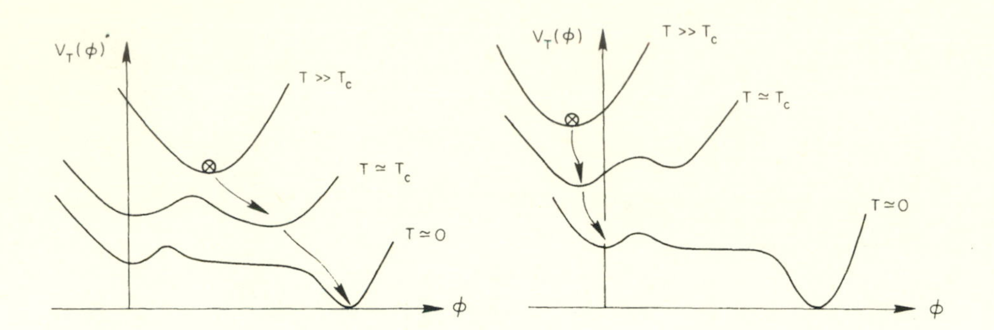

Fig. 8.6: Thermal evolution of a supersymmetric potential. (a) If the finite-temperature minimum of the potential occurs for $\langle\phi\rangle_T > 0$, it may smoothly evolve into the zero-temperature minimum of the potential; (b) insuring that $\langle\phi\rangle_T < 0$ solves the problem.

generally not at $\phi = 0$, and often occurs for $\langle\phi\rangle_T > 0$, in which case the high-temperature minimum can smoothly evolve to the zero-temperature minimum, thereby precluding the possibility of inflation [25] (see Fig. 8.6). There are at least two possible remedies. If the high-temperature minimum occurs for $\langle\phi\rangle_T \leq 0$, there will always be a barrier between the high-temperature and the low-temperature minimum, for which $\langle\phi\rangle \geq 0$. The other solution is to ignore the problem! Since the inflaton must be weakly coupled, it is likely that it was never in contact with the bath of thermal particles in the Universe, so that the initial value of ϕ is not determined by thermal considerations. The initial value of ϕ will take on different values in different regions of the Universe, and in some regions it will be near $\phi = 0$. Those regions will undergo inflation. (This, in fact, was the point of view taken by Holman, Ramond, and Ross above [24].)

A second problem that arises in supersymmetric models has to do with the gravitino, the supersymmetric partner of the graviton. Gravitinos are weakly interacting, long-lived particles (see Chapter 5). They are produced during reheating, with an abundance proportional to the reheat temperature (they never achieve their thermal abundance [26]). Unless the reheat temperature is less than about 10^9 GeV, decays of the gravitinos that are produced will adversely alter the light element abundances produced in primordial nucleosynthesis by dissociating ^{4}He and the other light elements produced [27].

Finally, in supersymmetric models where supersymmetry breaking is done with a Polonyi field,[30] the Polonyi field can be set into oscillations, just as the inflaton field, except that the Polonyi field oscillations do not decay because the Polonyi field is so weakly coupled [28]. Since the energy density in a coherently oscillating field behaves like non-relativistic matter, it will eventually come to dominate the Universe (long before the Universe should become matter-dominated).

As a final example of a model that successfully implements inflation, let us consider the very simple and elegant model suggested by Linde [29], which he calls "chaotic inflation." The scalar field is not part of any unified theory, and in fact its only purpose is to implement inflation. The potential for chaotic inflation is delightfully simple:[31]

$$V(\phi) = \lambda\phi^4. \tag{8.70}$$

The minimum of this potential is at $\phi = 0$, and so the potential clearly has nothing to do with SSB. In Linde's model the scalar field must initially be displaced from $\phi = 0$; we will denote the initial value of ϕ by ϕ_i. Linde envisions that the initial distribution of ϕ_i is "chaotic" (hence the name "chaotic inflation"), with ϕ_i taking on different values in different regions of the Universe.

Inflation with this potential is very easy to analyze. It is simple to show that the slow-roll approximation is valid for $\phi^2 \gtrsim \phi_e^2 = m_{Pl}^2/2\pi$. During the slow roll,

$$\begin{aligned} N(\phi \rightarrow \phi_e = m_{Pl}/\sqrt{2\pi}\,) &= \pi(\phi/m_{Pl})^2 - 1/2, \\ N_{\rm TOT} = N(\phi_i \rightarrow m_{Pl}/\sqrt{2\pi}\,) &= \pi(\phi_i/m_{Pl})^2 - 1/2, \\ \left(\frac{H^2}{\dot{\phi}}\right) &= 4\sqrt{2/3}\lambda^{1/2}N_\lambda{}^{3/2}. \end{aligned} \tag{8.71}$$

In order to obtain sufficient inflation, say 60 *e*-folds, ϕ_i must be greater than about $4.4\,m_{Pl}$; to obtain acceptable density perturbations, λ must be about 10^{-15}—as per usual. The model in its simplicity is not definite

---

[30]The Polonyi field is a very-weakly-interacting field (gravitational-strength interactions with ordinary matter) whose sole purpose in life is to break supersymmetry.

[31]In fact, the equally simple potential, $V(\phi) = m^2\phi^2/2$, works just as well. Such a potential corresponds to a massive scalar field. We leave it as an exercise for the reader to analyze such a potential. Answer: $m^2 = 2 \times 10^{-13} m_{Pl}^2$.

enough to discuss reheating. However, one can imagine coupling ϕ to other fields, in which case $T_{\rm RH} \simeq (\Gamma_\phi m_{Pl})^{1/2}$.[32]

This model illustrates one of the prerequisites for inflation to occur: the required smoothness of the inflationary patch. Remember for inflation to occur the energy density in the spatial-gradient terms must be subdominant:

$$(\nabla\phi)^2 \ll \lambda\phi_i^4. \tag{8.72}$$

If we take the dimension of the region over which ϕ varies by order unity to be L,

$$(\nabla\phi)^2 \sim (\phi_i/L)^2 \ll \lambda\phi_i^4 \quad \Rightarrow \quad L \gg 2\frac{\phi_i}{m_{Pl}}H^{-1}. \tag{8.73}$$

That is, ϕ must be smooth on a scale much greater than the Hubble radius—such a requirement doesn't sound very chaotic.[33]

From these examples of specific models that implement inflation we have learned a number of important lessons, lessons which generalize to become a "prescription" for successful inflation [17]. That prescription includes the following requirements:

• *Start Inflation*: The scalar field must be smooth in a large enough region so that the energy density and pressure associated with spatial gradients in ϕ are smaller than the potential energy. If the average value of ϕ within the would-be inflationary patch is ϕ_i, and if the region has spatial dimension L, this requirement implies

$$(\nabla\phi)^2 = \mathcal{O}[(\phi_i/L)^2] \ll V(\phi_i). \tag{8.74}$$

If this requirement is not met and the $(\nabla\phi)^2$ terms dominate, inflation will not occur.

• *Slow Roll/Sufficient Inflation*: In order to have a slow roll, the potential must have a flat region, where $|V''(\phi)| \lesssim 9H^2$ and $|V'(\phi)m_{Pl}/V(\phi)| \lesssim (48\pi)^{1/2}$. In addition, the interval of slow roll, $[\phi_i, \phi_e]$, must be long enough so that quantum fluctuations do not quickly "scoot" the field across the

---

[32]The coherent oscillation phase for this potential is somewhat different; for $V = \lambda\phi^4$ it follows that $\langle p_\phi\rangle = \langle\dot\phi^2/2 + V\rangle = \rho_\phi/3$ (like relativistic matter). Thus during the coherent oscillation phase $\rho_\phi \propto R^{-4}$.

[33]Linde [30] has a different view on this matter. He imagines that chaotic inflation begins at the Planck epoch with $\phi_i \simeq \lambda^{-1/4}m_{Pl}$, so that $V(\phi_i) \sim m_{Pl}^4$ (note that $N_{\rm TOT} \sim \pi\lambda^{-1/2} \sim 10^8$). Moreover, he assumes that $V(\phi_i)$ and $(\nabla\phi)_i^2$ are at best comparable, so that $L \sim \lambda^{-1/4}H^{-1} \sim 10^4 H^{-1}$. Therefore he argues that there will be regions of size greater than H^{-1} where $V(\phi_i) \gg (\nabla\phi)_i^2$.

flat part of the potential. That is, the quantum fluctuations should be a small perturbation to the classical trajectory of the field. Stated quantitatively, the classical change in ϕ in a Hubble time, $\Delta\phi_{\rm cl} \simeq -V'/3H^2$, should be much greater than the change in the quantum variance in ϕ in a Hubble time, $\Delta\phi_{\rm QM} = H/2\pi$. This requires that

$$2\pi \gg \frac{3H^3}{-V'}. \qquad (8.75)$$

We recognize $3H^3/V'$ as $H^2/\dot{\phi}$, which sets the size of density perturbations; thus, this condition is automatically satisfied provided that density perturbations of an acceptable size are achieved.

In addition, the number of e-folds of expansion must be larger than about 60 to solve the *smoothness* and *flatness* problems, cf. (8.26) and (8.36). This condition is generally very easy to satisfy.

• *Acceptable Density Perturbations*: The magnitude of the inflation-produced, adiabatic density perturbations must be less than about 10^{-4} on large scales in order to avoid excessively large fluctuations in the temperature of the CMBR. In addition, if the fluctuations produced by inflation are to lead to structure formation, they should have amplitude greater than about 10^{-5} or so. This leads to the condition that $H^2/\dot{\phi} \simeq 10^{-5}$ to 10^{-4} at about 50 e-folds before the end of inflation. This is by far the most difficult constraint to satisfy, and the fact that inflationary potentials must be exceeding flat is a result of meeting this requirement.

• *Sufficient Reheating*: The reheat temperature must be high enough so the Universe is radiation dominated in time for primordial nucleosynthesis, i.e., $T_{\rm RH} \gtrsim 1\,{\rm MeV}$—if not, the beautiful concordance between the predicted and observed light-element abundances will be upset. Using the fact that $T_{\rm RH} = (\Gamma_\phi m_{Pl})^{1/2}$, this implies that Γ_ϕ must be greater than about 10^{-25} GeV—a rather modest requirement. Baryogenesis provides a more challenging requirement. If the baryon asymmetry is produced by the out-of-equilibrium decays of superheavy bosons, the Universe must reheat to a temperature comparable to the mass of the superheavy boson, in order that these bosons be produced from the thermal bath after reheating. While it is difficult to be precise about the minimum mass of such a boson, the longevity of the proton would strongly suggest that any such boson which couples to light quarks must be more massive than about 10^{10} GeV. That would imply the reheat temperature must be more than 10^9 GeV.

It is possible that the baryon asymmetry is produced directly by the decay of the inflaton field itself, in which case, $n_B/s \simeq \epsilon T_{\rm RH}/m_\phi$, and a significantly lower reheat temperature can be tolerated. In fact, models have been suggested wherein baryogenesis occurs at temperatures as low as a few GeV [31].

The reader has probably noticed that the density perturbation requirement works at cross purposes with reheating. To obtain sufficiently small adiabatic density perturbations, the inflaton must be very weakly coupled, which tends to make the reheat temperature small, since $T_{\rm RH} \propto \Gamma_\phi^{1/2}$.

• *Unwanted Relics*: One of the prime motivations for inflation was to rid the Universe of an unwanted relic: superheavy monopoles. One must be careful that inflation does not lead to the production of any nasty relics itself. Certainly, GUT SSB should occur before or during inflation to avoid excessive monopole production. Moreover, one must worry about the production of other dangerous relics during reheating.

Of particular concern is any weakly interacting, stable (or long-lived) NR species. This is because the energy density of matter grows relative to radiation while the Universe is radiation-dominated, $\rho_{\rm NR}/\rho_R \propto R(t)$. Even a small contamination of stable, NR particles can be dangerous because of the long post-inflationary, radiation-dominated epoch: From reheating until the time the Universe *should* become matter dominated, $\rho_{\rm NR}/\rho_R$ grows by a factor of about $10^{18}(T_{\rm RH}/10^{10}\,{\rm GeV})$. Thus, the production of any stable, NR species must suppressed by at least this factor, lest the Universe become matter-dominated too early.[34] Achieving such a large suppression factor is not always an easy matter! Potentially dangerous NR relics include gravitinos, whose decays can also adversely modify the abundances of the light elements, and any field that might be set into oscillation, as the inflaton field is (e.g., the Poloyni field in some supersymmetric models).

• *Part of a Sensible Particle Physics Model*: Perhaps the most important, and most difficult, task in building a successful inflationary model is to ensure that the inflaton field is an integral part of a sensible model of particle physics. It is this aesthetic requirement that causes us to say that there is at present no truly compelling[35] model of inflation. Put another way, the inflaton field should spring forth from some grander theory, and

[34] If the Universe becomes MD at a temperature $T_{\rm MD}$, it reaches its present temperature of 3 K at an age of about $10\,{\rm Gyr}\,(T_{\rm MD}/10\,{\rm eV})^{-1/2}$. A Universe that becomes matter-dominated much before a temperature of about 10 eV would be unacceptably young today.

[35] *Compelling model* means a model that is "compelling" and "beautiful" to people other than those who proposed it!

not vice versa.

That is not to say that there is a dearth of inflationary models. Nothing could be further from the truth. There are workable models based upon grand unification [22, 23], supersymmetry and supergravity [24], non-minimally coupled scalar fields [32], induced gravity [33], higher-derivative gravity theories (e.g., $\mathcal{R}^2$ gravity) [34], theories with extra spatial dimensions (where the inflaton is related to the size of the extra, compactified dimensions) [35], and of course Linde's chaotic model. It is now apparent that inflation, which was originally so closely tied to spontaneous symmetry breaking, is a much more general early-Universe phenomenon, one that may, or may not, have anything to do with a cosmological phase transition. Stated in its full generality, inflation involves the dynamical evolution of a very weakly coupled scalar field that is initially displaced from the minimum of its potential. Unfortunately, at present, inflation remains a very attractive paradigm in search of a compelling model.

8.6 Cosmic No-Hair Theorems

Inflation is cosmologically attractive because it offers the possibility of rendering the present state of the Universe, in regions much larger than our present Hubble volume, insensitive to the initial state of the Universe. Given this fact, our treatment of inflation thus far must seem very paradoxical, as we have analyzed inflation in the context of a flat, FRW cosmological model. We will now address this central issue of inflation: Does it proceed from very general initial conditions and resolve the dilemma of initial data? Or put another way, is there a cosmic no-hair theorem? While a complete proof of such a theorem does not exist—in fact, a very simple class of counter examples exists (closed FRW models that recollapse before they can inflate)—we will show that the idea is very well motivated, and in the process we will justify our FRW treatment of the dynamics of inflation.

Lacking the most general cosmological solution to Einstein's theory, let us begin by considering the homogeneous but anisotropic cosmological models—the so-called Bianchi models, classified according to the Lie algebras of their isometries.[36] For these models the expansion rate of the

[36]The complete class of homogeneous models also includes the special case of the Kantowski-Sachs model, whose spatial hypersurfaces have the topology of $S^2 \times R^1$. Our discussion here also applies to these models.

Universe can be written as [36]

$$H^2 \equiv \frac{1}{3}\left(\frac{\dot{V}}{V}\right)^2 = (\dot{\bar{R}}/\bar{R})^2 = \frac{8\pi G\rho}{3} + F(X_1, X_2, X_3), \tag{8.76}$$

where X_i are the scale factors of the three principal axes of the Universe; $V = X_1X_2X_3$ is the "volume scale factor," the mean scale factor $\bar{R} \propto V^{1/3}$; and ρ is the total energy density including both the scalar field (ρ_ϕ) and any "ordinary stress energy" present, e.g., matter ($\rho_M \propto \bar{R}^{-3}$) and/or radiation ($\rho_R \propto \bar{R}^{-4}$). In such cosmological models, the different axes of the Universe expand at different rates. The function F accounts for the effect of the anisotropic expansion on the mean expansion rate. For a Bianchi I model, the functional form for F is very simple: $F \propto \bar{R}^{-6}$; however, in general F is very complicated. The crucial feature of F is that for all the Bianchi models F decreases at least as fast as $\bar{R}^{-2}$. The quantity F can be thought of as "the anisotropy energy density:" $8\pi G\rho_{\rm AN}/3 \equiv F$.

The FRW models are included in the Bianchi classes as special cases: $k > 0$, type IX; $k = 0$, types I and VII$_0$; $k < 0$, types V and VII$_h$. For the FRW models $X_1 = X_2 = X_3$ and $F = -k/R^2$ simply corresponds to the ordinary 3-curvature. Because of this, our discussion of anisotropic space-times also applies to highly negatively curved FRW models, with $F = |k|/R^2$.

The equation of motion for the inflaton field ϕ in a Bianchi cosmological model is the same as in a RW model,

$$\ddot{\phi} + 3H\dot{\phi} + V'(\phi) = 0. \tag{8.77}$$

(We have not included the $\Gamma_\phi\dot{\phi}$ term, as reheating will not play an essential role in our discussions.) As usual, we will assume that ϕ is constant and initially displaced from the minimum of its potential, $\phi = \phi_i \neq \sigma$; and we will also define $V(\phi_i) \equiv V_0$. So long as $\phi \simeq \phi_i$, the Universe is endowed with a non-zero, constant vacuum energy, or equivalently an effective cosmological term. Further, lacking a definite, standard model of inflation, we will assume that in the *absence* of anisotropy, ϕ would slowly roll down its potential, and while doing so the scale factor would grow by a factor of $\exp(N_{\rm TOT})$, where $N_{\rm TOT}$ is given by (8.26), and only depends upon microphysical quantities, e.g., $V(\phi)$, m_{Pl}, etc. That the field ϕ, in the *absence* of anisotropy, undergoes slow rollover, means the following conditions are satisfied:

$$|V''(\phi_i)| \lesssim 9H_V^2, \tag{8.78}$$

$$|V'(\phi_i)/m_{PL}| \lesssim \left(\frac{27}{4\pi}\right)^{1/2} H_V^2, \tag{8.79}$$

where $H_V^2 = 8\pi V_0/3m_{PL}^2$. We will study the inflating properties of the same model, i.e., same initial value of $\phi = \phi_i$ and same potential $V(\phi)$, in the presence of anisotropy.

The proof of the no-hair theorem for the Bianchi models involves three basic elements [36]: (i) showing that a cosmological model with non-zero vacuum energy (i.e., a cosmological constant) asymptotically evolves to a de Sitter state; (ii) verifying that, in the time it takes the space-time to become nearly de Sitter, ϕ does not roll down to the minimum of its potential, i.e., making sure that the vacuum energy does not "disappear" before the model begins to inflate; and (iii) making sure that any growing modes of anisotropy present before inflation do not "grow back" by the present epoch.

With regard to (i), Wald has proven, with the exception of some highly positively curved Bianchi IX models, that all Bianchi models with a positive cosmological constant become asymptotically de Sitter, in a time of order $(8\pi GV_0/3)^{-1/2}$ [37]. His result is very easy to understand. As the Universe expands, the anisotropy energy density F decreases (at least as fast as $\bar{R}^{-2}$); likewise, the energy density in ordinary stress energy decreases as some power of $\bar{R}$. The vacuum energy, however, remains constant. So eventually it must come to dominate the rhs of (8.76). When it does, $H^2 \simeq 8\pi GV_0/3$, so that $t = t_V \sim H_V^{-1} \simeq (8\pi GV_0/3)^{-1/2}$. Once the vacuum energy density begins to dominate, $\bar{R} \propto \exp(H_V t)$, and the anisotropy and ordinary stress energy density terms decrease exponentially relative to the vacuum energy, and the space-time rapidly approaches de Sitter. The exceptional cases are also easy to understand: For highly positively curved Bianchi IX, where $F = -k/R^2$, recollapse commences before vacuum-domination can begin.[37] We should also note that Wald's theorem

[37]Consider a closed Universe model comprised of vacuum energy (V_0) and radiation ($\rho_R = aT^4$). Assuming that the expansion is adiabatic, the entropy contained within a curvature volume, $S_{\rm curv} = (4\pi/3)(Rk^{-1/2})^3(4aT^3/3)$, remains constant and serves to quantify the degree of curvature—smaller $S_{\rm curv}$ corresponds to a higher degree of curvature. If we ignore vacuum energy for the moment, the temperature at the epoch of recollapse is determined by $H(T_{\rm min}) = 0$: $T_{\rm min} = (3\pi a)^{-1/6} S_{\rm curv}^{-1/3} m_{Pl}$. In order for

ensures that shortly after the onset of vacuum energy domination, the Universe is well approximated by a flat Robertson-Walker model, justifying this assumption in our analysis of inflationary scalar field dynamics (see Section 8.3).

Wald's theorem is not directly applicable to inflation, as in an inflationary Universe, the Universe has an effective cosmological constant only so long as $\phi \neq \sigma$, where σ is the zero-energy minimum of the potential. That brings us to the second element of the theorem, verifying that ϕ does not roll to the minimum of its potential before vacuum domination begins. *Provided* that the conditions for slow roll are satisfied, the $\ddot{\phi}$ term can be neglected, and the equation of motion for ϕ is given by

$$\dot{\phi} \simeq -\frac{V'}{3H}. \tag{8.80}$$

The conditions that must be satisfied to justify the neglect of the $\ddot{\phi}$ then are: $|V''| \lesssim 9H^2$ and $|V'/m_{PL}| \leq (27/4\pi)^{1/2}H^2$. During the pre-de Sitter phase $H \geq H_V$, because anisotropy, curvature, and ordinary energy density *serve only to speed up* the expansion rate. Thus, the conditions for slow rollover are necessarily satisfied with anisotropy or curvature present. The physical explanation is simple; the more rapid expansion rate leads to additional friction (larger $3H\dot{\phi}$ term) and hence slower evolution for ϕ.

Just how far does ϕ roll during the time it takes the Universe to become vacuum dominated? The answer is obtained by integrating (8.80):

$$\int_{\phi_i}^{\phi} \frac{d\phi}{-V'(\phi)} = \int_0^{t_V} \frac{dt}{3H}, \tag{8.81}$$

where $t = t_V$ is the time when the Universe becomes vacuum dominated, i.e., when V_0 becomes greater than or equal to all the other terms on the rhs of (8.76). To compute how far ϕ rolls before the Universe becomes vacuum dominated, let us assume: (i) that $\bar{R} \propto t^n$, so that $3H = 3nt^{-1}$;[38] and (ii) that $\Delta\phi/\phi_i$ is small, so that $-\int d\phi/V'(\phi) \simeq -\Delta\phi/V'(\phi_i)$ (we will verify the consistency of this assumption). We then find that the change in ϕ during the period $t = 0 \rightarrow t = t_V$ is:

$$\frac{\Delta\phi}{-V'(\phi_i)} \simeq \frac{1}{6n}\, t_V^2. \tag{8.82}$$

---

the Universe to recollapse before it becomes vacuum dominated, $aT_{\rm min}^4$ must be greater than V_0, or $S_{\rm curv} \lesssim (a/36\pi^2)^{1/4}(m_{Pl}^3/V_0^{3/4})$.

[38]It is simple to show that $\bar{R} \propto t^n$ pertains when the rhs of (8.76) varies as $\bar{R}^{-2/n}$.

By matching $\bar{R}$ and $\dot{\bar{R}}$ at the time of change over from power law growth of $\bar{R}$ to exponential growth we find that $t_V = n\ H_V^{-1}$. Using this fact we can rewrite $\Delta\phi$ in a very suggestive form:

$$\Delta\phi \simeq \frac{n}{2}\frac{-V'(\phi_i)}{3H_V}H_V^{-1}. \tag{8.83}$$

By comparing this expression to (8.80) (with $H = H_V$), we see that the change in ϕ during the pre-inflationary epoch is simply $(n/2)$ times the change in ϕ during the first Hubble time of the de Sitter epoch. Put another way, during the pre-inflationary epoch ϕ rolls down its potential only about the amount that it would during the first e-fold of inflation. The physical reason behind this fact is straightforward: The anisotropy energy density leads to an increased expansion rate and thereby increased "friction," and so the scalar field ϕ moves very little during the time required for the vacuum energy to become dominant. Part (ii) of the theorem is now complete.

Since the period of exponential expansion is finite, the space-time never reaches the asymptotic de Sitter state. After inflation ceases, growing modes of curvature, or anisotropy, which decrease more slowly than $\bar{R}^{-4}$ (during radiation domination), or $\bar{R}^{-3}$ (during matter domination), will grow relative to the energy density in the Universe. We must then ask if these growing modes grow enough by the present epoch for the Universe to again become anisotropic or curvature dominated. As we shall now show, if the number of e-folds of inflation is large enough to solve the smoothness and flatness problems, then the answer is no, completing the proof of the no-hair theorem for anisotropic cosmologies.

At the beginning of the de Sitter epoch the anisotropy energy density is, by definition, at most comparable to V_0. The fastest growing mode of anisotropy evolves as $\rho_{\rm AN} \propto \bar{R}^{-2}$, just like the curvature term in a FRW model. Calculating the value of $\rho_{\rm AN}$ today is essentially identical to calculating the size of an ordinary curvature term today, and we find that

$$\rho_{\rm AN}/\rho_0 \lesssim \exp(-2N_{\rm TOT})(M/T_{\rm RH})^{4/3}(T_{\rm RH}/10\ {\rm eV})^2(10\ {\rm eV}/3\,{\rm K}), \tag{8.84}$$

where the inequality follows because $\rho_{\rm AN}$ is at most comparable to V_0 at the onset of the de Sitter phase. To ensure that the level of anisotropy is small today, i.e., $\rho_{\rm AN}/\rho_0 \lesssim 1$, the number of e-folds of inflation must satisfy

$$N_{\rm TOT} \gtrsim N_{\rm min} = 53 + \frac{2}{3}\ln(M/10^{14}\,{\rm GeV}) + \frac{1}{3}\ln(T_{\rm RH}/10^{10}\,{\rm GeV}) \qquad (8.85)$$

—the very same requirement that pertains to solving the smoothness and flatness problems. Note too that this lower bound to $N_{\rm TOT}$ is *independent* of the level of initial anisotropy, or ordinary curvature. For the modes where F decreases even faster than $\bar{R}^{-2}$, the requirement on $N_{\rm min}$ is even less stringent. Since the number of e-folds of de Sitter expansion is virtually unaffected by the pre-de Sitter epoch, if the model, specified by $V(\phi)$ and ϕ_i, inflated sufficiently to solve the smoothness and flatness problems in the absence of anisotropy (or curvature), growing modes of anisotropy will be sufficiently reduced so that $\rho_{\rm AN}/\rho_0$ is still small today. Of course, just as with the 3-curvature, inflation only postpones the inevitable: If there were growing modes of anisotropy present before inflation, then they will eventually—in the exponentially distant future—come to be important again, at time $t = t_{\rm AN}$, where

$$t_{\rm AN} \gtrsim \exp(3N_{\rm TOT} - 3N_{\rm min})\ 10^{10}{\rm yr}. \qquad (8.86)$$

To summarize the anisotropic models, all but a small class, those highly positively curved Bianchi IX models, will inflate, become isotropic, and remain isotropic until the present epoch. That's the no-hair theorem for homogeneous models; what about the more general case of inhomogeneous models? First consider small perturbations from homogeneity [38]. Specifically, consider a perturbation that is well outside the horizon at the onset of inflation ($R\lambda \gg H_V^{-1}$). Suppose that in the absence of inflation it has comoving wavelength λ and horizon-crossing amplitude ε_H. During inflation this mode remains super-horizon sized (see Fig. 8.4). Because the mode is super-horizon sized during inflation the gauge-invariant quantity ζ that describes its amplitude is completely unaffected by inflation: It has the same value before and after inflation. Thus, when the mode eventually re-enters the horizon, long after inflation, it does so with the same amplitude that it would have had, had inflation not occurred: $(\delta\rho/\rho)_{\rm HOR} = \varepsilon_H$, where horizon crossing occurs when $R\lambda \simeq H^{-1}$.

Because of the inflationary phase, the physical wavelength of this perturbation will have grown by an additional factor of $\exp(N_{\rm TOT})$ compared to the non-inflationary case. Therefore, inflation postpones the re-entry into the horizon by an exponential amount. Consider a specific example, a mode that in the absence of inflation, would at the present epoch (specified by $T = 2.75$ K) have size 1 Mpc. That same mode, in an inflationary model, would have a physical size today $\lambda_{\rm phys} = \exp(N_{\rm TOT})$ Mpc! Thus, inflation does not reduce the level of inhomogeneity; rather it postpones the epoch when a given mode re-enters the horizon. And of course, we need not worry about the scales that are sub-horizon sized today (i.e., $\lambda \lesssim 3000$ Mpc), because they were sub-horizon sized during inflation and had their amplitudes set by the de Sitter-produced scalar field fluctuations.

Finally, some brief comments on the most general case—initially inhomogeneous space-times, with arbitrary inhomogeneity. Jensen and Stein-Schabes [39] have proven the analogue of Wald's theorem for negatively curved models and have conjectured that inhomogeneous models that contain a sufficiently large, negatively curved patch will undergo inflation (in that region). Starobinskii [40] has shown that inhomogeneous space-times that do indeed inflate are of the most general type (a metric that allows four completely arbitrary functions of space and time to be specified).

In sum, a finite period of inflation does not make the Universe forever isotropic, homogeneous, and flat. Nor do all initial space-times inflate—a very positively curved FRW model will recollapse before it becomes vacuum dominated and can inflate. However, a large class of initial space-times will inflate—the precise size of which depends upon the measure one chooses to use on the set of initial space-time geometries—and the process of inflation creates very large smooth and flat regions, which easily encompass our present Hubble volume. Not bad for 60 or so e-folds of work! While inflation does not forever render the space-time smooth and isotropic; it does postpone (exponentially) the inevitable: If our initial space-time was anisotropic and/or inhomogeneous, that anisotropy and/or inhomogeneity will eventually re-appear.

8.7 Testing the Inflationary Paradigm

Experiment and observation are the final arbiter of all theoretical speculation. The crowning glory of any theory is its ultimate verification by experiment and observation. No different for inflation. To date one of the few blemishes on early Universe cosmology is the lack of experimental and/or observational tests for many, if not most, of the very interesting theoretical speculations. Inflation, however, is an early Universe theory that is rapidly becoming amenable to test.

How does one go about testing a theory that lacks a specific model and involves physics at energy scales well beyond those that can be probed in the terrestrial laboratory? The key is to sieze upon the "inescapable" predictions of inflation. For any inflationary model, those seem to be:

• $\Omega_0 = 1.0$: To be more precise, a flat Universe today, with $R_{\rm curv} = R(t)|k|^{-1/2} = H_0^{-1}/|\Omega_0 - 1|^{1/2} \gg H_0^{-1}$.[39] Here, $\Omega_0 = \rho_{\rm TOT}/\rho_C$, where $\rho_{\rm TOT}$ includes all forms of energy density. As we have discussed, all viable inflationary models result in many more e-folds than are necessary to ensure that $\Omega_0 = 1.0$ today; moreover, since the *smoothness* and *flatness* problems require the same amount of inflation, if the amount of inflation were insufficient to guarantee that $\Omega_0 = 1.0$ today, the *smoothness* problem would not be solved.

• Harrison-Zel'dovich spectrum of constant-curvature fluctuations. The amplitude of the spectrum is model-dependent, but the form of the spectrum is a generic prediction.

• Spectrum of relic gravitational waves. De Sitter space-produced fluctuations in the metric tensor itself result in a stochastic background of relic gravitational waves with wavelengths from 1 km to 3000 Mpc (see Fig. 8.5).

These are the inescapable predictions; how can one test them?

(1) $\Omega_0 = 1.0$. In Chapter 1 we discussed the kinematical methods for determining Ω_0, including the Hubble diagram, the galaxy-number-count–red shift-test, etc. While no method has yet provided a definitive determination, several of these methods seem very promising, and results may be forthcoming in the not too distant future. In addition, provided that the Universe is matter dominated today, the prediction that $\Omega_0 = 1.0$ also implies that $H_0 t_0 = 2/3$. Here too, the data are less than definitive; taking the following reasonable ranges for H_0 and t_0: $H_0 = 40$ to $100\,{\rm km\,s^{-1}\,Mpc^{-1}}$ and $t_0 = 12$ to $20\,{\rm Gyr}$, $H_0 t_0$ is constrained to lie only in the interval [0.5, 2.0]. However, the situation could change dramatically, if, for example, H_0 were determined to be greater than $65\,{\rm km\,s^{-1}\,Mpc^{-1}}$, in which case one could conclude that $H_0 t_0 > 0.66$, in contradiction with the inflationary prediction.

Recalling that primordial nucleosynthesis restricts the baryonic contribution to be $\Omega_B \lesssim 0.15$, we see that inflation *requires* non-baryonic matter to be the dominant form of matter in the Universe. The simplest and most plausible form of such is relic WIMPs.[40] Prime candidates for the

---

[39]Since density perturbations correspond to fluctuations in the curvature, the density perturbations on the present Hubble scale imply that a very accurate measurement of Ω_0 would actually yield a value, $\Omega_0 = 1.0 \pm (\delta\rho/\rho)_{\rm HOR} = 1.0 \pm \mathcal{O}(10^{-5})$.

[40]**WIMP**© is a copyrighted trademark of the Chicago group, standing for **W**eakly **I**nteracting **M**assive **P**article.

non-baryonic dark matter include axions, massive neutrinos, neutralinos, and superheavy monopoles. Moreover, a number of experimental efforts are underway, or are being planned, to detect the relic WIMPs that should reside in our halo; for further discussion of such efforts, see, e.g. [41].

As we discussed in Chapter 1, dynamical observations strongly suggest that the amount of matter that is associated with bright galaxies contributes: $\Omega_{\lesssim 10-30} \simeq 0.2 \pm 0.1$. This appears to contradict the inflationary prediction. If this determination is accurate—and there is no reason to believe that it isn't—and inflation is not to be falsified, then there must be a sizeable component of mass density that is not associated with bright galaxies and is more smoothly distributed: $\Omega_{\rm SM} \simeq 0.8 \pm 0.1$. A simple resolution to this dilemma is illustrated by a Universe dominated by $90h^2$ eV neutrinos. Owing to their large velocities, perturbations in the neutrinos on scales $\lambda_\nu \lesssim 13h^{-2}$ Mpc would have been smoothed out by neutrino "free streaming" (see Chapter 9), so that one would expect neutrinos to be smooth today on scales smaller than λ_ν. Because of this, their contribution to Ω_0 would not be revealed by measurements of the amount of material that clusters on scales less than 10 to 30 Mpc. Another possibility is the existence of a substantial amount of material in faint galaxies that do not cluster with bright galaxies—this is referred to as biased galaxy formation; we will discuss this idea in the next Chapter. A number of even more exotic suggestions have been put forth: a relic cosmological term, or a sea of relativistic particles produced by the recent decay of an unstable WIMP.[41]

(2) Scale-invariant density perturbations. As we shall discuss in more detail in the next Chapter, density perturbations necessarily lead to temperature fluctuations in the CMBR, of roughly the same amplitude. On angular scales less than about 1° the temperature fluctuations may have been diminished by microphysical processes. However, since the size of the Hubble radius at decoupling corresponds to an angular size of only about 1° on the sky today, the temperature fluctuations on large-angular scales should accurately mirror the virgin, primeval density fluctuations. While the amplitude of the density fluctuations required to initiate structure formation in the Universe is not precisely determined, it seems certain that perturbations of amplitude greater than about 10^{-5} are required and that the CMBR fluctuations on large-angular scales are necessarily larger

---

[41]If the mass density of the Universe is dominated by a relic cosmological term, then $H_0t_0 > 2/3$, and if it is dominated by relativistic particles, $H_0t_0 < 2/3$. Moreover, if it is dominated by relativistic particles they must be of recent origin, otherwise the Universe would never have been matter dominated, so that density perturbations would never have been able to grow into the structures that exist today.

than a *few* $\times\, 10^{-6}$. Fluctuations of this amplitude on large, as well as small, angular scales are definitely within the reach of the next generation of experiments designed to probe the anisotropy of the CMBR.

As we shall discuss at length in the next Chapter, within the gravitational instability picture, structure formation is truly an "initial data" problem; the initial data being the quantity and composition of the matter in the Universe and the nature of the primeval density perturbations. Inflation provides the initial data: gaussian density perturbations with the Harrison-Zel'dovich spectrum and $\Omega_0 = 1.0$, with baryons providing about 10% of the mass density, and relic WIMPs providing the other 90%. Of course, the amplitude of the primeval fluctuations is not specified but must be determined by other considerations. With such initial data, structure formation reduces to two, or possibly three, very specific scenarios: hot dark matter, where the relic WIMPs are light neutrinos or a similar species; cold dark matter, where the relic WIMPs are just about anything else; and warm dark matter, where the relic WIMPs are keV mass particles. The initial data provided by inflation are definite enough that the resulting scenarios are specific enough to be falsified. At present the hot dark matter scenario does not fare well when compared with the observations and has been all but falsified. Cold dark matter seems to do a very good job, although the scenario is at variance with a couple of pieces of observational data. The possibility of warm dark matter has received little study. Structure formation will be the subject of the next Chapter.

(3) Relic background of gravitons. Inflation really makes two predictions here. First, that the thermal background of about 1 K gravitons should not be present, having been exponentially supercooled by inflation. The second prediction is the spectrum of gravitational radiation illustrated in Fig. 8.5, extending from wavelengths of about 1 km to 3000 Mpc! As previously mentioned, the limit to the quadrupole anisotropy of the CMBR constrains H/m_{Pl} to be less than about 10^{-4}. Since $\Omega_{\rm graviton}$ increases monotonically with λ (see Fig. 8.5) and $\Omega_{\rm graviton}(\lambda \simeq 3000\ {\rm Mpc}) \simeq (H/m_{Pl})^2$, the isotropy of the CMBR restricts $\Omega_{\rm graviton}$ to be less than about 10^{-8} on any scale. An additional constraint to $\Omega_{\rm graviton}$ is provided by accurate timing measurements of the millisecond pulsar: at present, $\Omega_{\rm graviton}(\lambda \sim {\rm pc}) \lesssim 3 \times 10^{-7}$ [42].[42] Moreover, the LIGO project (laser interferometer gravity wave observatory) and similar projects in Europe and

---

[42] A background of relic gravitational waves would introduce a stochastic term in the pulse arrival time of this very accurate "clock." The sensitivity of this unique gravity-wave detector continues to improve with time [42].

Australia when fully implemented should reach a sensitivity of $\Omega_{\rm graviton} \sim 10^{-11}$, and a beam in space could achieve a sensitivity of $\Omega_{\rm graviton} \sim 10^{-16}$ [43]. The spectrum of relic gravitational waves has the potential to provide a unique test of inflation and if detected could provide valuable information about the details of inflation: expansion rate during inflation and reheat temperature. It is fair to say that the detection of these relic gravitational waves provides a very great challenge, even to the most clever experimenters.

In discussing the testing of the inflationary Universe, we zeroed in on the three most fundamental predictions and were bold enough to refer to these predictions as "inescapable." Of course, theorists are a clever lot, and even these three predictions are not entirely inescapable. For example, with the right potential, it is possible for the density perturbations to deviate significantly from constant amplitude [17], or perhaps to be non-gaussian [18]. Several theorists have explored the idea of multiple inflation (even double-bubble inflation), with the spectrum of density perturbations being determined by two or more periods of inflation [44].[43] Some [45] have even raised this to an art form, "dialing in" the terms in the potential (or potentials) to create the desired spectrum ("designer spectra" so to speak!), with scalar fields rolling and rebounding on their potentials like pinballs in a pinball machine! While the density perturbations that arise due to de Sitter space-produced fluctuations in the inflaton field are always present at some level, other perturbations may also be produced, e.g., isocurvature perturbations in a single species (axions or even baryons), and it is possible that these could be the dominant perturbations [46]. When testing the inflationary paradigm it seems most fruitful to focus on the most fundamental and simplest predictions and only when and if they are falsified, to turn to the theorists for more exotic ideas!

8.8 Summary: A Paradigm in Search of a Model

Over the past decades many cosmologists, including Hoyle, Gold, Bondi, K. Sato, Gliner, Starobinskii, and Kazanas, to name a few, have appreciated and emphasized some of the very interesting cosmological aspects of a period of exponential expansion (for a historical perspective, see [47]). It was Guth's 1981 paper [7] that was the "shot heard round the world," and

[43]The trick here is to arrange for the final period of inflation to last less than 60 or so e-folds, so that some of the cosmologically interesting scales cross outside the horizon during previous periods of inflation.

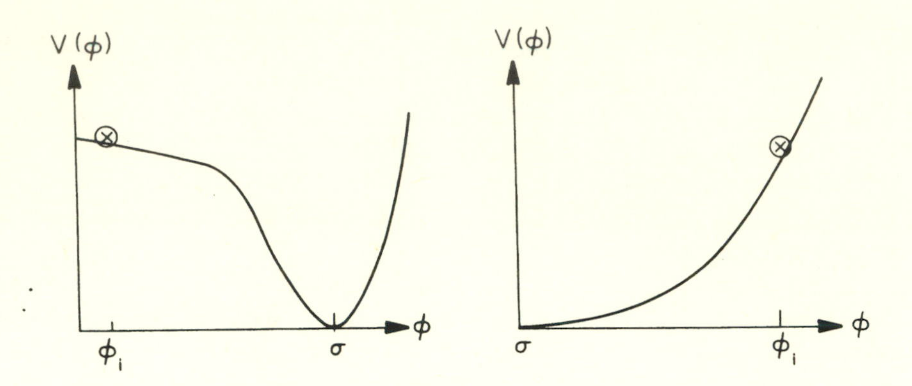

Fig. 8.7: Examples of inflationary potentials. All such potentials are necessarily very flat but do not necessarily have their minima at $\phi = 0$. It is crucial to inflation that, for one reason or another, ϕ be initially displaced from the minimum of its potential (indicated by ⊗).

that clearly spelled out how a period of exponential expansion, or inflation, might resolve the shortcomings of the standard cosmology. The idea of slow-rollover (or new) inflation made Guth's idea into a workable early Universe scenario. Today, inflation is a cosmological paradigm based upon the dynamical evolution of a very weakly coupled scalar field, which at some early but post-Planck epoch was displaced from the minimum of its potential (see Fig. 8.7). That paradigm, however, is still without a standard model for its implementation. Of course, that shortcoming should be viewed in light of the fact that our understanding of physics at energy scales well beyond that of the standard model of particle physics is still quite incomplete.

While in its original form inflation was associated with a cosmological, symmetry-breaking phase transition, it is now well appreciated that that need not be the case. Such an association is still perhaps the most attractive realization of inflation. While inflation has the potential to solve all of the purely cosmological conundrums of the standard cosmology, it does not address the puzzle of the cosmological constant. Moreover, that puzzle is a potential Achilles heel for the scenario: If some grand principle should be discovered that renders vacuum energy *at all times* impotent, inflation would lose the ultimate source of its power.

The implications of inflation for cosmology are manifold. For example, inflation implies that the Universe is many, many orders of magnitude larger than a simple adiabatic extrapolation would indicate (see Fig. 8.2). Moreover, all that we see today arose from "nothing," in the form of false vacuum energy. While our immediate neighborhood and well beyond should be smooth and flat if the Universe inflated, it could well be that the Universe on the very largest scales is very irregular, with regions of space-time inflating at different times, and some regions never inflating at all. Moreover, the "Universes" that evolve in different inflationary regions could be quite different. For example, if the C, CP violation required for baryogenesis arises due to SSB, then the sign of the baryon asymmetry in different regions would presumably be different, resulting in a *globally* baryon-symmetric Universe. In addition, some unified theories (particularily supersymmetric and superstring theories) are plagued with the existence of numerous degenerate vacuum states. Inflation could solve the riddle of why our vacuum is $SU(3)_C \otimes SU(2)_L \otimes U(1)_Y$—different inflationary regions have different vacuum states.[44] If inflation is indeed a part of the early history of the Universe, the global structure of the Universe is likely to be quite rich indeed.

Several people have addressed the space-time evolution of an isolated, spherical region of false vacuum within an otherwise empty Universe [48]. This problem poses a paradox: An observer inside the bubble would expect the bubble to grow exponentially due to the false-vacuum energy density, while an observer outside the bubble would expect the bubble to shrink as the pressure within it is negative and the pressure outside is zero. The resolution is quite interesting. If the false-vacuum region is larger than a certain critical size, a bubble-like region of space expands exponentially; as viewed by an observer in the empty space-time, it collapses into a black hole and becomes causally disconnected (see Fig. 8.8). Moreover, Guth and his collaborators [49] have addressed the issue of producing additional inflationary bubbles in the laboratory today, an event that they have shown is classically forbidden but that can occur via quantum mechanical tunnelling.

Linde and Starobinskii [50] have further explored the effects of quantum mechanical fluctuations during inflation and have shown that inflationary bubbles within bubbles can result; and due to the role of quantum fluctuations, most of the physical volume of the Universe should always be inflating away! The physical mechanism behind this is simple: If the

[44]Since it is possible that the realization of physical law is different in different inflationary regions, inflation may, God forbid, provide some rational basis for the anthropic principle, as inflation provides a multitude of "Universes" from which to choose.

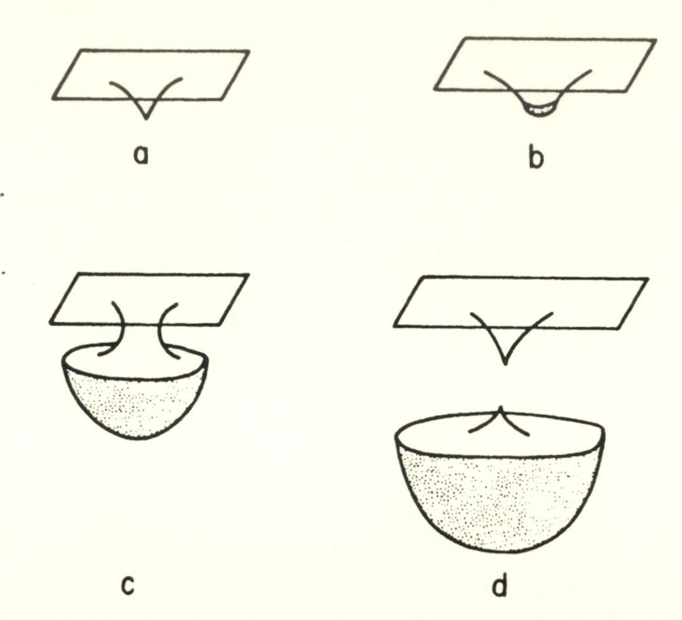

Fig. 8.8: The space-time evolution of a bubble of false vacuum in an otherwise empty space-time. If the initial region is sufficiently large, it expands exponentially and becomes causally disconnected from the surrounding space-time. As viewed from outside, the region shrinks and becomes a black hole (from Guth, et al. [48]).

quantum change in ϕ in a Hubble time, $\Delta\phi_{\rm QM} \sim H/2\pi$, is larger than the classical change in ϕ, $\Delta\phi_{\rm cl} \sim V'/3H^2$, ϕ is just as likely to roll up the potential as to roll down the potential, and so in some regions ϕ will climb the potential, prolonging inflation. The condition $\Delta\phi_{\rm QM} > \Delta\phi_{\rm cl}$ is satisfied in Linde's chaotic model provided that $\phi_i \gtrsim \lambda^{-1/6} m_{Pl} \sim 300\ m_{Pl}$.

Many issues in inflationary cosmology remain unsettled, not the least of which is the confrontation between inflation and cosmological observations. Other pressing issues include the following.

• Who is ϕ? How does the inflaton field fit into the grander scheme? Models have been put forth wherein the inflaton is reponsible for GUT SSB, or for the SSB of supersymmetry, where ϕ is associated with the compactification of extra spatial dimensions, where ϕ is a composite field made out of preonic fermion fields, where ϕ is related to the scalar curvature in a higher-derivative theory of gravity, or as in Linde's model, where ϕ is just some random scalar field. Of course, a not-unrelated issue is why is ϕ so weakly coupled? Surely the answers to these two questions must be related.

• What determines ϕ_i? In the original models of inflation the initial value of ϕ was determined by thermal considerations. In many, if not most, inflationary models ϕ is so weakly coupled that it is oblivious to the thermal bath of particles in the Universe. Moreover, it should be clear that so long as ϕ is displaced from the minimum of its potential, the existence of an initial thermal sea of particles, or anything else for that matter, is irrelevant! That is, the pre-inflationary phase need not be radiation dominated.

• The quantum mechanics of inflation. While we have shown that the semi-classical treatment of the evolution of ϕ is very well justified during inflation, could it be that the quantum mechanics of the inflaton field at early times determines its initial value? Is some aspect of quantum cosmology, or quantum gravity, key to understanding the initial conditions for inflation? We will discuss quantum cosmology in Chapter 11.

8.9 References

1. J. Yang, et al., *Ap. J.* **281**, 493 (1984).

2. C. B. Collins and S. W. Hawking, *Ap. J.* **180**, 317 (1973).

3. C. W. Misner, *Ap. J.* **151**, 431 (1968); in *Magic Without Magic*, ed. J. Klauder (Freeman, San Francisco, 1972).

4. R. Penrose, in *General Relativity: An Einstein Centenary Survey*, eds. S. W. Hawking and W. Israel (Cambridge Univ. Press, Cambridge, 1979).

5. J. B. Hartle and S. W. Hawking, *Phys. Rev. D* **28**, 2960 (1983).

6. A. Vilenkin, in *Particles and the Universe*, eds. G. Lazarides and Q. Shafi (North-Holland, Amsterdam, 1986).

7. A. H. Guth, *Phys. Rev. D* **23**, 347 (1981).

8. A. D. Linde, *Phys. Lett.* **108B**, 389 (1982).

9. A. Albrecht and P. J. Steinhardt, *Phys. Rev. Lett.* **48**, 1220 (1982).

10. A. H. Guth and E. J. Weinberg, *Nucl. Phys.* **B212**, 321 (1983); S. W. Hawking, I. Moss, and J. Stewart, *Phys. Rev. D* **26**, 2681 (1982).

11. A. Albrecht, P. J. Steinhardt, M. S. Turner, and F. Wilczek, *Phys. Rev. Lett.* **48**, 1437 (1982); L. Abbott, E. Farhi, and M. Wise, *Phys. Lett.* **117B**, 29 (1982); A. Dolgov and A. D. Linde, *Phys. Lett.* **116B**, 329 (1982).

12. T. Bunch and P. C. W. Davies, *Proc. Roy. Soc. London* **A360**, 117 (1978).

13. A. Vilenkin and L. Ford, *Phys. Rev. D* **26**, 1231 (1982); A. D. Linde, *Phys. Lett.* **116B**, 335 (1982).

14. W. Fischler, B. Ratra, and L. Susskind, *Nucl. Phys.* **B259**, 730 (1985).

15. The results discussed here are from J. M. Bardeen, P. J. Steinhardt, and M. S. Turner, *Phys. Rev. D* **28**, 679 (1983). The spectrum of density perturbations produced during inflation has also been calculated independently by A. H. Guth and S.-Y. Pi, *Phys. Rev. Lett.* **49**, 1110 (1982); A. A. Starobinskii, *Phys. Lett.* **117B**, 175 (1982); and S. W. Hawking, *Phys. Lett.* **115B**, 295 (1982).

16. E. R. Harrison, *Phys. Rev. D* **1**, 2726 (1970); Ya. B. Zel'dovich, *Mon. Not. Roy. Astron. Soc.* **160**, 1p (1972).

17. P. J. Steinhardt and M. S. Turner, *Phys. Rev. D* **29**, 2162 (1984).

18. T. J. Allen, B. Grinstein, and M. Wise, *Phys. Lett.* **197B**, 66 (1987).

19. M. S. Turner and L. M. Widrow, *Phys. Rev. D* **37**, 3428 (1988).

20. The results shown here are from M. T. Ressell and M. S. Turner, *Phys. Rev. D*, in press (1989). Graviton production (and superadiabatic amplification) in the early Universe was first analyzed by L. P. Grishchuk, *Sov. Phys. JETP* **40**, 409 (1975). In the context of inflation, graviton production was first studied by V. A. Rubakov, M. Sazhin, and A. Veryaskin, *Phys. Lett.* **115B**, 189 (1982); R. Fabbri and M. Pollock, *Phys. Lett.* **125B**, 445 (1983); B. Allen, *Phys. Rev. D* **37**, 2078 (1988); and L. Abbott and M. Wise, *Nucl. Phys.* **B244**, 541 (1984).

21. M. S. Turner and L. M. Widrow, *Phys. Rev. D* **37**, 2743 (1988).

22. S.-Y. Pi, *Phys. Rev. Lett.* **52**, 1725 (1984).

23. Q. Shafi and A. Vilenkin, *Phys. Rev. Lett.* **52**, 691 (1984).

24. R. Holman, P. Ramond, and G. G. Ross, *Phys. Lett.* **137B**, 343 (1984).

25. B. Ovrut and P. J. Steinhardt, *Phys. Lett.* **133B**, 161 (1983); L. Jensen and K. A. Olive, *Nucl. Phys.* **B263**, 731 (1986).

26. L. L. Krauss, *Nucl. Phys.* **B227**, 556 (1983).

27. J. Ellis, J. E. Kim, and D. Nanopoulos, *Phys. Lett.* **145B**, 181 (1984); M. Yu. Khlopov and A. D. Linde, *Phys. Lett.* **138B**, 265 (1984).

28. G. D. Coughlan, W. Fischler, E. W. Kolb, S. Raby, and G. G. Ross, *Phys. Lett.* **131B**, 59 (1983).

29. A. D. Linde, *Phys. Lett.* **129B**, 177 (1983).

30. A. D. Linde, *Rep. Prog. Phys.* **47**, 925 (1984); *Comments on Astrophys.* **10**, 229 (1985); *Prog. Theo. Phys.* **85**, 279 (1985).

31. S. Dimopoulos and L. Hall, *Phys. Lett.* **196B**, 135 (1987).

32. K. Maeda, *Phys. Rev. D*, in press (1989).

33. B. L. Spokoiny, *Phys. Lett.* **147B**, 39 (1984); F. S. Accetta, D. Zoller, and M. S. Turner, *Phys. Rev. D* **31**, 3046 (1985).

34. A. A. Starobinskii, *Phys. Lett.* **91B**, 99 (1980); M. B. Mijić, M. S. Morris, and W.-M. Suen, *Phys. Rev. D* **34**, 2934 (1986).

35. Q. Shafi and C. Wetterich, *Phys. Lett* **129B**, 387 (1983); *ibid* **152B**, 51 (1985).

36. M. S. Turner and L. M. Widrow, *Phys. Rev. Lett.* **57**, 2237 (1986); L. Jensen and J. Stein-Schabes, *Phys. Rev. D* **34**, 931 (1986).

37. R. M. Wald, *Phys. Rev. D* **28**, 2118 (1983).

38. J. Frieman and M. S. Turner, *Phys. Rev. D* **30**, 265 (1984).

39. L. Jensen and J. A. Stein-Schabes, *Phys. Rev. D* **35**, 1146 (1987).

40. A. A. Starobinskii, *JETP Lett.* **37**, 66 (1983).

41. J. Primack, D. Seckel, and B. Sadoulet, *Ann. Rev. Nucl. Part. Sci.* **38**, 751 (1988).

42. L. A. Rawley, et al., *Science* **238**, 761 (1987); J. Taylor, et al., in preparation (1989).

43. See, e.g., K. S. Thorne, in *300 Years of Gravitation*, eds. S. W. Hawking and W. Israel (Cambridge Univ. Press, Cambridge, 1989), p. 330.

44. J. Silk and M. S. Turner, *Phys. Rev. D* **35**, 419 (1986); L. A. Kofman, A. D. Linde, and J. Einasto, *Nature* **326**, 48 (1987).

45. D. Salopek, J. R. Bond, and J. M. Bardeen, *Phys. Rev. D* **40**, 1753 (1989).

46. Isocurvature axion fluctuations have been discussed by D. Seckel and M. S. Turner, *Phys. Rev. D* **32**, 3178 (1985); A. D. Linde, *JETP Lett.* **40**, 1333 (1984); A. D. Linde, *Phys. Lett.* **158B**, 375 (1985). Isocurvature baryon number fluctuations have been discussed by M. S. Turner, A. Cohen, and D. Kaplan, *Phys. Lett.* **216B**, 20 (1989).

47. D. Lindley, Fermilab preprint, unpublished (1985).

48. V. A. Berezin, V. A. Kuzmin, and I. I. Tkachev, *Phys. Lett.* **120B**, 91 (1983); S. K. Blau, E. I. Guendelman, and A. H. Guth, *Phys. Rev. D* **35**, 1747 (1987); K. Sato, et al., *Phys. Lett.* **108B**, 103 (1982); *Prog. Theor. Phys.* **65**, 1443 (1981); *ibid* **66**, 2287 (1981); *ibid* **68**, 1979 (1982).

49. E. Farhi and A. H. Guth, *Phys. Rev. D*, in press (1989).

50. The question of what fraction of the Universe remains in a de Sitter state goes all the way back to "old inflation," where most of the physical volume of the Universe remained in the de Sitter state (Guth and Weinberg [10]). Vilenkin [*Phys. Rev. D* **27**, 2848 (1983)] showed that due to quantum fluctuations the same is true for new inflation—although the consequences are not deleterious. Linde [*Mod. Phys. Lett.* **1A**, 81 (1986); *Phys. Lett.* **175B**, 395 (1986)] and Starobinskii [in *Fundamental Interactions* (MGPI Press, Moscow, 1984), p. 55] have shown that the same is true for chaotic inflation, and Linde [*Intl. J. Mod. Phys.* **2A**, 561 (1987)] has explored some of the consequences of the "eternally reproducing" Universe.

9

STRUCTURE FORMATION

9.1 Overview

On small scales the Universe today is very lumpy. For example, the density within a galaxy is about 10^5 the average density of the Universe, and that within a cluster of galaxies is about 10^2 to 10^3 times the average density in the Universe. Of course on very large scales, say much larger than 100 Mpc, the Universe is smooth, as evidenced by the isotropy of the CMBR, the isotropy of the x-ray background, and number counts of radio sources.

The surface of last scattering for the 2.75 K background radiation is the Universe at $180{,}000(\Omega_0 h^2)^{-1/2}$ yr after the bang, when the temperature $T \simeq 0.26$ eV, and the scale factor $R \equiv R_0/1100 \simeq 9.1 \times 10^{-4}$.[1] Thus the CMBR is a fossil record of the Universe at that early epoch. The 2.75 K background radiation is very isotropic, $\delta T/T \lesssim \mathcal{O}(10^{-4})$ on angular scales ranging from $1'$ to $180°$ (see Fig. 1.6). There is a calculable relationship between $\delta T/T$ and $\delta\rho/\rho$, which depends upon the nature of density perturbations present, but typically

$$\left(\frac{\delta\rho}{\rho}\right)_{dec} = \text{const} \times \left(\frac{\delta T}{T}\right) \lesssim \mathcal{O}(10^{-2} - 10^{-3}), \tag{9.1}$$

where "const" is $\mathcal{O}(10$ to $100)$. Thus the isotropy of the CMBR implies that the Universe at decoupling was very smooth. (We will discuss the rela-

[1] In discussing structure formation it is very convenient to set $R_0 \equiv 1$; we will do so throughout this Chapter. Although decoupling, recombination, and freeze in of a residual ionization are three distinct events whose red shifts depend upon Ω_0 and Ω_B (see Section 3.4), for simplicity, we will assume that all occur at a red shift $1 + z_{dec} \equiv 1100$.

tionship between anisotropies in the CMBR and density inhomongeneities in Section 9.6.)

So the Universe was very smooth, and is very lumpy—how did it get to "here" from "there?" For the past decade or so cosmologists have had a general picture that describes how this took place: Small, primeval density inhomogeneities grew via the Jeans, or gravitational, instability into the large inhomogeneities we observe today, i.e., galaxies, clusters of galaxies, superclusters, voids, etc. The growth of density inhomogeneities *can* begin as soon as the Universe is matter-dominated; and this is the case in a Universe dominated by non-interacting relic WIMPs. Baryonic inhomogeneities cannot begin to grow until after decoupling because until then, baryons are tightly coupled to the photons. After decoupling, when the Universe is matter-dominated and baryons are free of the pressure support provided by photons, density inhomogeneities in the baryons and any other matter components grow as

$$\frac{\delta\rho}{\rho} \propto \begin{cases} R & \delta\rho/\rho \lesssim 1 \quad \text{(linear regime)} \\ R^n \; (n \gtrsim 3) & \delta\rho/\rho \gtrsim 1 \quad \text{(non-linear regime).} \end{cases} \tag{9.2}$$

The cosmic scale factor $R(t)$ has grown by slightly more than a factor of 10^3 since decoupling, and the isotropy of the CMBR is consistent with perturbations as large as 10^{-2} to 10^{-3} at decoupling. Thus, it is possible for the large perturbations we see today to have grown from small perturbations present at decoupling. The basic gravitational instability picture is at least self-consistent and is a generally accepted part of the standard cosmology.

One would like to understand the formation of structure in detail. In order to do so one needs to know the initial conditions at the time structure formation began. The formation of structure[2] began when the Universe became matter-dominated, $t_{\rm EQ} \simeq 4.4 \times 10^{10}(\Omega_0 h^2)^{-2}$ sec, $R_{\rm EQ} \simeq 4.3 \times 10^{-5}(\Omega_0 h^2)^{-1}$, and $T_{\rm EQ} \simeq 5.5(\Omega_0 h^2)$ eV, as this is the time density perturbations present in the matter component of the Universe can begin to grow. Thus the time of matter–radiation equality is the initial epoch for structure formation. In order to fill in the details of structure formation one needs the "initial data" for that epoch. The initial data required include: (1) the total amount of non-relativistic matter in the Universe, quantified by Ω_0; (2) the composition of the Universe, as quantified by the

[2]The formation of structure is often simply referred to as galaxy formation; strictly speaking, that term should be reserved for details of how galaxies themselves form.

Suspect Relic	Mass	Origin t, T	Abundance cm^{-3}
Invisible Axion	10^{-5}eV	10^{-30}sec, 10^{12}GeV	10^{9}
Light Neutrino	30 eV	1 sec, 1 MeV	100
Photino—Gravitino	keV	10^{-4}sec, 100 MeV	10
Photino—Sneutrino—Neutralino—Axino—Heavy Neutrino	GeV	10^{-3}sec, 10 MeV	10^{-5}
Magnetic Monopoles	10^{16}GeV	10^{-34}sec, 10^{14}GeV	10^{-21}
Pyrgons—Maximons—Newtorites	10^{19}GeV	10^{-43}sec, 10^{19}GeV	10^{-24}
Quark Nuggets	$\simeq 10^{15}$g	10^{-5}sec, 300 MeV	10^{-44}
Primordial Black Holes	$\gtrsim 10^{15}$g	$\gtrsim 10^{-12}$sec, $\lesssim 10^{3}$GeV	$\lesssim 10^{-44}$

Table 9.1: WIMP candidates for the dark matter. The cosmic abundance required for closure density is $n_{\rm WIMP} \simeq 1.05h^2 \times 10^{-5}\mathrm{cm}^{-3}/m_{\rm WIMP}(\mathrm{GeV})$.

fraction of critical density, Ω_i, contributed by the various components ($i =$ baryons, relic WIMPs, cosmological constant, relativistic particles, etc.); and (3) the spectrum and type (i.e., "adiabatic" or "isocurvature") of primeval density perturbations. Given these initial data one can imagine constructing a detailed scenario of structure formation, e.g., by numerical simulation—one that can be tested by comparing to the Universe we observe today.

The importance of specifying the initial data for the structure formation problem cannot be overemphasized. Without knowledge of the initial data, a detailed picture cannot be formulated. Moreover, "inverting" present observations of the Universe to infer the initial data does not seem feasible because of all the astrophysical filtering that has taken place. Recent speculations about the earliest history of the Universe have provided crucial and well motivated "hints" as to the appropriate initial data, hints that have led to a renaissance in the study of structure formation. These important hints include: $\Omega_0 = 1$—from inflation; $0.014 \lesssim \Omega_B \lesssim 0.15$ and $\Omega_{\rm WIMP} \sim 0.9$—from inflation, primordial nucleosynthesis, and particle relics; adiabatic, or possibly isocurvature, density inhomogeneities with the Harrison-Zel'dovich spectrum—from inflation; and cosmic strings as the seed for structure formation—from phase transitions.

The list of suspect relics that could provide closure density and thus be the dark matter is long: axions, light neutrinos, neutralinos, superheavy monopoles, primordial black holes, etc. (see Table 9.1). As we shall see, so far as structure formation goes, there are two limiting cases: relics with large internal velocity dispersion, "hot dark matter," which only includes light neutrinos; and relics with very small internal velocity dispersion, "cold dark matter," which includes essentially ever other candidate. Taken with the hints for the seed perturbations that initiate structure formation, the result is four well motivated scenarios: hot and cold dark matter with adiabatic (or possibly isocurvature) perturbations, and hot and cold dark matter with cosmic-string–induced perturbations. In addition to being well motivated, these early-Universe inspired scenarios are sufficiently specific to be falsified!

At this point the reader may wish to re-read Chapter 1, an overview of the observed Universe, to remind one what it is that structure formation is supposed to produce. While our treatment of structure formation in this Chapter is rather lengthy, it is by no means exhaustive. We refer the interested reader to the other excellent treatments of the subject that exist; see [1].

9.2 Notation, Definitions, and Preliminaries

It is convenient to discuss the density field of the Universe in terms of the density contrast,

$$\delta(\vec{\mathbf{x}}) \equiv \frac{\delta\rho(\vec{\mathbf{x}})}{\overline{\rho}} = \frac{\rho(\vec{\mathbf{x}}) - \overline{\rho}}{\overline{\rho}}, \tag{9.3}$$

and to express the density contrast $\delta(\vec{\mathbf{x}})$ in a Fourier expansion:

$$\delta(\vec{\mathbf{x}}) = \sum_{l,m,n=-\infty}^{\infty} \exp(-i\vec{\mathbf{k}}\cdot\vec{\mathbf{x}})\delta_k \rightarrow \frac{V}{(2\pi)^3}\int_{\rm Vol} \delta_k \exp(-i\vec{\mathbf{k}}\cdot\vec{\mathbf{x}})d^3k,$$

$$\delta_k = V^{-1} \int_{\text{Vol}} \delta(\vec{x}) \exp(i\vec{k} \cdot \vec{x}) d^3x. \tag{9.4}$$

Here $\bar{\rho}$ is the average density of the Universe, periodic boundary conditions have been imposed, and $V \equiv L^3$ is the volume of the fundamental cube. The discrete spectrum of modes is labeled by integers l, m, and n, and the wavenumber $\vec{k}$ is related to l, m, and n by: $k_x = 2\pi l/L$, $k_y = 2\pi m/L$, and $k_z = 2\pi n/L$. In the limit that L is large compared to any length scale in the problem, the passage to the continuum description is well justified. Moreover, in the continuum limit it is convenient to eliminate the volume factors in (9.4) by a re-definition of δ_k: $\delta_k \to V\delta_k$. In this Chapter we will use the continuum description and eliminate the volume factors by such a re-definition.[3] Further, since $\delta(\vec{x})$ is a scalar quantity, one can use either comoving or physical coordinates in the Fourier expansion; we will always use comoving coordinates.

A particular Fourier component is characterized by its amplitude $|\delta_k|$ and its comoving wavenumber k. Since $\vec{x}$ and $\vec{k}$ are (comoving) coordinate quantities, the physical distance and physical wavenumber are related to the comoving distance and wavenumber by

$$dx_{\text{phys}} = R(t)dx, \qquad k_{\text{phys}} = k/R(t).$$

The wavelength of a perturbation is related to the wavenumber by

$$\lambda \equiv 2\pi/k, \qquad \lambda_{\text{phys}} = R(t)\lambda. \tag{9.5}$$

Because we have choosen to normalize the scale factor by $R_0 = 1$, $k_{\text{phys}} = k$ and $\lambda_{\text{phys}} = \lambda$ today.

Before we go on, we should mention some of the assumptions implicit in our description of density fluctuations. Strictly speaking, our expansion is valid only in spatially flat models (curvature radius $R_{\text{curv}} = R(t)k^{-1/2} = \infty$); however, at early times the effect of spatial curvature is small and can usually be neglected. In any case, it can always be neglected for modes with $\lambda_{\text{phys}} \ll R_{\text{curv}}$. (In spatially curved space-times, plane waves are replaced by generalized solutions to the Helmholtz equation.) Second, the quantity $\delta(\vec{x})$ is not a gauge invariant quantity. In fact, it is always possible to make $\delta(\vec{x}) = 0$ by a suitable gauge transformation: $x^\mu \to x^\mu + \varepsilon^\mu$,

[3] Periodic boundary conditions and a Fourier series expansion of the density field are routinely used in numerical simulations of structure formation.

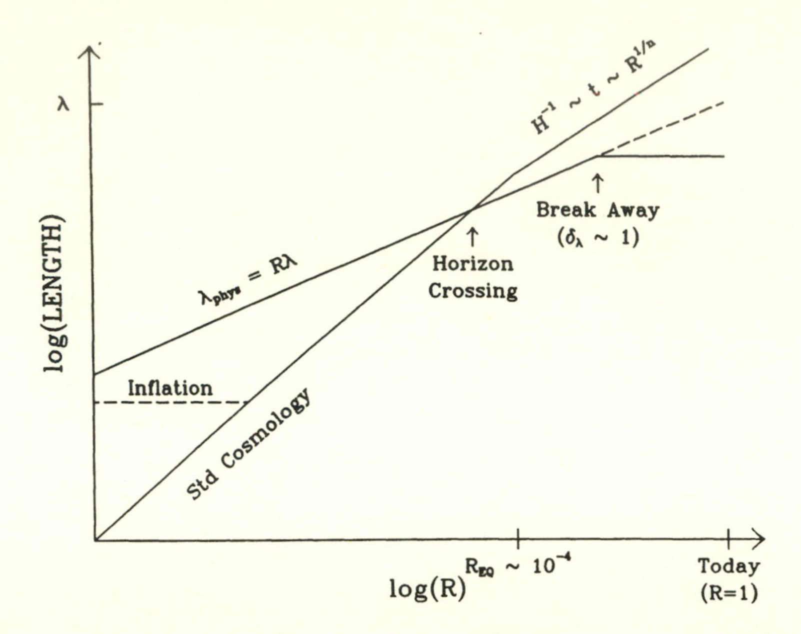

Fig. 9.1: The physical size of a given perturbation mode, $\lambda_{\rm phys}$, and the Hubble radius, H^{-1}, as a function of the cosmic scale factor. For $R(t) \propto t^n$, the Hubble radius $H^{-1} \propto R^{1/n}$. Note that $\lambda_{\rm phys}$ begins super-horizon sized and then crosses inside the horizon. So long as $\delta_k \lesssim 1$, $\lambda_{\rm phys}$ increases as $R(t)$; when a perturbation becomes non-linear, $\delta_k \gtrsim 1$, it separates from the general expansion and thereafter maintains approximately constant physical size. It is conventional and very convenient to label perturbations that have become non-linear by the present, by the size ($= \lambda$) they would have today, had they *not* become non-linear.

$g \to g + \mathcal{O}(\varepsilon)$, where $\varepsilon^\mu \simeq \mathcal{O}(\delta\, kt/R)$. For sub-horizon-sized perturbations, i.e., $\lambda_{\rm phys} \ll H^{-1} \sim t$, this fact is of little consequence, as a Newtonian analysis suffices. However, for super-horizon-sized perturbations, i.e., $\lambda_{\rm phys} \gg H^{-1}$, this fact can be troublesome, as the gauge transformation necessary can be very small, $\varepsilon \sim (t/\lambda_{\rm phys})\delta \ll \delta$, and care must be taken in discussing the evolution of super-horizon-sized perturbations. This issue will be addressed in some detail in the next Section.

In the linear regime ($\delta_k < 1$), the physical size of a given density perturbation grows with the expansion: $\lambda_{\rm phys} = R(t)\lambda$. The use of a comoving label (k or λ) for a given mode is very convenient because it does not change with time—i.e., the same physical perturbation is characterized by the same comoving label. It is also useful (and conventional) to characterize a density perturbation by the (invariant) rest mass M within a sphere of radius $\lambda/2$:

$$M \equiv \frac{\pi}{6}\lambda^3_{\rm phys}\rho_{\rm matter} \simeq 1.5 \times 10^{11} M_\odot(\Omega_0 h^2)\lambda^3_{\rm Mpc}, \tag{9.6}$$

where as usual Ω_0 is the fraction of critical density in NR particles. For reference, a galactic mass perturbation ($10^{12} M_\odot$ including the dark matter) corresponds to

$$\lambda_{GAL} \simeq 1.9 \text{ Mpc}(\Omega_0 h^2)^{-1/3}. \tag{9.7}$$

Of course, the physical size of a galaxy is much less than 1.9 Mpc, more like 30 kpc; this is because galactic-sized perturbations have already gone non-linear ($\delta_k > 1$). Once a perturbation enters the non-linear regime, it "separates" from the the general expansion, i.e., ceases expanding, and becomes a self-bound gravitational system. The scale 1.9 Mpc, then, is the size a perturbation containing $10^{12} M_\odot$ would have today had it not already gone non-linear and separated from the general expansion of the Universe. The evolution of the physical size of a density perturbation is shown in Fig. 9.1.

9.2.1 Spherical collapse model

To develop a feel for how a perturbation "breaks away" from the expansion, consider the evolution of a spherical overdense region in an otherwise flat, homogeneous, matter-dominated FRW model.[4] Specifically, let $\rho(r) = \rho > \overline{\rho}$ for $r \leq r_1$; $\rho(r) = \text{const} < \overline{\rho}$ for $r_1 \leq r \leq r_2$; and $\rho(r) = \overline{\rho}$ for $r \geq r_2$ (where const is chosen so that the average density interior to $r = r_2$ is just $\overline{\rho}$). The overdense region behaves like a portion of a closed universe model, so that within the overdense region

$$H^2 = \frac{8\pi G\rho}{3} - \frac{k}{R^2}. \tag{9.8}$$

On the other hand, well outside the overdense region the model behaves like a flat universe, with

$$H^2 = \frac{8\pi G\overline{\rho}}{3}, \quad \overline{R}(t) \propto t^{2/3}, \quad \overline{\rho} = \frac{1}{6\pi G t^2}. \tag{9.9}$$

[4] We could equally well consider an overdense region in an otherwise open, or closed, FRW model. We leave these interesting cases as an exercise for the reader.

The scale factor in the overdense region will expand to a maximum size and will then subsequently recollapse. In terms of the Hubble constant, scale factor, and value of Ω at some early epoch (labeled by subscript i), the epoch of maximum expansion is specified by, cf. (3.40–3.42),

$$
\begin{aligned}
R_{\rm MAX}/R_i &= \Omega_i/(\Omega_i - 1), \\
H_i t_{\rm MAX} &= \frac{\pi}{2}\frac{\Omega_i}{(\Omega_i - 1)^{3/2}} \simeq \frac{\pi}{2}(R_{\rm MAX}/R_i)^{3/2}, \\
\rho_{\rm MAX} &= (R_i/R_{\rm MAX})^{3/2}\rho_i = \frac{3\pi}{32Gt_{\rm MAX}^2}.
\end{aligned} \tag{9.10}
$$

Taking the initial epoch to be sufficiently early so that $\rho_i \gg k/R_i^2$, $\Omega_i \simeq 1$, and $\rho_i \simeq \overline{\rho}(t_i)$, it follows that the density contrast at the time of maximum expansion is

$$
\frac{\rho(t_{\rm MAX})}{\overline{\rho}(t_{\rm MAX})} = \frac{9\pi^2}{16} \simeq 5.55 \quad \text{or} \quad \delta(t_{\rm MAX}) = 4.55. \tag{9.11}
$$

If we compare the density in the overdense region with that in the rest of the Universe at early times, $t \ll t_{\rm MAX}$, we find that $\delta = 1 - \Omega^{-1} \propto \overline{R}(t)$; i.e., the density contrast grows as the scale factor. At the time of maximum expansion, the Universe has expanded by a factor of $(t_{\rm MAX}/t_i)^{2/3} = (3\pi/4)^{2/3}\Omega_i/(\Omega_i - 1) \simeq 1.77\delta_i^{-1}$; i.e., *had* the density contrast continued to increase linearly with $\overline{R}$, the value of δ at maximum expansion would be 1.77.

After reaching its maximum size, the overdense region will recollapse and virialize.[5] Once the overdense region virializes, $KE = -W_{\rm GRAV}/2$, where KE is the kinetic energy and $W_{\rm GRAV} = -\text{const} \times G/r$ is the gravitational potential energy. (For a constant density configuration, const $= -3M^2/10$; M is the mass of the lump.) Assuming that dissipative forces are not important, the total energy of the overdense region remains constant during virialization. Before it virializes $KE = 0$, so that $E_{\rm TOT} = W_{\rm GRAV} = -\text{const} \times G/r_{\rm MAX}$. After virialization, the total energy is, $E_{\rm TOT} = W_{\rm GRAV}/2 = -\text{const} \times G/2r_{\rm VIR}$. Equating the pre- and

---

[5] As we shall discuss later, the physical mechanism of dissipationless virialization is the time-varying gravitational field of the collapsing lump, and virialization occurs in a few dynamical times, i.e., $\tau_{\rm VIR} \sim (G\rho_{\rm MAX})^{-1/2}$. This process is known as "violent relaxation" [2].

post-virialization energies we find that $r_{\rm VIR} = r_{\rm MAX}/2$. That is, after virialization, the lump that separates from the expansion has shrunk by a factor of 2, increasing its density by a factor of 8.

9.2.2 The power spectrum and related quantities

Returning to the Fourier expansion of the density field, any statistical quantity one might wish to compute can be specified in terms of the *power spectrum* $|\delta_k|^2$.[6] In the absence of a specific model for the spectrum of primeval fluctuations, it is often assumed that

$$|\delta_k|^2 = AVk^n, \tag{9.12}$$

that is, a featureless power law (here V is the volume of the fundamental cube).

First consider the *rms* density fluctuation defined by,

$$\frac{\delta\rho}{\rho} = \langle\delta(\vec{\mathrm{x}})\delta(\vec{\mathrm{x}})\rangle^{1/2}, \tag{9.13}$$

where $\langle\cdots\rangle$ indicates the average over all space. A bit of Fourier algebra and the use of Parseval's theorem yields:

$$\left(\frac{\delta\rho}{\rho}\right)^2 = V^{-1}\int_0^\infty \frac{k^3|\delta_k|^2}{2\pi^2}\frac{dk}{k}. \tag{9.14}$$

where we have also made use of the fact that the power spectrum is isotropic, i.e., depends only upon $k = |\vec{\mathrm{k}}|$.[7] Evidently, the contribution to $(\delta\rho/\rho)^2$ from a given logarithmic interval in k, $dk/k = d\ln k \sim 1$, is given by

$$\left(\frac{\delta\rho}{\rho}\right)_k^2 \approx \Delta^2(k) \equiv V^{-1}\frac{k^3|\delta_k|^2}{2\pi^2}. \tag{9.15}$$

---

[6]This statement is only rigorously true for *gaussian* random fluctuations, i.e., for a density field whose origin involves a gaussian random process. For gaussian fluctuations the phases of the different Fourier modes are uncorrelated, referred to as "random phases." Inflation-produced fluctuations are gaussian.

[7]For the sake of accuracy we have included the volume factor in this and subsequent expressions which involve averaging over all space. The reader may wish to follow the advice of one of the authors to ignore the various factors of V; they play no crucial role. Moreover, the less ambitious reader may also wish to ignore all factors of π, the so-called small-circle approximation.

The fluctuation power per logarithmic interval, or *variance*, will occur often in our discussions, and so we have denoted it by $\Delta^2(k)$.

Now consider $(\delta M/M)$, the *rms* mass fluctuation on a given mass scale. This is what most people mean when they refer to the density contrast on a given mass scale. Mechanically, one would measure $(\delta M)_{rms}$ as follows: Take a volume V_W, which on average contains mass M, place it at all points throughout space, measure the mass within it, and then compute the *rms* mass fluctuation. Although it is simplest to choose a spherical volume V_W with a sharp surface, to avoid surface effects one often wishes to smear the surface. This is done by using a *window function* $W(r)$, which smoothly defines a volume V_W and mass $M = \bar{\rho} V_W$, where

$$V_W = 4\pi \int_0^\infty r^2 W(r) dr. \tag{9.16}$$

The *rms* mass fluctuation on the mass scale $M \equiv \bar{\rho} V_W$ is given in terms of the density contrast and window function by

$$\left(\frac{\delta M}{M}\right)^2 = \left\langle \left(\frac{\delta M}{M}\right)^2 \right\rangle \equiv \left\langle \left(\int \delta(\vec{\mathbf{x}} + \vec{\mathbf{r}}) W(\vec{\mathbf{r}}) d^3r\right)^2 / V_W^2 \right\rangle. \tag{9.17}$$

After some simple Fourier algebra, it follows that

$$\left(\frac{\delta M}{M}\right)^2 = \frac{1}{V_W^2} \int \Delta^2(k) |W(k)|^2 \frac{dk}{k}. \tag{9.18}$$

Notice that the *rms* mass fluctuation is given in terms of an integral over the variance $\Delta^2(k)$.

First consider the *top hat* window function, which has a sharply defined edge:

$$\begin{aligned}
W(r) &= \begin{cases} 1 & r \le r_0 \\ 0 & r > r_0, \end{cases} \\
V_W &= 4\pi r_0^3/3, \\
W_k &= 4\pi r_0^3 \left[\frac{\sin kr_0}{(kr_0)^3} - \frac{\cos kr_0}{(kr_0)^2}\right] \\
&\rightarrow 4\pi r_0^3 \begin{cases} [1 - (kr_0)^2/10]/3 & kr_0 \ll 1 \\ -\cos kr_0/(kr_0)^2 & kr_0 \gg 1. \end{cases}
\end{aligned} \tag{9.19}$$

The Fourier transform of the top hat window function is shown in Fig. 9.2. For this window function,

$$\left(\frac{\delta M}{M}\right)^2_{r_0} = 9\int_0^\infty \Delta^2(k)\left[\frac{\sin kr_0}{(kr_0)^3} - \frac{\cos kr_0}{(kr_0)^2}\right]^2 \frac{dk}{k}. \tag{9.20}$$

Taking $|\delta_k|^2 = Ak^n$, we find numerically that

$$\left(\frac{\delta M}{M}\right)_{r_0} \simeq (1.4 \text{ to } 4.2)\Delta(k = r_0^{-1}), \tag{9.21}$$

for $0.8 > n > -2.9$. Moreover, since $|W_k|^2 \propto k^{-4}$ for $k \gg r_0^{-1}$, we see that the contribution of short wavelength modes (large k) is negligible only if $n < 1$. This is the aforementioned unpleasant effect of the sharp edge of this simplest window function. Provided that $1 > n > -3$, $(\delta M/M)$ is given by an integral over all wavelengths longer than about r_0 and is roughly equal to the *rms* variance $\Delta(k = r_0^{-1})$. Using a top hat window function, Davis and Peebles [3] find for the CfA red shift survey that $(\delta M/M) = 1$ for a sphere of radius of $r_0 = 8h^{-1}$ Mpc.

Next consider the gaussian window function, which is both simple and smooth:

$$\begin{aligned} W(r) &= \exp(-r^2/2r_0^2), \\ V_W &= (2\pi)^{3/2} r_0^3, \\ W_k &= V_W \exp(-k^2 r_0^2/2) \end{aligned} \tag{9.22}$$

(see Fig. 9.2). For the gaussian window function,

$$\left(\frac{\delta M}{M}\right)^2 = \int_0^\infty \Delta^2(k)\exp(-k^2 r_0^2)\frac{dk}{k}. \tag{9.23}$$

We immediately see that for a gaussian window function the contribution of short wavelength fluctuations is exponentially cut off, and so it is never a problem. Taking the usual power law form for the power spectrum, we find

$$\left(\frac{\delta M}{M}\right)^2 = \frac{1}{2}\Gamma\left(\frac{n+3}{2}\right)\Delta^2(k = r_0^{-1}), \tag{9.24}$$

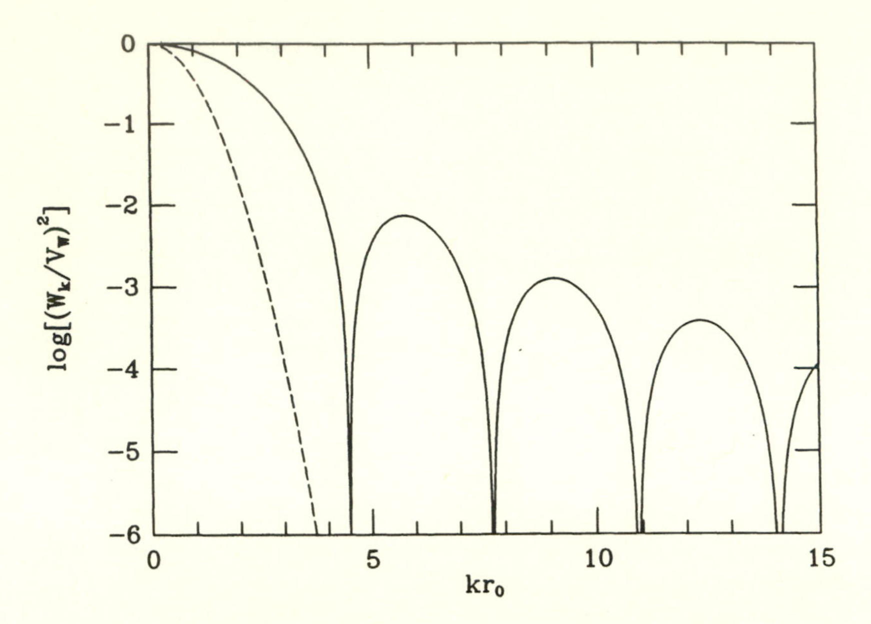

Fig. 9.2: The Fourier transform squared of the top hat window function (solid curve), cf. (9.19), and the gaussian window function (broken curve), cf. (9.22).

where $\Gamma(m) = (m-1)!$. Once again we see that $(\delta M/M)^2$ is related to an integral over modes with wavelengths longer than order r_0, and provided that $n > -3$, the *rms* mass fluctuation on the scale r_0 is given in terms of the variance Δ evaluated on that scale. That the short wavelength modes do not contribute makes physical sense: Fluctuations with wavelengths less than r_0 average to zero in a volume of size r_0.

We have just discovered that the *rms* mass fluctuation is related to the variance by $(\delta M/M)_\lambda \simeq \Delta(\lambda^{-1})$, where the precise numerical factors depend upon the form of the window function and of the power spectrum. Further, we have discovered that the contribution of fluctuations on the scale λ to the *rms* density fluctuation is also determined by $\Delta(\lambda^{-1})$. As asserted, the power spectrum, or equivalently the variance $\Delta^2(k) \equiv V^{-1}k^3|\delta_k|^2/2\pi^2$, is all important in describing the density field and its properties. Finally, taking the usual power law form for $|\delta_k|^2$, and remembering that $k \sim \lambda^{-1} \propto M^{-1/3}$, we see that

$$(\delta\rho/\rho)_M \equiv (\delta M/M) \sim k^{3/2}|\delta_k| \propto M^{-1/2-n/6}. \qquad (9.25)$$

Since $(\delta\rho/\rho) \propto |\delta_k|$, it is quite common to see $|\delta_k|$ and $(\delta\rho/\rho)$ used interchangeably.

Next, consider the mass autocorrelation function $\xi(r)$, defined by

$$\xi(\vec{r}) \equiv \langle \delta(\vec{x}+\vec{r})\delta(\vec{x}) \rangle. \tag{9.26}$$

Again, after some simple Fourier algebra, we obtain

$$\xi(\vec{r}) = (2\pi)^{-3}V^{-1}\int |\delta_k|^2 \exp(-i\vec{k}\cdot\vec{r})d^3k, \tag{9.27}$$

$$|\delta_k|^2 = V\int \xi(\vec{r})\exp(i\vec{k}\cdot\vec{r})d^3r. \tag{9.28}$$

The power spectrum is the Fourier transform of the mass autocorrelation function (and vice versa). Moreover, it follows that $\xi(\vec{r}=0) = (\delta\rho/\rho)^2$. Based upon the isotropy of the Universe, we expect that $\xi(\vec{r}) = \xi(r)$ where $r = |\vec{r}|$. If we suppose that $|\delta_k|^2 = AVk^n$, then for $-3 < n < -1$,

$$\begin{aligned} \xi(r) &= \frac{Ar^{-n-3}}{2\pi^2}\int_0^\infty u^{n+1}\sin u\, du \\ &= \frac{Ar^{-n-3}}{2\pi^2}\Gamma(n+2)\sin[(n+2)\pi/2]. \end{aligned} \tag{9.29}$$

While we do not have the means to measure the *mass* autocorrelation function directly, the galaxy autocorrelation function (or galaxy–galaxy correlation function) has been measured, cf. (1.16),

$$\xi_{\rm GG}(r) \simeq (r/5h^{-1}\,{\rm Mpc})^{-1.8}. \tag{9.30}$$

The galaxy autocorrelation function is defined as

$$\xi_{\rm GG}(r) = \langle \delta n(\vec{x}+\vec{r})\delta n(\vec{x}) \rangle. \tag{9.31}$$

If we adopt the *seemingly* reasonable proposition that the number density of galaxies $n(\vec{x})$ is proportional to the mass density $\rho(\vec{x})$, $n(\vec{x}) = {\rm const}\times\rho(\vec{x})$, it then follows directly that the mass autocorrelation function and the galaxy–galaxy correlation function should be identical.

Furthermore, the galaxy autocorrelation function also has a very physical interpretation: It measures the joint probability, over random and averaged over all space, of finding two galaxies separated by a distance r (in volume elements δV_1 and δV_2):

$$\langle \delta P_{12} \rangle \equiv \bar{n}^2 \delta V_1 \delta V_2 [1+\xi(r)]. \tag{9.32}$$

Or stated another way, $\overline{n}\delta V[1+\xi(r)]$ is the probability of finding a galaxy at a distance r from a given galaxy.

Since galaxies are identified by the light they emit, another way of stating the assumption that $\rho(\vec{x}) = \text{const} \times n(\vec{x})$ is "light traces mass," i.e., the amount of light is proportional to the amount of mass (with a "constant" constant of proportionality). This is a very strong and not well tested assumption. Moreover, there is some indication that it is not strictly true.

If it were strictly true, then the unit amount of light selected to quantify the mass density should be irrelevant. For example, one could use as the "unit of light" a cluster rather than a galaxy, and if light faithfully traced mass, the cluster–cluster correlation function would also provide us with the mass autocorrelation function. That is, under the assumption of light traces mass, the cluster–cluster and galaxy–galaxy correlation functions should be identical since they both are determined by the same underlying mass density. However, as we discussed in Chapter 1, there is some evidence that the galaxy–galaxy and cluster–cluster correlation functions are very different, apparently belying the assumption that light faithfully traces mass:

$$\xi_{\rm CC}(r) \simeq (r/25h^{-1}\,{\rm Mpc})^{-1.8} \simeq 20\xi_{\rm GG}(r). \tag{9.33}$$

Moreover, the cluster–cluster correlation function depends upon the "richness class" of the clusters being considered.[8]

As we emphasized in Chapter 1, the issue of whether or not the cluster–cluster correlation function has been reliably measured is still unsettled; the discrepancy between $\xi_{\rm CC}$ and $\xi_{\rm GG}$ is intriguing as it seems to undermine the simple assumption of light traces mass. For the moment we will take $\xi(r)$ to be determined by the galaxy autocorrelation function, and we will return to the important issue of whether or not mass traces light when discussing cold dark matter.

Next, consider the following integral over $\xi(r)$:

$$J_3(R) \equiv \frac{1}{4\pi}\int_0^R \xi(r)d^3r = \int_0^R r^2\xi(r)dr. \tag{9.34}$$

[8] Abell classified clusters by their "richness" (0, 1, 2, $\cdots$, 5), which is a measure of the number of galaxies in a cluster.

In terms of the variance $\Delta(k)$, $J_3(R)$ is given by

$$J_3(R) = R^3 \int_0^\infty \Delta^2(k) \left[\frac{\sin kR}{(kR)^3} - \frac{\cos kR}{(kR)^2} \right] \frac{dk}{k} \tag{9.35}$$

—just the integral of $\Delta^2(k)$ times the window function (more precisely, $W_k/3V_W$) for a top hat of radius R. Using the form for $\delta M/M$ appropriate to a top hat window function and approximating W_k by a step function, we see that

$$\left(\frac{\delta M}{M} \right)_R^2 = \frac{3J_3(R)}{R^3}. \tag{9.36}$$

That is, $J_3(R)$ is directly related to the *rms* mass fluctuation on the scale R. Although $J_3(R)$ involves an integral over all scales, it is weighted such that it is proportional to $(\delta M/M)_R$.

Based upon measurements of the galaxy–galaxy correlation function [3], $J_3(R)$ is determined to be $270h^{-3}\,\mathrm{Mpc}^3$ on the scale of $10h^{-1}$ Mpc, and $600h^{-3}\,\mathrm{Mpc}^3$ on the scale of $30h^{-1}$ Mpc. Using this data we can infer that

$$\begin{aligned} (\delta M/M)_{\lambda \simeq 10h^{-1}\,\mathrm{Mpc}} &\simeq 0.9, \\ (\delta M/M)_{\lambda \simeq 30h^{-1}\,\mathrm{Mpc}} &\simeq 0.25. \end{aligned} \tag{9.37}$$

Therefore, a scale just smaller than about $10h^{-1}$ Mpc separates the scales that have already gone non-linear from those that are yet to go non-linear. This is consistent with the determination of $(\delta M/M) = 1$ for a spherical top hat window of radius $8h^{-1}$ Mpc.

By taking $R \to \infty$ we obtain an interesting integral constraint to the correlation function:

$$\begin{aligned} J_3 &\equiv J_3(R \to \infty) = \frac{V^{-1}}{4\pi(2\pi)^3} \int |\delta_k|^2 e^{-ikr} d^3k d^3r \\ &= \frac{V^{-1}}{4\pi} \int |\delta_k|^2 \delta^3(\vec{\mathbf{k}}) d^3k = V^{-1} \left. \frac{|\delta_k|^2}{4\pi} \right|_{k=0}. \end{aligned} \tag{9.38}$$

Taking as usual $|\delta_k|^2 \propto k^n$ we see that provided $n > 0$ (i.e., $\delta\rho/\rho$ decreases faster than $M^{-1/2}$),

$$J_3 \equiv \int_0^\infty r^2 \xi(r) dr = 0. \tag{9.39}$$

This constraint to $\xi(r)$ is really just a statement of mass conservation: If mass (galaxies) are clustered on small scales, then on large scales they must be "anti-clustered" to conserve the total amount of mass (number of galaxies). On scales $r \lesssim 5h^{-1}$ Mpc, $\xi(r)$ is positive and greater than unity; therefore, on some larger scale $\xi(r)$ must become negative. Whether or not, and on what scale, the galaxy–galaxy correlation function becomes negative is an important probe of the power spectrum. The observational side of this issue is quite unsettled at the moment.

We have just discussed how $J_3(R)$ can be used to infer $\delta M/M$. In principle, one can also directly use $\xi(r)$ to infer the power spectrum. In practice, it is somewhat more difficult. The power spectrum and $\xi(r)$ are related by

$$|\delta_k|^2 = V\int \xi(r)\exp(i\vec{\mathbf{k}}\cdot\vec{\mathbf{r}})d^3r = V\frac{4\pi}{k}\int r\xi(r)\sin kr\,dr. \tag{9.40}$$

The galaxy correlation function is well determined only on scales 100 kpc $\lesssim r \lesssim 10h^{-1}$ Mpc; to invert $\xi(r)$ to obtain the power spectrum we need to know $\xi(r)$ on all scales. However, we can make a stab at determining $|\delta_k|^2$ from (9.40) by approximating $\xi(r)$ as $\xi(r) = (r/r_0)^{-1.8}$ on all scales, where $r_0 \simeq 5h^{-1}$ Mpc. Doing so, we find

$$\begin{aligned}\Delta^2(k) &\equiv V^{-1}\frac{k^3|\delta_k|^2}{2\pi^2} = \frac{2}{\pi}(kr_0)^{1.8}\int_0^\infty u^{-0.8}\sin u\,du \\ &= \frac{2}{\pi}(kr_0)^{1.8}\,\Gamma(0.2)\sin(0.1\pi) \simeq 0.9\,(kr_0)^{1.8}.\end{aligned} \tag{9.41}$$

Taking $k \simeq r_0^{-1} = (5h^{-1}\,\mathrm{Mpc})^{-1}$, the scale on which $\xi(r) \simeq 1$, our approximate inversion implies that

$$\left(\frac{\delta M}{M}\right)_{r_0\simeq 5h^{-1}\,\mathrm{Mpc}} \simeq \Delta(k = r_0^{-1}) \simeq 1. \tag{9.42}$$

This result should not come as much of a surprise: One would have expected that $\delta M/M \sim 1$ on the scale where the clustering is just becoming non-linear, and this result is consistent with the previous estimates based upon $J_3(R)$ and the direct determination that $(\delta M/M)_{8h^{-1}\,\mathrm{Mpc}} = 1$. The difficulty of inverting $\xi(r)$ to obtain the power spectrum illustrates the utility of $J_3(R)$: The factor of r^2 in the integral weights $J_3(R)$ in such a way that it is directly related to $(\delta\rho/\rho)_R$, whereas $\xi(r)$ is related to the power on all scales.

9.2.3 Curvature and isocurvature perturbations

We have set up the machinery for describing density perturbations, and in the next Section we will discuss the evolution thereof in some detail. Before we do, we will discuss the two different kinds of primeval fluctuations: curvature (or adiabatic) and isocurvature (or isothermal).

As we will emphasize in the next Section, when discussing the evolution of density perturbations, there are two qualitatively different regimes: (1) The early regime, when the perturbation is outside the horizon, $\lambda_{\rm phys} \gtrsim H^{-1}$, during which causal, microphysical processes are impotent and unimportant; (2) the late regime, when the perturbation is inside the horizon, during which causal, microphysical processes can be important, and the evolution of the perturbation can be well described with a Newtonian analysis (see Fig. 9.1).

For modes that cross inside the horizon before matter–radiation equality, $\lambda \lesssim \lambda_{\rm EQ} = 13\,(\Omega_0 h^2)^{-1}\,{\rm Mpc}$, the horizon-crossing epoch ($\lambda_{\rm phys} = t$) is given by

$$
\begin{aligned}
T_{\rm HOR}(\lambda) &= 63\, g_{*S}^{1/3} g_*^{-1/2} \lambda_{\rm Mpc}^{-1}\,{\rm eV}, \\
t_{\rm HOR}(\lambda) &= 6.1 \times 10^8\, g_{*S}^{-2/3} g_*^{1/2} \lambda_{\rm Mpc}^{2}\,{\rm sec}.
\end{aligned}
\tag{9.43}
$$

While for those that cross inside the horizon after matter–radiation equality, the horizon-crossing epoch is given by

$$
\begin{aligned}
T_{\rm HOR}(\lambda) &= 950\,(\Omega_0 h^2)^{-1} \lambda_{\rm Mpc}^{-2}\,{\rm eV}, \\
t_{\rm HOR}(\lambda) &= 2.6 \times 10^7\,(\Omega_0 h^2) \lambda_{\rm Mpc}^{3}\,{\rm sec}.
\end{aligned}
\tag{9.44}
$$

At late times, $t \gtrsim t_{\rm EQ}$, practically all the interesting fluctuation modes are well within the horizon and correspond to "Newtonian" density perturbations that can be so analyzed. However, at very early times, $t \ll t_{\rm EQ}$, when the cosmologically interesting scales were well outside the horizon, there are two distinct and orthogonal types of primeval fluctuations: curvature (or adiabatic) fluctuations and isocurvature (or isothermal) fluctuations.

Curvature fluctuations are "honest-to-God" fluctuations in the energy density ($\delta\rho \neq 0$) that, as we shall see in the next Section, can be characterized in a gauge-invariant manner as fluctuations in the local value of the spatial curvature. By the equivalence priciple, all components of the energy density participate in such perturbations:

$$\delta \equiv \frac{\delta n_B}{n_B} = \frac{\delta n_X}{n_X} = \frac{\delta s}{s} \quad \left(\Rightarrow \frac{\delta T}{T} = \frac{1}{3}\delta \right), \tag{9.45}$$

where s is the entropy density, subscript B refers to baryons, and subscript X to any other species X. Because both the radiation and matter species participate in the fluctuation, $\delta(n_X/s) = \delta n_X/s - n_X \delta s/s^2 = 0$. That is, the fluctuation in the local number density of any species relative to the entropy density, i.e., the fluctuation in the number per comoving volume, vanishes; hence the name adiabatic. Note too that the magnitude of the temperature fluctuation is one third that in any species.

On the other hand, isocurvature fluctuations are not "honest-to-God" fluctuations in the energy density ($\delta\rho = 0$) and are not characterized by a fluctuation in the local curvature. Rather, they correspond to fluctuations in the form of the local equation of state. The simplest example is provided by spatial fluctuations in the number of some species X per comoving volume, e.g., baryons or axions:

$$\delta(n_X/s) = f(\vec{\mathbf{x}}) \neq 0. \tag{9.46}$$

This corresponds to a spatial variation in the equation of state since the local pressure depends not only upon the density ρ but also upon the composition, e.g., the number of X's in a comoving volume.

So long as an isocurvature mode is super-horizon sized it must be characterized by $\delta\rho = 0$, because causality precludes the re-distribution of energy density on scales larger than the horizon. As we shall see, once an isocurvature mode becomes sub-horizon sized, fluctuations in the local pressure can "push" energy density around and convert an isocurvature fluctuation into an energy density perturbation. Thereafter, the distinction between isocurvature and curvature perturbations is not important.

Using the fact that $\delta\rho$ must remain equal to zero for super-horizon modes, we can show that isocurvature fluctuations are not *truly* isothermal (although the terms are routinely used interchangeably). Consider an isocurvature fluctuation in a species X that is non-relativistic; fluctuations in the local number density of X's, n_X, correspond to energy density fluctuations in ρ_X. For simplicity write the total energy density as a component due to X and a component due to radiation: $\rho = \rho_X + \rho_R$, where $\rho_R = \text{const} \times T^4$. The requirement that $\delta\rho$ vanishes implies that there *must* be compensating fluctuations in ρ_R:

$$\begin{aligned}\delta\rho &= \delta(m_X n_X + \text{const} \times T^4) = 0 \\ \Rightarrow \frac{\delta T}{T} &= -\frac{1}{4}\frac{\rho_X}{\rho_R}\frac{\delta n_X}{n_X}.\end{aligned} \tag{9.47}$$

The size of the compensating temperature fluctuations is determined by ρ_X/ρ_R and $\delta n_X/n_X$. Recall that n_X/s measures the number of X's in a comoving volume element. Let the fluctuation in the number of X's in a comoving volume be characterized by $(\delta_X)_i$,

$$(\delta_X)_i \equiv \frac{\delta(n_X/s)}{(n_X/s)}. \tag{9.48}$$

While the mode in question is well outside the horizon the fluctuation in the local number of X's in a comoving volume cannot change—causality precludes X's from being transported across super-horizon-sized distances. Since $s \propto T^3$, we can relate $(\delta_X)_i$, $\delta n_X/n_X$, and $\delta T/T$:

$$(\delta_X)_i = \frac{\delta n_X}{n_X} - 3\frac{\delta T}{T}. \tag{9.49}$$

Combining this relation with (9.47) we find that

$$\begin{aligned}\frac{\delta T}{T} &= -\frac{(\delta_X)_i}{4}\left(\frac{\rho_X/\rho_R}{1 + 0.75\rho_X/\rho_R}\right), \\ \frac{\delta n_X}{n_X} &= (\delta_X)_i + 3\frac{\delta T}{T}.\end{aligned} \tag{9.50}$$

At early times when $\rho_R \gg \rho_X$, the compensating temperature fluctuation is small:

$$\frac{\delta T}{T} \simeq -\frac{1}{4}\frac{\rho_X}{\rho_R}(\delta_X)_i \ll (\delta_X)_i; \tag{9.51}$$

hence the name isothermal. This is easy to understand: At early times it does not take much of a temperature fluctuation to compensate for an energy density fluctuation in the X's since $\rho_R \gg \rho_X$. In addition, we see that the fluctuation in the local number density of X's is just $\delta n_X/n_X \simeq (\delta_X)_i$.

On the other hand, at late times, as the Universe becomes matter-dominated, ρ_X becomes comparable to, and then greater than, ρ_R, and the temperature fluctuation becomes comparable to $(\delta_X)_i$:

$$\frac{\delta T}{T} \to -\frac{(\delta_X)_i}{3} \quad \text{for} \quad \rho_X \gtrsim \rho_R. \tag{9.52}$$

In addition, the fluctuation in the local number density of X's becomes smaller than $(\delta_X)_i$:

$$\frac{\delta n_X}{n_X} \to \frac{4}{3}(\delta_X)_i \frac{\rho_R}{\rho_X} \quad \text{for} \quad \rho_X \gtrsim \rho_R. \tag{9.53}$$

That this must occur is also clear: As ρ_X becomes greater than ρ_R, the only way $\delta\rho = 0$ can be maintained is for $\delta n_X/n_X$ to decrease.

Once an isocurvature mode enters the horizon, $\lambda_{\rm phys} \lesssim H^{-1}$, energy density (and X's) can be transported on the scale of the size of the fluctuation, and the energy constraint $\delta\rho = 0$ no longer applies. Physically, the transport of energy density occurs because of the local variations in pressure that arise due to the local variation in the equation of state: Although isocurvature modes enter the horizon with $\delta\rho = \delta\rho_R + \delta\rho_X = 0$, the compositional variations, here $\delta n_X/n_X$, give rise to pressure variations since $p = p_R + p_X$ ($p_R = \rho_R/3$ and $p_X \simeq 0$) and $\delta p \simeq \delta\rho_R/3$. Because microphysical processes become important once a mode crosses inside the horizon, the analysis becomes more complex; however, the end result is easy to state.

First, consider isocurvature modes that cross inside the horizon before matter–radiation equality. As the fluctuation crosses inside the horizon, $\delta n_X/n_X \simeq (\delta_X)_i$ and $\delta T/T = -(\rho_X/\rho_R)(\delta_X)_i/4$. The photon fluctuation becomes a sound wave in the baryon-photon plasma. The perturbation in the X's, $\delta n_X/n_X$, maintains a constant amplitude (about $(\delta_X)_i$) until matter domination commences. As the radiation energy density becomes sub-dominant, the fluctuation in the number density of X's becomes a density perturbation of amplitude $\delta\rho/\rho \simeq (\delta_X)_i$, which, after matter domination begins, behaves just like its adiabatic counterpart.

The situation for isocurvature modes that enter the horizon after matter domination begins is slightly different. Just before horizon crossing, $\delta T/T \simeq (\delta_X)_i/3$ and $\delta n_X/n_X \simeq (4/3)(\delta_X)_i(\rho_R/\rho_X)$. The temperature fluctuation remains imprinted upon the photons, giving rise to an anisotropy in the CMBR temperature (to be discussed in Section 9.6). As the mode crosses inside the horizon, the fluctuation in the number density of X's grows to a size $\delta n_X / n_X \simeq (\delta_X)_i$, because of the pressure gradients that have developed due to the compositional variations. After horizon crossing, the X fluctuation behaves just as an X density perturbation of amplitude $(\delta_X)_i$, and it continues to grow thereafter, just like its adiabatic counterpart. As one can see, after horizon crossing, the difference between adiabatic and isocurvature becomes irrelevant.

Adiabatic (curvature) fluctuations either are present initially as fluctuations in the curvature (and energy density) or can be produced at early times provided that some mechanism (like inflation) allows sub-horizon-sized scales to become super-horizon-sized (see Figs. 8.4 and 9.1). In inflationary models, de Sitter space produced fluctuations in the inflaton field give rise to adiabatic fluctuations (see Section 8.4). Isocurvature fluctuations can also be produced during inflation, e.g., as fluctuations in the local number density of baryons or axions (see Section 10.4). More generally, isocurvature fluctuations are produced by causal, microphysical processes that cannot transport energy across super-horizon scales but that can perturb the local equation of state. These fluctuations in the local equation of state ultimately lead to density perturbations after the mode enters the horizon and the Universe becomes matter-dominated (as described above). Cosmic strings provide another such example. During the phase transition that leads to the production of these linelike defects, some of the energy density in radiation is converted into cosmic string. Later on, cosmic string becomes the seed for structure formation.

9.3 The Evolution of Density Inhomogeneities: The Standard Lore

As we have stated several times now, it is widely believed that the structure we observe in the Universe today is the end result of the gravitational amplification of small primeval perturbations in an otherwise smooth Universe. In this Section we will discuss the theory of gravitational instability, beginning with the Newtonian theory and working our way to the full general relativistic theory. We will start by considering the simplest possible form of gravitational instability, the Jeans instability in a non-expanding, perfect fluid. We will then include the effects of the expansion. The three important scales that arise in the Newtonian analysis—the baryon Jeans mass, the collisional-damping scale, and the collisionless-damping scale—will then be discussed. We then proceed to the full general-relativistic analysis that is necessary to treat perturbations that are larger than the horizon. Finally, as a reward for the readers who survive this long Section, we provide an executive summary of important results.

9.3.1 Jeans analysis in a non-expanding fluid

The Eulerian equations of Newtonian motion describing a perfect fluid are

$$\begin{aligned}
&\frac{\partial \rho}{\partial t} + \vec{\nabla} \cdot (\rho \vec{v}) = 0, \\
&\frac{\partial \vec{v}}{\partial t} + (\vec{v} \cdot \vec{\nabla})\vec{v} + \frac{1}{\rho}\vec{\nabla} p + \vec{\nabla}\phi = 0, \\
&\nabla^2 \phi = 4\pi G \rho.
\end{aligned} \tag{9.54}$$

Here ρ is the matter density, p the matter pressure, $\vec{v}$ the local fluid velocity, and ϕ the gravitational potential.

The simplest solution is the static one where the matter is at rest ($\vec{v}_0 = 0$) and uniformly distributed in space ($\rho_0 = \text{const}$, $p_0 = \text{const}$).[9] Throughout this Section, we will denote unperturbed quantities with the subscript 0. The reader should take note of this, since previously we reserved the subscript 0 for the values of various quantities at the present epoch. Consider perturbations about this static solution, expanded as follows

$$\begin{aligned}
\rho &= \rho_0 + \rho_1, \\
p &= p_0 + p_1, \\
\vec{v} &= \vec{v}_0 + \vec{v}_1, \\
\phi &= \phi_0 + \phi_1.
\end{aligned} \tag{9.55}$$

The pressure is related to the energy density by the equation of state. We will consider adiabatic perturbations, that is, perturbations for which there are no spatial variations in the equation of state. The (adiabatic)

[9]We follow Jeans in assuming that for such a configuration the gravitational force vanishes, i.e., $\nabla \phi_0 = 0$. Of course, this contradicts the Poisson equation (the Jeans swindle).

sound speed, v_s^2, is defined as

$$v_s^2 \equiv \left(\frac{\partial p}{\partial \rho}\right)_{\text{adiabatic}}, \tag{9.56}$$

and since by assumption there are no spatial variations in the equation of state,

$$v_s^2 = \frac{p_1}{\rho_1}. \tag{9.57}$$

To first order, the small perturbations, ρ_1, $\vec{\mathbf{v}}_1$, and ϕ_1, satisfy the perturbed versions of (9.54):

$$\frac{\partial \rho_1}{\partial t} + \rho_0 \vec{\nabla} \cdot \vec{\mathbf{v}}_1 = 0,$$

$$\frac{\partial \vec{\mathbf{v}}_1}{\partial t} + \frac{v_s^2}{\rho_0}\vec{\nabla}\rho_1 + \vec{\nabla}\phi_1 = 0,$$

$$\nabla^2 \phi_1 = 4\pi G \rho_1. \tag{9.58}$$

Equations (9.58) can be combined to form a single, second-order differential equation for ρ_1:

$$\frac{\partial^2 \rho_1}{\partial t^2} - v_s^2 \nabla^2 \rho_1 = 4\pi G \rho_0 \rho_1. \tag{9.59}$$

Solutions to (9.59) are of the form

$$\rho_1(\vec{\mathbf{r}}, t) = \delta(\vec{\mathbf{r}}, t)\rho_0 = A \exp\left[-i\vec{\mathbf{k}} \cdot \vec{\mathbf{r}} + i\omega t\right] \rho_0, \tag{9.60}$$

and ω and $\vec{\mathbf{k}}$ satisfy the dispersion relation

$$\omega^2 = v_s^2 k^2 - 4\pi G \rho_0, \tag{9.61}$$

where $k \equiv |\vec{\mathbf{k}}|$.

If ω is imaginary, there will be exponentially growing modes; if ω is real, the perturbations will simply oscillate as sound (compressional) waves. From (9.61), it is clear that for k less than some critical value, ω will be imaginary. This critical value is called the Jeans wavenumber, k_J, and is given by

$$k_J = \left(\frac{4\pi G \rho_0}{v_s^2}\right)^{1/2}. \tag{9.62}$$

For $k^2 \ll k_J^2$, ρ_1 grows (or decays) exponentially on the dynamical timescale

$$\tau_{\rm dyn} = ({\rm Im}\ \omega)^{-1} \simeq (4\pi G\rho_0)^{-1/2}. \tag{9.63}$$

It is useful to define the Jeans mass, the total mass contained within a sphere of radius $\lambda_J/2 = \pi/k_J$:

$$M_J \equiv \frac{4\pi}{3}\left(\frac{\pi}{k_J}\right)^3 \rho_0 = \frac{\pi^{5/2}}{6}\frac{v_s^3}{G^{3/2}\rho_0^{1/2}}. \tag{9.64}$$

Perturbations of mass less than M_J are stable against gravitational collapse, while those of mass greater than M_J are unstable.

There is a simple physical explanation for the fact that the scale λ_J separates gravitationally stable from gravitationally unstable modes. The time scale for gravitational collapse is given by $\tau_{\rm grav} \simeq (G\rho_0)^{-1/2}$, which is just the dynamical time scale. On the other hand, the "pressure response" time scale is given by the size of the perturbation divided by the sound speed: $\tau_{\rm pressure} \sim \lambda/v_s$. If $\tau_{\rm pressure}$ exceeds $\tau_{\rm grav}$, the gravitational collapse of a perturbation can occur before pressure forces can respond to restore hydrostatic equilibrium; $\tau_{\rm pressure} \gtrsim \tau_{\rm grav}$ occurs for $\lambda \gtrsim v_s/(G\rho_0)^{1/2} \sim \lambda_J$.

The classical Jeans analysis is not directly applicable to cosmology because the expansion of the Universe is not taken into account and, further, because the analysis is Newtonian. For modes of wavelength less than that of the horizon, i.e., $\lambda_{\rm phys} \ll H^{-1}$, a Newtonian analysis suffices so long as the expansion is taken into account.

9.3.2 Jeans analysis in an expanding fluid

When the expansion of the Universe is taken into account, the unperturbed solution to (9.58) is

$$\rho_0 = \rho_0(t_0)R^{-3}(t), \qquad \vec{\mathbf{v}}_0 = \frac{\dot{R}}{R}\vec{\mathbf{r}}, \qquad \vec{\nabla}\phi_0 = \frac{4\pi G\rho_0}{3}\vec{\mathbf{r}}, \tag{9.65}$$

where $R(t)$ satisfies the usual Friedmann equation.[10]

[10] Here it is not necessary to invoke the Jeans swindle for the unperturbed solution; but for scales $|\vec{\mathbf{r}}| \gtrsim H^{-1}$, the "velocity" $|\vec{\mathbf{v}}|$ exceeds that of light. However, we remind the reader that the Newtonian Jeans analysis given here is valid only for perturbation scales that are sub-horizon sized.

The first order perturbations in ρ, $\vec{\mathbf{v}}$, and ϕ satisfy the following set of equations:

$$\frac{\partial \rho_1}{\partial t} + 3\frac{\dot{R}}{R}\rho_1 + \frac{\dot{R}}{R}(\vec{\mathbf{r}} \cdot \vec{\nabla})\rho_1 + \rho_0 \vec{\nabla} \cdot \vec{\mathbf{v}}_1 = 0,$$
$$\frac{\partial \vec{\mathbf{v}}_1}{\partial t} + \frac{\dot{R}}{R}\vec{\mathbf{v}}_1 + \frac{\dot{R}}{R}(\vec{\mathbf{r}} \cdot \vec{\nabla})\vec{\mathbf{v}}_1 + \frac{v_s^2}{\rho_0}\vec{\nabla}\rho_1 + \vec{\nabla}\phi_1 = 0,$$
$$\vec{\nabla}^2 \phi_1 = 4\pi G \rho_1. \tag{9.66}$$

If we define $\delta(\vec{\mathbf{r}}, t) = \rho_1/\rho_0$, and if we expand δ, $\vec{\mathbf{v}}_1$, and ϕ_1 in Fourier integrals

$$\Psi(\vec{\mathbf{r}}, t) = (2\pi)^{-3} \int \Psi_k(t) \exp\left[\frac{-i\vec{\mathbf{k}} \cdot \vec{\mathbf{r}}}{R(t)}\right] d^3r \tag{9.67}$$

($\Psi = \delta$, $\vec{\mathbf{v}}_1$, ϕ_1), equations (9.66) simplify considerably:

$$\dot{\delta}_k - \frac{i\vec{\mathbf{k}}}{R} \cdot \vec{\mathbf{v}}_k = 0,$$
$$d(R\vec{\mathbf{v}}_k)/dt - i\vec{\mathbf{k}}v_s^2\delta_k - i\vec{\mathbf{k}}\phi_k = 0,$$
$$\phi_k = -\frac{4\pi G \rho_0}{k^2} R^2 \delta_k. \tag{9.68}$$

Here we recognize $\vec{\mathbf{k}}$ as the comoving wavenumber, and $\vec{\mathbf{x}} = \vec{\mathbf{r}}/R$ as the comoving coordinate. Moreover, it is useful to decompose the perturbed velocity field into its rotational ($\vec{\mathbf{v}}_\perp$) and irrotational ($\vec{\mathbf{v}}_\parallel$) pieces:

$$\vec{\mathbf{v}}_1 = \vec{\mathbf{v}}_\parallel + \vec{\mathbf{v}}_\perp, \qquad \vec{\mathbf{v}}_k = \vec{\mathbf{v}}_\parallel(\vec{\mathbf{k}}) + \vec{\mathbf{v}}_\perp(\vec{\mathbf{k}}), \tag{9.69}$$

where

$$\nabla \cdot \vec{\mathbf{v}}_\perp = 0, \qquad \nabla \times \vec{\mathbf{v}}_\parallel = 0,$$
$$\vec{\mathbf{k}} \cdot \vec{\mathbf{v}}_\perp(\vec{\mathbf{k}}) = 0, \qquad \vec{\mathbf{k}} \cdot \vec{\mathbf{v}}_\parallel(\vec{\mathbf{k}}) = |\vec{\mathbf{k}}||\vec{\mathbf{v}}_\parallel(\vec{\mathbf{k}})|. \tag{9.70}$$

Note that $\vec{\mathbf{v}}_\parallel$ can be written as the gradient of some scalar field, while $\vec{\mathbf{v}}_\perp$ can be written as the curl of some vector field. The first-order equations governing the perturbations become

$$d[R\vec{v}_\perp(\vec{k})]/dt = 0 \quad \Rightarrow \quad \vec{v}_\perp \propto R(t)^{-1},$$
$$\vec{v}_\parallel(\vec{k}) = \frac{R}{ik}\dot{\delta}_k + \frac{\text{const}}{R(t)},$$
$$\ddot{\delta}_k + 2\frac{\dot{R}}{R}\dot{\delta}_k + \left(\frac{v_s^2k^2}{R^2} - 4\pi G\rho_0\right)\delta_k = 0. \tag{9.71}$$

We immediately see that the rotational modes are not coupled to density perturbations and that they decay as R^{-1}.[11]

Henceforth, we will be interested in only the irrotational modes, which do couple to the density perturbations. For brevity we will drop the subscripts $\|$ and k. From the equation for δ we see that in a non-expanding Universe (i.e., $\dot{R} = 0$) Jeans' result is recovered when we recognize $|\vec{k}|/R$ as the physical wavenumber. It is also clear that the qualitative behavior of the solution depends upon whether $v_s^2k^2/R^2$ is larger, or smaller, than $4\pi G\rho_0$. In an expanding Universe as well, the Jeans wavenumber, $k_J^2 \equiv 4\pi G\rho_0 R^2/v_s^2$, separates the gravitationally stable and unstable modes.[12]

For short-wavelength modes, $k \gg k_J$, the perturbations oscillate as a sound wave

$$\delta(t) \sim \exp(\pm i\omega t), \tag{9.72}$$

where

$$\omega = \frac{v_s k}{R(t)(1-n)}, \tag{9.73}$$

and it has been assumed that $R(t) \propto t^n$. In general, the exact solution is given by a Bessel function of some type, and the amplitude of the sound wave slowly decreases. The exact solution depends upon whether the Universe is matter or radiation dominated, and how v_s evolves.

For $k \ll k_J$ there are unstable (growing mode) solutions. In this limit the pressure gradient term can be neglected, and if we further assume the spatially flat ($k = 0$), matter-dominated FRW model where $\dot{R}/R =$

---

[11] The rotational modes evolve so as to conserve angular momentum: $L \sim (\rho_0 r^3) r v_1 =$ const, from which it follows that they must decay as $v_1 \propto r^{-1} \propto R^{-1}$.

[12] It is sometimes useful to write the differential equation for δ in terms of the conformal time variable η ($d\eta = dt/R(t)$, $\eta = (1-n)^{-1}t/R(t)$ + const if $R \propto t^n$): $\delta'' + (R'/R)\delta' + (v_s^2k^2 - 4\pi G\rho_0 R^2)\delta = 0$, where prime denotes $d/d\eta$.

$(2/3)t^{-1}$ and $\rho_0 = (6\pi G t^2)^{-1}$, (9.71) becomes

$$\ddot{\delta} + \frac{4}{3t}\dot{\delta} - \frac{2}{3t^2}\delta = 0, \qquad k \ll k_J. \tag{9.74}$$

This equation has two independent solutions, a growing mode, δ_+, and a decaying mode, δ_-, with time dependence given by

$$\delta_+(t) = \delta_+(t_i)\left(\frac{t}{t_i}\right)^{2/3}; \quad \delta_-(t) = \delta_-(t_i)\left(\frac{t}{t_i}\right)^{-1}. \tag{9.75}$$

The time t_i is some convenient but arbitrary time chosen to normalize the solution. Here we see the key difference between the Jeans instability in the static regime (e.g., within a galaxy) and in the expanding Universe: The expansion of the Universe slows the exponential growth of the instability and results in power-law growth for unstable modes.

As expected, since the equation governing δ is second order, there are two solutions. A given perturbation is expressed as a linear combination of $\delta_+(t)$ and $\delta_-(t)$; at late times, only the projection onto the growing mode will be important. Physically, the decaying mode corresponds to a perturbation with initial overdensity and velocity arranged so that the initial velocity perturbation eventually "undoes" the density perturbation.

Thus far we have assumed that there is only one component to the mass density. If there are several components, $\rho = \sum_i \rho_i$ (i = baryons, photons, neutrinos, exotic particle relics, etc.), the equation describing the evolution of the perturbation in the NR component i is

$$\ddot{\delta}_i + 2\frac{\dot{R}}{R}\dot{\delta}_i + \left[\frac{v_{si}^2 k^2}{R^2}\delta_i - 4\pi G\rho_0 \sum_j \epsilon_j \delta_j\right] = 0, \tag{9.76}$$

where $\epsilon_j \equiv \rho_j/\rho_0$ is the fraction of the total mass density contributed by species j.[13]

Consider a two-component model with NR species i (e.g., i = baryons, or a weakly interacting, relic species) and photons during the radiation-dominated era. In this case $\dot{R}/R = 1/2t$, $\epsilon_j \ll 1$. First, consider the

[13]Since the derivation of (9.76) relied upon a Newtonian analysis, one might question whether (9.76) is valid in a radiation-dominated era. Equation (9.76) is valid for all sub-horizon-sized perturbations in any non-relativistic species, so long as the usual Friedmann equation for the expansion rate is used.

evolution of perturbations that are Jeans unstable, $k \ll k_J$. Supposing that the photons are smooth ($\delta_\gamma = 0$), (9.76) becomes

$$\ddot{\delta}_i + \frac{1}{t}\dot{\delta}_i = 0. \tag{9.77}$$

In this case the solution is $\delta_i(t) = \delta_i(t_i)[1 + a\ln(t/t_i)]$, so only a perturbation with an "initial velocity," $\dot{\delta}(t_i) \neq 0$, can actually grow, and only logarithmically at that.

Next, consider the evolution of small perturbations during a curvature-dominated epoch. When the Universe is curvature dominated, $R \propto t$ and $4\pi G\rho_0 = \frac{3}{2}\Omega(t)H^2 \ll t^{-2}$, so that (9.76) becomes

$$\ddot{\delta} + \frac{2}{t}\dot{\delta} + \frac{3}{2t^2}\Omega(t)\delta \simeq \ddot{\delta} + \frac{2}{t}\dot{\delta} = 0. \tag{9.78}$$

The solution in this regime is $\delta(t) = a_1 + a_2 t^{-1}$. That is, small perturbations cease growing and become frozen in.[14]

The growth of linear perturbations in both a radiation-dominated and curvature-dominated Universe is inhibited. The physical reason behind this fact is simple. The classical Jeans instability, which is characterized by exponential growth, is moderated by the expansion of the Universe. In a matter-dominated epoch, perturbations grow as a power law. In a radiation- or curvature-dominated epoch, the expansion rate is faster than what it would be if there were only matter present, and so the growth of perturbations is further moderated—in fact, quenched.

9.3.3 The baryon Jeans mass

Now let us consider the Jeans mass for baryonic perturbations. For simplicity, we will assume that the Universe consists only of photons ($g_* = 2$) and baryons. For such a model the unperturbed energy density is $\rho_0 = \rho_B + \rho_\gamma = mn_B + \pi^2T^4/15$. It is useful to express ρ_B in terms of the baryon number $B \equiv n_B/s = 3.8 \times 10^{-9}(\Omega_B h^2)$ and the entropy density $s = 4\pi^2T^3/45$: $\rho_B = msB = 4\pi^2 mBT^3/45$. The epoch of matter–radiation equality for this model occurs for $R_{\rm EQ} \simeq 4.3 \times 10^{-5}(\Omega_B h^2)^{-1}$,

[14]We leave as an exercise for the reader a matter-dominated, flat Universe with two matter components, one of which is smooth ($\epsilon_1 = \rho_1/\rho_{\rm TOT} = r$, $\delta_2 = 0$). In this case $\delta_1 \propto t^{\alpha_\pm}$, where $\alpha_\pm = -(1/6)[1 \pm (1 + 24r)^{1/2}]$; and the unstable mode grows more slowly than $t^{2/3}$.

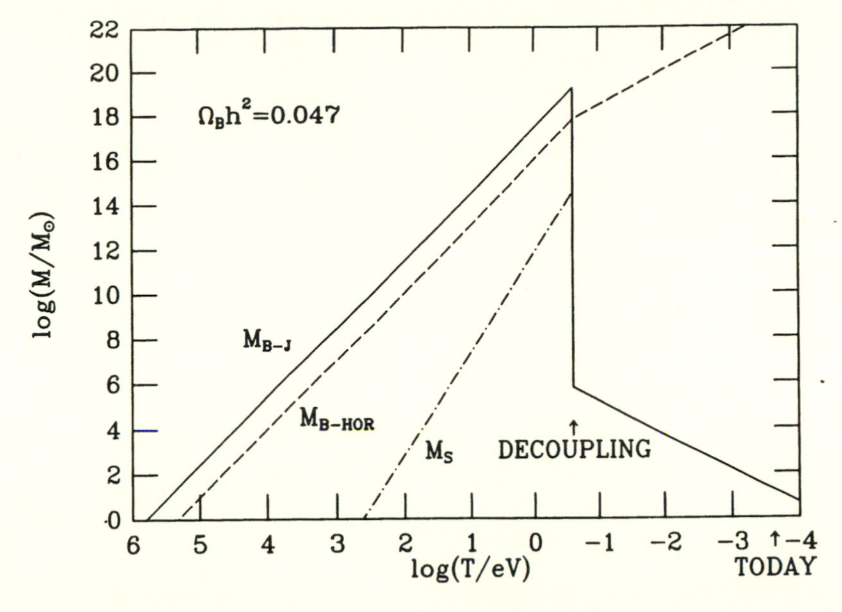

Fig. 9.3: The Jeans mass in baryons (solid line) and the mass in baryons within the horizon (dashed line) as a function of temperature for a baryon-dominated model where decoupling and matter–radiation equality occur simultaneously (i.e., $\Omega_0 = \Omega_B$ and $\Omega_B h^2 \simeq 0.047$). Also shown is the evolution of the Silk mass, $M_S = \pi\lambda_S(t)^3\rho_0/6$, where $\lambda_S(t)$ is given by (9.95).

while the epoch of recombination occurs for $R_{rec} \simeq 9.1 \times 10^{-4}$. For simplicity let us assume that recombination and the advent of matter domination occur simultaneously, which corresponds to $\Omega_B h^2 \simeq 0.047$.[15]

First, consider the Jeans mass in the radiation-dominated era (i.e., before recombination). Before recombination the baryons are tightly coupled to the photons. The fact that the Universe is radiation-dominated implies that the pressure in the baryon–photon plasma is supplied by the photons, and so $v_s^2 = 1/3$. The physical Jeans wavenumber is given by

$$k_{J-\text{phys}} = \frac{k_J}{R(t)} = \left(\frac{4\pi G\rho_0}{v_s^2}\right)^{1/2} = \left(\frac{4\pi^3}{5}\right)^{1/2} \frac{T^2}{m_{Pl}}. \tag{9.79}$$

[15]Unlike most of our readers, one of the authors can still remember when it was fashionable to talk about a baryon-dominated Universe.

The Jeans mass in *baryons* during the pre-recombination era is

$$M_{B-J} = \frac{4\pi}{3}\rho_B \left(\frac{\pi}{k_{J-\text{phys}}}\right)^3 \simeq 5.4 \times 10^{18} \, (\Omega_B h^2) T_{\text{eV}}^{-3} M_\odot, \tag{9.80}$$

where T_{eV} is the temperature in eV.[16]

Let us compare the baryon Jeans mass to the baryon mass within the horizon. The mass inside the horizon depends upon the baryon density and the distance to the horizon, $d_H \sim t$:

$$M_{B-HOR} \equiv \frac{4\pi}{3}\rho_B t^3 = \frac{t^3}{(\pi/k_{J-\text{phys}})^3} M_{J-B}. \tag{9.81}$$

Expressing $k_{J-\text{phys}}$ in terms of $t^{-1} = (32\pi G\rho_0/3)^{1/2}$,

$$\frac{\pi}{k_{J-\text{phys}}} = \left(\frac{8}{3}\right)^{1/2} \pi v_s t, \tag{9.82}$$

the baryon Jeans mass can be written in terms of the baryon mass within the horizon,

$$\frac{M_{B-J}}{M_{B-HOR}} = \left(\frac{8}{3}\right)^{3/2} (\pi v_s)^3 \simeq 26. \tag{9.83}$$

During the radiation-dominated era, the baryonic Jeans mass is larger than the mass of baryons within the horizon, and so the only potentially unstable modes cannot be treated with a Newtonian analysis.

After recombination, matter is decoupled from the radiation. The pressure support is provided only by non-relativistic hydrogen atoms, and the sound speed is: $v_s^2 = (5/3)(T_B/m) = (5/3)T^2/mT_{rec}$.[17] In the post-recombination era, $\rho_0 \simeq \rho_B = (4\pi^2/45)mBT^3$, and the Jeans mass is given by (9.64) with $\rho_0 = \rho_B$:

$$M_{B-J} \quad = \quad \frac{25}{12\sqrt{3}} \frac{\pi^{3/2}}{B^{1/2}} \frac{m_{Pl}^3}{m^2} \left(\frac{T}{T_{rec}}\right)^{3/2}$$

[16]Because we are ignoring the contribution of neutrinos to the mass density, the formulae in this Section differ from their counterparts in the Appendix by factors and powers of $g_*/2 \simeq 1.68$.

[17]Once the baryons lose thermal contact with the photons, the temperature of the baryons decreases as R^{-2}, rather than R^{-1}. Therefore, the temperature of the baryons, T_B, is related to the photon temperature by $T_B = (R_{rec}/R)T = T^2/T_{rec}$.

$$\simeq 1.3 \times 10^5 (\Omega_B h^2)^{-1/2} \left(\frac{z}{1100}\right)^{3/2} M_\odot. \qquad (9.84)$$

The precipitous decrease in M_{B-J} owes to the enormous decrease in pressure when the baryons decouple from the photons. This dramatic drop is crucial for structure formation: Only after recombination can fluctuations in the baryons on sub-horizon-sized scales begin to grow. Before recombination, the baryons cannot freely move through the photon plasma to collapse. The fact that the value of the baryon Jeans mass just after decoupling is about that of a globular cluster has intrigued cosmologists for decades. The evolution of the Jeans mass in baryons is shown in Fig. 9.3.

We have thus far assumed that the stress energy in the Universe is accurately described as a perfect fluid. This is not always a good approximation, and when it breaks down there can be dissipational effects. We will now consider two important such effects: collisionless damping (or free streaming) and collisional (or Silk) damping.

9.3.4 Collisionless damping

Perturbations in a nearly collisionless component (e.g., neutrinos, axions, etc.) are subject to Landau damping, also known as collisionless phase mixing or free streaming. Until perturbations become Jeans unstable and begin to grow when $t = t_{\rm EQ}$, collisionless particles can stream out of overdense regions and into underdense regions, in the process smoothing out inhomogeneities. This effect has not been taken into account thus far, as we have assumed a perfect fluid. In order to take this effect into account properly, one must integrate the Boltzmann equation that describes the collisionless component. Here we shall just estimate the scale of collisionless damping.

Once a species decouples from the plasma, it simply travels in free fall in the expanding Universe. As we have before, we may choose the particle's motion to be along $d\phi = d\theta = 0$, so that the motion of a freely propagating particle is simply $R(t)dr = v(t)dt$. The coordinate distance traversed by a free-streaming particle from some initial time t_i to time t is just

$$\lambda_{FS}(t) \equiv r(t) - r(t_i) = \int_{t_i}^{t} \frac{v(t')}{R(t')} dt'. \qquad (9.85)$$

Of course the physical distance traveled is simply $R(t)$ times the comoving distance. Note the similarity of this expression to that for the horizon distance, cf. (2.21) to (2.23).

We are interested in the free streaming that takes place before the Jeans instability becomes operative, at $t = t_{\rm EQ}$. Thus, the comoving free-streaming scale is given by

$$\lambda_{FS} = \int_0^{t_{\rm NR}} \frac{1}{R(t')}dt' + \int_{t_{\rm NR}}^{t_{\rm EQ}} \frac{v(t')}{R(t')}dt', \tag{9.86}$$

where we have split the integral into two pieces: the relativistic regime, when $v \simeq 1$; and the non-relativistic regime, when $v \lesssim 1$. Since the species is freely propagating, its momentum simply red shifts with the expansion, $p \propto R^{-1}$; once it becomes non-relativistic, $p = mv$ and so $v \propto R^{-1}$. Therefore, we can write λ_{FS} as

$$\lambda_{FS} \simeq 2\frac{t_{\rm NR}}{R_{\rm NR}} + \int_{t_{\rm NR}}^{t_{\rm EQ}} \frac{R_{\rm NR}}{R^2(t')}dt', \tag{9.87}$$

where we have assumed that the Universe is radiation-dominated while the particle is relativistic, i.e., $t_{\rm NR} \lesssim t_{\rm EQ}$; this is almost always the case. In the radiation-dominated era, $t = t_{\rm NR}(R/R_{\rm NR})^2$, so that

$$\begin{aligned}\lambda_{FS} &\simeq 2\frac{t_{\rm NR}}{R_{\rm NR}} + \int_{t_{\rm NR}}^{t_{\rm EQ}} \frac{R_{\rm NR}}{R^2(t')}dt' \\ &\simeq \left(\frac{t_{\rm NR}}{R_{\rm NR}}\right)\left[2 + \ln(t_{\rm EQ}/t_{\rm NR})\right].\end{aligned} \tag{9.88}$$

A species X becomes NR when $T_X \simeq m_X/3$, where we must remember that for a species that is weakly interacting, T_X is likely to be less than the photon temperature T. Using these facts we can calculate $t_{\rm NR}$ and $R_{\rm NR}$:

$$\begin{aligned}t_{\rm NR} &\simeq 1.2\times 10^7 \left(\frac{\rm keV}{m_X}\right)^2 \left(\frac{T_X}{T}\right)^2 {\rm sec}, \\ R_{\rm NR} &\simeq 7.1\times 10^{-7}\left(\frac{\rm keV}{m_X}\right)\left(\frac{T_X}{T}\right), \\ t_{\rm EQ}/t_{\rm NR} &\simeq \left[\frac{m_X}{17(\Omega_0 h^2)(T_X/T){\rm eV}}\right]^2,\end{aligned} \tag{9.89}$$

where T_X/T is the ratio of the temperature of species X to that of the photons; e.g., for a neutrino species $T_X/T = (4/11)^{1/3}$. It then follows that

$$\lambda_{FS} \simeq 0.2 \text{ Mpc} \left(\frac{m_X}{\text{keV}}\right)^{-1} \left(\frac{T_X}{T}\right) \left[\ln\left(\frac{t_{\text{EQ}}}{t_{\text{NR}}}\right) + 2\right]. \tag{9.90}$$

Using the fact $\Omega_X h^2 = 30(m_X/\text{keV})(T_X/T)^3$ for a two-component fermion species, the free-streaming length can be written in a very suggestive way:

$$\lambda_{FS} \simeq 30(\Omega_X h^2)^{-1}(T_X/T)^4 \text{ Mpc}, \tag{9.91}$$

where $\ln(t_{\text{EQ}}/t_{\text{NR}})$ has been set equal to 3. For a fixed value of $\Omega_X h^2$, earlier decoupling results in a smaller value for T_X/T and thereby a smaller free-streaming length.

Consider a light neutrino species; $T_\nu/T \simeq 0.71$, and so the characteristic scale for free streaming is

$$\lambda_{FS-\nu} \simeq 20 \text{ Mpc} \left(\frac{m_\nu}{30\text{eV}}\right)^{-1}, \quad M_{FS-\nu} \simeq 4 \times 10^{14} \left(\frac{m_\nu}{30\text{eV}}\right)^{-2} M_\odot, \tag{9.92}$$

where we have used $M(\lambda) \simeq 1.5 \times 10^{11} M_\odot(\Omega_0 h^2)\lambda_{Mpc}^3$ and $\Omega_\nu h^2 \simeq m_\nu/91$ eV. A more accurate determination of $\lambda_{FS-\nu}$ using the Boltzmann equation gives a number more like 40 Mpc for a neutrino mass of 30 eV [8]. The neutrino free-streaming mass can be written in terms of fundamental quantities as $M_{FS-\nu} \simeq m_{Pl}^3/m_\nu^2$, which is precisely the same form as for the Chandrasekhar mass ($m_N \rightarrow m_\nu$).

9.3.5 Collisional damping

Since baryons are not collisionless, they do not suffer collisionless damping. However, perturbations in the photon–baryon plasma do suffer damping around the time of recombination, because the assumption of a perfect fluid breaks down. As decoupling is approached, the photon mean free path becomes larger, and photons can diffuse out of overdense regions into underdense regions, thereby damping inhomogeneities in the photon–baryon plasma.[18] In this case the operative word is not free streaming, but rather photon diffusion, and the effect is known as Silk damping. A careful treatment of Silk damping requires using the Boltzmann equations

[18] The mean free path of the photons, $\lambda_\gamma = (n_e\sigma_T)^{-1}$, is always much larger than that of an electron, $\lambda_e = (n_\gamma\sigma_T)^{-1}$, since $n_\gamma \sim 10^{10} n_e$. Therefore, photon diffusion is much more important than matter diffusion, and the diffusing radiation "drags" the matter with it. Isocurvature baryon fluctuations are damped only by matter diffusion, since the associated radiation perturbations are small.

to follow the photons and matter through decoupling. Here, we will simply estimate the scale for Silk damping.

For $T \ll m_e$ the photon mean free path is

$$\lambda_\gamma \equiv (X_e n_e \sigma_T)^{-1} \simeq 1.3 \times 10^{29} X_e^{-1} R^3 (\Omega_B h^2)^{-1} \mathrm{cm}, \tag{9.93}$$

where σ_T is the Thomson cross section and X_e is the electron ionization fraction. Well before decoupling $X_e \simeq 1$; around the time of decoupling $X_e \sim 0.1$; thereafter, the residual ionization freezes in at a value, $X_e \simeq 3 \times 10^{-5}(\Omega_0^{1/2}/\Omega_B h)$. The perfect fluid approximation clearly breaks down for any perturbation of wavelength $\lambda \lesssim \lambda_\gamma$, and photon streaming should completely damp any such perturbation. At decoupling, $R \equiv 9.1 \times 10^{-4}$ and λ_γ corresponds to a comoving scale of about $0.03(\Omega_B h^2)^{-1}$ Mpc.

Moreover, in a time Δt, a photon suffers $\Delta t/\lambda_\gamma(t)$ collisions and thereby undergoes a random walk characterized by a mean coordinate distance squared of

$$(\Delta r)^2 \simeq \frac{\Delta t}{\lambda_\gamma(t)} \frac{\lambda_\gamma(t)^2}{R(t)^2}. \tag{9.94}$$

Summing up the mean coordinate distances squared until the time of decoupling, we obtain an estimate for the photon diffusion length squared,

$$\lambda_S^2 = \int_0^{t_{dec}} dt \frac{\lambda_\gamma}{R^2(t)}. \tag{9.95}$$

We shall define this coordinate distance squared to be the square of the Silk damping scale. (The evolution of $\lambda_S(t)$ as a function of time is shown in Fig. 9.3.)

Assuming that the Universe is matter-dominated around the epoch of decoupling (although not necessarily by baryons), the evolution of the scale factor is given by $R(t) = [t/2.1 \times 10^{17}(\Omega_0 h^2)^{-1/2}\,\mathrm{sec}]^{2/3}$. Using this fact we can evaluate λ_S^2:

$$\lambda_S^2 = \frac{3}{5} \frac{t_{dec} \lambda_\gamma(t_{dec})}{R_{dec}^2}; \tag{9.96}$$

that is, up to a factor of $\sqrt{3/5}$, the total diffusion length is just the characteristic photon diffusion length at the epoch of decoupling—essentially all of the Silk damping occurs at decoupling. As we discussed in Chapter 3, the precise epoch of decoupling and recombination depends upon Ω_0 and Ω_B. As we have throughout, we will take $R_{dec}^{-1} = 1100$, at which time

$X_e \sim 0.1$. Using these values we find

$$\begin{aligned}\lambda_S &= 3.5(\Omega_0/\Omega_B)^{1/2}(\Omega_0 h^2)^{-3/4}\,\mathrm{Mpc},\\ M_S &= 6.2\times 10^{12}(\Omega_0/\Omega_B)^{3/2}(\Omega_0 h^2)^{-5/4}\,M_\odot. \end{aligned} \tag{9.97}$$

More accurate estimates of the Silk mass are a factor of about 3 smaller [9]. It is intriguing that the Silk scale is close to that of a cluster of galaxies.

We should emphasize that the scales we have calculated here are somewhat arbitrary. For instance, does one define the Jeans mass as the mass contained within a sphere of radius $2\pi/k_J$, $1/k_J$, or π/k_J, etc? Since the Jeans mass is proportional to the cube of the length chosen, conventions concerning 2's, 2π's, etc. can be very important. The free-streaming scale and the Silk scale are merely estimates for the scales below which there is appreciable (exponential) damping. A definitive analysis of the damping and the growth of perturbations is best addressed by numerical simulation, and the results of some of these simulations will be reviewed in the next two Sections.

9.3.6 Small fluctuations in general relativity

The analysis of the evolution of density perturbations to this point has been Newtonian. For modes that are well within the horizon, $\lambda_{\rm phys} \ll H^{-1}$, the Newtonian analysis suffices (and, as one would expect, is exact in the limit $\lambda_{\rm phys}/H^{-1} \to 0$). The Newtonian analysis clearly has its limitations; on distance scales $r \gtrsim H^{-1}$, the Newtonian expansion velocity exceeds c and ϕ_0 exceeds c^2; and retardation effects have not been taken into account in computing ϕ. In short, the analysis goes to hell for $\lambda_{\rm phys} \gg H^{-1}$! Further, as we discussed in the previous Section, for modes well within the horizon a density perturbation is a "density perturbation," and the gauge ambiguities associated with the gauge non-invariance of $\delta\rho/\rho$ do not cloud the issue. However, for super-horizon-sized modes, the gauge non-invariance of $\delta\rho/\rho$ must be reckoned with, and the difference between isocurvature and curvature type perturbations is very important.

As we emphasized in the previous Section, it is very natural to divide the discussion of the evolution of a density perturbation into two regimes. (1) The early regime, when the perturbation is super-horizon sized, $\lambda_{\rm phys} \gtrsim H^{-1}$ or $k/HR \lesssim 1$, and microphysics (e.g., pressure forces) are impotent. While subtleties arise due to the gauge non-invariance of $\delta\rho/\rho$, the evolution of the perturbation in this regime is basically kinematic—the evolution of a ripple in the fabric of space-time. We shall see that the evolution in this regime can be summarized very simply: the constancy of a gauge-invariant quantity $\zeta \equiv \delta\rho/(\rho_0 + p_0)$. (2) The late regime, when the perturbation is sub-horizon sized, $\lambda_{\rm phys} \lesssim H^{-1}$ or $k/HR \gtrsim 1$, and microphysics is important. During the sub-horizon sized regime, the Newtonian analysis discussed above is applicable. Fig. 9.1 illustrates the evolution of the physical size of a perturbation, showing its size relative to the Hubble radius decreasing with time.

In order to treat the evolution of a density perturbation during the early phase one needs a full, general-relativistic analysis. To begin, let us illustrate in a simple, geometric way the behavior of perturbations outside the horizon by considering perturbations of the spatially flat ($k = 0$) FRW model. Recall that the Friedmann equation for the unperturbed $k = 0$ model is

$$H^2 = 8\pi G\rho_0/3 \qquad (k = 0). \tag{9.98}$$

Now consider a similar FRW model, one with the same expansion rate H, but one that has higher density, $\rho = \rho_1$, and is therefore positively curved:[19]

$$H^2 = \frac{8\pi G\rho_1}{3} - \frac{k}{R^2} \qquad (k > 0). \tag{9.99}$$

We shall always compare these two models when their expansion rates are equal—this choice is called the uniform Hubble flow condition. We immediately see that the density contrast between the two models is given in terms of the curvature of the closed model by

$$\delta \equiv \frac{\rho_1 - \rho_0}{\rho_0} = \frac{k/R^2}{8\pi G\rho_0/3}. \tag{9.100}$$

The evolution of δ has been reduced to that of the curvature k/R^2 relative to the energy density ρ_0. So long as δ is small, equivalently $k/R^2 \lesssim 8\pi G\rho_0$, the scale factors for the two models are essentially equal (fractional difference of order δ). In a matter-dominated Universe, $\rho \propto R^{-3}$, while in a radiation-dominated Universe $\rho \propto R^{-4}$, so

$$\delta \propto \frac{R^{-2}}{\rho} \propto \begin{cases} R^2 & \text{radiation dominated} \\ R & \text{matter dominated.} \end{cases} \tag{9.101}$$

---

[19] Of course, the same analysis can also be applied to an underdense model, in which case the spatial curvature is negative.

Recall that in a matter-dominated Universe $R \propto t^{2/3}$, while in a radiation-dominated Universe $R \propto t^{1/2}$, so

$$\delta = \delta_i \begin{cases} t/t_i & \text{radiation dominated} \\ (t/t_i)^{2/3} & \text{matter dominated.} \end{cases} \tag{9.102}$$

This simple model illustrates several important points about the evolution of super-horizon-sized perturbations. (1) The geometric character of a density perturbation, and why honest-to-God density perturbations are referred to as curvature fluctuations. (2) That what is actually relevant is the difference in the evolution of the perturbed model as compared to an unperturbed, reference model. (3) That the freedom in the choice of the reference model is equivalent to a gauge choice, so that in general, δ will be gauge dependent. As usual, when confronted with gauge ambiguity, the correct thing to do is to ask a physical question, one whose answer cannot depend upon the gauge. Here, the perturbed and reference models are compared by matching their expansion rates.

To formulate a more quantitative picture of the evolution of perturbations outside the horizon it is necessary to solve the perturbed Einstein equations. This program was started (and essentially finished) in 1946 by E. M. Lifshitz [4]. Here we will simply outline the steps involved, state the key results, and refer the reader to several excellent sources for details [1, 5]. The Lifshitz analysis relies upon selecting a gauge, finding the solutions in that gauge, and then identifying the gauge modes. Another approach is to formulate the problem in terms of gauge-invariant quantities; Bardeen [6] has developed such a formalism, and we refer the interested reader to his important work.

In the Lifshitz analysis the metric $g_{\mu\nu}$ is taken to be the spatially flat FRW metric, $g^0_{\mu\nu}$, as a background, plus small metric perturbations $h_{\mu\nu}$:[20]

$$g_{\mu\nu} = g^0_{\mu\nu} + h_{\mu\nu}. \tag{9.103}$$

In addition, the gauge freedom can be exploited to choose a synchronous gauge: $h_{00} = h_{i0} = 0$.[21] If we are interested only in scales larger than the horizon, microphysical processes such as shear, heat conduction, and bulk viscosity can be ignored, and the perturbed stress tensor is described simply in terms of the perturbed energy density, $\rho = \rho_0 + \rho_1$, pressure,

[20]In Chapter 2 we defined $h_{ij} = -g_{ij}$ to be the spatial part of the FRW metric. Here, $h_{\mu\nu}$ is the metric perturbation. The two different definitions should not be confused.

[21]This, unfortunately, does not exhaust all of the gauge freedom.

$p = p_0 + p_1$, and matter velocity $U^\mu = U_0^\mu + U_1^\mu$.[22] We will again assume that the equation of state is everywhere the same; i.e., we will not consider isocurvature fluctuations, only curvature fluctuations. The perturbation equation that must be solved is $\delta R_{\mu\nu} - (1/2)\delta\left[g_{\mu\nu}\mathcal{R}\right] = 8\pi G\delta T_{\mu\nu}$, or equivalently

$$\delta R_{\mu\nu} = 8\pi G\delta T_{\mu\nu} - 4\pi G\delta\left[g_{\mu\nu}\mathcal{T}\right], \tag{9.104}$$

where $\mathcal{T}$ is the trace of the stress tensor, $\mathcal{T} = \rho - 3p = (\rho_0 - 3p_0) + (\rho_1 - 3p_1)$. Now we will outline the steps involved in solving the perturbed Einstein equations.

(1) Calculate $\delta\Gamma^\mu_{\nu\alpha}$ to first order in $h_{\mu\nu}$. For example, recalling that for the unperturbed FRW metric $\Gamma^0_{ij} = -(1/2)\partial g_{ij}/\partial t$, it follows that

$$\delta\Gamma^0_{ij} = -\frac{1}{2}\frac{\partial h_{ij}}{\partial t}. \tag{9.105}$$

(2) Calculate $\delta R_{\mu\nu}$ to first order in $h_{\mu\nu}$. For example,

$$\delta R_{00} = \frac{1}{2}\ddot{h} + \frac{\dot{R}}{R}\dot{h}, \tag{9.106}$$

where $h \equiv h^\mu{}_\mu = h^k{}_k = -\sum_{k=1}^3 h_{kk}/R^2$ is the trace of the metric perturbation. As expected, $\delta R_{\mu\nu}$ involves second derivatives and products of first derivatives. Recall that for the unperturbed FRW metric $R_{00} = -3(\ddot{R}/R)g_{00}$.

(3) Calculate $\delta T_{\mu\nu}$. To first order

$$\delta T_{\mu\nu} = -p_1 g^0_{\mu\nu} + (\rho_1 + p_1)U_{0\mu}U_{0\nu} - p_0 h_{\mu\nu} + (\rho_0 + p_0)(U_{0\mu}U_{1\nu} + U_{1\mu}U_{0\nu}). \tag{9.107}$$

(4) Calculate $\delta[g_{\mu\nu}\mathcal{T}]$:

$$\delta\left[g_{\mu\nu}\mathcal{T}\right] = g^0_{\mu\nu}(\rho_1 - 3p_1) + h_{\mu\nu}(\rho_0 - 3p_0). \tag{9.108}$$

(5) Substitute the results from (1)–(4) into (9.104). The result is three equations: the 00, the ij, and the $0i$ components.[23] For example, the 00

[22] The first-order perturbation of the identity $g_{\mu\nu}U^\mu U^\nu = 1$ gives $g_{00}U_0^0 U_1^0 = 0$ when we recall that $U_0^\mu = (1,\ 0,\ 0,\ 0)$, and so $U_1^0 = 0$. Therefore, U_1^μ has only non-vanishing spatial components, denoted by $\vec{\mathbf{U}}_1$.

[23] Recall that for the unperturbed FRW metric, only two independent equations follow from the field equations; the $0i$ component gives $0 = 0$.

component of (9.104) gives

$$\ddot{h} + 2H\dot{h} = 8\pi G(\rho_1 + 3p_1). \tag{9.109}$$

As in the analysis of the FRW models, it is useful to use energy-momentum conservation, $T^{\mu\nu}{}_{;\nu} = 0$, in place of one of the field equations. To first order, the perturbed equation of energy-momentum conservation gives

$$\dot{\rho}_1 + 3H(\rho_1 + p_1) + (\rho_0 + p_0)\left[-\dot{h}/2 + \vec{\nabla}\cdot\vec{\mathbf{U}}_1\right] = 0. \tag{9.110}$$

Equations (9.104) to (9.110) govern the first-order perturbations of the metric.

The metric tensor can always be decomposed into its trace, a transverse, traceless tensor, and a vectorial piece. Thus, the complete set of coupled equations above describes three types of perturbations. For example, the perturbations in the transverse, traceless part of $h_{\mu\nu}$ describe the propagation of gravitational waves, and these modes do not couple to $\vec{\mathbf{U}}_1$, ρ_1, or p_1.[24] The vectorial part of $h_{\mu\nu}$ couples only to the rotational part of the matter velocity $\vec{\mathbf{U}}_1$, where just as in the Newtonian case, $\vec{\mathbf{U}}_1 = \vec{\mathbf{U}}_\perp + \vec{\mathbf{U}}_\parallel$. The evolution of $\vec{\mathbf{U}}_\perp$ is given by $|\vec{\mathbf{U}}_\perp| \propto [(1+v_s^2)R^4\rho_0]^{-1}$, which for $\rho_0 \propto R^{-3}$, reduces to the previous Newtonian result. The vectorial part of $h_{\mu\nu}$ is given by an integral over $\vec{\mathbf{U}}_\perp(\vec{\mathbf{x}},t)$.

Of interest to us are the scalar (density) perturbations, which correspond to perturbations in the trace of the metric perturbation (defined as h). These perturbations couple to ρ_1, p_1, and $\vec{\mathbf{U}}_\parallel$. For brevity, we will hereafter omit the subscript $\parallel$, and use $\vec{\mathbf{U}}_1$ to denote $\vec{\mathbf{U}}_\parallel$. As before, we will express the perturbed quantities in a Fourier expansion

$$h,\ \rho_1,\ p_1,\ \vec{\mathbf{U}}_1 \propto \exp(-i\vec{\mathbf{k}}\cdot\vec{\mathbf{x}}). \tag{9.111}$$

For the density perturbation (scalar) modes the perturbed Einstein equations are

$$\ddot{h} + 2H\dot{h} - 3H^2(1+3v_s^2)\delta = 0,$$

[24]It is straightforward to show that the transverse, traceless (TT) components of $h_{\mu\nu}$ satisfy the same equation of motion as a minimally coupled scalar field: $\ddot{\psi} + 3H\dot{\psi} + k^2\psi/R^2 = 0$, where $\psi = (h^{TT})^i{}_j$.

$$\dot{\delta} + (1 + p_0/\rho_0)(\theta - \dot{h}/2) + 3H(v_s^2 - p_0/\rho_0)\delta = 0,$$
$$\dot{\theta} + (2 - 3v_s^2)H\theta - \frac{k^2 v_s^2}{R^2(1 + p_0/\rho_0)}\delta = 0, \tag{9.112}$$

where $\theta \equiv \vec{\nabla} \cdot \vec{\mathbf{U}}_1 = -i\vec{\mathbf{k}} \cdot \vec{\mathbf{U}}_1$, and $\delta = \rho_1/\rho_0$. These equations may be greatly simplified by assuming that $p_0/\rho_0 = v_s^2 \equiv p_1/\rho_1$, defining $\phi \equiv \theta/H$, and using $y \equiv \ln R$ as the time variable:

$$h'' + \frac{1}{2}(1 - 3v_s^2)h' - 3(1 + 3v_s^2)\delta = 0,$$
$$\delta' + (1 + v_s^2)(\phi - h'/2) = 0,$$
$$\phi' - \frac{1}{2}(9v_s^2 - 1)\phi = 0, \tag{9.113}$$

where prime denotes $d/dy = H^{-1}d/dt$, and the last term in (9.112) has been neglected, as it is sub-dominant for super-horizon-sized modes.

This set of equations is equivalent to a single, fourth-order equation. Therefore, we expect four independent solutions. This is somewhat surprising since our Newtonian analysis indicated that there are only two physical modes: one growing and one decaying. However, not all of the four modes are physical—two correspond to pure gauge modes (the residual gauge freedom in synchronous gauge).

To illustrate a gauge mode, consider solutions where $\delta = \phi = 0$ and h_{ij} is defined in terms of $R^2(t)$ and spatial derivatives of $\xi(\vec{\mathbf{x}})$, with $\xi(\vec{\mathbf{x}})$ a function of position only:

$$h_{ij}(\vec{\mathbf{x}}, t) = R^2(t)\left[\xi_{i,j} + \xi_{j,i}\right]. \tag{9.114}$$

We will now show that such a perturbation does not describe a physical perturbation, and that such a perturbation can be removed by a simple coordinate transformation of the unperturbed metric, $g^0_{\mu\nu}$, without loss of synchronicity. To wit, consider a coordinate transformation $x^\mu \to x'^\mu = x^\mu - \epsilon^\mu(x)$. For small ϵ^μ,

$$\frac{\partial x'^\mu}{\partial x^\nu} = \delta^\mu{}_\nu - \frac{\partial \epsilon^\mu(x)}{\partial x^\nu},$$
$$\frac{\partial x^\nu}{\partial x'^\mu} = \delta^\nu{}_\mu + \frac{\partial \epsilon^\nu(x)}{\partial x^\mu} + \cdots. \tag{9.115}$$

The coordinate transformation will result in a new metric $g'_{\mu\nu}(x')$. For small ϵ^μ we can expand $g'_{\mu\nu}(x)$ about $g'_{\mu\nu}(x')$:

$$g'_{\mu\nu}(x) = g'_{\mu\nu}(x') + \frac{\partial g_{\mu\nu}(x)}{\partial x^\lambda}\epsilon^\lambda(x) + \cdots . \tag{9.116}$$

The new metric $g'_{\mu\nu}(x')$ is related to $g_{\mu\nu}(x)$ by the usual transformation rules for a tensor,

$$\begin{aligned} g'_{\mu\nu}(x) &= g_{\mu\nu}(x) + g_{\lambda\nu}(x)\frac{\partial\epsilon^\lambda(x)}{\partial x^\mu} + g_{\lambda\mu}(x)\frac{\partial\epsilon^\lambda(x)}{\partial x^\nu} \\ &\quad + \frac{\partial g_{\mu\nu}(x)}{\partial x^\lambda}\epsilon^\lambda(x) + \cdots \\ &= g_{\mu\nu}(x) + \epsilon_{\mu;\nu} + \epsilon_{\nu;\mu}. \end{aligned} \tag{9.117}$$

Since $h_{\mu\nu} = g_{\mu\nu}(x) - g^0_{\mu\nu}(x)$, we can see that the "perturbation" (9.114) is not a true physical perturbation but rather simply represents a coordinate transformation of the unperturbed metric given by $\epsilon^0 = 0$ and $\epsilon^i(\vec{x}, t) = R^2(t)\xi^i(\vec{x})$. This is an example of a gauge mode.

The set of equations (9.113) for $[\delta,\ \phi\ , h,\ h']$ has four power-law solutions,

$$[\delta,\ \phi,\ h,\ h'] \propto \chi_i t^{\lambda_i}, \tag{9.118}$$

where $\lambda_1 = 0$, $\lambda_2 = -1$, $\lambda_3 = (2 + 6v_s^2)/(3 + 3v_s^2)$, and $\lambda_4 = (9v_s^2 - 1)/(3 + 3v_s^2)$, and the χ_i are given by

$$\begin{aligned} \chi_1 &= [0,\ 0,\ 1,\ 0], \\ \chi_2 &= [(1+v_s^2)/2,\ 0,\ 1,\ -3(1+v_s^2)/2], \\ \chi_3 &= [(1+v_s^2)/2,\ 0,\ 1,\ (1+3v_s^2)], \\ \chi_4 &= [v_s^2(1+v_s^2)(9v_s^2-1),\ (3v_s^2+1/2)(1-v_s^2)(9v_s^2-1), \\ &\quad\ 2(1+3v_s^2)(1+v_s^2),\ (1+3v_s^2)(1+v_s^2)(9v_s^2-1)]. \end{aligned} \tag{9.119}$$

For the matter-dominated case ($v_s^2 = 0$),

$$\begin{array}{ll} \lambda_1 = 0, & \chi_1 = [0,\ 0,\ 1,\ 0], \\ \lambda_2 = -1, & \chi_2 = [1/2,\ 0,\ 1,\ -3/2], \\ \lambda_3 = 2/3, & \chi_3 = [1/2,\ 0,\ 1,\ 1], \\ \lambda_4 = -1/3, & \chi_4 = [0,\ -1/2,\ 2,\ -1]; \end{array} \tag{9.120}$$

while for the radiation-dominated case ($v_s^2 = 1/3$),

$$\begin{aligned} &\lambda_1 = 0, && \chi_1 = [0,\ 0,\ 1,\ 0], \\ &\lambda_2 = -1, && \chi_2 = [2/3,\ 0,\ 1,\ -2], \\ &\lambda_3 = 1, && \chi_3 = [2/3,\ 0,\ 1,\ 2], \\ &\lambda_4 = 1/2, && \chi_4 = [8/9,\ 2,\ 16/3,\ 16/3]. \end{aligned} \tag{9.121}$$

The first two modes are pure gauge modes. Therefore, in the synchronous gauge, the physical growing and decaying modes for super-horizon sized perturbations are

$$\begin{aligned} &\delta_+(t) = \delta_+(t_i)(t/t_i)^{2/3}, && \delta_-(t) = \delta_-(t_i)(t/t_i)^{-1/3} && (v_s^2 = 0), \\ &\delta_+(t) = \delta_+(t_i)(t/t_i), && \delta_-(t) = \delta_-(t_i)(t/t_i)^{1/2} && (v_s^2 = 1/3). \end{aligned} \tag{9.122}$$

Note that these are the same growing-mode solutions that we obtained in our uniform Hubble flow analysis.

So far we have neglected perturbations in the the equation of state, i.e., isocurvature (or isothermal) perturbations. Such perturbations do not grow while they are outside the horizon—after all, they are not honest-to-God wrinkles in space-time. Of course, we should remember that this is a gauge-dependent statement, true in synchronous gauge. What is true and gauge-invariant is that an isocurvature perturbation eventually becomes a density perturbation of the same amplitude after horizon crossing (see the previous Section).

Finally, we should mention again the very useful gauge-invariant quantity introduced in Chapter 8, $\zeta \equiv \delta\rho/(\rho_0 + p_0)$.[25] The evolution of ζ is particularily simple and *independent* of the background space-time. For super-horizon-sized modes the evolution is: ζ = const for $\lambda_{\rm phys} \gtrsim H^{-1}$. In discussing the amplitude of the inflation-produced, adiabatic density perturbations, we used the constancy of ζ to compute the amplitude of a given mode when it re-enters the horizon: When a mode re-enters the horizon, the value of ζ, which is equal to $(1+w)^{-1}(\delta\rho/\rho)_{\rm HOR}$, is given by the value of ζ when the mode crossed outside the horizon during inflation. The evolution of ζ embodies the simplicity that we expected to be associated with the "kinematic" evolution of super-horizon-sized perturbations.

[25]Since $\delta\rho$ is not gauge-invariant, ζ is not manifestly gauge-invariant; ζ takes on this simple form at horizon crossing in the uniform Hubble constant gauge, see [7].

9.3.7 Executive summary

To finish, let us sumarize the highlights of this rather lengthy review of the standard lore.

- For sub-horizon-sized perturbations a Newtonian treatment of the evolution of perturbations suffices. There are two physical modes, a growing and a decaying mode. During a radiation- or curvature-dominated epoch, perturbations do not grow. During a matter-dominated epoch perturbations on scales larger than the Jeans scale grow as $\delta \propto R \propto t^{2/3}$. Perturbations on scales smaller than the Jeans scale oscillate as acoustic waves, and depending upon the evolution of the sound speed, the amplitude of the acoustic wave may decrease. Before recombination, the baryon Jeans mass is larger than the horizon mass, by a factor of about 30; after recombination the baryon Jeans mass drops precipitously to about $10^5 M_\odot$.

- As we discussed in Section 9.2, there are two types of perturbations: *curvature* (or *adiabatic*) perturbations and *isocurvature* (or *isothermal*) perturbations. For super-horizon-sized modes the distinction is very important: Curvature perturbations correspond to honest-to-God energy density fluctuations (fluctuations in the spatial curvature), while isocurvature perturbations do not and are characterized by $\delta\rho = 0$. Isocurvature fluctuations correspond to local variations in the equation of state, i.e., pressure variations, that eventually lead to density perturbations of the same amplitude once they enter the horizon. Unlike their adiabatic counterparts, super-horizon-sized isothermal perturbations do not grow whilst they are outside the horizon.

- Collisionless phase mixing damps perturbations on scales smaller than the free-streaming scale, $\lambda_{FS} \simeq few \times (t_{NR}/R_{NR})$, where the subscript NR denotes the value of the quantity at the epoch when the species became non relativistic. Taking the particle's mass to be m_X and the ratio of its temperature to the photon temperature to be T_X/T, the free-streaming scale is roughly $\lambda_{FS} \simeq 1$ Mpc $(m_X/\text{keV})^{-1}(T_X/T)$.

- Due to photon diffusion, adiabatic fluctuations in the baryons are strongly damped on scales smaller than the Silk scale. This damping occurs primarily just as the photons and baryons decouple. The Silk scale is given by $\lambda_S \simeq 3.5(\Omega_0/\Omega_B)^{1/2}(\Omega_0 h^2)^{-3/4}$ Mpc.

- For modes that are super-horizon sized the subtleties of the gauge non-invariance of $\delta\rho$ are important, and a full general relativistic treatment is required. There are two physical modes, a decaying mode and a growing mode, and two pure gauge modes. In synchronous gauge, the growing mode evolves as $\delta \propto R^2 \propto t$ (radiation dominated) and $\delta \propto R \propto t^{2/3}$ (matter dominated). Alternatively, the evolution of super-horizon-sized modes can be described by the constancy of the quantity $\zeta = \delta\rho/(\rho_0 + p_0)$.

9.4 The Spectrum of Density Perturbations

We know how to describe the spectrum of density perturbations and how a given mode evolves. Now the question arises as to when we should specify the spectrum. From our discussion of the evolution of density perturbations, it seems very sensible to specify the amplitude of a given mode just as it crosses inside the horizon, before any microphysical processing can occur. Specifying the amplitude of a mode at horizon crossing not only evades the subtleties associated with the gauge non-invariance of $\delta(\vec{x}, t)$, but also has a simple Newtonian interpretation: The amplitude at horizon crossing directly corresponds to the amplitude of the perturbation in the gravitational potential, cf. (9.68). Moreover, for $\lambda_{\rm phys} \gtrsim H^{-1}$, in the synchronous gauge, δ_k evolves in such a way that $\delta\phi_k$ remains constant. Of course, when one specifies the fluctuation spectrum by the amplitude of each mode at horizon crossing, the amplitudes for different modes are specified at *different* times. However, it is easy to relate the spectrum at a fixed time to that at horizon crossing.

In the absence of a definite model for the primeval fluctuations, one often supposes that the spectrum is a featureless power law,

$$\left(\frac{\delta\rho}{\rho}\right)_{\rm HOR} \simeq \frac{k^{3/2}|\delta_k|}{\sqrt{2}\pi} = A M^{-\alpha}, \tag{9.123}$$

where α and n ($|\delta_k|^2 \propto k^n$) are related by $\alpha = 1/2 + n/6$. As we discussed in the previous Chapter, the inflationary prediction is the Harrison-Zel'dovich spectrum of constant-curvature perturbations: $\alpha = 0$ ($n = -3$).

Aside from the inflationary prediction, what can we say about α? Roughly speaking, the formation of galaxies requires that $(\delta\rho/\rho)_{HOR} \simeq 10^{-4\pm1}$ on the scale of a galaxy, i.e., $10^{12} M_\odot$. The measured isotropy of the CMBR constrains $(\delta\rho/\rho)_{HOR}$ on the scale of the present horizon (about $10^{22} M_\odot$) to be less than about 10^{-4}. If the primeval spectrum has no long wavelength cutoff, then these two facts imply that the slope α must be greater than about -0.1.

Now consider the perturbations on scales much less than $10^{12} M_\odot$. If a perturbation crosses the horizon with amplitude greater than order unity, black hole formation is inevitable. Overdense regions will behave like small closed-Universe models and will recollapse before pressure forces can respond to prevent black hole formation. Black holes less massive than order 10^{15}g will have evaporated via the Hawking process before the present epoch. However, primordial black holes more massive than 10^{15}g will still be with us today, and those of mass 10^{15}g will be evaporating now and radiating γ rays. If $(\delta\rho/\rho)_{HOR} \gtrsim \mathcal{O}(1)$ on a scale larger than 10^{15}g, there would be far too many black holes with us today to be consistent with the observed γ background and the mass density of the Universe. These facts constrain α to be less than about 0.2, provided that the spectrum is not cut off at short wavelengths. Since the long and short wavelength constraints restrict α to be in the small interval $[-0.1,\ 0.2]$, the $\alpha = 0$, constant-curvature spectrum is clearly singled out (if the primeval spectrum is featureless).

While it is most natural to specify the perturbation amplitude at horizon crossing, it is sometimes most convenient to specify the spectrum of perturbations at a single, fixed time, e.g., at $t = t_{\rm EQ}$ when structure formation commences. To discuss the relationship between $(\delta\rho/\rho)_t$ and $(\delta\rho/\rho)_{\rm HOR}$ for modes that are still outside the horizon at time t, one must, of course, specify a gauge. The most commonly used gauge is the synchronous gauge (used in the previous Section). Super-horizon-sized modes at time t have yet to achieve their horizon-crossing amplitude. Using the fact that $\delta\rho/\rho$ grows as R^n (n = 1-matter dominated and n = 2-radiation dominated) and the fact that the mass in non-relativistic particles within the horizon $M_{\rm HOR}$ grows as R^m ($m = 3/2$–matter dominated, and $m = 3$–radiation dominated), it is simple to show that

$$(\delta\rho/\rho)_t = \left(\frac{M}{M_{\rm HOR}(t)}\right)^{-2/3} (\delta\rho/\rho)_{\rm HOR}. \tag{9.124}$$

Thus, for scales that are still outside the horizon at time t,

$$(\delta\rho/\rho)_t = AM^{-\alpha-2/3}. \tag{9.125}$$

Note for the constant-curvature spectrum ($\alpha = 0$), $(\delta\rho/\rho)_t \propto M^{-2/3}$.

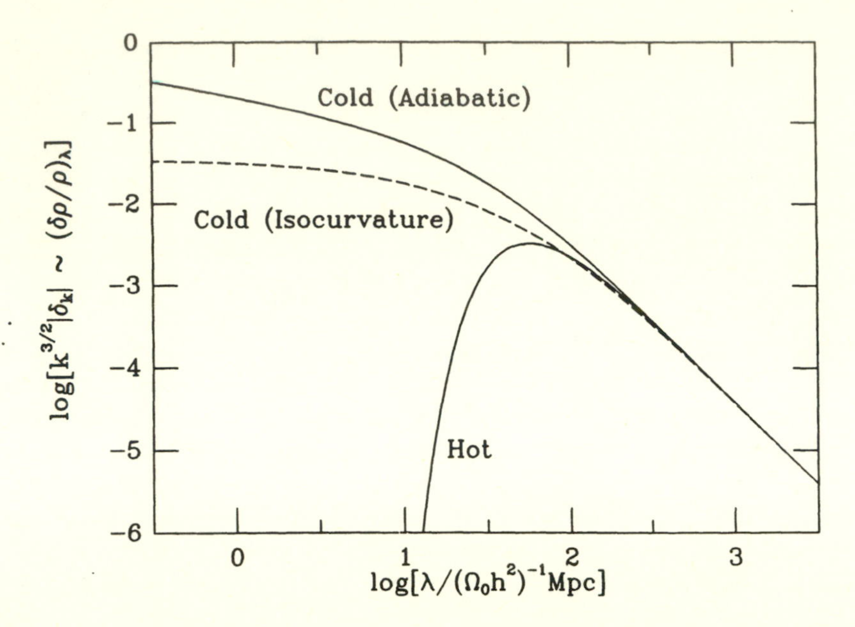

Fig. 9.4: The fully processed spectrum, $k^{3/2}|\delta_k|$, specified at some time after the epoch of matter–radiation equality, for hot and cold dark matter with constant-curvature, adiabatic fluctuations; and for cold dark matter with isocurvature, Harrison-Zel'dovich fluctuations, cf. (9.127, 9.128). For all three cases, $k^{3/2}|\delta_k| \propto k^2 \propto M^{-2/3}$ for $\lambda \gg \lambda_{EQ}$. The overall normalization is arbitrary.

The relationship between $(\delta\rho/\rho)_t$ and $(\delta\rho/\rho)_{\rm HOR}$ for scales that are sub-horizon sized at time t depends upon any microphysical processing that may have taken place once they enter the horizon. Given the primeval spectrum, i.e., $(\delta\rho/\rho)_{HOR}$, we can use the results of the previous Section to infer the processed spectrum.

A particularly convenient time to specify the processed spectrum is at $t \simeq t_{\rm EQ}$, when structure formation begins. For sake of a concrete example, let us assume that the primeval spectrum is the Harrison-Zel'dovich spectrum. (If not, the slopes of the various regions of the processed spectrum can be obtained by changing those given here by α.) Scales that cross inside the horizon before $t_{\rm EQ}$, i.e., those with

$$\lambda \lesssim \lambda_{\rm EQ} \simeq \frac{ct_{\rm EQ}}{R_{\rm EQ}} \simeq 13(\Omega_0 h^2)^{-1}\ {\rm Mpc}, \qquad (9.126)$$

at best grow logarithmically from horizon-crossing until $t \simeq t_{\rm EQ}$. On scales greater than $\lambda_{\rm EQ}$, $(\delta\rho/\rho)_{\rm EQ} = M^{-2/3}(\delta\rho/\rho)_{\rm HOR}$, as discussed above. Finally, we need to take into account the damping of sub-horizon-sized perturbations by the streaming of matter particles out of the perturbations before gravity acts to hold them together. For relic WIMPs, the damping scale is given by the free-streaming scale, $\lambda_D \simeq \lambda_{\rm FS}$; while for baryons, the damping scale is given by the Silk scale, $\lambda_D \simeq \lambda_S$. Roughly speaking then, the processed spectrum is characterized by $(\delta\rho/\rho)_{\rm EQ} \simeq 0$ for $\lambda \lesssim \lambda_D$; $(\delta\rho/\rho)_{\rm EQ} \simeq$ const $\times$ $(\delta\rho/\rho)_{\rm HOR}$ for $\lambda_D \lesssim \lambda \lesssim \lambda_{\rm EQ}$; and $(\delta\rho/\rho)_{\rm EQ} = (\delta\rho/\rho)_{\rm HOR}(\lambda/\lambda_{\rm EQ})^{-2}$ for $\lambda \gtrsim \lambda_{\rm EQ}$. The processed spectrum is shown in Fig. 9.4.

From the epoch of matter domination until decoupling, perturbations in relic WIMPs can grow ($\delta\rho/\rho \propto R \propto t^{2/3}$); however, perturbations in the baryons cannot, since they are still tightly coupled to the photons. After decoupling, the baryons are free of the pressure support provided by the photons, and they quickly fall into the potential wells formed by the WIMP perturbations. By a few expansion times after decoupling, the baryon perturbations catch up with the WIMP perturbations (provided that $\Omega_{\rm WIMP} \gg \Omega_B$), and then perturbations in both components grow together (see Fig. 9.5).

After matter–radiation equality, perturbations on all scales grow together, $\delta\rho/\rho \propto t^{2/3}$; and thereafter, the shape of the spectrum remains the same, with its overall amplitude increasing as $R(t)$. If $\Omega_0 < 1$, the growth of density perturbations ceases when $R \simeq \Omega_0$. When $(\delta\rho/\rho)$ becomes of order unity on a scale, perturbations of that mass form bound systems that separate from the general expansion. As we discussed in Section 9.2, in the context of the spherical-collapse model, perturbations "break away" from the general expansion when $\delta \simeq 4.6$,[26] and they collapse further during virialization, by about a factor of 2, increasing their density by a factor of about 8. For the Harrison-Zel'dovich spectrum (or any spectrum that decreases with increasing mass scale), the first scale on which structures form is set by λ_D (see Fig. 9.4).

There are two limiting cases; the first is when $\lambda_D \simeq \lambda_{\rm EQ} \simeq 13(\Omega_0 h^2)^{-1}$ Mpc, which occurs for relic WIMPs that become NR at $t \simeq t_{\rm EQ}$—of the candidates in Table 9.1, this applies only to light neutrinos. In this case the damping scale is much greater than a galactic mass; it is closer to the

[26] This can be stated in a somewhat more useful way; in discussing the spherical-collapse model we found that the epoch of maximum expansion is characterized by $\delta_i(R_{MAX}/R_i) \simeq 1.77$; that is, estimating perturbation growth as being linear, break away occurs when "δ" is about 1.77.

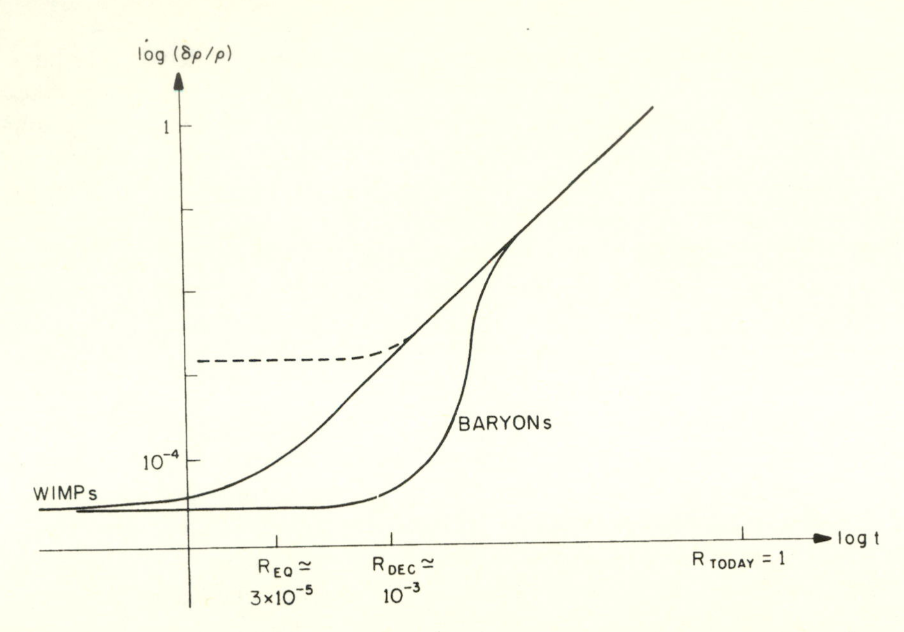

Fig. 9.5: The evolution of $(\delta\rho/\rho)$ in the baryon and WIMP components. The perturbations in the WIMPs begin to grow at the epoch of matter–radiation equality; however, the perturbations in the baryons cannot begin to grow until just after decoupling. After decoupling baryons fall into the WIMP potential wells, and within a few expansion times the baryon perturbations "catch up" with the WIMP perturbations.

mass of a supercluster. This limiting case is known as *hot dark matter.* The processed spectrum for hot dark matter is given precisely by [8]

$$\begin{aligned} |\delta_k|^2 &= Ak^{1+6\alpha}10^{-2(k/k_\nu)^{1.5}} \\ &= Ak^{1+6\alpha}\exp[-4.61(k/k_\nu)^{1.5}], \end{aligned} \tag{9.127}$$

where the neutrino damping scale $k_\nu = 0.16(m_\nu/30\,\mathrm{eV})\,\mathrm{Mpc}^{-1}$ ($\lambda_\nu = 40(m_\nu/30\,\mathrm{eV})^{-1}$ Mpc), A provides the overall normalization of the power spectrum, and $(\delta\rho/\rho)_{\rm HOR}$ is assumed to be proportional to $M^{-\alpha}$.

The second limiting case is when $\lambda_D \lesssim 1$ Mpc (i.e., $\lambda_D \ll \lambda_{\rm EQ}$), which occurs for relic WIMPs that become non-relativistic long before the epoch of matter-domination. In this case the first bound structures to form are of galactic mass (or smaller). This limiting case is known as *cold dark matter.* Assuming that λ_D is negligibly small, the processed spectrum for cold dark matter is given precisely by [10]:[27]

$$|\delta_k|^2 = \frac{Ak^{1+6\alpha}}{(1+\beta k + \omega k^{1.5} + \gamma k^2)^2}, \tag{9.128}$$

$\beta = 1.7(\Omega_0 h^2)^{-1}$ Mpc, $\omega = 9.0(\Omega_0 h^2)^{-1.5}\,\text{Mpc}^{1.5}$, $\gamma = 1.0(\Omega_0 h^2)^{-2}$ Mpc^2.

There is, of course, an intermediate case, $\lambda_D \simeq 1$ Mpc, corresponding to $T_X/T \sim 0.4$ and $m_X \sim$ keV, known as *warm dark matter*; however, we will not discuss this possiblity here (for details, see [12]). In this case the first objects to form are necessarily of galactic mass. In the next Section we will briefly review the course of structure formation in the *hot* and *cold* dark matter scenarios.

We should note that while the shape of the processed spectrum is precisely predicted in all four cases, the overall normalization is not. It must ultimately be determined by observational data, e.g., $\xi_{GG}(r)$, CMBR anisotropies, or the peculiar velocity field. Of course, there is always the possibility that a very specific inflation scenario could precisely predict A.

9.5 Two Stories: Hot and Cold Dark Matter

• *The Hints*: As we have tried to emphasize throughout this Chapter, structure formation is basically an initial-data problem; and in the past decade or so there has been great progress in the study of structure formation due to "hints" about the correct initial data that derive from theories of the earliest history of the Universe. The necessary initial data are the quantity and composition of matter in the Universe and the nature of the primeval inhomogeneities. Inflation (and good taste) argue for $\Omega_0 = 1$; moreover, primordial nucleosynthesis restricts Ω_B to be less than 0.15—barring a highly non-standard epoch of nucleosynthesis (see Chapter 4). Together, these facts suggest that the bulk of the material in the Universe today should be non-baryonic, and we have a very long list of candidate relics whose present energy density can provide closure density (see Table 9.1). As we have discussed in the two previous Chapters, both inflation and cosmic strings provide attractive possibilities for the origin of the primeval inhomogeneity. Inflation predicts the Harrison-Zel'dovich spectrum of adiabatic density perturbations—and in some models, isocurva-

[27] For isocurvature fluctuations, the processed spectrum is, $|\delta_k|^2 = Ak^{1+6\alpha}/[1+(\beta k + \omega k^{1.5} + \gamma k^2)^{1.24}]^{1.61}$, where $\beta = 15(\Omega_0 h^2)^{-1}$ Mpc, $\omega = 0.8(\Omega_0 h^2)^{-1.5}\,\text{Mpc}^{1.5}$, and $\gamma = 31.4(\Omega_0 h^2)^{-2}\,\text{Mpc}^2$ [11].

ture axion or baryon fluctuations. Cosmic string loops (or possibly string-produced wakes) can act as seeds for galaxy formation. Thus, there are at least four well motivated scenarios inspired by early Universe microphysics: inflation-produced, constant-curvature adiabatic (or isocurvature) fluctuations with hot or cold dark matter; and cosmic strings with hot or cold dark matter. Because the first two possibilities have become very well developed scenarios, we will focus our attention on them. But before we do, we should address a very important question.

• *Why Not Baryons?*—Why, one might ask, isn't a baryon-dominated Universe one of the well motivated possibilities? Well, there are several reasons.[28] In a baryon-dominated Universe density perturbations cannot begin to grow until decoupling, $R_{dec} \simeq 9.1 \times 10^{-4}$, and they cease growing when the Universe becomes curvature dominated, $R \simeq \Omega_B$. Therefore, the total growth factor for perturbations is only about $10^3\Omega_B$; for comparison, in an $\Omega_0 = 1$, WIMP-dominated Universe, perturbations can begin to grow as soon as the Universe becomes matter dominated, $R = R_{\rm EQ} \simeq 4.3 \times 10^{-5}h^{-2}$. Therefore, the growth factor in a WIMP-dominated Universe is larger by a factor of about $20h^2/\Omega_B \sim 200h^2$. Because of this, in a baryon-dominated Universe one needs fluctuations whose amplitude at horizon crossing is that much larger. The CMBR anisotropies predicted by the hot and cold dark matter scenarios are already close to the observational upper limits, so an $\Omega_B = 0.15$ model is certainly in dire straits, and even an $\Omega_B = 1$ model is in difficulty (see Section 9.6). If one wishes to entertain the possibility of an $\Omega_B = 1$ model, one has to sidestep the bound from primordial nucleosynthesis *and* explain why only about 1% of the baryons are luminous, and where all the dark baryons reside. While there are strong theoretical arguments against a baryon-dominated Universe, if push came to shove, theorists could likely live with a baryon-dominated Universe. For example, in Chapter 4 we mentioned two unconventional scenarios for getting around the nucleosynthesis constraint, and anisotropies in the CMBR can be significantly reduced provided that the primeval inhomogeneities are isocurvature and that the Universe is re-ionized after decoupling.

9.5.1 Hot dark matter

The processed spectrum for hot dark matter, cf. (9.127), is shown in Fig. 9.4. The crucial feature of the spectrum is the short wavelength cut-

[28]The perceptive reader will realize that the case against a baryon-dominated Universe must not be airtight, otherwise a single good reason would suffice.

off, $\lambda_D \simeq 40(m_\nu/30\text{eV})^{-1}$ Mpc, due to neutrino free streaming. Because of this cutoff, the first structures to form are of this size, roughly that of a supercluster ($10^{15} M_\odot$ or so).[29] Moreover, since this is such a large scale, collapse on this scale must have occurred rather recently ($z \lesssim 3$). Zel'dovich [13] has argued that when perturbations on the scale λ_D go non-linear, they do so in a highly non-spherical way, and that the resulting structures should be one-dimensional objects, resembling pancakes or blini.[30] Once a pancake forms and goes non-linear in *one* of its dimensions, the baryons within it can collide with one another and dissipate their gravitational energy.[31] Thereby, the baryons in the pancake can fragment and condense into smaller (e.g., galaxy-sized) objects. The neutrinos, being so weakly interacting, do not collide with one another or the baryons, cannot dissipate their gravitational energy, and therefore cannot collapse into more tightly bound objects. Thus they should remain less condensed, with the resulting structure resembling a baryon–neutrino sandwich (the baryons are the "meat"). Of course, some slowly moving neutrinos may subsequently be captured by the baryon-dominated galaxies. In a phrase, in a hot dark matter Universe the structure is said to form "from the top down."

Several groups [14, 15] have numerically simulated structure formation in a neutrino-dominated Universe.[32] In all of the simulations one can see a cell-like structure at late times (see Fig. 9.6). The cell-like structure traces to the existence of a feature in the processed spectrum: the damping scale. The large-scale structure in these simulations (scales $\gtrsim 10$ Mpc) seems to some to be qualitatively similar to the voids and filamentary structure seen in some of the red shift surveys, e.g., the CfA slices of the Universe, cf. Fig. 1.12. That's the good news. Now the bad news; in order to reproduce the observed galaxy–galaxy correlation function, the epoch of "pancaking" must be made to happen yesterday—that is, at a very low red shift ($z \lesssim 1$). This fact is difficult to reconcile with the many galaxies with

[29]Because of the short-wavelength cutoff, the slope of the spectrum is relatively unimportant, except in determining very-large-scale structure.

[30]A baryon-dominated Universe with adiabatic perturbations will behave in a similar way, with a slightly smaller cutoff scale (λ_S); in fact, the pancake scenario was first discussed in this context by Zel'dovich [13].

[31]We should emphasize that the crucial aspect of pancaking is the formation of a caustic surface in one dimension; this can, and generally does, occur before the pancake as a whole has gone non-linear.

[32]We remind the reader of the Oxford English Dictionary definition of ***simulation***: 1. The act, or practice, of simulating with intent to deceive; false pretense. 2. A false assumption or display; an imitation of something.

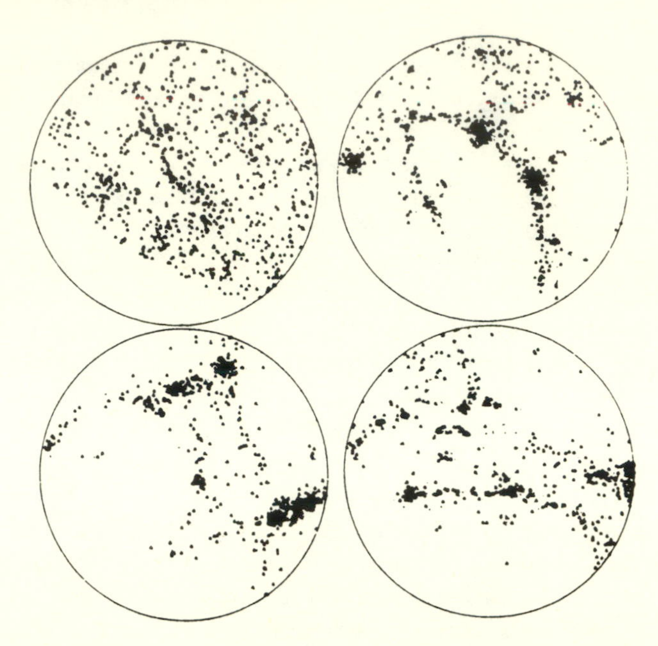

Fig. 9.6: Equal area projections (onto the celestial sphere) of the galaxy distribution on the northern sky made from the CfA catalogue of galaxies (upper left) and from red shift catalogues produced from numerical simulations of a neutrino-dominated Universe with $\Omega_0 = 1$ and $h = 0.54$, in which galaxy formation began at a red shift of $z = 2.5$ (from [16]). Note how the structure in the simulations is overdeveloped as compared to the CfA galaxies.

red shifts greater than 1, and QSO's with red shifts greater than 3, that are observed. Furthermore, the fact that the baryons are shock heated when they fall into the pancakes is problematic. The baryons may be heated to such a temperature that they are too hot to collapse, or remain so hot that they radiate prodigious amounts of x rays.

That the epoch of pancaking needs to be so recent in order to match the observed galaxy–galaxy correlation function should not come as a great surprise. The pancake scale is order 40 $(m_\nu/30\text{eV})^{-1}$ Mpc, while the galaxy correlation function indicates that the scale on which non-linearity is occurring today is $r_0 \simeq 5h^{-1}$ Mpc. Thus it is necessary that pancaking should have been very recent if structure is not to "over develop" by the present. The fundamental problem in the hot dark matter scenario is a mismatch in scales: $\lambda_D \simeq 40(m_\nu/30\text{eV})^{-1} \gg r_0 \simeq 5h^{-1}$ Mpc. Recall that $\Omega_\nu h^2 \simeq m_\nu/91$ eV, so that λ_D can also be expressed as $13(\Omega_\nu h^2)^{-1}$ Mpc. The simulations can be made to agree relatively well with the observed galaxy correlation function if λ_D is close to r_0, which requires that $\Omega_\nu h \sim 2$—not a very attractive proposition. Another idea proposed to resuscitate the neutrino-dominated Universe is to suppose that the density field of the Universe is better traced by clusters, and that the cluster–cluster correlation function is reliably described by $\xi_{CC} = (r/r_0)^{-1.8}$ where $r_0 = 25h^{-1}$ Mpc; here too, the mismatch between r_0 and λ_D is alleviated.

A general comment about numerical simulations of hot dark matter is in order; the most detailed simulations of structure formation include only neutrinos interacting through gravity, and thus cannot address the crucial question: Where are the baryons and how do galaxies form? In a hot dark matter model the formation of galaxies is necessarily a rather complicated process, involving the hydrodynamics and thermodynamics of shocked material. In order to compare their results to the Universe we see, the simulators have to make plausible assumptions as to how matter and light are correlated. In discussing the apparent failure of a neutrino-dominated Universe one should keep this fact in mind.

Hot dark matter has another interesting feature. Since neutrinos are fermions, their total number in any object is limited by the phase space available: $N_\nu \lesssim p^3 r^3$ (p is the typical neutrino momentum and r is the characteristic size of the object). From this, an upper bound to the mass of an object composed solely of neutrinos follows: $M_{\rm max} \lesssim m_\nu^4 \sigma^3 r^3$, where r and σ are the characteristic size and velocity dispersion of the object ($GM_{\rm max}/r \sim \sigma^2$). Turning this around, *if* neutrinos are to be the dominant mass component in a gravitationally bound object, one can place a *lower* limit to the neutrino mass [17],

$$m_\nu \gtrsim 120 \left(\frac{\sigma}{100\,{\rm km\,sec^{-1}}}\right)^{-1/4} \left(\frac{r}{\rm kpc}\right)^{-1/2} {\rm eV}. \qquad (9.129)$$

For a galaxy like ours, $r \simeq 10$ kpc and $\sigma \simeq 220\,{\rm km\,sec^{-1}}$; this translates to the limit, $m_\nu \gtrsim 30$ eV, which is in the ball park of the neutrino mass required to provide closure density.

For dwarf spheroidal galaxies (whose masses are as small as $10^6 M_\odot$ and include the satellite galaxies of the Milky Way, Draco, Carina, and Ursa Minor) the constraint is much more stringent, more like $m_\nu \gtrsim 300$ to 500 eV, which is *not* consistent with the cosmological bound to the mass of a neutrino species. If such objects have appreciable amounts of dark matter—which is still an open question because of the difficulty of measuring the tiny velocity dispersions of such objects, $\mathcal{O}(10\,{\rm km\,sec^{-1}})$—it can't be neutrinos. Of course, it could just be dark baryons, as such objects contribute only a tiny fraction of the total baryon density.

The so-called Tremaine-Gunn bound [17] is a kinematic constraint; dynamics also play a role. Around the epoch of galaxy formation, say $z \simeq$ 2 to 3, neutrinos are smooth on scales smaller than $\lambda_d \simeq 40(m_\nu/30\text{eV})^{-1}$ Mpc, and so are not likely to find their way into structures as small as galaxies, or perhaps even clusters. This aspect of hot dark matter could be a blessing in disguise: That neutrinos are likely to be smooth on scales less than 30 Mpc or so could reconcile the theoretical prejudice for $\Omega_0 = 1$ with the observational reality that the amount of material clustered on scales less than 10 to 30 Mpc is only 0.2 ± 0.1 of critical. If this is the case, the dark matter in galaxies and clusters of galaxies must be either dark baryons or the small fraction of neutrinos that did get captured.

In sum, hot dark matter has several attractive features. The large-scale structure seen in the simulations is qualitatively similar to the voids and filamentary structures that seem to be found in some galaxy surveys. If neutrinos stay smooth on scales of less than 10 to 30 Mpc, that would reconcile $\Omega_0 = 1$ with $\Omega_{\lesssim 10-30} \simeq 0.2 \pm 0.1$. However, in the hot dark matter picture it is difficult to achieve galaxy formation at a sufficiently early epoch, say $z \sim 2$ to 3, without having large-scale structure become over developed by the present time, a problem that traces to the mismatch of the damping scale and the present scale of non-linearity. As we will discuss in the next Section, the predicted level of CMBR anisotropy on the angular scale of $4.5'$ is very close to the present upper limit. Hot dark matter (with adiabatic) perturbations seems to be down for the eight count, but not out yet. In fact, it is a certainty that if the mass of one of the known neutrino species were determined to be of order 30 eV, hot dark matter would become viable in short order!

9.5.2 Cold dark matter

The processed spectrum for cold dark matter and constant-curvature, adiabatic fluctuations is shown in Fig. 9.4. In this case the damping scale is irrelevant (much less than 1 Mpc), while the initial shape of the spectrum is crucial. As a result of the small growth of perturbations that does take place between horizon crossing and the epoch of matter–radiation equality for scales less than $\lambda_{EQ} \simeq 13(\Omega_0 h^2)^{-1}$ Mpc, $\delta\rho/\rho$ increases as one goes to smaller scales. Thus, the first objects to form are of sub-galactic size. As objects break away from the expansion, they virialize into gravitationally bound objects through the process of *violent relaxation* [2]. The underlying mechanism of violent relaxation is the chaotic gravitational field of a collapsing, non-spherical self-gravitating system. The time- and space-varying gravitational field provides the means for individual, non-interacting particles to change their energies and to become well mixed in phase space. After only a few dynamical times ($\tau \sim (G\rho)^{-1/2}$) the result is a virialized distribution of matter whose phase space distribution is roughly Maxwellian and whose density varies as r^{-2}—just like galactic halos. Such a configuration is often referred to as an isothermal sphere, because it corresponds to the spherically symmetric, isothermal solution for a self-gravitating system of particles. If the WIMPs and baryons have different masses, then isothermal is somewhat of a misnomer since the two species will necessarily have different "temperatures." In addition, baryonic matter can, through dissipative processes, lose energy and condense even further into the cores of the isothermal spheres that form. If the galactic halos have angular momentum, then after dissipation baryons will wind up in disk-like structures.

As larger and larger scales go non-linear, bigger structures form through tidal interaction and mergers of smaller objects. In contrast to hot dark matter, structure formation in a cold dark matter model proceeds hierarchically, "from the bottom up." Were it not for dissipative processes that allow baryons to sink to the cores of the WIMP halos that they are emersed in, one might expect the formation of ever larger objects at the expense of the previous generation of smaller objects. The ability of baryonic matter to undergo dissipation allows objects of astrophysical size to condense out as individual and distinct entities. Moreover, since the dissipative processes are well known and understood—e.g., collisional excitation of atoms and molecules, and Compton scattering off the CMBR—the sizes of the objects that can cool rapidly enough to form at a given epoch can be calculated, and the properties of these objects, e.g, internal density and gravitational potential, can be predicted.[33] In the cold dark matter scenario, the predicted properties of the different kinds of galaxies that form agree quite well with observation [19].

The formation of structure in a cold dark matter Universe has been addressed in a beautiful series of numerical simulations carried out by Davis, Efstathiou, Frenk, and White [20]. As before, the simulations consist only of cold dark matter particles interacting via gravity in an expand-

[33]It is important to mention that the mass of a galaxy is generally thought to be unrelated to any feature in the spectrum of density perturbations; instead, it is believed that it is determined by the amount of gas that can cool and condense at a given epoch; see [18].

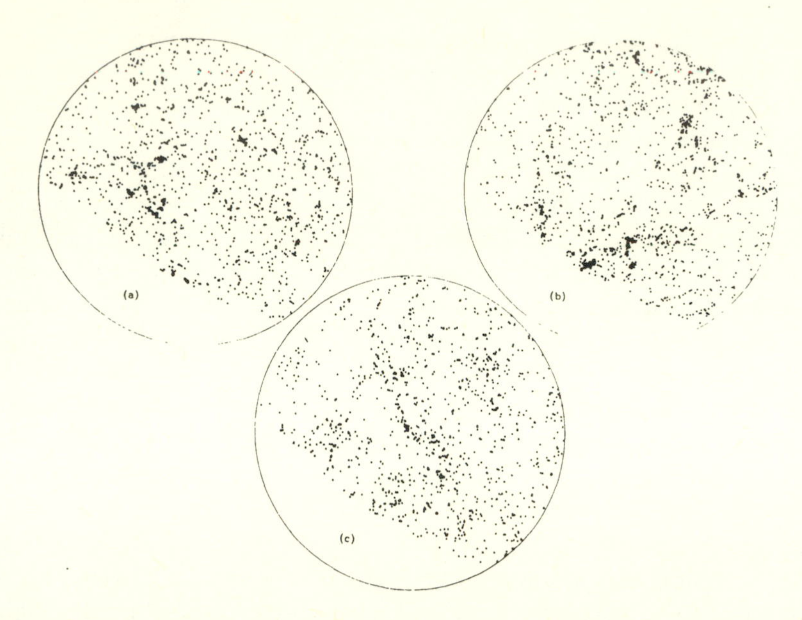

Fig. 9.7: Equal area projections (onto the celestial sphere) of the galaxy distribution on the northern sky made from the CfA red shift catalogue (shown in c), and from red shift catalogues produced from numerical simulations of a cold dark matter model with $\Omega_0 h \sim 0.2$ (a and b). In (b), the observer was chosen to be located near a prominent cluster (from [20]).

ing Universe. Thus the resulting structures cannot be directly identified with galaxies; however, they should be able to be identified with objects that have not undergone significant dissipation, such as galactic halos and clusters of galaxies. (Of course, even this comparison with the observed Universe depends in a critical way upon how one assigns light to mass.) Some of their results, and for reference, a projection of the CfA red shift survey, are shown in Fig. 9.7.

Making the simplest assumption, that light faithfully traces mass, Davis and his colleagues find a good match between the correlation function for their simulations and that of the observed Universe, provided that $\Omega_0 h \sim 0.2$.[34] In so doing they find that the epoch of collapse of galactic-

[34] The evolution of perturbations in an Einstein-de Sitter model are "scale free;" the only scale in a cold dark matter (or hot dark matter) model is set by $\lambda_{EQ} \simeq 13(\Omega_0 h^2)^{-1}$ Mpc, cf. (9.127) or (9.128). In the simulations, the present epoch is determined by when the slope of the correlation function (which changes with time) is -1.8. Comparing the correlation scale, $r_0 \simeq 5h^{-1}$ Mpc, with the only scale in the simulation, λ_{EQ}, then fixes

sized objects is relatively recent, $z \sim$ *few*, which is at least consistent with present observations but could prove troublesome if numerous galaxies and/or QSO's are discovered with red shifts in excess of 3. Of course, the fact that the simulations require $\Omega_0 h \sim 0.2$ is not particularly pleasing since our prejudice is for $\Omega_0 = 1$ and present determinations of h are in the range 0.4 to 1. The situation for flat models is actually somewhat worse; when the pairwise velocities of galaxies in the simulations are examined, it is found that they are excessive in flat models. That is, open models with $\Omega_0 h \sim 0.2$ are preferred.

It is also interesting to note that cold dark matter models "predict" that the galaxy–galaxy correlation function should become negative on scales larger than about $15(\Omega_0 h^2)^{-1}$ Mpc; we discussed the necessity of such in Section 9.2, as well as the fact that there is no definitive determination of whether or not the correlation function does become negative.

Because the formation of structures like halos, clusters of galaxies, etc. in cold dark matter models primarily involves gravitational forces alone, it is relatively straightforward to compare the simulations directly to the observed Universe. In such a comparison, cold dark matter does quite well, nicely accounting for the internal densities and gravitational potentials of galaxy halos, the number density of galaxies and clusters, and the like.

The shortcomings of cold dark matter are twofold. The first is its preference for $\Omega_0 h \sim 0.2$. Not only does galaxy clustering seem to require such, but with cold dark matter there is no reason that the ratio of baryons to WIMPs should vary in structures whose formation does not involve dissipation. That means the dynamical estimates for Ω_0 based upon galactic rotation curves and cluster dynamics should reflect the global value of Ω_0—supporting the view that $\Omega_0 \simeq 0.2 \pm 0.1$. We will return to discuss this dilemma, dubbed the Ω problem, in a later Section. For now, it suffices to say that a key assumption underlying the comparison of the numerical simulations and the actual observations is that of light faithfully tracing mass.

The second shortcoming of cold dark matter involves large-scale structure. The amplitude of the cluster–cluster correlation function in cold dark matter simulations falls at least a factor of 3 short of reproducing the measured value. In addition, giant voids like that in Boötes are not found, nor is it clear that the froth-like structure apparently seen in the slices of the Universe survey, cf. Fig. 1.12, is present. The present discrepancy between cold dark matter and the observed large-scale structure is

---

the value of $\Omega_0 h$.

difficult to quantify since our knowledge of such is less certain and more qualitative. In the next Section we will discuss the peculiar velocity field and CMBR anisotropies predicted by cold dark matter.

In closing, cold dark matter is a very successful scenario for structure formation: It is well motivated, has great predictive power, and is able to reproduce most of the features of the observed Universe. Whether or not it will be falsified remains to be seen; at least for the moment it serves as a useful paradigm for structure formation.

9.6 Probing the Primeval Spectrum

If light faithfully traces mass, then the distribution of bright galaxies provides us with a probe of the distribution of the mass density of the Universe. However, that is a very bold assumption, and as we have discussed, there is some evidence that it is not correct. In this Section we will discuss two other probes of the underlying mass distribution of the Universe: the peculiar velocity field and anisotropies of the CMBR. Both rely on the fact that matter is not uniformly distributed. Inhomogeneities in the mass density result in a (Newtonian) gravitational potential that is spatially varying. Therefore test particles will be accelerated relative to the cosmic rest frame, developing peculiar velocities; and photons traveling from the last-scattering surface from different directions will suffer different gravitational red shifts, giving rise to temperature anisotropies in the CMBR.

9.6.1 The peculiar velocity field

First let us discuss the peculiar velocity field. In our Newtonian analysis of the evolution of density perturbations we decomposed the peculiar velocity of a particle $\vec{\mathbf{v}}_1$ into its irrotational ($\vec{\mathbf{v}}_{\|}$) and rotational ($\vec{\mathbf{v}}_{\perp}$) pieces, and we discovered that the rotational piece did not couple to density inhomogeneities and decays as R^{-1}. On the other hand, the irrotational piece does couple to the density field of the Universe,

$$\vec{\mathbf{v}}_{\|}(k) = \frac{R}{ik}\dot{\delta}_k + \frac{\mathbf{const}}{R} \tag{9.130}$$

where the constant piece, like its rotational counterpart, decays away. Using equations (9.68) we can solve for the peculiar velocity,

$$\vec{\mathbf{v}}_1(\vec{\mathbf{r}},t) = \frac{\textbf{const}}{R} + \frac{2}{3}(\Omega_0 H_0)^{-1}\frac{d\ln\delta}{d\ln R}\,\vec{\mathbf{g}}(\vec{\mathbf{r}},t), \tag{9.131}$$

where the *peculiar acceleration* field $\vec{\mathbf{g}}$ is given by

$$\vec{\mathbf{g}}(\vec{\mathbf{r}},t) = -G\int\frac{\delta\rho(\vec{\mathbf{r}}\,',t)(\vec{\mathbf{r}}-\vec{\mathbf{r}}\,')d^3r'}{|\vec{\mathbf{r}}-\vec{\mathbf{r}}\,'|^3}, \tag{9.132}$$

and the first term represents the decaying piece of both the rotational and irrotational components of $\vec{\mathbf{v}}_1$. In writing (9.132) we have used the fact that $\dot{\delta}_k = H(d\ln\delta_k/d\ln R)\delta_k$. For a flat model and modes $k \ll k_J$, $\delta_k \propto R$, so that $d\ln\delta_k/d\ln R = 1$. For open models, $d\ln\delta_k/d\ln R$ is well approximated by $\Omega_0^{0.60}$. The expression for $\vec{\mathbf{v}}_1$ can also be written in terms of the density contrast alone,

$$\vec{\mathbf{v}}_1(\vec{\mathbf{r}},t) = \frac{\textbf{const}}{R} + \frac{H_0}{4\pi}\frac{d\ln\delta}{d\ln R}\,\vec{\mathbf{h}}(\vec{\mathbf{r}},t), \tag{9.133}$$

where

$$\vec{\mathbf{h}}(\vec{\mathbf{r}},t) \equiv \frac{\vec{\mathbf{g}}}{G\rho_0} = -\int\frac{\delta(\vec{\mathbf{r}}\,',t)(\vec{\mathbf{r}}-\vec{\mathbf{r}}\,')d^3r'}{|\vec{\mathbf{r}}-\vec{\mathbf{r}}\,'|^3}. \tag{9.134}$$

The physical content of (9.132) is manifest: The peculiar velocity is related to the gradient of the perturbed part of the gravitational potential times a time of order the Hubble time. From these two expressions for $\vec{\mathbf{v}}_1$ we can learn several important facts. First, that the contribution to $\vec{\mathbf{v}}_1$ from perturbations on the scale λ is roughly

$$(\vec{\mathbf{v}}_1)_\lambda \sim \frac{\lambda}{H_0^{-1}}\Omega_0^{0.6}(\delta\rho/\rho)_\lambda. \tag{9.135}$$

Provided that $(\delta\rho/\rho)_\lambda$ decreases sufficiently rapidly, faster than λ^{-1} (or $M^{-1/3}$), the contribution of very large scales ($\lambda \gtrsim H_0^{-1}$) will be negligible, thereby justifying our Newtonian treatment of $\vec{\mathbf{v}}_1$. Second, since the non-decaying part of the peculiar velocity is supported by inhomogeneities in the mass density, $\vec{\mathbf{v}}_1$ is directly related to the density field and can be used to probe it. Finally, the decaying part of $\vec{\mathbf{v}}_1$ arises due to non-gravitational effects (e.g., explosive hydrodynamics early on) and thereafter decays. Presumably this piece of $\vec{\mathbf{v}}_1$ is small today, and hereafter we will ignore it.

Expression (9.133) provides a very useful relationship between the peculiar velocity, Ω_0, and the density contrast. If one were confident that one knew the value of Ω_0, then (9.133) relates the underlying density contrast to $\vec{v}_1(\vec{r}, t)$. That is, by measuring $\vec{v}_1$, e.g., by using galaxies as test particles, one could, in principle, infer the density field.

Conversely, as we mentioned in Chapter 1, one can use (9.133) to measure Ω_0. The velocity of the Local Group relative to the cosmic rest frame has been determined from the dipole anisotropy of the CMBR: $600 \pm 50\,\mathrm{km\,sec^{-1}}$ toward Hydra-Centaurus [21]. A red shift survey of our nearby volume (e.g., that provided by the IRAS sample of infrared selected galaxies) together with the assumption that $\delta\rho/\rho = \delta n_{GAL}/n_{GAL}$, allows one to compute $\vec{h}$, independent of ρ_0. From the CMBR determination of $\vec{v}_1$, by evaluating $\vec{h}$, and by exploiting the Ω_0 dependence of $d\ln\delta/d\ln R$, one can infer the value of Ω_0. The dipole of the distribution of IRAS galaxies points in approximately the same direction (within 10° to 20°) as the dipole of the CMBR—which is what one expects if the inhomogeneous mass distribution (on scales where $\delta\rho/\rho$ is linear) is responsible for the motion of the Local Group [22]. As we mentioned in Chapter 1, the recent determinations of Ω_0 using this technique and the IRAS sample of galaxies give a value of Ω_0 close to unity and seem to indicate that the integral for $\vec{h}$ converges at a distance of about $40h^{-1}$ Mpc [23].

Returning to expression (9.130), which relates the Fourier components of $\vec{v}_1$ to those of the density contrast, we can express the statistical properties of the peculiar velocity field in terms of δ_k. For example, using (9.130) we can compute the *rms* peculiar velocity,

$$\begin{aligned} v^2 &\equiv \langle \vec{v}_1(\vec{x}) \cdot \vec{v}_1(\vec{x}) \rangle = V^{-1} \int_0^\infty \frac{k^3 |\vec{v}_{||}(k)|^2}{2\pi^2} \frac{dk}{k} \\ &= V^{-1} H_0^2 \left(\frac{d\ln\delta}{d\ln R} \right)^2 \int_0^\infty \frac{|\delta_k|^2 dk}{2\pi^2}, \end{aligned} \tag{9.136}$$

and the *rms* value of the peculiar velocity averaged over a volume of size r, defined by a window function $W(r)$,

$$\begin{aligned} v^2(r) &\equiv \left\langle \left(\int \vec{v}_1(\vec{r} + \vec{x}) W(r) d^3r \right)^2 / V_W^2 \right\rangle \\ &= V^{-1} H_0^2 \left(\frac{d\ln\delta}{d\ln R} \right)^2 \int_0^\infty \frac{|\delta_k|^2 |W_k|^2 dk}{2\pi^2 V_W^2}. \end{aligned} \tag{9.137}$$

Aside from the prefactors, these expressions are very similar to the analogous expressions for $\delta\rho/\rho$, with one important exception: The integrand here is proportional to $V^{-1}|\delta_k|^2/2\pi^2 = \Delta^2(k)/k^3$ rather than $\Delta^2(k)/k$. The averaged properties of the peculiar velocity involve a different moment of the power spectrum, one that weights larger scales more heavily. If we suppose that the power spectrum $|\delta_k|^2 \propto k^n$, it follows that $v(r) \propto r^{-(n+1)/2}$; for the Harrison-Zel'dovich spectrum, on scales much greater than $\lambda_{EQ} \simeq 13$ Mpc, $n = 1$, so that it is expected that $v(r) \propto r^{-1}$ on large scales.

Determining the peculiar velocity field affords a great opportunity for probing the underlying mass distribution *directly*. When probing the mass distribution with the correlation properties of bright galaxies, one has to make some assumption about the relationship of mass and light. In contrast, when using the peculiar velocity field, galaxies are merely being used as test particles to measure the local gradient of the gravitational field.[35] Of course, it is not easy to determine the peculiar velocity of a galaxy other than our own. In order to determine the peculiar velocity of a distant galaxy, one must independently measure the velocity of the galaxy—using its red shift z—and the distance to the galaxy d—e.g., using the Tully-Fisher or Faber-Jackson relations. The line-of-sight component of the peculiar velocity is then given by, $v_{pec} = (zc - H_0 d)$. The difficulties associated with this method are clear: (i) Very reliable distance indicators are required; (ii) the typical peculiar velocity of a galaxy is independent of its location, which implies that the size of the peculiar velocity relative to the galaxy's Hubble velocity decreases with distance—and because the relative errors in determining distances remain constant at best, the relative errors in determining the peculiar velocity increase with distance; and (iii) only the line-of-sight component of the peculiar velocity is determined.

In spite of the difficulties involved, several groups have measured the volume-averaged, peculiar velocity in our local neighborhood on scales $r \simeq 10h^{-1}$ to $60h^{-1}$ Mpc [24]. Because of the way in which real observations are made, it is not a simple matter to compare the observational results with the theoretical predictions. For example, the observers do not sample the volume with a window function, or even uniformly, and the errors in the peculiar velocity depend upon r (typically increasing with r as discussed above). The values obtained for $v(r)$ on scales from $10h^{-1}$

---

[35] This statement is not entirely accurate, as galaxies do not uniformly sample space. If one had the luxury of being able to select galaxies at random throughout a given volume, this statement would be precisely true.

to $60h^{-1}$ Mpc range from a few $100\,\mathrm{km\,sec^{-1}}$ to $600\,\mathrm{km\,sec^{-1}}$, with one of the largest measurements, $600\,\mathrm{km\,sec^{-1}}$, that of [25], associated with the largest volume, about $(60h^{-1}\,\mathrm{Mpc})^3$. The best—but not the simplest—way to compare these important observations to theory is to take a numerical realization of a model, e.g., hot or cold dark matter, and use the same protocol as the observers do to compute the same quantities. This is only beginning to be done.

To crudely summarize the present situation, we mention that the predicted *rms* average peculiar velocity for hot dark matter with $\Omega_0 = 1$, constant-curvature perturbations, and a gaussian window function is [26]

$$v(r) \simeq 200\,(r/25h^{-1}\,\mathrm{Mpc})^{-1}h^{-5/3}\,\mathrm{km\,sec^{-1}}, \tag{9.138}$$

and for cold dark matter it is [27]

$$v(r) \simeq 160\,(r/25h^{-1}\,\mathrm{Mpc})^{-1}h^{-3/4}\,\mathrm{km\,sec^{-1}}. \tag{9.139}$$

The predicted peculiar velocities for hot or cold dark matter *seem* to be significantly lower than the observations,[36] but one should keep in mind that these predictions are for the *rms* average over the entire Universe, that the predicted values depend upon the precise details of how the models are normalized, that these predictions cannot be compared in a simple way to the observations, and that the observational situation is far from being settled.

The model that seems to best reproduce the observations of [25] is the so-called Great Attractor model, in which there is a $10^{16}M_\odot$ mass concentration about $40h^{-1}$ Mpc from the Milky Way, in the direction of Hydra-Centaurus [29]. The monopole field of the Great Attractor accounts for the relatively large and roughly uniform peculiar velocity of the surrounding volume of radius about $60h^{-1}$ Mpc. In the context of cold dark matter the probability of such a mass concentration developing nearby is very small. While the implications of the present measurements of the peculiar velocity field are not completely clear, the importance of the peculiar velocity field as probe of the density field is!

[36] The same is true for the predicted peculiar velocities for cosmic strings with hot or cold dark matter [28].

9.6.2 Anisotropies in the CMBR

Temperature fluctuations in the CMBR arise due to five distinct physical effects: (i) our peculiar velocity with respect to the cosmic rest frame; (ii) fluctuations in the gravitational potential on the last scattering surface; (iii) fluctuations intrinsic to the radiation field itself on the last-scattering surface; (iv) the peculiar velocity of the last-scattering surface; and (v) damping of anisotropies if the Universe should be re-ionized after decoupling.[37] The first effect gives rise to a dipole anisotropy, as we have just discussed above. The second effect, known as the Sachs-Wolfe effect [32], is the dominant contribution to the anisotropy on large-angular scales, $\theta \gg 1°$. The last three effects provide the dominant contributions to the anisotropy on small-angular scales, $\theta \ll 1°$. That the angular scale 1° divides the large-scale and small-scale anisotropy traces to the fact that the Hubble scale at decoupling subtends an angle of

$$\theta_{dec} = 0.87° \, \Omega_0^{1/2}(z_{dec}/1100)^{-1/2}. \qquad (9.140)$$

Thus, small-angular scales correspond to comoving lengths that are sub-horizon sized at decoupling, while large-angular scales correspond to comoving lengths that are super-horizon sized at decoupling.

First consider the temperature fluctuations on large-angular scales that arise due to the Sachs-Wolfe effect. These anisotropies are simple to compute because they probe length scales that were super-horizon sized at decoupling, and therefore insensitive to microphysical processes. Furthermore, because of their insensitiviy to microphysical processing, they provide a probe of the "virgin" spectrum of primeval fluctuations. In the synchronous gauge and for a flat cosmology, the large-angle fluctuations

[37] Additional fluctuations can arise if the line of sight to the last-scattering surface intersects a cloud of very hot electrons, e.g., the hot gas associated with many rich clusters. CMBR photons are "up scattered" by inverse Compton scattering, with the number of photons being conserved. The resulting distortion can be described by a change in the equivalent black body temperature as a function of wavelength: $\Delta T/T = y[x/\tanh(x/2) - 4]$ (valid for $x^2 y$, $y \lesssim 1$), where $x = E_\gamma/T$, $y = (T_e/m_e)\tau$, and the optical depth $\tau = n_e \sigma_T c$ (n_e and T_e are the electron temperature and number density). On the Rayleigh-Jeans side, $\delta T/T \simeq -2y$, i.e., a "dip" in the temperature. This is known as the Sunyaev-Zel'dovich effect [30] and has been detected in the direction of several rich clusters [31].

in the CMBR temperature are given by

$$\begin{aligned}\frac{\delta T(\hat{\mathbf{x}})}{T} &= -\frac{R_0^2 H_0^2}{2(2\pi)^3}\int k^{-2}\delta_k \exp(-i\vec{\mathbf{k}}\cdot\vec{\mathbf{x}})d^3k \\ &= \frac{1}{3}\phi_1(\vec{\mathbf{x}},t_0) \sim (\delta\rho/\rho)_{\lambda\sim H_0^{-1}},\end{aligned} \tag{9.141}$$

where the vector $\vec{\mathbf{x}}$ points to the last-scattering surface and has length $|\vec{\mathbf{x}}| = 2H_0^{-1}$ (the distance to the horizon). The second expression for $\delta T/T$ follows from the first by using the relationship between ϕ_k and δ_k, cf. (9.68). During the post-recombination epoch $\delta_k \propto R$ and $H_0^2 \propto R^{-3}$, so that the quantity $H^2R^2\delta_k$ remains constant. Therefore, $\phi_1(\vec{\mathbf{x}},t_0)$ corresponds to the perturbation in the Newtonian potential on the last-scattering surface. Based upon Newtonian intuition one would have "guessed" that $\delta T/T$ should be given by $\phi_1(\vec{\mathbf{x}},t)$; when all the general relativistic effects are taken into account, the result is one third of that.

It is often convenient to expand $\delta T/T$ in spherical harmonics,

$$\frac{\delta T(\hat{\mathbf{x}})}{T} = \sum_{l=2}^{\infty}\sum_{m=-l}^{m=+l} a_{lm}Y_{lm}(\theta,\phi), \tag{9.142}$$

where the angles θ and ϕ are spherical angles on the sky. The dipole component has been left out; as we have discussed, it arises due to our peculiar velocity.[38] Using this expansion it follows that

$$\langle|a_{lm}|^2\rangle = V^{-1}\frac{H_0^4}{2\pi}\int_0^\infty \frac{dk}{k^2}|\delta_k|^2[j_l(kx)]^2, \tag{9.143}$$

where j_l is the spherical Bessel function of order l, and the average is over all observation positions in the Universe.[39] Taking the power spectrum to be $|\delta_k|^2 = AVk^n$, the integral over $[j_l(kx)]^2$ can be performed, with the result that

$$\langle|a_{lm}|^2\rangle = \frac{AH_0^{n+3}}{16}\frac{\Gamma(3-n)}{\Gamma[(4-n)/2]^2}\frac{\Gamma[(2l+n-1)/2]}{\Gamma[(2l+5-n)/2]}. \tag{9.144}$$

---

[38] Moreover, it is not possible to separate an intrinsic dipole anisotropy from our peculiar velocity.

[39] Of course, the spatial average of any of the a_{lm} vanishes identically.

Remembering that $(\delta M/M)_R^2 \sim (k^3|\delta_k|^2)|_{k\sim R^{-1}} \sim AR^{-(n+3)}$, we see that up to a numerical factor $\langle|a_{lm}|^2\rangle^{1/2}$ is equal to the value of $(\delta\rho/\rho)$ on the present horizon scale.

We expect the spectrum of density perturbations on scales greater than $\lambda_{EQ} \simeq 13(\Omega_0 h^2)^{-1}$ Mpc to preserve its initial form (for Harrison-Zel'dovich, $n = 1$). Therefore, we can relate the quadrupole anisotropy, for example, to the *rms* mass fluctuation on some scale r greater than λ_{EQ}. Taking $n = 1$ and a top hat window function to define $(\delta M/M)_R$, we find that

$$\langle|a_{2m}|^2\rangle = \frac{\pi}{3}(H_0 R)^4 \left(\frac{\delta M}{M}\right)_R^2 . \tag{9.145}$$

Furthermore, using $R = 30h^{-1}$ Mpc and $(\delta M/M)_R \simeq 1/4$, as determined by $J_3(R)$ (see Section 9.2), we find

$$\langle|a_{2m}|^2\rangle^{1/2} \simeq 2 \times 10^{-5}. \tag{9.146}$$

The present 90% confidence level upper limit to $\langle|a_{2m}|^2\rangle^{1/2}$ is slightly smaller than 10^{-4} [33]. Since the very same density inhomogeneities lead to both structure formation and anisotropies in the CMBR, either can be used to normalize the spectrum—and to predict the other.

CMBR anisotropy experiments often measure the *rms* temperature difference between two antennas separated by an angle α, averaged over some substantial fraction of the sky: $\Delta T/T \equiv \langle(T_1 - T_2)^2\rangle^{1/2}/T$. The theoretical expectation for $\Delta T/T$ is conveniently expressed in terms of the autocorrelation function of $\delta T(\hat{\mathbf{x}})/T$:

$$\begin{aligned} C_\sigma(\alpha) &\equiv \left\langle \frac{\delta T(\hat{\mathbf{x}}_1)}{T}\frac{\delta T(\hat{\mathbf{x}}_2)}{T}\right\rangle, \\ \left(\frac{\Delta T}{T}\right)^2 &= 2\left[C_\sigma(0) - C_\sigma(\alpha)\right], \end{aligned} \tag{9.147}$$

where the averaging is done over the entire sky and all observation positions in the Universe.[40] The predicted autocorrelation function for two antennas

[40] In order to reduce systematic effects not associated with anisotropy intrinsic to the CMBR, experimenters often use a triple beam arrangement, measuring $\Delta T/T = \langle[T_0 - (T_1 + T_2)/2]^2\rangle^{1/2}/T$, where subscript 0 denotes the central beam, and subscripts 1 and 2 denote beams displaced by an angle α to either side of the central beam. In this case, it is simple to show that $(\Delta T/T)^2 = 1.5C_\sigma(0) - 2C_\sigma(\alpha) + 0.5C_\sigma(2\alpha)$.

whose angular response is well modeled by a gaussian of angular width σ is given by[41]

$$\left\langle \frac{\delta T(\hat{\mathbf{x}}_1)}{T}\frac{\delta T(\hat{\mathbf{x}}_2)}{T}\right\rangle = \frac{1}{4\pi}\sum_{l=2}^{\infty}(2l+1)\langle|a_{lm}|^2\rangle P_l(\hat{\mathbf{x}}_1\cdot\hat{\mathbf{x}}_2)\exp[-(l+1/2)^2\sigma^2], \tag{9.148}$$

where $\hat{\mathbf{x}}_1\cdot\hat{\mathbf{x}}_2 = \cos\alpha$, and P_l is the Legendre function of order l. Note that the finite beam width, which smears out any information on angular scales less than about σ, effectively cuts off the sum at $l \sim \sigma^{-1}$.

The CMBR anisotropy on small-angular scales is more difficult to compute as these angular scales correspond to comoving length scales that were sub-horizon sized at decoupling, so that microphysical processing is very important. In order to calculate the small-scale anisotropy one must solve the coupled Boltzmann equations for the evolution of matter and radiation perturbations through the epoch of decoupling.

In general, the dominant effect of the five previously mentioned is the intrinsic anisotropy of the radiation field itself, with the Sachs-Wolfe and peculiar velocity of the last-scattering surface being sub-dominant, and reionization being applicable only if the Universe is subsequently re-ionized. Roughly speaking, for adiabatic perturbations $\delta T/T \sim \delta\rho_B/3\rho_B$ at decoupling. Moreover, if there is a substantial WIMP component present, $\delta\rho_{\rm WIMP}/\rho_{\rm WIMP}$ will be larger than $\delta\rho_B/\rho_B$ as sub-horizon-sized perturbations in the WIMPs begin growing when the Universe becomes matter dominated. Since $R_{dec}/R_{\rm EQ} \simeq 21(\Omega_0 h^2)$, one would expect that

$$\delta\rho_B/\rho_B \sim 0.05(\Omega_0 h^2)^{-1}(\delta\rho_{\rm WIMP}/\rho_{\rm WIMP}) \tag{9.149}$$

at decoupling. As a crude estimate then, one would predict

$$\begin{aligned}(\delta T/T)_\theta &\simeq \frac{1}{3}(\delta\rho/\rho)_\lambda \quad \text{(baryons only)},\\ (\delta T/T)_\theta &\simeq \frac{(\Omega_0 h^2)^{-1}}{60}(\delta\rho/\rho)_\lambda \quad \text{(WIMP - dominated)},\end{aligned} \tag{9.150}$$

where the comoving wavelength λ corresponds to the angular scale on the sky, $\theta = 34.4''(\Omega_0 h)(\lambda/\text{Mpc})$.

To make precise predictions, one has to integrate the Boltzmann equations, taking into account all the sources of anisotropy and the washing out

[41]Specifically, the response function is $dR(\phi,\theta) = \exp(-\theta^2/2\sigma^2)\theta d\theta d\phi/2\pi\sigma^2$, where θ and ϕ are spherical polar angles and σ is assumed to be much less than 1.

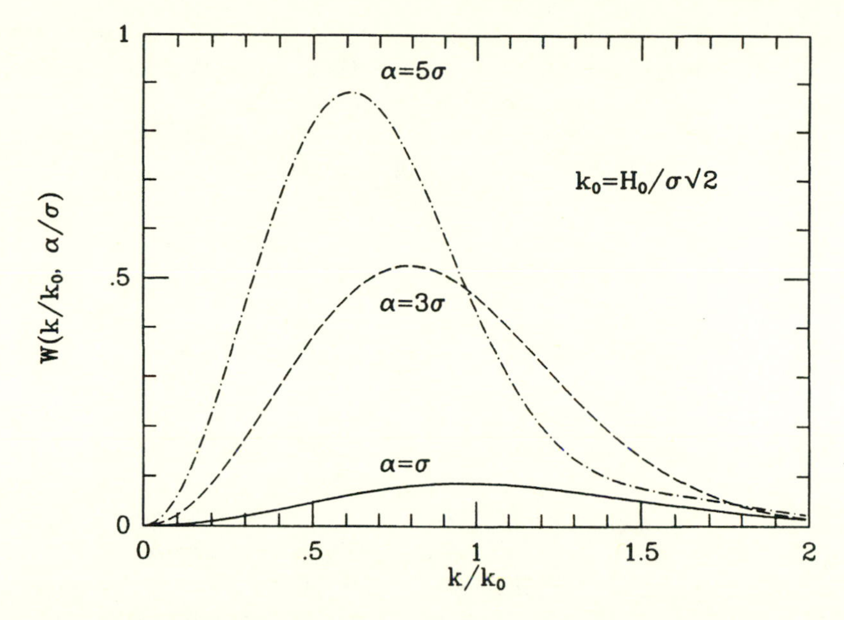

Fig. 9.8: The effective "window function" $W(k/2^{-1/2}H_0\sigma^{-1}, \alpha/\sigma)$ for a two-beam experiment to measure $\Delta T/T$, where the separation of the two antennas $\alpha = 1$, 3, and 5 σ. The function $W = \exp(-2k^2H_0^{-2}\sigma^2)[1 - J_0(\sqrt{2}kH_0^{-1}\alpha)]$, cf. (9.152).

of anisotropy on small-angular scales ($\theta \ll 8'$) due to the finite thickness of the last-scattering surface.[42] Once the Fourier transform of the temperature fluctuation on the last scattering surface, $\delta_T(\mu, k)$, is obtained, it is relatively straightforward to calculate the autocorrelation function of the CMBR temperature on the sky today in terms of $|\delta_T(\mu, k)|^2$, α, and σ:

$$C_\sigma(\alpha) = \left\langle \frac{\delta T(\hat{\mathbf{x}}_1)}{T} \frac{\delta T(\hat{\mathbf{x}}_2)}{T} \right\rangle = \frac{V^{-1}}{4\pi^2} \int_0^\infty k^2 dk \int_{-1}^1 d\mu |\delta_T(\mu, k)|^2$$

$$\times \exp[-4k^2H_0^{-2}(1-\mu^2)\sigma^2] J_0[2kH_0^{-1}(1-\mu^2)^{1/2}\alpha], \quad (9.151)$$

where $\hat{\mathbf{x}}_1 \cdot \hat{\mathbf{x}}_2 = \cos\alpha$, α is the angular separation of the two antennas, σ is the gaussian beam width of the antennas, $\mu = \cos\beta = \hat{\mathbf{x}} \cdot \hat{\mathbf{k}}$, β is the angle

[42]Because decoupling is not instantaneous, the last-scattering surface has a finite width. Estimating that $\Delta z_{dec}/z_{dec} \sim 0.1$, cf. Section 3.4, this corresponds to a comoving scale and angle on the sky of about 15 Mpc and $8'$, respectively. Owing to this fact, any measurement of the CMBR temperature effectively averages over an angular scale of about $8'$, smearing out any information on smaller scales.

between the direction to the last-scattering surface and the wavenumber $\vec{k}$, and J_0 is the zeroth-order Bessel function. (The Fourier transform of the temperature fluctuation $\delta_T(\mu, k)$ can depend upon the angle β.)

The prediction for the *rms* temperature fluctuation measured in an experiment with two antennas, each of beam width σ, separated by angle α, is

$$\begin{aligned}\left(\frac{\Delta T}{T}\right)^2 &= \frac{V^{-1}}{2\pi^2}\int k^2 dk \int_{-1}^{1} d\mu |\delta_T(\mu, k)|^2 \exp[-4k^2 H_0^{-2}(1-\mu^2)\sigma^2] \\ &\quad \times \left\{1 - J_0[2kH_0^{-1}(1-\mu^2)^{1/2}\alpha]\right\}. \end{aligned} \tag{9.152}$$

Note that in the integral expression for $(\Delta T/T)^2$ the power spectrum $|\delta_T(\mu, k)|^2$ is weighted by two factors that act like a window function: The exponential factor, which arises because of smearing due to the finite beam width and cuts off the contribution of wavelengths shorter than σH_0^{-1}; and the $[1 - J_0]$ factor, which arises due to the separation of the two antennas and cuts off the contribution of all wavelengths longer than αH_0^{-1}. Thus, roughly speaking, $\Delta T/T$ receives contributions from scales σH_0^{-1} to αH_0^{-1} (angular scales on the sky from σ to α). Furthermore, one should remember that due to the finite thickness of the last-scattering surface, $\delta_T(\mu, k)$ itself should decrease sharply on scales shorter than 15 Mpc (angles smaller than about 8′). The "window function" for $\Delta T/T$ is shown in Fig. 9.8.

While the large-angle anisotropy probes the fluctuation spectrum only on scales comparable to H_0^{-1}, the small-angle anisotropy has the potential to probe it on scales closer to those relevant to structure formation. However, the relationship between the small-angle anisotropy and the power spectrum of the density field is much more complicated, as it involves detailed microphysics. Moreover, small-angle anisotropy can be reduced by re-ionization effects: If the ionization fraction should be increased to a value close to unity after decoupling (e.g., due to the ionizing effects of a very early population of QSO's), the photon mean free path does not become of order *ct* until

$$1 + z_{dec} = 7.5 X_e^{-2/3}(\Omega_0/\Omega_B)^{2/3}(\Omega_0 h^2)^{-1/3}. \tag{9.153}$$

That is, with sufficient re-ionization, the Universe can become opaque at a red shift of order 10. Due to this, anisotropy on scales up to the horizon scale at the epoch of final decoupling, or $\theta \simeq 10.5° X_e^{1/3}(\Omega_0/\Omega_B)^{-1/3}(\Omega^2 h)^{1/3}$, can be significantly reduced.

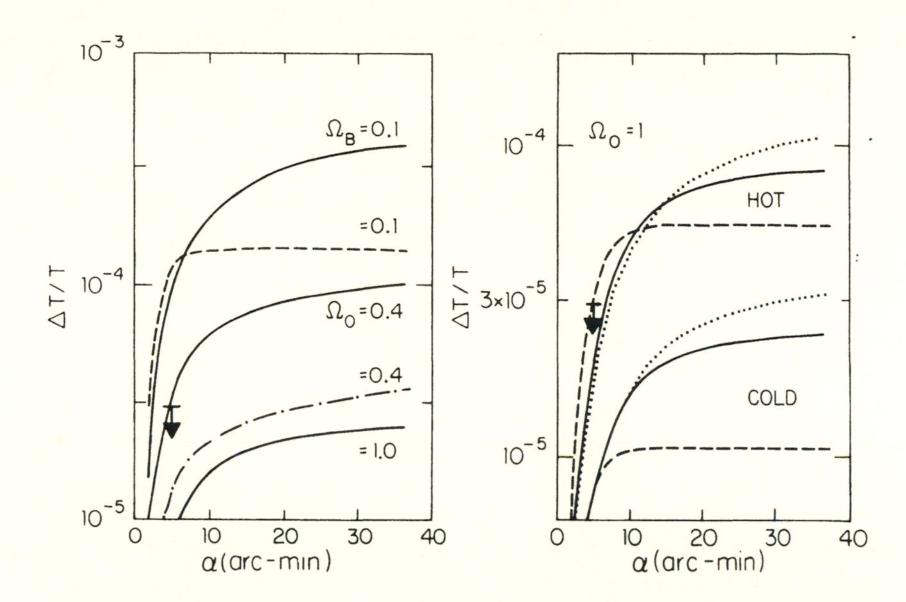

Fig. 9.9: The predicted small-scale anisotropy for hot and cold dark matter models and for a baryon-dominated model as a function of angle in a triple-beam experiment with antenna beam widths of 1.5′. Also indicated is the upper limit to the anisotropy on 4.5′. For the solid curves the Harrison-Zel'dovich spectrum ($n = 1$) is assumed; for the broken curves, a primeval spectrum with $n = 4$; and for the dotted curves, $n = 0$. On the left are shown a baryon-dominated model with $\Omega_B = 0.1$, and cold dark matter models with $\Omega_0 = 0.4$ and $\Omega_0 = 1$. On the right are shown hot and cold models with $\Omega_0 = 1$. For all models, $h = 1/2$, with the exception of the broken dotted curve ($h = 3/4$) (from [34]).

With these facts in mind, we show in Fig. 9.9 the predicted anisotropy as a function of angular scale for hot and cold dark matter models and for a baryon-dominated model, as well as the upper limit to the anisotropy on the angular scale of 4.5′, $\delta T/T < 3 \times 10^{-5}$, based upon the measurements of Uson and Wilkinson [35] and Readhead, et al. [36]. The hot dark matter predictions are near the upper limit, while the cold dark matter predictions can be as much as an order of magnitude below. These fine-scale anisotropy measurements rule out any baryon-dominated model without substantial re-ionization, independent of the value of Ω_B [37].

The CMBR is a powerful probe of the Universe at decoupling and perhaps even later, e.g., through the Sunyaev-Zel'dovich effect or if the Universe was re-ionized. The large-angle anisotropy probes the density field of the Universe on scales comparable to $H_0^{-1} \sim 3000h^{-1}$ Mpc, where microphysical processing is unimportant. Large-angle anisotropy measurements provide the means of determining the "virgin" spectrum of fluctuations. Small-angle anisotropy measurements probe the Universe on scales closer to those relevant for structure formation, $\lambda \sim 10$'s Mpc, where microphysical processing can be a very significant effect. Together, the large- and small-angle anisotropy of the CMBR offers a unique probe of the spectrum of density fluctuations and a stringent test of any scenario of structure formation.

9.7 The Ω Problem

This brings us to a pressing and very significant problem: the fact that dynamical observations indicate that $\Omega_{\lesssim 10-30} \simeq 0.2 \pm 0.1$, while our theoretical prejudices strongly favor $\Omega_0 = 1.0$. As we discussed in Chapter 1, the dynamical determinations involve measuring the amount of mass associated with bright galaxies on scales less than 10 to 30 Mpc. However, a component of mass density that is smoothly distributed on these scales would not be reflected in $\Omega_{\lesssim 10-30}$. In the cold dark matter scenario the Ω problem is particularly acute as on the scales of galactic halos and larger, baryons (and the associated light) should be a good tracer of the mass, since on scales this large there is no mechanism to separate baryons and WIMPs.

Because of the strong theoretical preference for $\Omega_0 = 1$, based not only on inflation, but also on the fact that the growth of density perturbations in an open Universe is smaller by a factor of Ω_0^2, the Ω problem has received a great deal of attention. A number of solutions have been proposed. All basically involve the same idea: The existence of a "smooth" component of mass density that is less clustered on scales of 10 to 30 Mpc than bright galaxies and provides the additional mass density required to bring Ω_0 to 1: $\Omega_{\rm SM} = 1 - \Omega_{\lesssim 10-30} \simeq 0.8 \pm 0.1$. The presence of this component would not have been detected in dynamical measurements of the mass density previously discussed, as they have probed only smaller scales. Of course, its presence would be revealed by the global, kinematic techniques discussed in Chapters 1 and 2, e.g., the magnitude red shift, angle red shift, and galaxy-count red shift tests, and these measurements are still inconclusive. Suggestions for the smooth component include, first, a relic cosmological term (i.e., $\Lambda = 3\Omega_{SM} H_0^2$), which is absolutely smooth, and corresponds to a uniform energy density $\rho_{\rm VAC} = \Lambda/8\pi G$ throughout space [39]. As we have previously emphasized, theory provides no prediction for the present vacuum energy density other than the preposterous one,

$\rho_{\rm VAC} = m_{Pl}^4$. A relic cosmological term has very little effect on the growth of density inhomogeneities, and in fact, a cold dark matter model with $\Omega_{\rm WIMP} \simeq 0.2$ and $\Omega_\Lambda = 0.8$ provides a very good match to the observed Universe [20]. As we discussed in Chapter 2, a model with $\Lambda \neq 0$ is older than its matter-dominated counterpart: $H_0t_0 \geq 2/3$.

Second, relativistic (or very fast-moving) particles, produced by the recent (red shift $z \sim 2$ to 5) decay of an unstable WIMP species, e.g., a 100 eV neutrino species that decays to a light neutrino and a massless scalar [40]. Very fast moving particles cannot, because of their high speeds, cluster. It is necessary that the relativistic particles be of recent origin and not "primordial." Were they present at early times when the Universe was radiation-dominated, the fact that they dominate the mass density today would imply that the Universe never became matter dominated, and so small density perturbations would never have been amplified by the Jeans instability. In a model with unstable dark matter the Universe becomes matter dominated at an earlier epoch, $R'_{EQ} = (\Omega_{NR}/\Omega_R)R_{EQ}$ and then becomes radiation dominated again when $R = \Omega_{NR}/\Omega_R$, resulting in the same growth factor for density perturbations as in a matter-dominated, flat model (here Ω_R is the fraction of critical density contributed today by relativistic particles, and Ω_{NR} is that of non-relativistic particles). Of course, a model with $\Omega_R \sim 0.8$ is younger than its matter-dominated counterpart: $1/2 < H_0t_0 \leq 2/3$, which could be problematic.

Third, a string network comprised of tangled strings, or of very fast-moving, non-intercommuting strings, neither of which can cluster [41]. Unlike the "ordinary" strings discussed in Chapter 7, the string networks envisioned here would never achieve a scaling solution, and the energy density in string would evolve as: $\rho_{\rm string} \propto R^{-2}$ (tangled network) or $\rho_{\rm string} \propto (Rt)^{-1}$ (fast moving strings). In order that such strings not come to dominate the mass density of the Universe before the present epoch, the string tension $\mu \sim \eta^2$ and corresponding symmetry breaking scale η must be very much smaller than the preferred values for ordinary cosmic strings ($G\mu \sim 10^{-6}$ and $\eta \sim 10^{16}$ GeV): $G\mu \lesssim 10^{-30}$ and $\eta \lesssim 10^4$ GeV (tangled network); and $G\mu \lesssim 10^{-20}$ and $\eta \lesssim 10^9$ GeV (fast moving strings). If such strings exist, they must have been created in a "relatively" recent phase transition. For either string scenario, once the strings dominate the mass density the scale factor $R(t)$ evolves as t—just like a curvature-dominated Universe, and for this reason the growth of density perturbations in such models is very similar to an $\Omega_0 < 1$ model (suppressed by a factor of Ω_0^2 or more) [42].

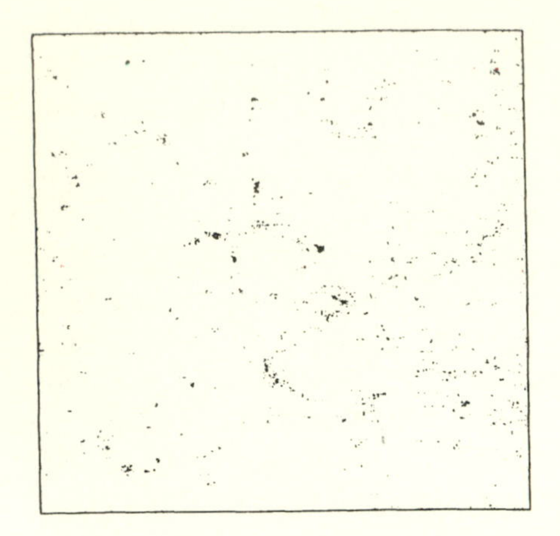
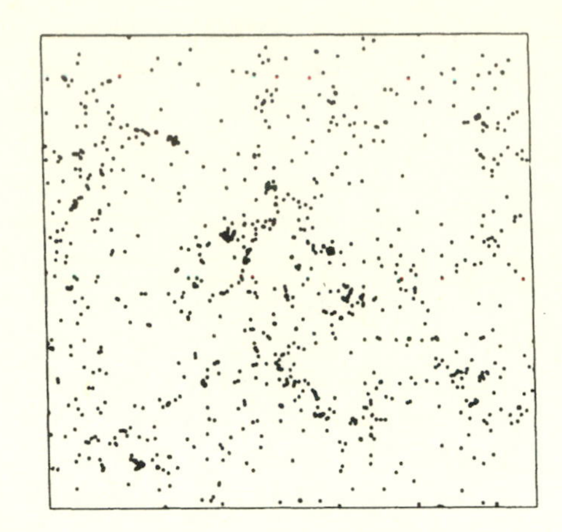

Fig. 9.10: The projected distribution of biased mass points ($\nu_{\rm th} = 2.5$) in an $\Omega_0 = 1$ cold dark matter numerical simulation (right) and all the mass points (left). Notice how much more strongly clustered is the biased sample of mass points (from [20]).

Fourth, a more smoothly distributed population of galaxies that are either too faint to be seen or never lit up [43]. This interesting possiblity is known as "biased" galaxy formation. Biased galaxy formation is an intriguing idea that relies upon some of the fundamental statistical properties of the density field itself. To understand how biasing works, consider the local density field smoothed over a window function appropriate to a galaxy (window size $\sim$ Mpc),

$$\delta(\vec{\mathbf{x}}) \equiv \frac{\langle\delta\rho(\vec{\mathbf{x}})\rangle_W}{\overline{\rho}} = (V_W\overline{\rho})^{-1} \int \delta\rho(\vec{\mathbf{r}} + \vec{\mathbf{x}}) W(\vec{\mathbf{r}}) d^3r. \tag{9.154}$$

If the origin of the density field involves a gaussian-random process (as in inflationary scenarios), then the amplitude of galactic-sized perturbations $\delta(\vec{\mathbf{x}})$ is gaussian distributed, with an *rms* value and variance given by $\overline{\delta} = (\delta M/M)_{\rm GAL}$, cf. (9.17). The probability of having a density contrast δ at a given point in space is proportional to $\exp(-\delta^2/2\overline{\delta}^2)$. In some regions of space the value of $\delta(\vec{\mathbf{x}})$ exceeds $\overline{\delta}$; i.e., some galactic-sized perturbations are particularly overdense. These perturbations will necessarily collapse and form galaxies before the more common-sized perturbations (those with $\delta \simeq \overline{\delta}$).

Roughly speaking then, the 3σ high-density peaks ($\delta = 3\overline{\delta}$) form into galaxies first, followed by the 2σ peaks, and so on. That is, different sized perturbations, for definiteness labeled by $\nu = \delta/\bar{\delta}$, form into galaxies at different epochs. Now consider the clustering properties of the different "classes" of galaxies that form. Because of non-linear interactions between different Fourier components, the clustering properties depend upon ν. Roughly speaking, the correlation function of galaxies that are formed from density perturbations of size ν is given by

$$\xi_\nu(r) \simeq \frac{\nu^2}{1.7(1 + z_{\rm collapse})}\xi(r), \tag{9.155}$$

where $\xi(r)$ is the autocorrelation function of the underlying mass distribution, and $z_{\rm collapse}$ is the red shift of collapse of the ν peaks. That is, the galaxies that form from the high density peaks are more strongly correlated than the mass density itself (see Fig. 9.10). Why this should occur can be illustrated by a simple example.

Imagine a sinusoidal wave with wavelength $\lambda \sim$ Mpc, whose amplitude δ varies in space and is gaussian distributed about the mean value $\bar{\delta}$. Regions where the amplitude exceeds $\bar{\delta}$ by a significant factor are exponentially rare. Now imagine superimposing that wave on a longer wavelength sinusoid. The overall amplitude of the two waves has different statistical properties: It is easier to exceed a given threshold in amplitude, say $\nu_{\rm th} \gg 1$, by riding on one of the crests of the longer wavelength perturbation, and so peaks that exceed the threshold $\nu_{\rm th}$ will be preferentially found on the crests of the underlying longer wavelength wave. This is the origin of the stronger correlations among the high peaks.

A more "Earthly" analogy is sometimes used: Mountains of height greater than 25,000 ft are strongly clustered in the Himalayas. This traces to the fact that the Himalayas lie in the midst of the long-wavelength crest of the long-wavelength "perturbation" created by the collision of the continents of India and Eurasia.

The above discussion involves only elementary statistics. Now for the astrophysics. Suppose for some reason that the formation of a bright galaxy is a threshold phenomenon, i.e., that bright galaxies correspond to peaks $\nu \geq \nu_{\rm th}$, with faint galaxies corresponding to the more common $\nu \leq \nu_{\rm th}$ peaks. Our knowledge of the Universe is based upon the biased sample of bright galaxies ($\nu \geq \nu_{\rm th}$) whose clustering properties are very different from the underlying mass distribution. Several interesting facts follow upon adopting the biasing hypothesis. (1) Since $\xi_{GG}(r)$ is determined by the clustering properties of bright galaxies, the autocorrelation function of the underlying mass density is a factor of order $\nu_{\rm th}^2$ smaller, and the true value of $|\delta_k|$ is a factor of order $\nu_{\rm th}$ smaller than that inferred from $\xi_{GG}(r)$. That is, the mass distribution is less lumpy than the distribution of bright galaxies leads us to believe! (2) There is a large population of more smoothly distributed faint (or failed?) galaxies that is yet to detected. The ratio of the bright to faint galaxies depends upon $\nu_{\rm th}$; for $\nu_{\rm th} \sim 2$ to 3, the ratio is 3 to 5. (3) Because the "faint" galaxies are more numerous and more smoothly distributed, biasing reconciles the $\Omega_0 = 1$ model with the values of Ω_0 determined from measurements of the mass associated with bright galaxies.

The biasing hypothesis resolves the Ω problem in a very clever way. The population of failed galaxies provides the "smoother" component of mass density that accounts for $\Omega_{SM} = 1 - \Omega_{\rm bright} \simeq 0.8$. While the population of more common galaxies that formed from the "1 σ" peaks are fainter, they should have the same mix of WIMPs and baryons. The only difference between these galaxies and the bright galaxies is that for some reason (astrophysical or otherwise) they are not bright today. If we write $\overline{\rho} = \overline{n}_{GAL} M_{GAL}$, as we did in Chapter 1, biasing solves the Ω problem not by changing the mass associated with a galaxy, M_{GAL}, but rather, by increasing the number density of galaxies $\overline{n}_{GAL}$, by a factor of 3 to 5 because the faint galaxies have thus far gone undetected.

Biasing also affects other dynamical determinations of Ω_0. Recall in the last Section we discussed how our own peculiar velocity can be used to infer Ω_0, by exploiting the relationship between the peculiar velocity, Ω_0, and the density contrast, cf. (9.133). The density contrast is determined by counting bright galaxies—which are "overclustered" as compared to the mass density—and so leads to an inferred value of Ω_0 that is smaller than the true value.

The idea of biasing is very plausible and suggests a most interesting resolution to the Ω problem. Biasing has also been suggested as the explanation for the discrepancy between the cluster–cluster and galaxy–galaxy correlation functions; presumably rich clusters also correspond to very rare density peaks [44].

Many important questions about biasing remain to be answered. What is the physical mechanism associated with biasing? For example, could it be gravitational, or is it that the early population of galaxies somehow "poisons" the environment and stunts the development of the later forming, more common galaxies? If there are indeed large numbers of faint galaxies (three or four for every bright galaxy), they must have some observational effects: Do they fill the voids? Do they act as gravitational lenses? Have they already been detected as the QSO absorption line systems? Furthermore, if biasing is correct, then smaller scale dynamical measurements should continue to yield values for Ω_0 that are less than 1, while global determinations should yield unity. We refer the reader interested in more details about the statistical basis of biasing to [45].

9.8 Epilogue

The hints that follow from theories of the very early Universe have helped to focus the efforts of those studying structure formation. We have at present two rather detailed stories—hot and cold dark matter with inflation produced adiabatic density perturbations—and two stories under development—cosmic string with hot and cold dark matter [46]. Neither of the two detailed stories provides a completely satisfactory picture. At present, cold dark matter does a better job, and provides an attractive (if only temporary)[43] paradigm for structure formation.

One thing seems certain: The origin of structure in the Universe most likely traces back to events that took place during the earliest moments of the Universe, and involved fundamental physics that today is still very speculative. If cold dark matter should fall by the wayside, we should not be too worried, as early Universe theorists are far from running out of interesting and attractive ideas. There is another side to this lesson. Since the origins of structure formation seem to be so intimately related to early Universe microphysics, sorting out the details of structure formation provides a potentially powerful and unique probe of that interesting and very distant epoch.

Lest we become overly impressed with our recent progress, we should remind ourselves that our eventual understanding of the details of structure formation may involve a bold departure from the two stories and ideas discussed here. The discussion of structure formation usually focuses solely on the role of gravity, with astrophysical fireworks (energy produced by the galaxies themselves by nuclear and other processes) being ignored. During the very early stages of structure formation (before any astrophysical objects form) this is probably a very good approximation; but once a few galaxies light up, the energy released by astrophysical processes within them may play an important, or even primary, role. This point has been particularly emphasized by Ostriker and others [47]. Moreover, many of the

[43]One of the authors is willing to wager a large amount of hard currency that cold dark matter proves to be correct. The other author is willing to keep the money until the winner is determined.

mechanisms suggested for biasing galaxy formation involve astrophysical processes. One of the most interesting and most speculative early-Universe inspired scenarios for structure formation involves the explosive release of prodigious amounts of electromagnetic energy when superconducting cosmic strings exceed their critical currents and quench [48].

While we have focused on the role of primeval perturbations, i.e., those produced very early on, it could be that the necessary perturbations arise due to physical processes that occur rather late, e.g., processes that result in black hole formation (masses 10^{-6} to $10^{6} M_{\odot}$), with these black holes then acting as the seeds for structure formation [49]. It has also been suggested that a post-recombination phase transition could have triggered structure formation by the production of cosmological-scale domain walls [50]. As Hamlet is reported to have said,

> There are more things in heaven and earth, Horatio,
> Than are dreamt of in your philosophy.

Although much progress has been made toward understanding structure formation in the last decade, a "standard model" of structure formation still does not exist. Vital to developing such is more observational and/or experimental data. And that much-needed experimental and/or observational aid may be coming soon! Consider the relic WIMP hypothesis; experiments are underway, or are being planned, to detect the relic photinos, neutralinos, axions, or even magnetic monopoles that may comprise the halo of our galaxy [51]. Moreover, experiments to search for supersymmetric particles—including the one that could be the dark matter—are ongoing at CERN's $p\bar{p}$ collider (S$p\bar{p}$S) and $e^{\pm}$ collider (LEP), at Fermilab's $p\bar{p}$ collider (the Tevatron), and at SLAC's $e^{\pm}$ linear collider (SLC). The Superconducting Super Collider (SSC), a 20 TeV on 20 TeV pp collider, will be built in Waxahachie, Texas, in the next decade. On the astronomical side, the Hubble Space Telescope and the Gamma-Ray Observatory should be launched soon, and a host of new, large ground-based telescopes are planned or under construction. In addition, space-based x-ray (AXAF) and infrared (SIRTF) observatories are planned. All of these new observatories should help to reveal the Universe at large red shift (large here means $z \sim 2$ to 10). A ten-year effort to produce a red shift survey of 10^6 galaxies (a map of the Universe out to a red shift of about 0.1) using a dedicated 2.5 m telescope is in the early planning stages [52]. Several new CMBR anisotropy experiments (including the COBE satellite) are underway and should ultimately achieve sensitivities of $\delta T/T \sim$

few $\times 10^{-6}$. Efforts to understand the peculiar velocity field and to clarify the cluster–cluster correlation function continue. The next decade should provide exciting developments in the area of structure formation, and in the process we will gain valuable information about fundamental physics and the earliest history of the Universe.

9.9 References

1. P. J. E. Peebles, *The Large-Scale Structure of the Universe* (Princeton Univ. Press, Princeton, 1980); S. Weinberg, *Gravitation and Cosmology* (Wiley, New York, 1972), Chapter 15; L. D. Landau and E. M. Lifshitz, *The Classical Theory of Fields* (Pergamon Press, Oxford, 1975); G. Efstathiou and J. Silk, *Fund. Cosmic Phys.* **9**, 1 (1983); Ya. B. Zel'dovich and I. Novikov, *Relativistic Astrophysics, Vol. II* (The Univ. of Chicago Press, Chicago, 1983); J. R. Primack, in *Proc. of the Int'l School of Physics "Enrico Fermi," Course 92*, ed. N. Cabibbo (North-Holland, Amsterdam, 1984), p. 137; G. Blumenthal, in *The Sante Fe TASI-87*, eds. R. Slansky and G. West (WSPC, Singapore, 1988).

2. D. Lynden-Bell, *Mon. Not. Roy. Astron. Soc.* **136**, 101 (1967); F. H. Shu, *Ap. J.* **225**, 83 (1978).

3. M. Davis and P. J. E. Peebles, *Ap. J.* **267**, 465 (1983).

4. E. M. Lifshitz, *J. Phys. (Moscow)* **10**, 116 (1946); E. M. Lifshitz and I. M. Khalatnikov, *Adv. Phys.* **12**, 185 (1963).

5. W. H. Press and E. T. Vishniac, *Ap. J.* **239**, 1 (1980).

6. J. M. Bardeen, *Phys. Rev. D* **22**, 1882 (1980).

7. J. M. Bardeen, P. J. Steinhardt, and M. S. Turner, *Phys. Rev. D* **28**, 679 (1983).

8. J. R. Bond and A. S. Szalay, *Ap. J.* **276**, 443 (1983).

9. G. Efstathiou and J. Silk, *Fund. Cosmic Phys.* **9**, 1 (1983).

10. P. J. E. Peebles, *Ap. J.* **263**, L1 (1983); M. Davis, et al., *Ap. J.* **292**, 371 (1985).

11. G. Efstathiou and J. R. Bond, *Mon. Not. Roy. Astron. Soc.* **218**, 103 (1986).

12. J. R. Bond, A. S. Szalay, and M. S. Turner, *Phys. Rev. Lett.* **48**, 1636 (1982).

13. Ya. B. Zel'dovich, *Astrofizika* **6**, 319 (1970); *Astron. Astrophys.* **5**, 84 (1970).

14. J. Centrella and A. Melott, *Nature* **305**, 196 (1982).

15. S. D. M. White, C. Frenk, and M. Davis, *Ap. J.* **274**, L1 (1983); *ibid* **287**, 1 (1983).

16. S. D. M. White, in *Inner Space/Outer Space*, eds. E. W. Kolb, et al. (The Univ. of Chicago Press, Chicago, 1986), p. 228.

17. S. Tremaine and J. Gunn, *Phys. Rev. Lett.* **42**, 407 (1979).

18. J. Silk, *Ap. J.* **211**, 638 (1977); M. J. Rees and J. P. Ostriker, *Mon. Not. Roy. Astron. Soc.* **179**, 541 (1977).

19. G. R. Blumenthal, S. M. Faber, J. R. Primack, and M. Rees, *Nature* **311**, 517 (1984).

20. M. Davis, G. Efstathiou, C. Frenk, and S. D. M. White, *Ap. J.* **292**, 371 (1985); *ibid* **327** 507 (1988); *Ap. J. Suppl.* **57**, 241 (1985); *Nature* **317**, 595 (1985); *ibid* **330**, 451 (1987).

21. P. M. Lubin, G. L. Epstein, and G. F. Smoot, *Phys. Rev. Lett.* **50**, 616 (1983); P. M. Lubin, et al., *Ap. J.* **298**, L1 (1985); D. J. Fixsen, E. S. Chang, and D. T. Wilkinson, *Phys. Rev. Lett.* **50**, 620 (1983); I. A. Strukov, et al., *Sov. Astron. Lett.* **13**, 65 (1987).

22. A. Meiksen and M. Davis, *Astron. J.* **91**, 191 (1986); A. Yahil, et al., *Ap. J.* **301**, L1 (1986).

23. M. A. Strauss and M. Davis, in *Large-Scale Motions in the Universe*, eds. V. Rubin and G. V. Coyne (Pontifical Scientific Academy, Vatican City, 1988), p. 256; A. Yahil, *ibid*, p. 219.

24. A. Dressler, et al., *Ap. J.* **313**, L37 (1987); D. Lynden-Bell, et al., *Ap. J.* **326**, 19 (1988); C. A. Collins, et al., *Nature* **320**, 506 (1986); M. Aaronson, et al., *Ap. J.* **302**, 536 (1986).

25. A. Dressler, et al., *Ap. J.* **313**, L37 (1987); D. Lynden-Bell, et al., *Ap. J.* **326**, 19 (1988).

26. N. Kaiser, *Ap. J.* **273**, L17 (1983); N. Vittorio and J. Silk, *Ap. J.* **293**, L1 (1985); N. Vittorio and M. S. Turner, *Ap. J.* **316**, 475 (1987).

27. N. Vittorio and J. Silk, *Ap. J.* **293**, L1 (1985); N. Vittorio and M. S. Turner, *Ap. J.* **316**, 475 (1987).

28. See, e.g., E. Bertschinger, *Ap. J.* **324**, 5 (1988).

29. D. Lynden-Bell, et al., *Ap. J.* **326**, 19 (1988).

30. R. Sunyaev and Ya. B. Zel'dovich, *Comm. Astrophys. Sp. Phys.* **4**, 173 (1972); *Mon. Not. Roy. Astron. Soc.* **190**, 413 (1980). For a very nice pedagogical treatment, see J. Bernstein and S. Dodelson, *Phys. Rev. D*, in press (1990).

31. M. Birkinshaw, G. F. Gull, and H. Hardebeck, *Nature* **309**, 34 (1984).

32. R. K. Sachs and A. M. Wolfe, *Ap. J.* **147**, 73 (1967).

33. P. M. Lubin, G. L. Epstein, and G. F. Smoot, *Phys. Rev. Lett.* **50**, 616 (1983); D. J. Fixsen, E. S. Chang, and D. T. Wilkinson, *Phys. Rev. Lett.* **50**, 620 (1983); A. A. Klypin, et al., *Sov. Astron. Lett.* **13**, 104 (1987); P. M. Lubin, et al., *Ap. J.* **298**, L1 (1985).

34. J. Silk, in *Inner Space/Outer Space*, eds. E. W. Kolb, et al. (The Univ. of Chicago Press, Chicago, 1986), p. 143.

35. J. Uson and D. Wilkinson, *Nature* **312**, 427 (1984).

36. A. C. S. Readhead, et al., *Ap. J.*, in press (1989).

37. M. Wilson and J. Silk, *Ap. J.* **243**, 14 (1981); P. J. E. Peebles, *Ap. J.* **315**, L73 (1987); A. C. S. Readhead, et al., *Ap. J.*, in press (1989).

38. N. Vittorio and J. Silk, *Ap. J.* **285**, L39 (1984); J.R. Bond and G. Efstathiou, *ibid* **285**, L44 (1984).

39. M. S. Turner, G. Steigman, and L. L. Krauss, *Phys. Rev. Lett.* **52**, 2090 (1984); P. J. E. Peebles, *Ap. J.* **284**, 439 (1984).

40. D. A. Dicus, E. W. Kolb, and V. L. Teplitz, *Phys. Rev. Lett.* **39**, 168 (1977); M. S. Turner, G. Steigman, and L. L. Krauss, *Phys. Rev. Lett.* **52**, 2090 (1984); M. Fukugita and T. Yanagida, *Phys. Lett.* **144B**, 386 (1984); G. Gelmini, D. N. Schramm, and J. Valle, *Phys. Lett.* **146B**, 311 (1984).

41. A. Vilenkin, *Phys. Rev. Lett.* **53**, 1016 (1984).

42. M. S. Turner, *Phys. Rev. Lett.* **54**, 252 (1985).

43. N. Kaiser, *Ap. J.* **284**, L9 (1985); in *Inner Space/Outer Space*, eds. E. W. Kolb, et al. (The Univ. of Chicago Press, Chicago, 1986), p. 258; J. M. Bardeen, *ibid*, p. 212.

44. N. Kaiser, *Ap. J.* **284**, L9 (1985).

45. J. M. Bardeen, J. R. Bond, N. Kaiser, and A. S. Szalay, *Ap. J.* **304**, 15 (1986).

46. N. Turok and R. Brandenberger, *Phys. Rev. D* **33**, 2175 (1986); R. Brandenberger, et al., *Phys. Rev. Lett.* **59**, 2371 (1987); R. J. Scherrer, *Ap. J.* **320**, 1 (1987); A. Melott and R. J. Scherrer, *Nature*, **328**, 691 (1987); R. J. Scherrer, A. Melott, and E. Bertschinger, *Phys. Rev. Lett.* **62**, 379 (1989); E. Bertschinger, *Ap. J.* **316**, 489 (1987).

47. J. P. Ostriker and L. Cowie, *Ap. J.* **243**, L127 (1981); S. Ikeuchi, *Publ. Astr. Soc. Jpn.* **33**, 211 (1981); C. J. Hogan, *Mon. Not. Roy. Astron. Soc.* **202**, 1101 (1983).

48. J. P. Ostriker, C. Thompson, and E. Witten, *Phys. Lett.* **180B**, 231 (1986).

49. B. J. Carr, *Nucl. Phys.* **B252**, 81 (1985).

50. C. Hill, J. N. Fry, and D. N. Schramm, *Comm. Nucl. Part. Phys.* **19**, 25 (1989).

51. J. Primack, D. Seckel, and B. Sadoulet, *Ann. Rev. Nucl. Part. Sci.* **38**, 751 (1988).

52. *The Million Red Shift Project*, a consortium of astronomers at The University of Chicago and Princeton University (1989).

10

AXIONS

Many of the most interesting theoretical speculations about physics beyond the standard model involve energy scales that cannot be probed in terrestrial laboratory experiments. Fortunately, the Heavenly Laboratory—in the guise of the early Universe and various contemporary astrophysical sites—has proven to be invaluable in studying physics in regimes beyond the reach of terrestrial laboratories. In Chapter 4 we discussed some of the interesting constraints on particle properties that follow from primordial nucleosynthesis, in Chapter 5 the dazzling array of astrophysical and cosmological constraints to the properties of neutrinos, and in Chapter 7 the stringent astrophysical constraints to the flux of relic monopoles. There is probably no better example of the use of the Heavenly Laboratory to probe fundamental physics than the axion, and we will devote this Chapter to the multitude of astrophysical and cosmological implications of axions. The axion is a prediction of the most elegant solution to the strong-CP problem of quantum chromodynamics (QCD), Peccei-Quinn symmetry. In the Peccei-Quinn scheme, the mass of the axion is essentially a free parameter that could plausibly be anywhere the range of 10^{-12}eV to 1 MeV. Laboratory experiments have probed only two or three of these eighteen orders of magnitude; the Heavenly Laboratory has probed some fourteen orders of magnitude, and leaves open only two windows on the axion. Moreover, if the axion does indeed exist, it has important astrophysical and cosmological implications, e.g., providing closure density.

10.1 The Axion and the Strong-CP Problem

QCD is a remarkable theory and is almost universally believed to be *the theory* of the strong interactions. Aside from technical difficulties associated with actually calculating the spectrum of states in the theory, QCD has but one serious blemish: the strong-CP problem [1]. The problem arises from the fact that non-perturbative effects violate CP, T, and P, and unless suppressed would lead to an electric dipole moment for the neutron which is in excess of experimental limits by some ten orders of magnitude.

Owing to the existence of non-trivial, vacuum gauge configurations, non-Abelian gauge theories have a rich vacuum structure. These degenerate vacuum configurations are characterized by distinct homotopy classes that cannot be continuously rotated into one another and are classified by the topological winding number n associated with them,

$$n = \frac{ig^3}{24\pi^2}\int d^3x \ \mathrm{Tr}\, \epsilon_{ijk} A^i(\vec{x})A^j(\vec{x})A^k(\vec{x}), \tag{10.1}$$

where g is the gauge coupling, A^i is the gauge field, and the temporal gauge ($A^0 = 0$) has been used. The correct vacuum state of the theory is a superposition of all the degenerate states $|n\rangle$,

$$|\Theta\rangle = \sum_n \exp(-in\Theta)|n\rangle, \tag{10.2}$$

where *a priori* Θ is an arbitrary parameter in the theory which must be measured. The state $|\Theta\rangle$ is referred to as "the Θ-vacuum." By appropriate means the effects of the Θ-vacuum can be recast into a single, additional non-perturbative term in the QCD Lagrangian,

$$\begin{aligned} \mathcal{L}_{\rm QCD} &= \mathcal{L}_{\rm PERT} + \bar{\Theta}\frac{g^2}{32\pi^2}G^{a\mu\nu}\tilde{G}_{a\mu\nu}, \\ \bar{\Theta} &= \Theta + \mathrm{Arg}\,\det\mathcal{M}, \end{aligned} \tag{10.3}$$

where $G^{a\mu\nu}$ is the gluon field strength tensor, $\tilde{G}^{a\mu\nu}$ is the dual of the field strength tensor,[1] and $\mathcal{M}$ is the quark mass matrix. Note that the effective Θ term in the theory involves both the bare Θ term and the phase of the quark mass matrix. Since the $G\tilde{G}$ term is a total derivative, it affects neither the equations of motion nor the perturbative aspects of the theory. However, the existence of such a term in the QCD Lagrangian violates CP, T, and P, and leads to a neutron electric dipole moment of order $d_n \simeq 5\times 10^{-16}\bar{\Theta}$ e cm. The present experimental bound to the electric dipole moment of the neutron, $d_n \lesssim 10^{-25}$ e cm, constrains $\bar{\Theta}$ to be less

[1] The dual of the field strength tensor $\tilde{G}^{a\mu\nu} = \epsilon^{\mu\nu}{}_{\alpha\beta}G^{a\alpha\beta}$, and $G\tilde{G} \propto \vec{\mathbf{E}}_{\rm color}\cdot\vec{\mathbf{B}}_{\rm color}$.

than (or of order of) 10^{-10}. Why is the $\bar{\Theta}$ parameter in QCD so small? This is the strong-CP problem.

Before going on to discuss the axion, some general comments about the strong-CP problem are in order. The unwanted, non-perturbative term in the Lagrangian arises due to two separate and independent effects: the Θ structure of the pure QCD vacuum and electroweak effects involving quark masses. In the limit that one or more of the quark species are massless the $G\tilde{G}$ term has no physically measurable effects: The Θ term can be rotated away by a chiral rotation of the quark fields, and there is no strong-CP problem. In the absence of a massless quark species (which seems to be the case in our world, but see [2]), the effective $G\tilde{G}$ term is made of two unrelated contributions that have no reason to cancel.

One might be tempted to ignore this mysterious topological contribution to the QCD Lagrangian on grounds that one has no need for it or in the hope that its absence will be understood at some future date. This is not a particularly good thing to do since the Θ structure of the QCD vacuum has at least one beneficial feature: the resolution of the $U(1)_A$ puzzle. In the absence of the $G\tilde{G}$ term one would expect four Nambu-Goldstone bosons when the $U(2)_L \otimes U(2)_R$ global symmetry (of a massless, two flavor QCD theory) is spontaneously broken by nonperturbative QCD effects. These Nambu-Goldstone bosons are the π and η mesons, the η meson being the Goldstone mode of the spontaneously broken $U(1)_A$ global symmetry. When non-zero u and d quark masses are taken into account one can show that the mass of the η must satisfy $m_\eta \leq \sqrt{3}m_\pi$, which is badly contradicted by reality [3]. The existence of the Θ vacuum structure of QCD corrects this erroneous prediction and solves the $U(1)_A$ problem.

So it seems likely that the Θ structure of the QCD vacuum and the strong-CP problem are to be taken seriously. The most compelling solution is the one proposed by Peccei and Quinn in 1977 [4]. The crux of their idea is to make $\bar{\Theta}$ a dynamical variable, which is driven to zero by the action of its classical potential. This feat is accomplished by introducing an additional global, chiral symmetry, now known as PQ (or Peccei-Quinn) symmetry, which is spontaneously broken at a scale f_{PQ}. Weinberg and Wilczek pointed out that because $U(1)_{PQ}$ is a spontaneously broken global symmetry, there must be a Nambu-Goldstone boson, "the axion," associated with it [5]. Because $U(1)_{PQ}$ suffers from a chiral anomaly,[2] the axion is not massless but acquires a small mass of order $\Lambda^2_{\rm QCD}/f_{PQ}$. Moreover,

[2]While PQ symmetry is a classical symmetry of the Lagrangian, quantum-mechanical effects, in the form of triangle diagrams, break the symmetry.

due to the anomaly a term in the QCD Lagrangian of the form

$$\mathcal{L}_{\rm QCD} = \cdots + C_a \frac{a}{f_{PQ}} \frac{g^2}{32\pi^2} G_a^{\mu\nu} \tilde{G}^a_{\mu\nu} \tag{10.4}$$

arises, where a is the axion field and C_a is a model-dependent constant. Note the similarity of this term to the previously discussed $\bar{\Theta}$ term. These two terms amount to a potential for the axion field, which is minimized by

$$\langle a \rangle = -\frac{\bar{\Theta} f_{PQ}}{C_a}. \tag{10.5}$$

Moreover, for this value of $\langle a \rangle$ the coefficient of the offending $G\tilde{G}$ term vanishes! Expanding the axion field about its vacuum expectation value $\langle a \rangle$, the axion part of the QCD Lagrangian is

$$\mathcal{L}_{\rm axion} = \frac{1}{2}\partial^\mu a \partial_\mu a + C_a \frac{a}{f_{PQ}} \frac{g^2}{32\pi^2} G_a^{\mu\nu} \tilde{G}^a_{\mu\nu} \tag{10.6}$$

(where the axion's other interactions have yet to be included). Note that the $\bar{\Theta}$ parameter has been effectively replaced by the dynamical axion field, and the axion acquires a mass due to the non-perturbative $G\tilde{G}$ term.

In an axion model then, the price for resolving the strong-CP problem is the existence of an additional, spontaneously broken global symmetry[3] and its associated pseudo-Goldstone boson—the axion. *A priori* the mass of the axion (or equivalently the PQ symmetry breaking scale) is arbitrary: All values solve the strong-CP problem equally well. Taking f_{PQ} to be somewhere between 100 GeV and 10^{19} GeV, the axion mass lies between about 1 MeV and 10^{-12}eV—a span of some eighteen or so orders of magnitude to search.

10.1.1 Properties of the axion

So much for the high-brow theory and philosophy! If we are to search for the axion, we must know how it couples to ordinary matter. As alluded to above, an axion model has one basic free parameter: the axion mass, or

[3]Such a global symmetry often arises in supersymmetric and superstring-inspired models in any case.

equivalently the PQ symmetry-breaking scale. They are related by

$$m_a = \frac{\sqrt{z}}{1+z}\frac{f_\pi m_\pi}{f_{PQ}/N} \simeq 0.62\,\text{eV}\frac{10^7\,\text{GeV}}{f_{PQ}/N}, \tag{10.7}$$

where $z \equiv m_u/m_d \simeq 0.56$, $m_u \simeq 5$ MeV is the mass of the up quark, $m_d \simeq 9$ MeV is the mass of the down quark, $m_\pi = 135$ MeV and $f_\pi = 93$ MeV are the pion mass and decay constant, and N is the *color* anomaly of the PQ symmetry.[4] The axion field a is related to the $\bar{\Theta}$ angle by

$$a = (f_{\rm PQ}/N)\bar{\Theta}. \tag{10.8}$$

The effective Lagrangian for the interaction of the axion with ordinary matter (nucleons, electrons, and photons) is

$$\begin{aligned} \mathcal{L}_{\rm int} &= i\frac{g_{\rm aNN}}{2m_N}\partial_\mu a(\bar{N}\gamma^\mu\gamma_5 N) + i\frac{g_{\rm aee}}{2m_e}\partial_\mu a(\bar{e}\gamma^\mu\gamma_5 e) \\ &\quad + g_{\rm a\gamma\gamma}\, a\,\vec{\mathbf{E}}\cdot\vec{\mathbf{B}}; \end{aligned} \tag{10.9}$$

the corresponding Feynman diagrams are shown in Fig. 10.1.[5]

The axion couplings $g_{\rm aii}$ are given by

$$\begin{aligned} g_{\rm aee} &= \left[\frac{X_e}{N} + \frac{3\alpha^2}{4\pi}\left(\frac{E_{PQ}}{N}\ln(f_{PQ}/m_e) - 1.95\ln(\Lambda_{\rm QCD}/m_e)\right)\right] \\ &\quad \times\frac{m_e}{f_{PQ}/N} \\ g_{\rm a\gamma\gamma} &= \frac{\alpha/2\pi}{f_{PQ}/N}(E_{PQ}/N - 1.95) \\ g_{\rm ann} &= [(-F_{A0} + F_{A3})(X_u/2N - 0.32) \\ &\quad +(-F_{A0} - F_{A3})(X_d/2N - 0.18)]\frac{m_N}{f_{PQ}/N} \\ g_{\rm app} &= [(-F_{A0} - F_{A3})(X_u/2N - 0.32) \\ &\quad +(-F_{A0} + F_{A3})(X_d/2N - 0.18)]\frac{m_N}{f_{PQ}/N} \end{aligned} \tag{10.10}$$

---

[4]Normalization conventions and notation differ; see [6].

[5]If there is only one Nambu-Goldstone boson in the problem, the pseudo-vector coupling may be written as a pseudo-scalar coupling, e.g., $ig_{\rm aNN}a(\bar{N}\gamma_5 N)$, by means of a suitable phase rotation of the fermion fields. For further discussion, see [7].

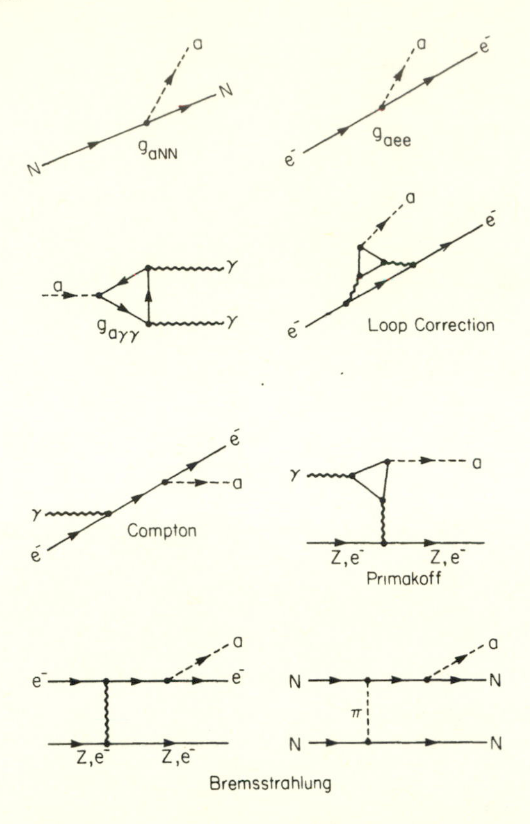

Fig. 10.1: Axion couplings to ordinary particles and the dominant axion emission processes in stars. The "pion–axion" conversion process, $N + \pi \leftrightarrow N + a$, corresponds to the nucleon–nucleon, axion bremsstrahlung diagram, without the lower nucleon line.

where E_{PQ} is the *electromagnetic* anomaly of the PQ symmetry, $1.95 = 2(4+z)/3(1+z)$, $0.32 = 1/2(1+z)$, $0.18 = z/2(1+z)$, $\alpha \simeq 1/137$ is the fine structure constant, $F_{A0} \simeq -0.75$ is the axial-vector, isoscalar part of the pion–nucleon coupling, $F_{A3} \simeq -1.25$ is the axial-vector, isovector part of the pion–nucleon coupling, $m_e = 0.511$ MeV is the mass of the electron, and $m_N \simeq 940$ MeV is the mass of the nucleon. The quantities X_i $(i = u,\, d,\, e)$ are the PQ charges of the u and d quarks and the electron. Depending upon the PQ charge of the electron, the axion can be classified as one of two generic types: hadronic [8], the case where $X_e = 0$ (no tree-level coupling to the electron); or DFSZ [9], the case where all the X_i are of order unity.[6] Even hadronic axions couple to the electron through a

[6]In the original model of Dine, Fischler, and Srednicki [9], $N = 6$, $X_e/N = \cos^2\beta/3$,

small loop correction, proportional to α^2, arising from the anomalous two-photon coupling of the axion (see Fig. 10.1). The axion–electron coupling is of great importance in determining the astrophysical effects of the axion, and for this reason the the astrophysical constraints to DFSZ and hadronic axions must be discussed separately.

Note that all the axion couplings are proportional to $1/(f_{PQ}/N)$, or equivalently, m_a: the smaller the axion mass, or the larger the PQ-SSB scale, the more weakly the axion couples. The coupling of the axion to the photon arises through the electromagnetic anomaly of the PQ symmetry and allows the axion to decay to two photons, with a lifetime,

$$\tau_a = 6.8 \times 10^{24}\,\mathrm{sec}\frac{(m_a/\mathrm{eV})^{-5}}{[(E_{PQ}/N - 1.95)/0.72]^2}. \tag{10.11}$$

The coupling of the axion to two photons depends upon the ratio of the electromagnetic to the color anomaly; when the axion is incorporated into the simplest GUTs, $E_{PQ}/N = 8/3$ and $(E_{PQ}/N - 1.95) \simeq 0.72$.[7]

The axion–nucleon coupling arises from two roughly equal contributions: the tree-level coupling of the axion to up and down quarks and a contribution that arises due to axion–pion mixing (both the axion and pion are Nambu-Goldstone bosons with the same quantum numbers, and the physically propagating states mix). This means that even a hadronic axion that does not couple to light quarks at tree level[8] still has a coupling to nucleons comparable to that of a DFSZ axion.

While we have taken the view here that *a priori* the axion mass (or symmetry-breaking scale) is an arbitrary parameter to be determined by experiment, that viewpoint belies the history of the axion. The original axion proposed by Peccei and Quinn was based upon a PQ symmetry-breaking scale equal to that of the weak scale ($f_{PQ} \sim 250$ GeV), leading to an axion mass of about 200 keV. Such an axion was quickly ruled out by a combination of direct experimental searches and astrophysical arguments. The most sensitive laboratory searches involve looking for rare Kaon and

---

$X_u = 1 - \cos 2\beta$, and $X_d = 1 + \cos 2\beta$. Here β parameterizes the ratio of the "up" and "down" PQ vacuum expectation values.

[7]However, it is possible that E_{PQ}/N could have a different value, even 2, in which case its two-photon coupling ($E_{PQ}/N - 1.95 \simeq 0.05$) would be strongly suppressed [10]. We should keep this fact in mind, as it will be of some importance when discussing the astrophysical effects of a hadronic axion.

[8]This was the case with the original "invisible" axion (see Kim [8]), which only coupled to one very heavy, exotic quark.

quarkonium decays: $K^+ \to \pi^+ + a$; $J/\psi \to a + \gamma$; and $\Upsilon \to a + \gamma$. The absence of such decays implies that if the axion exists, its mass must be less than about 10 keV (or $f_{PQ} \gtrsim 10^3$ GeV) [11]. The most restrictive astrophysical limit to the mass of the original axion followed from considering the effect of axion emission on stellar evolution, particularly red-giant stars. These astrophysical arguments also ruled out the original axion [12].

After the original axion was ruled out, the "invisible axion" was invented [8, 9]. The invisible axion is characterized by a symmetry-breaking scale much larger than 250 GeV, mass much less than 200 keV, and interactions that are necessarily extremely weak (recall $g_{aii} \propto m_a \propto f_{PQ}^{-1}$). And of course, it goes without saying that once the weak scale had been ruled out, all educated bets as to the PQ-SSB scale were off![9]

10.2 Axions and Stars

The rather uncomplicated life of a star involves only the simple struggle to regain the state of nuclear statistical equilibrium (NSE) lost during primordial nucleosynthesis (see Chapter 4). The far-from-equilibrium condition is reflected in the enormous nuclear free energy associated with the primordial composition of the Universe (for every 10 atoms, roughly 9 H atoms, 1 He atom, and a trace of D, ^{3}He, and ^{7}Li). Given the short timescale intrinsic to nuclear interactions, it might be a great surprise that stars live as long as they do: A star like our sun burns hydrogen for 10^{10} yr. However NSE is like innocence: Once lost, it is not easily regained.

The reason for stellar longevity is simple; the rate at which a star can liberate its nuclear free energy is not controlled by nuclear reaction rates, but rather by the rate at which the nuclear energy liberated can be transported through the star and radiated into the vacuum of space. Under the conditions that exist in a typical star, say our sun, the mean free path of a photon is only about one cm (!), and the time required for a photon liberated at the center of the sun to make its way to the surface is of order 10^7 yr. The enormous opacity of ordinary matter to photons of course traces to the strength of the electromagnetic interactions: The Thomson cross section is very large, $\sigma_T \simeq 0.67 \times 10^{-24}$ cm^2. Stellar longevity then is

[9] Very recently it has been argued that wormhole effects might cause the wave function of the Universe to be very highly peaked at $\bar{\Theta} = \pi$ (a CP-conserving value). If correct, this would solve the strong-CP problem and obviate the need for the axion [13]. In any case, Nature may still provide us with an axion, as PQ symmetry seems to be generic to supersymmetric and superstring inspired models.

explained by the large interaction cross section of the photon with ordinary matter.[10]

The existence of a light (i.e., compared to typical stellar temperatures, $T \sim$ keV to MeV), weakly interacting particle has the potential to accelerate the evolutionary process of stars of all types by more efficiently transporting energy out of the star, thereby shortening stellar lifetimes. To carry away the free energy liberated by the nuclear reactions in a star effectively, the hypothetical "super coolant" must interact weakly enough so that it streams out without being hindered by numerous interactions, but strongly enough so that such particles are produced in sufficient numbers to carry away large amounts of energy. As a rough rule of thumb, the most efficient coolant will have an interaction length comparable to the size of the object.[11] Nature has provided us with at least three potential super coolants—the three neutrino species—and contemporary theorists have postulated another, the axion.

Before turning to the axion, let us orient ourselves by considering neutrino cooling in stars. Because of the nature of the weak interaction, neutrino cross sections are highly temperature sensitive, proportional to $G_F^2 T^2$. In ordinary main-sequence stars the neutrino luminosity $\mathcal{L}_\nu$ is proportional to T_c^8, whereas as we shall see shortly, the photon luminosity $\mathcal{L}_\gamma \propto T_c^{1/2}$ (T_c is the central temperature of the star). Only in stars hotter than about 10^8 K does neutrino cooling begin to compete with photon cooling; for these stars (oxygen- and silicon-burning stars and beyond) neutrino emission is the dominant cooling mechanism, and as a result the timescale for these burning phases is greatly reduced as neutrinos can just stream out (O burning timescale is of order 10^5 yr; Si burning timescale is of order seconds). In fact, long ago it was argued that the existence of carbon-burning stars places a limit to G_F; had G_F been a factor of three or so larger, the evolution timescale for C stars would have been greatly reduced due to neutrino emission, so much so that C burning stars would evolve so quickly that they would not exist long enough to be observed [15]. The approximate timescales for the various stellar burning phases are summarized in Fig. 10.2.

The most dramatic example of the importance of neutrino cooling is SN 1987A. Thermal neutrinos carried away the bulk of the 3×10^{53} erg

---

[10]For a textbook discussion of stars and stellar evolution, see [14].

[11]An interaction length much smaller than the object means that cooling will be regulated by the slow process of diffusion, while an interaction length much larger than the object means that the coolant interacts so weakly that its production rate is necessarily very small.

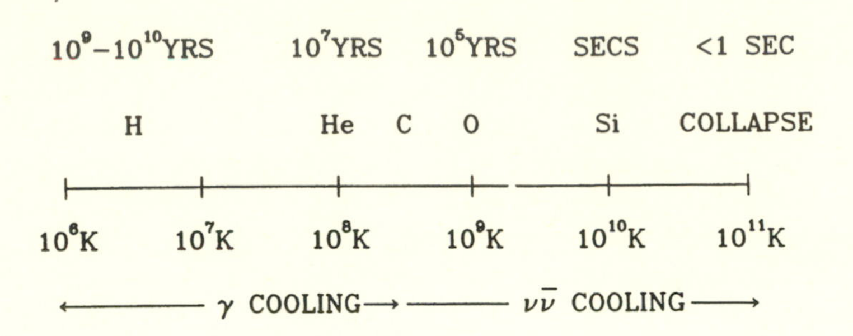

Fig. 10.2: Timescales and temperatures for the main stages of stellar nucleosynthesis.

of binding energy of the newly formed neutron star in about 10 sec—as indicated by the detection of nineteen neutrino events [16, 17], and as predicted for years by theorists. This observation can be used to limit the number of neutrino species. If the number were nine, say, rather than three, cooling would have proceeded three times as rapidly (in about 3 sec), in contradiction to observation.

The effect of axion emission on stars is clearly the acceleration of their evolution and shortening of their lifetimes.[12] In main-sequence stars and red giants, the primary axion emission processes are the Compton-like process, $\gamma + e^- \to a + e^-$; and axion bremsstrahlung, $e^- + Z \to a + e^- + Z$, both of which are proportional to $g_{aee}^2 \propto m_a^2$. Of lesser importance (unless g_{aee} vanishes at tree level, as it does for a hadronic axion) is the Primakoff process $\gamma + Z(\text{ or } e^-) \to a + Z(\text{ or } e^-)$. In stars of low mass ($M \lesssim 0.2 M_\odot$) emission through the axio-electric effect (the analogue of the photo-electric effect) is also important [18]. Feynman diagrams for the stellar axion emission processes are shown in Fig. 10.1.

It is convenient to express the rate at which axions carry away energy from stars, $\dot{\varepsilon}_a$, in erg per gram per second. For the Compton-like processs $\dot{\varepsilon}_a$ is given in terms of the production cross section σ, electron number density n_e, photon number density n_γ, mass density ρ, and axion energy E_a, by:

$$\dot{\varepsilon}_a(C) = n_e n_\gamma \frac{\langle E_a \, \sigma |v| \rangle}{\rho}, \tag{10.12}$$

where brackets indicate thermal averaging. The total axion luminosity of

[12]The importance of this fact for constraining the properties of the axion was first pointed out in [12].

a star is related to $\dot{\varepsilon}_a$ by

$$\mathcal{L}_a = \int_{\rm star} \dot{\varepsilon}_a \, dM, \qquad (10.13)$$

where dM is the stellar mass element.

In the regime where electrons are non-degenerate and non-relativistic (relevant conditions for most stellar interiors), and where the axion is relativistic, the axion emission rates for the Compton-like process, the bremsstrahlung process, and the Primakoff process are given by [19]

$$\begin{aligned}
\dot{\varepsilon}_a(C) &= \frac{n_e}{\rho}\frac{40\alpha}{\pi^2} g_{aee}^2 \zeta(6) \frac{T^6}{m_e^4} \\
&\simeq 1.8\,\mu_e^{-1} (T/10^7\,{\rm K})^6 \\
&\quad \times [g_{aee}/(Nm_e/f_{\rm PQ})]^2 (m_a/{\rm eV})^2\,{\rm erg\,g^{-1}\,sec^{-1}}; \qquad (10.14)\\
\dot{\varepsilon}_a(B) &= \frac{n_e n_Z}{\rho}\frac{64Z^2\alpha^2}{15(2\pi)^{3/2}}\frac{T^{5/2}}{m_e^{7/2}} g_{aee}^2 \\
&\simeq 0.15\,(\mu_e\mu')^{-1} (\rho/{\rm g\,cm^{-3}})(T/10^7\,{\rm K})^{5/2} \\
&\quad \times [g_{aee}/(Nm_e/f_{\rm PQ})]^2 (m_a/{\rm eV})^2\,{\rm erg\,g^{-1}\,sec^{-1}}; \qquad (10.15)\\
\dot{\varepsilon}_a(P) &= \frac{n_Z}{\rho}\frac{2Z^2\alpha}{\pi} g_{a\gamma\gamma}^2\, T^4 \,[6\zeta(4)[\ln 2 - 0.5 - \ln(\omega/T)] + 7.74] \\
&\simeq 0.2\mu'^{-1} [(E_{PQ}/N - 1.95)/0.72]^2 (T/10^7\,{\rm K})^4 \\
&\quad \times [1 - 0.72\ln(\omega/T)]\,(m_a/{\rm eV})^2\,{\rm erg\,g^{-1}\,sec^{-1}}; \qquad (10.16)
\end{aligned}$$

where n_Z is the number density of nuclear species ^{Z}A, $\zeta(4) = \pi^4/90$, $\zeta(6) = \pi^6/945$, and ω is the plasma mass of the photon under the ambient conditions. Here $\mu_e = [\Sigma_i X_i Z_i/A_i]^{-1}$ is the mean molecular weight per electron, $\mu' = [\Sigma_i X_i Z_i^2/A_i]^{-1}$ is the mean molecular weight, weighted by Z_i, and X_i is the mass fraction of nuclear species ^{Z}A.[13]

10.2.1 Axions and the sun

To begin, consider the sun. At the center of the sun the temperature is about 1.6×10^7 K, and nuclear reactions ($4p \rightarrow {}^4{\rm He} + 2e^+ + 2\nu_e + \gamma$) liberate

[13]For pure hydrogen, $X_H = 1$, $\mu_e = \mu' = 1$; while for pure ^{4}He, $X_{He} = 1$, $\mu_e = 2$ and $\mu' = 1$.

free energy at the rate of a few erg g^{-1} sec^{-1}. On the other hand, DFSZ axion emission via the Compton process carries away energy at the rate of

$$\dot{\epsilon}_a(C) \sim 10(m_a/\mathrm{eV})^2 \, \mathrm{erg\, g^{-1}\, sec^{-1}}. \tag{10.17}$$

Roughly speaking then, *if* a DFSZ axion of mass greater than about 1 eV existed, axions would carry energy away from the center of the sun faster than nuclear reactions could generate it. That statement bears closer scrutiny!

The thermal time constant of a star like the sun is only about 10^7yr; this is the time required for the star to radiate away its thermal energy reserves and is known as the Kelvin-Helmholtz time. Thus, an axion luminosity greater than the rate at which nuclear energy is released can be tolerated only on a short timescale ($\lesssim 10^7$yr or so). Put another way, if an axion of mass greater than about 1 eV suddenly came into being, the sun would have to "adjust itself" on a timescale shorter than 10^7 yr (relative to an "axionless" model) to re-establish energy balance. As we shall see, in a hypothetical star in which one "turns on" axion emission, the star contracts to raise its temperature and nuclear energy liberation rate to balance axion losses. In the process it would also raise its photon luminosity, and as a result of both axion emission and enhanced photon emission its lifetime would be shortened. On the basis of this we see that the observable that allows one to constrain axion emission from the sun is its age at a given epoch in its evolution. A convenient means of specifying the epoch is the ^{4}He content. For $m_a \gtrsim 1$ eV, a sun with our sun's ^{4}He abundance would be *younger* than our sun is known to be.

Let us consider the sun in slightly more detail [20]. As we have discussed, the photon luminosity of the sun is determined by the opacity of solar material. Just by analyzing hydrostatic equilibrium and energy transport in stars like the sun (i.e., stars less massive than about 2 $M_\odot$), Chandrasekhar [14] has derived a remarkable formula (the so-called luminosity formula) that relates the photon luminosity of a star to its central temperature,[14]

$$\mathcal{L}_\gamma \propto (G\mu^7)M^5 T_c^{1/2}, \tag{10.18}$$

[14]The derivation of this formula is straightforward: Hydrostatic equilibrium implies that $T \sim GM\mu/R$ (R is the radius of the star). Radiative transfer implies that $\mathcal{L}_\gamma \sim R^3T^4/\tau_{\rm diff}$, where $\tau_{\rm diff} \sim R^2/\lambda$ is the photon diffusion time, and $\lambda = (\kappa\rho)^{-1}$ is the mean free path (κ is the opacity). For stars like the sun, $\kappa \propto \rho T^{-3.5}$ and the radiative enrtopy per nucleon is $T^3/(\rho/m_N) \sim G^3 m_N^4 \mu^3 M^2$. Taken together, the equations of hydrostatic equilibrium and radiative transfer lead to (10.18).

where M is the mass of the star, T_c is the central temperature, and μ is the mean molecular weight per particle, $\mu = (\rho/m_N)/(n_e + \sum_Z n_Z) = [\sum_Z (X_i/A_i)(Z+1)]^{-1}$.

Energy balance requires that the energy liberated by nuclear reactions,

$$Q_{\rm nuc} = \int_{\rm star} \dot{\varepsilon}_{\rm nuc}\, dM, \tag{10.19}$$

be equal to the photon (plus axion) luminosity. In general the nuclear energy liberation rate is very temperature dependent and can be parameterized as $\dot{\varepsilon}_{\rm nuc} \propto \rho T^n \propto T^{n+3}$, where for the sun $n \simeq 4$.[15] In the absence of axion emission, energy equilibrium in a star requires the photon luminosity to be balanced by the rate of nuclear energy generation: $\mathcal{L}_\gamma^0 = Q_{\rm nuc}^0$ (0 will denote values for a star without axion emission).

Now consider how axion emission would modify the structure of this star. The additional luminiosity in axions, $\mathcal{L}_a$, must be compensated for by an increased energy generation rate, and energy balance now reads $\mathcal{L}_\gamma + \mathcal{L}_a = Q_{\rm nuc}$, where $Q_{\rm nuc} = Q_{\rm nuc}^0 + \delta Q_{\rm nuc}$, $\mathcal{L}_\gamma = \mathcal{L}_\gamma^0 + \delta\mathcal{L}_\gamma$. Using Chandrasekhar's luminosity formula, we see that $\delta\mathcal{L}_\gamma/\mathcal{L}_\gamma^0 = 0.5(\delta T_c/T_c^0)$. Since $\dot{\varepsilon}_{\rm nuc} \propto T^{n+3}$, we have $\delta Q_{\rm nuc}/Q_{\rm nuc}^0 = 7(\delta T_c/T_c^0)$. Using the perturbed energy balance equation in the form $\delta\mathcal{L}_\gamma + \mathcal{L}_a = \delta Q_{\rm nuc}$, and expressing $\mathcal{L}_a$ as a fraction of the unperturbed nuclear energy generation rate ($\mathcal{L}_a = \epsilon Q_{\rm nuc}^0$), we find that $\delta\mathcal{L}_\gamma/\mathcal{L}_\gamma^0 + \epsilon = \delta Q_{\rm nuc}/Q_{\rm nuc}^0$, or $\delta T_c/T_c^0 = \epsilon/6.5$. Using this expression for $\delta T_c/T_c^0$, we can compute the change in photon luminosity, nuclear energy generation rate, and radius

$$\begin{aligned}
\delta T_c/T_c^0 &= \epsilon/6.5 \qquad (\epsilon \equiv \mathcal{L}_a/Q_{\rm nuc}^0), \\
\delta\mathcal{L}_\gamma/\mathcal{L}_\gamma^0 &= \epsilon/13, \\
\delta Q_{\rm nuc}/Q_{\rm nuc}^0 &= 14\epsilon/13, \\
\delta R/R_0 &= -\epsilon/6.5,
\end{aligned} \tag{10.20}$$

where we have assumed that $R \propto T_c^{-1}$—which follows from hydrostatic equilibrium.

As advertised, a star perturbed by axion emission restores equilibrium by contracting, increasing its temperature and nuclear energy generation rate, and in the process, increasing its photon luminosity. Suppose that

[15]For stars like our sun the radiative entropy per baryon, which is proportional to T^3/ρ, is constant throughout the sun, so that $\rho \propto T^3$.

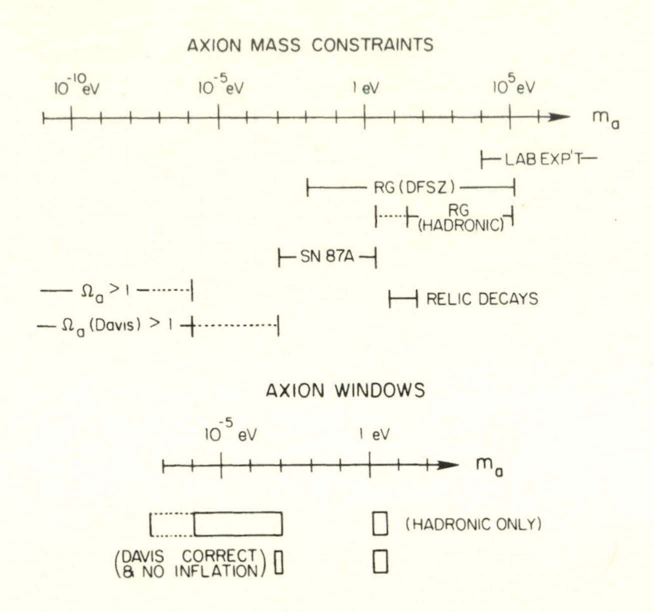

Fig. 10.3: Summary of laboratory, astrophysical, and cosmological constraints to the axion mass, and the two remaining windows on the axion.

$\epsilon = 1/2$; then the central temperature increases by about 8%, the total luminosity increases by about 54%, and the evolutionary timescale, proportional to $Q_{\rm nuc}^{-1} \propto (1 + \delta Q_{\rm nuc}/Q_{\rm nuc}^{0})^{-1}$, decreases by a factor of 20/13. This strongly suggests that $\epsilon \gtrsim 1/2$, or so, is inconsistent with our knowledge of the sun.[16]

An even more sensitive barometer for stellar axion emission is the rate of ^{8}B neutrino emission (the high energy neutrinos that have been detected by Davis' ^{37}Cl experiment and by the KII detector; for a discussion of solar neutrinos, see [21]). The rate of emission of these high-energy neutrinos is proportional to T_c^p ($p \simeq 13$, depending upon which quantities in the stellar model are held fixed). Using our previous formula for δT_c we see that the ^{8}B neutrino flux would increase by almost a factor of 3 (for $\epsilon = 1/2$), exacerbating an already large discrepancy. In the end, the simple limit provided by $\dot{\varepsilon}_a \lesssim \dot{\varepsilon}_{\rm nuc}$ is more than justified!

We should mention that the germanium double beta decay experiment of Avignone, et al. [22] provides a similar limit to the mass of a DFSZ

[16]To make such an argument rigorous one must also consider the very strong compositional dependence of stellar models, reflected in $\mathcal{L}_\gamma \propto \mu^7$; see [20].

axion. Axions emitted by the sun would be detected by their interactions with electrons in their Ge detector: $e^- + a \rightarrow e^- + \gamma$. Their data are inconsistent with the predicted flux of DFSZ axions if $m_a \gtrsim 1$ eV.

To summarize then, the age of the sun and the observed solar neutrino flux precludes a DFSZ axion more massive than about 1 eV (equivalently, $f_{PQ}/N \lesssim 10^7$ GeV). For a hadronic axion the reasoning is identical, but it is the Primakoff process that is most important, and the mass constraint for a hadronic axion is $m_a \lesssim 20\,\mathrm{eV}/[(E_{PQ}/N - 1.95)/0.72]$. Moreover, these limits do not apply to an axion of mass greater than 10 keV or so, since the production of such axions would be severely suppressed because the temperature at the center of the sun is only a few keV [12].

10.2.2 Axions and red giants

The discussion above should provide the reader with the flavor of stellar limits to the axion mass. They all rely upon the fact that axion emission modifies stellar evolution in such a way as to significantly affect an observable, usually the lifetime of the star. Now let us turn to the most stringent stellar evolution limits that exist at present. These limits are provided by the evolution of red giant stars—helium-burning stars whose central temperatures reach 10^8 K and whose central densities are 10^2 to $10^4\,\mathrm{g\,cm^{-3}}$. Because hadronic and DFSZ axions couple very differently to electrons, the constraints that follows for each are very different.

The constraint to the mass of the hadronic axion is based upon the helium-burning lifetimes of red giant stars. As we discussed above, when axion emission is taken into account the central temperature of the star is necessarily increased to satisfy the extra energy being carried away by axions, and this accelerates the evolution and shortens the lifetime. The helium-burning phase of a red giant lasts of order 10^8 yr—too long for *most* astronomers to observe. However, when one observes a cluster of stars (say M67, for example), the number of helium-burning red giants one sees is determined by the length of time red giants spend burning helium—the shorter the time, the fewer that will be seen. Raffelt and Dearborn [23] argue that a hadronic axion of mass greater than about $2\,\mathrm{eV}/[(E_{PQ}/N - 1.95)/0.72]$ would reduce the helium-burning timescale by more than an order of magnitude. In turn, this would reduce the number of red giants one would expect to observe by the same factor—in severe contradiction with the number of helium-burning red giants seen in the cluster M67.

The red giant limit for the DFSZ axion is based upon a slightly more subtle, dynamical argument. Before helium ignition occurs in the core of a red giant, the ^{4}He core is supported by electron degeneracy pressure. This is a very dangerous condition because any increase in temperature is not accompanied by a similar increase in pressure.[17] Once ^{4}He starts to burn (via the triple-α process), the temperature increases, which causes the rate of ^{4}He burning to increase, and so on. Runaway nuclear burning continues until thermal pressure support becomes dominant. The brief period of thermal runaway is referred to as the helium flash.[18] Before the helium flash, hydrogen continues to burn just outside the helium core. As the helium core grows in mass, its radius decreases (for degenerate matter $R \propto M^{-1/3}$), and the accompanying release of gravitational binding energy heats the core. Eventually the helium core becomes hot enough for the triple-α process to burn helium to carbon. The effect of axion cooling is to decrease the temperature rise in the helium core associated with the contraction, and according to Dearborn, et al. [24], for a DFSZ axion of mass greater than about 10^{-2}eV helium ignition never takes place.

Let us consider their argument in slightly more detail. As the helium core increases in mass, the size of the core contracts (MR^3 = const), thereby releasing gravitational energy:

$$\dot{E}_{\rm grav} \simeq \frac{d}{dt}\left(\frac{GM^2}{R}\right) \propto M^{4/3}\dot{M}. \tag{10.21}$$

In the absence of axion cooling, the dominant cooling mechanism for the approximately isothermal core is neutrino emission (because of the long mean free paths of electrons in degenerate matter, degenerate matter is almost always isothermal). For our purposes, let us assume that axion emission dominates, which is the case for an axion of sufficiently large mass to "screw up" the helium flash. The Compton-like process is the dominant axion emission process in the helium core, and the energy radiated in axions is, cf. (10.14), $\mathcal{L}_a \sim m_a^2 M T^6$. Energy equilibrium, i.e., $\dot{E}_{\rm grav} = Q_a$, determines the temperature of the core:

$$T_c \sim m_a^{-1/3} M^{1/18} \dot{M}^{1/6}. \tag{10.22}$$

[17]We remind the reader that in an ordinary star, the simple physical fact that the pressure is proportional to the temperature stabilizes nuclear burning, as any increase in temperature is accompanied by an increase in pressure, which causes the star to expand and thereby cool—Nature's stellar thermostat!

[18]A helium flash should not be confused with a hot flash.

Note, the larger the axion mass, the *lower* the temperature of the core; this differs from the usual case where the existence of axions actually causes the star to *raise* its temperature to compensate for axion emission. Of course, in the present situation that is not possible since nuclear reactions are not yet occurring in the ^{4}He core. One can easily appreciate how a sufficiently massive axion can prevent the core from reaching the temperature required to ignite helium.

Because the temperatures in the cores of red giant stars are of order 10 keV or so, emission of axions more massive than about 100 keV is severely suppressed, and so these arguments do not apply to axions more massive than about 200 keV (which of course are precluded by laboratory searches). The red giant limits to the axion mass are shown in Fig. 10.3. We now turn to the most stringent astrophysical test of the axion: SN 1987A.

10.2.3 Axions and SN 1987A

SN 1987A not only confirmed astrophysicists' more cherished beliefs about type II (core collapse) supernovae, but also provided a unique laboratory for the study of the properties of ordinary neutrinos and hypothetical particles such as right-handed neutrinos and axions. This supernova in the LMC was triggered by the gravitational collapse of the blue super giant Sanduleak -69 202 (thought to be a $15\,M_\odot$ star). Shortly after the gravitational collapse and hydrodynamic bounce of the $1.4\ M_\odot$ Fe core, the central temperature of the newly formed neutron star was about 30 MeV and the central density was about 8×10^{14} g cm^{-3}. During the catastrophic collapse of the Fe core about 3×10^{53} erg of binding energy were liberated, and according to the standard picture, this energy was radiated in thermal neutrinos of all three types. The neutrino mean free path within the core is much smaller than the size of the core (about 10 km). Thus, even neutrinos are "trapped" in the core and must diffuse out. In analogy to the photosphere of a star, neutrinos are radiated from a neutrino sphere ($R\sim 15$ km, $\rho\sim 10^{12}$g cm^{-3}, $T\sim 4$ MeV), defined as the surface beyond which a neutrino, on average, suffers about one scattering in escaping the star.

Neutrino emission is characterized by two phases. The first is powered by residual accretion and hydrodynamic contraction of the outer core and lasts 1 to 2 sec. The second phase is powered by the diffusion of heat trapped in the inner-core region and lasts about 5 to 10 sec, the timescale for neutrino diffusion from the inner core to the neutrino sphere. The energies associated with the two phases are comparable, and

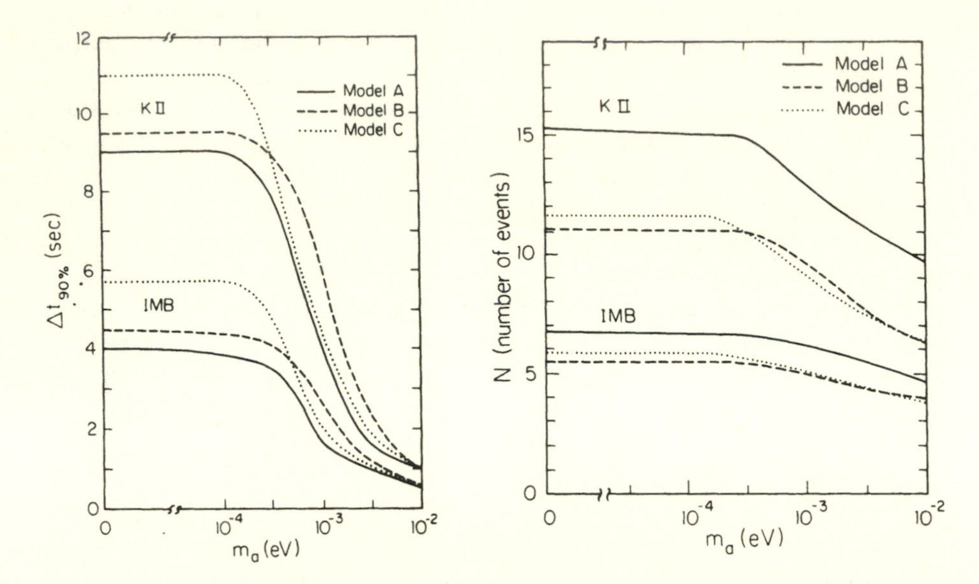

Fig. 10.4: The characteristic length of the predicted neutrino burst in the KII and IMB detectors, $\Delta t(90\%)$, and the predicted number of neutrino events, as a function of axion mass, for three numerical models of axion cooled, nascent neutron stars. The IMB burst consisted of eight events over about 6 sec, and the KII consisted of eleven events over about 12 sec (from [26]).

accordingly, one expects a neutrino burst of order 5 to 10 sec. This standard picture was dramatically confirmed [25] by the detection of nineteen anti-electron neutrinos by the Kamiokande II (KII) [16] and the Irivine-Brookhaven-Michigan (IMB) [17] water Cherenkov detectors, which interestingly enough were originally built to search for nucleon decay!

If the axion exists, it should play an important role in the cooling of this nascent neutron star. The observable effect of axion cooling would be the shortening of the neutrino burst. Fortunately there are nineteen beautiful neutrino events spread out over about 12 sec that detail the cooling of the neutron star.

Under the conditions that existed in the post-collapse core, nucleon–nucleon axion bremsstrahlung (NNAB) is the dominant axion emission process (see Fig. 10.1). The relevant axion coupling for this process is that to nucleons, which we should recall is insensitive to the axion type. In terms of the matrix element squared, $|\mathcal{M}|^2$, the axion emission rate is given by

$$\dot{\varepsilon}_a(\text{NNAB}) = \rho^{-1}\int d\Pi_1 d\Pi_2 d\Pi_3 d\Pi_4 d\Pi_a\,(2\pi)^4 \times S|\mathcal{M}|^2\delta^4(p_1+p_2-p_3-p_4-p_a) \times E_a f_1 f_2(1-f_3)(1-f_4), \qquad (10.23)$$

where $d\Pi_i = d^3p_i/(2\pi)^3 2E_i$, the labels $i = 1$ to 4 refer to the incoming (1,2) and outgoing (3,4) nucleons, $i = a$ denotes the axion, S is the symmetry factor for identical particles in the initial and final states, $|\mathcal{M}|^2$ is summed over initial and final nucleon spins, and the nucleon phase space distribution functions are $f_i = [\exp(E_i/T - \mu_i/T) + 1]^{-1}$. The emission rate is relatively easy to evaluate in the fully degenerate or non-degenerate regimes; however, the nucleons in the core are semi-degenerate, that is $\epsilon_{\text{FERMI}} \sim T$. In addition, since the post-collapse core has roughly equal numbers of neutrons and protons, three bremsstrahlung processes are important: $nn \to nn + a$, $pp \to pp + a$, and $np \to np + a$. The axion emission rate, for all three processes and arbitrary nucleon degeneracy, has been evaluated numerically, and it is found that the analytical rate for non-degenerate conditions provides a good approximation to the actual rate for the semi-degenerate conditions that exist in the hot neutron star [27]:

$$\dot{\varepsilon}_a(\text{NNAB}) \simeq 2.4\times 10^{25}\,(\rho/8\times 10^{14}\,\text{g cm}^{-3}) \times (T/30\,\text{MeV})^{3.5}(m_a/\text{eV})^2\,\text{erg g}^{-1}\,\text{sec}^{-1},$$
$$\mathcal{L}_a \simeq 1.4M_\odot\,\dot{\varepsilon}_a \simeq 10^{59}\,(m_a/\text{eV})^2\,\text{erg sec}^{-1}. \qquad (10.24)$$

For simplicity, it has been assumed that there are equal numbers of neutrons and protons, and that $g_{ann} = g_{app} = 0.5m_N/(f_{PQ}/N) \simeq 7.8\times 10^{-8}\,(m_a/\text{eV})$. Note that for an axion mass of about 10^{-3}eV, the axion luminosity is about that in thermal neutrinos: $\mathcal{L}_\nu \simeq (E \sim 3\times 10^{53}\,\text{erg})/(\Delta t \simeq \textit{few}\ \text{sec}) \sim 10^{53}\,\text{erg sec}^{-1}$. Thus one would expect axion emission to *significantly* affect the cooling of the neutron star associated with SN 1987A for $m_a \gtrsim 10^{-3}$eV—provided that axions just stream out of the neutron star.

Axions less massive than about 0.02 eV, once radiated, do freely stream out of the nascent neutron star without interacting and thereby accelerate its cooling. Qualitatively, one would expect axion emission to shorten

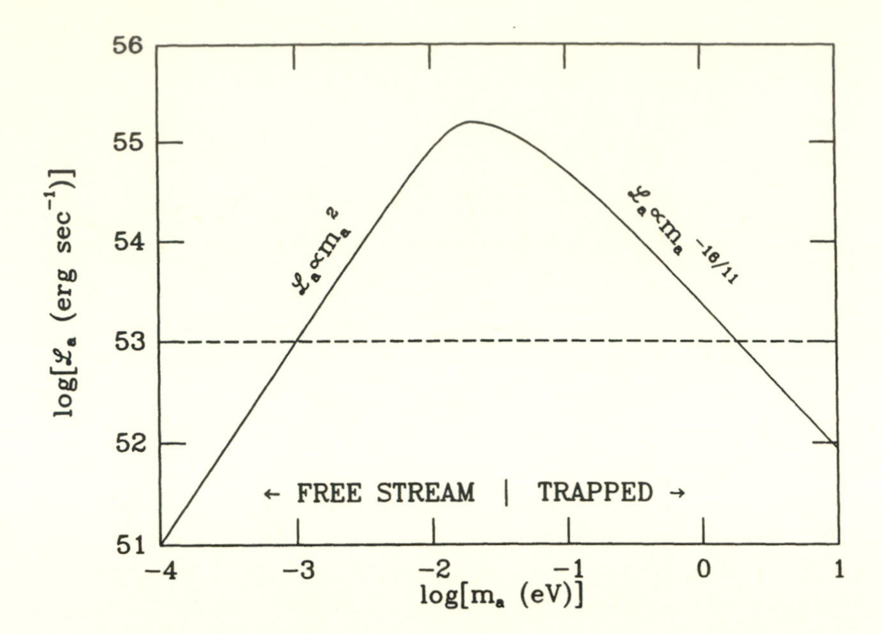

Fig. 10.5: The axion luminosity from SN 1987A as a function of axion mass based upon a simple, analytical model. In the free-streaming regime $\mathcal{L}_a \propto m_a^2$, while in the trapped regime $\mathcal{L}_a \propto m_a^{-16/11}$. For an axion mass between about 10^{-3} and 2 eV, the axion luminosity is unacceptably large.

the duration of the neutrino pulse.[19] Axion cooling has been incorporated into numerical models of the initial cooling of the nascent neutron star, and from these axion-cooled, numerical models the resulting neutrino flux and the predicted response of the KII and IMB detectors has been calculated [26]: expected number of events; and expected burst duration, $\Delta t(90\%)$, the time required for the number of events to achieve 90% of its final value. The quantity $\Delta t(90\%)$ proves to be the most sensitive indicator of axion emission. Axion emission tends to cool the inner core rapidly, depleting the energy reserve that powers the second part of the burst. This effect is clearly seen in Fig. 10.4, where $\Delta t(90\%)$ is plotted as a function of m_a. Axion emission has virtually no effect on $\Delta t(90\%)$ until a mass of $\sim 3\times10^{-4}$eV, and by an axion mass of 10^{-2}eV the duration of the neutrino burst has dropped to less than 1 sec (the timescale for the first phase of

[19]Of course axion emission, which proceeds predominantly from the high temperature, high density inner core, does not *directly* affect neutrino emission, which proceeds from the neutrino sphere (in the outer core). Axion emission does, however, "drain" the same energy reserves that power neutrino emission.

the burst). For comparison, for an axion mass of 10^{-2}eV, the expected number of neutrino events has only dropped from ten to eight for KII and from six to four for IMB (see Fig. 10.4). Based upon the burst duration, one can safely say that the cooling of the neutron star associated SN 1987A precludes an axion of mass 10^{-3} to 0.02 eV.[20]

For axion masses greater than about 0.02 eV, axions interact sufficiently strongly that they do not simply stream out. In this case the mean free path for axion absorption (e.g., $n + p + a \rightarrow n + p$) is smaller than the radius of the neutron star; a typical axion produced in the core will be absorbed before it can escape, and axions are said to be "trapped." Just as neutrinos are radiated from a neutrino sphere, axions of mass greater than 0.02 eV should be radiated from an axion sphere, with temperature T_a. In the trapping regime, the axion luminosity is proportional to T_a^4. With increasing axion mass (and hence interaction strength), the axion sphere moves outward and therefore has a lower temperature: $T_a \propto m_a^{-4/11}$. Thus, for $m_a \gtrsim 0.02$ eV the axion luminosity *decreases* with increasing axion mass (see Fig. 10.5); for comparison, recall that in the free-streaming regime the axion luminosity *increases* as m_a^2, cf. (10.24). For sufficiently large axion mass the effect of axion cooling becomes acceptable. Simple analytical models indicate that for an axion mass of about 2 eV or greater the axions are so strongly trapped that their presence is equivalent to less than a couple of additional neutrino species and is therefore consistent with the observations of KII and IMB [28].

In sum, the observed neutrino cooling of SN 1987A precludes an axion of either type in the mass range, 10^{-3} to 2 eV, providing the most stringent astrophysical constraint. To be sure there are residual uncertainties—the equation of state at supernuclear densities and the possible existence of an unusual state of matter at the core of the neutron star, e.g., pion condensate, quark matter, or even strange matter [29]. However, the physics of the cooling of the nascent neutron star is so simple, and the anti-electron neutrinos detected provide such a clear and complete record of the cooling of the neutron star, that this test of the existence of the axion is both a very stringent and a very reliable one. Moreover, it beautifully illustrates the power of unique, contemporary astrophysical laboratories to probe fundamental physics at scales well beyond the reach of the terrestrial laboratory.

[20]One might wonder if a finite mass for the electron neutrino could lengthen an axion-shortened neutrino burst. A mass of 20 eV or so *might* work just fine for the KII events; however, because the energies of the IMB events are much larger on average, a mass of 30 to 50 eV would be required to lengthen the IMB burst, a value precluded by the KII data and laboratory experiments.

10.3 Axions and Cosmology

The cosmological history of axions begins with PQ symmetry breaking, which occurs at a temperature $T \sim f_{\rm PQ}$. In most axion models PQ symmetry breaking occurs when a complex scalar field $\vec{\phi}$, which carries PQ charge, develops a vacuum expectation value. This occurs in the usual way discussed in Chapter 7: At temperatures below $T \sim f_{\rm PQ}$ the potential for $\vec{\phi}$ is of the "wine bottle" form

$$V(\vec{\phi}) = \lambda(|\vec{\phi}|^2 - f_{\rm PQ}^2/2)^2, \tag{10.25}$$

which is minimized for $\langle|\vec{\phi}|\rangle = f_{\rm PQ}/\sqrt{2}$ (see Fig. 10.6).[21] When the $U(1)$ global PQ symmetry is broken, the argument of $\langle\vec{\phi}\rangle$ is left undetermined; that is, $V(\vec{\phi})$ is independent of $\bar{\Theta} \equiv \arg\langle\vec{\phi}\rangle$. The massless $\bar{\Theta}$ degree of freedom is the axion: $a = (f_{\rm PQ}/N)\bar{\Theta}$.

At high temperatures, $T \gg \Lambda_{\rm QCD}$, the axion is massless; however, at low temperatures, the axion develops a mass due to QCD instanton effects. Because instantons are inherently large objects, size set by $\Lambda_{\rm QCD}^{-1}$, their effects are strongly suppressed at high temperatures. The temperature dependence of the axion mass is approximately given by [30]:

$$m_a(T) \simeq 0.1\, m_a(T=0)\,(\Lambda_{\rm QCD}/T)^{3.7}, \tag{10.26}$$

valid for $\pi T/\Lambda_{\rm QCD} \gg 1$. Roughly speaking then, the axion is massless from $T \sim f_{\rm PQ}$ until $T \sim \Lambda_{\rm QCD}$.

The axion mass is easy to visualize in models where $N = 1$. In terms of $V(\vec{\phi})$ one can think of the instanton effects as tipping the potential by a small angle, $\mathcal{O}(m_a^2/f_{\rm PQ}^2)$, so that the tipped potential develops a minimum at the CP-conserving value $\bar{\Theta} = 0$ and the axion develops a mass m_a. The axion degree of freedom then, is periodic in $\bar{\Theta} = a/f_{PQ}$, with a potential $V(a) = m_a^2 f_{PQ}^2(1 - \cos[a/f_{PQ}]) \simeq (1/2)m_a^2 a^2 + \cdots$.

In axion models where $N > 1$, the situation is more interesting and cannot be illustrated by the tipping of a wine bottle [31]. In such models, there is a residual discrete Z_N symmetry, and there are actually N degenerate and distinct CP-conserving minima of the axion potential at $\bar{\Theta} = 2\pi n$ $(n = 0, 1, 2, \cdots, N-1)$. As we learned in Section 7.2, the existence of these disconnected and degenerate vacuum states implies that

[21] The value of the coupling constant λ will be of no importance in our discussions here.

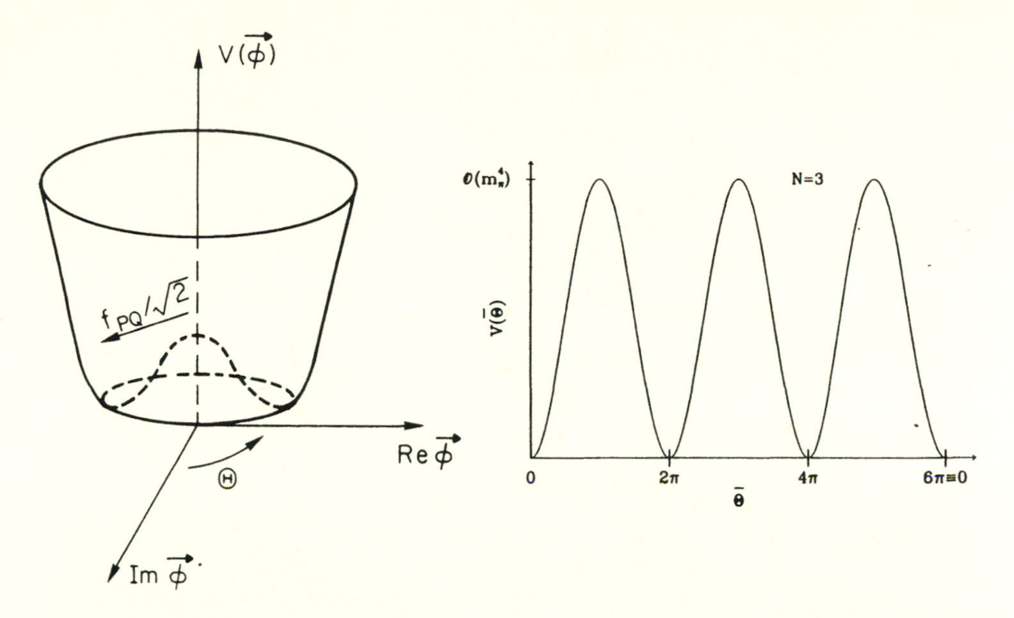

Fig. 10.6: The potential $V(\vec{\phi})$: left panel—just after PQ SSB; and right panel—$V(\bar{\Theta})$ for $T \ll \Lambda_{\rm QCD}$ and $N = 3$.

axionic domain walls, separating regions of space in adjacent minima, can develop when $T \sim \Lambda_{\rm QCD}$. In the case that $N > 1$, the axion potential is of the form

$$V(a) = m_a^2(f_{\rm PQ}/N)^2\,(1 - \cos[a/(f_{\rm PQ}/N)])\,, \tag{10.27}$$

where the overall factor of $m_a^2(f_{\rm PQ}/N)^2$ follows from the fact that the curvature of the potential at any of its minima is m_a^2. Following our analysis of domain walls in Chapter 7, we can estimate the properties of an axionic domain wall. The surface energy density, η, of an axionic wall of thickness Δ receives contributions from a potential term and a spatial gradient term:

$$\begin{aligned} \eta &\sim V(a = \pi f_{\rm PQ}/N)\Delta + [(\pi f_{\rm PQ}/N)/\Delta]^2\Delta \\ &\sim m_a^2(f_{\rm PQ}/N)^2\,\Delta + (\pi f_{\rm PQ}/N)^2/\Delta. \end{aligned} \tag{10.28}$$

The surface energy density η is minimized when the two contributions are comparable, which occurs for

$$\begin{aligned} \Delta &\sim m_a^{-1} \simeq 10^{-5}\,(m_a/{\rm eV})^{-1}\,{\rm cm}, \\ \eta &\sim m_a(f_{\rm PQ}/N)^2 \simeq 10^{9}\,(m_a/{\rm eV})^{-1}\,{\rm g\,cm^{-2}}. \end{aligned} \tag{10.29}$$

Since the $\bar{\Theta}$ degree of freedom is massless at very high temperatures, no value of $\bar{\Theta}$ is singled out by energetics, and $\bar{\Theta}$ will necessarily take on different randomly chosen values in different causally distinct regions of the Universe. When $T \sim \Lambda_{\rm QCD}$, the axion develops a mass due to QCD instanton effects, and the axion's potential is minimized for $\bar{\Theta} = 0$ mod(2π). If $N > 1$, there are N distinct minima of the axion's potential, and a network of axionic walls will necessarily form. Such domain walls are just as disastrous as other cosmological domain walls, and must be avoided! The problem of axionic domain walls is clearly side-stepped if $N = 1$, or if the Universe inflated before or during PQ SSB, so that $\bar{\Theta}$ is constant within the observable Universe.[22]

The observant reader is probably wondering what happens to $\bar{\Theta}$, which was last seen initially displaced from $\bar{\Theta} = 0$ by some random value, and has probably also noticed that the spontaneous breaking of the $U(1)_{\rm PQ}$ symmetry should lead to axionic (global) strings. We will return to these issues shortly—both lead to the production of relic axions.

The topic of relic particles from the early Universe is a most interesting one—especially when the relic is the axion! In Chapter 5 we discussed the making of a thermal relic, and in Chapter 7 the making of non-thermal topological relics. Relic axions arise due to both thermal and non-thermal processes. In fact, relic axions arise due to three different and distinct processes: as thermal relics, through the usual "freeze-out" process [33]; as non-thermal relics, through coherent production due to the initial misalignment of the axion field [34]; and as non-thermal relics, through the decay of axionic strings [35]. Each of these three processes can be the dominant production mechanism, depending upon the axion mass and whether or not the Universe ever underwent inflation (see Fig. 10.7). Let us consider the three processes in turn.

10.3.1 Thermal production

As the result of near thermal equilibrium that existed in the early Universe, all kinds of interesting particles—possibly including axions—were present in great abundance. For axions the important creation and destruction processes are photo-production $\gamma + Q \to Q + a$ (Q is a quark) and "pion–axion conversion" $N + \pi \to N + a$ (N is a nucleon) (see Fig. 10.1).

[22] Axion domain walls can also be avoided if the different minima are gauge-equivalent. This occurs in models where the Z_N symmetry coincides with, or lies within, the center of the gauge group. In this case the Z_N is effectively a "discrete gauge symmetry" [32].

In Chapter 5 we derived the Boltzmann equation governing the abundance of a particle species created and destroyed in pairs. However, to lowest order in their small coupling, axions are produced singly. Assuming that the processes relevant to regulating the abundance of axions are of the form $a + 1 \leftrightarrow 2 + 3$, where 1, 2, and 3 are other particle species that are in thermal equilibrium, it is easy to derive the Boltzmann equation governing the abundance of axions, i.e., the analogue of (5.24):

$$Y' = -\left(\frac{\Gamma_{\rm ABS}}{xH}\right)(Y - Y_{\rm EQ}); \tag{10.30}$$

$$Y = \frac{n_a}{s}; \quad Y_{\rm EQ} = \frac{n_a^{\rm EQ}}{s} \simeq \frac{0.278}{g_*}; \quad \Gamma_{\rm ABS} = n_1\langle\sigma|v|\rangle_{\rm ABS};$$

where $x = m_N/T$ (m_N = nucleon mass, or any other convenient energy scale), and brackets indicate thermal averaging. Further, we have assumed that the axion is relativistic during the epochs of interest, so that $n_a^{\rm EQ} = \zeta(3)T^3/\pi^2$. Note the similarity of this equation to (5.24), where the controlling reaction rate is the annihilation rate.

The solution to (10.30) is simple to write down:

$$\frac{1 - Y(x)/Y_{\rm EQ}}{1 - Y(0)/Y_{\rm EQ}} = \exp\left[-\int_0^x \frac{\Gamma_{\rm ABS}}{x'H}\,dx'\right]. \tag{10.31}$$

Since we are interested in the question of whether a thermal population of axions is ever created, let us assume that $Y(0) = 0$, in which case

$$Y(x) = Y_{\rm EQ}\left\{1 - \exp\left[-\int_0^x \frac{\Gamma_{\rm ABS}}{x'H}\,dx'\right]\right\}. \tag{10.32}$$

From the form of (10.32) it is clear that axions will reach their equilibrium abundance provided that $\Gamma_{\rm ABS}/H$ was greater than unity for at least an expansion time (in an expansion time $\Delta x/x \simeq |\Delta T/T| \sim 1$). As we shall see, for all the processes that create and destroy axions $\Gamma_{\rm ABS}/H$ increases with x for small x (high temperature), achieves its maximum value, and then decreases with x for large x (low temperatures). Long before thermal axions become non-relativistic (if ever), their interactions freeze out ($\Gamma_{\rm ABS}/H \lesssim 1$), and their relic abundance freezes in. Therefore, the relic abundance of thermal axions is given by

$$Y_\infty \simeq \frac{0.278}{g_{*F}}\left\{1 - \exp\left[-\int_0^\infty \frac{\Gamma_{\rm ABS}}{x'H}\,dx'\right]\right\}, \tag{10.33}$$

where g_{*F} is the value of g_* at the freeze-out temperature.

The most important process for creating a thermal population of axions is pion–axion conversion: $N + \pi \leftrightarrow N + a$. Since nucleons come into existence only after the quark/hadron transition, which occurs at a temperature of about 200 MeV, they are non-relativistic, with an equilibrium abundance of[23] $n_N \simeq (mT)^{3/2}e^{-x}$. The cross section for pion–axion conversion is

$$\langle\sigma|v|\rangle_{\rm ABS} \sim \frac{m_N^2}{(f_{\rm PQ}/N)^2}\left(\frac{T}{m_N}\right)^2 m_\pi^{-2}, \qquad (10.34)$$

where we have used the fact that $g_{a\rm NN} \simeq m_N/(f_{\rm PQ}/N)$. The axion absorption rate is $\Gamma_{\rm ABS} = n_N\langle\sigma|v|\rangle_{\rm ABS} \sim (T^3/m_\pi^2)[m_N/(f_{\rm PQ}/N)]^2 x^{-1/2}\exp(-x)$. The all-important ratio of the absorption rate to expansion is then

$$\frac{\Gamma_{\rm ABS}}{H} \sim (m_a/10^{-4}{\rm eV})^2 x^{-3/2}\exp(-x). \qquad (10.35)$$

At low temperatures $\Gamma_{\rm ABS}/H$ is exponentially cut off by the small number of nucleons present.

At high temperatures, $T \gtrsim 200$ MeV, there are no nucleons or pions, just a quark/gluon plasma, and the dominant axion production process is photo-production (or gluon-production): $a+Q \leftrightarrow Q+\gamma$ (or $+G$). Provided that the quark species is relativistic, the cross section for this process is $\langle\sigma|v|\rangle_{\rm ABS} \sim \alpha[m_Q/(f_{\rm PQ}/N)]^2/T^2$, and $\Gamma_{\rm ABS} \sim \alpha T[m_Q/(f_{\rm PQ}/N)]^2$ (where $m_Q/(f_{\rm PQ}/N)$ is the quark–axion coupling). The ratio of the absorption rate to the expansion rate is given by

$$\frac{\Gamma_{\rm ABS}}{H} \sim x(m_Q/{\rm GeV})^2(m_a/0.1\,{\rm eV})^2, \qquad (10.36)$$

valid for $T \gtrsim m_Q$.

At high temperatures $\Gamma_{\rm ABS}/H$ varies as x, and at low temperatures as $x^{-3/2}e^{-x}$, achieving its maximum value just after the quark/hadron transition. To estimate the relic abundance of thermal axions we can simply integrate forward from the time of the quark/hadron transition:

$$Y_\infty = \frac{0.278}{g_{*F}}\left[1 - \exp\left[-(m_a/10^{-4}{\rm eV})^2\exp(-x_{\rm qh})/x_{\rm qh}^{5/2}\right]\right] \qquad (10.37)$$

[23] We are ignoring the non-zero chemical potential associated with nucleons, which ensures that today $\eta \simeq (4$ to $7) \times 10^{-10}$. At the temperatures of interest the abundance of nucleon–antinucleon pairs is greater than this. Only for $T \lesssim 35$ MeV is the baryon chemical potential important in determining the number of nucleons present.

where we have approximated $\int_{x_{\rm qh}}^{\infty} \exp(-x)x^{-5/2}\,dx$ by $\exp(-x_{\rm qh})/x_{\rm qh}^{5/2}$, and $x_{\rm qh}$ is the value of $x = m_N/T$ at the quark/hadron transition. From this expression we see that a nearly equilibrium abundance of axions is produced by thermal processes provided that the axion mass is greater than 10^{-3} to 10^{-2} eV. For an axion mass of much less than 10^{-3} eV, axions interact so weakly that a thermal population never results. The actual abundance of relic, thermal axions depends slightly upon the freeze-out temperature, through g_{*F}. As in Chapter 5, given Y_∞, we can compute the relic number density of thermal axions,

$$n_a = s_0 Y_\infty \simeq \frac{83\,{\rm cm}^{-3}}{g_{*F}/10}. \tag{10.38}$$

Not surprisingly, their relic abundance is comparable to that of light neutrinos or the CMBR photons. Their present contribution to the mass density is

$$\Omega_{\rm TH} h^2 = \frac{m_a}{130\,{\rm eV}} \left(\frac{10}{g_{*F}}\right). \tag{10.39}$$

If thermal axions are to provide closure density the axion mass must be about $130h^2$eV. However, an axion of such mass would decay in less than the age of the Universe and is excluded by the astrophysical constraints discussed in the previous Section.[24] While thermal axions cannot close the Universe, their decays may be detectable as we shall discuss in Section 10.5.

10.3.2 Misalignment production

Axions are also produced by a very interesting and highly non-thermal process involving the relaxation of the $\bar{\Theta}$ angle. Today $\bar{\Theta}$ is anchored at the CP-conserving value $\bar{\Theta} = 0$. However, as we have just discussed, the axion mass, which arises due to instanton effects, is very temperature dependent, cf. (10.26), and at very high temperatures the axion is essentially massless. That means that no special value of $\bar{\Theta}$ is specified by dynamics and all values of $\bar{\Theta}$ are equally palatable! Therefore, the initial value of $\bar{\Theta}$, denoted by $\bar{\Theta}_1$, must be chosen by some stochastic process, and so there is no reason to expect $\bar{\Theta}_1$ to be zero. Thus, at early times, there is every reason to

[24]If $h \sim 1/2$, a hadronic axion of mass 30 eV or so, with $E_{PQ}/N \simeq 2$, could possibly avoid the astrophysical constraint and live long enough to provide closure density.

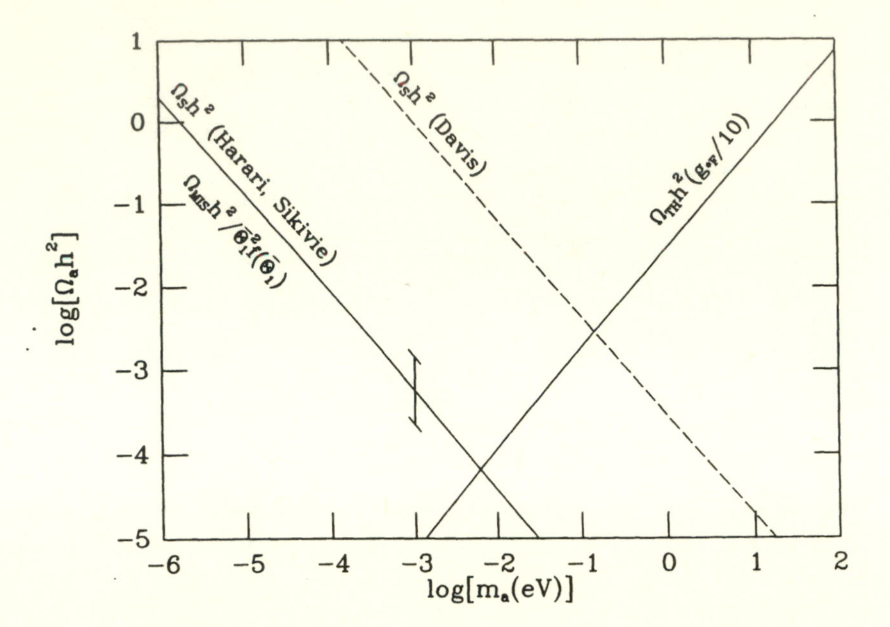

Fig. 10.7: Summary of $\Omega_a h^2$ as a function of axion mass for the three production processes: thermal (TH) (10.39); misalignment (MIS) (10.49); and axionic string decay (S) (10.56, 10.55). The theoretical uncertainty in $\Omega_{\rm MIS}$, $(10^{\pm 0.4})$ is indicated by the error bar [see (10.49)].

expect that the axion field is misaligned with the minimum of its potential ($\bar\Theta = 0$).

When the axion mass does "turn on" around a temperature of $T \sim \Lambda_{\rm QCD}$, and becomes comparable to the expansion rate of the Universe, the axion field will start to roll toward $\bar\Theta = 0$, and of course will overshoot $\bar\Theta = 0$. Thereafter, it will oscillate. These cosmic oscillations of the axion field correspond to a zero-momentum condensate of axions and are just like the coherent oscillations of the inflaton field discussed in Chapter 8—with one important exception—the axion oscillations do not decay.

The evolution of $\bar\Theta$ is easy to calculate: It is just that of a scalar field initially displaced from the minimum of its potential! Near the minima of the axion potential, the potential is $V(\bar\Theta) \simeq m_a^2(f_{PQ}/N)^2\bar\Theta^2/2$ [see (10.27)], and the axion Lagrangian density can be written in terms of $\bar\Theta$ as:

$$\mathcal{L} = (f_{\rm PQ}/N)^2\left[\dot{\bar\Theta}^2/2 - m_a^2\bar\Theta^2/2\right] \qquad (10.40)$$

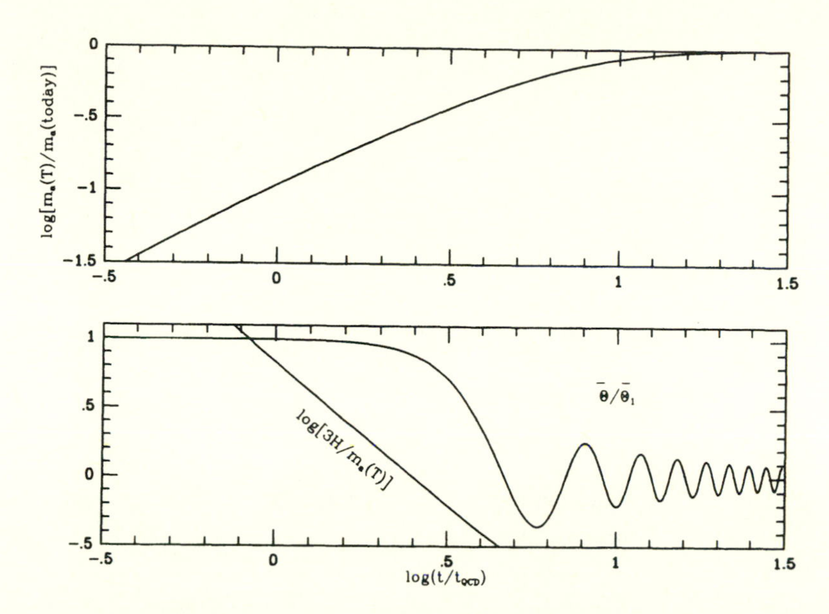

Fig. 10.8: The evolution of $m_a(T)$ and $\bar{\Theta}$ for $m_a = 10^{-10}$eV, where $t_{\rm QCD} \equiv t(T = 200 \text{ MeV}) \simeq 1.4 \times 10^{-5}$sec. A functional form that smoothly interpolates between $m_a(T \gg \Lambda_{\rm QCD})$, cf. (10.26), and $m_a(T=0)$ has been used.

where we have assumed that $\bar{\Theta}$ is homogeneous.[25]

The equation of motion for $\bar{\Theta}$ can be obtained in the usual way by varying the action, $S = \int d^4x\, R^3 \mathcal{L}$,

$$\ddot{\bar{\Theta}} + 3H\dot{\bar{\Theta}} + m_a^2(T)\bar{\Theta} = 0. \qquad (10.41)$$

Note, because the decay width of the axion is so small, the $\Gamma_a\dot{\bar{\Theta}}$ term need not be included. The new wrinkle with this familiar equation is the temperature dependence of the axion mass.

At high temperatures, $T \gg \Lambda_{\rm QCD}$, $m_a \simeq 0$, and the solution to (10.41) is $\bar{\Theta} = \bar{\Theta}_1 = \text{const.}$[26] Once the axion mass becomes greater than the expansion rate, $m_a(T) \gtrsim 3H \simeq 5g_*^{1/2}T^2/m_{Pl}$, $\bar{\Theta}$ begins to oscillate with angular frequency $m_a(T)$ (see Fig. 10.8). When the coherent $\bar{\Theta}$ oscillations commence, we can replace $\dot{\bar{\Theta}}^2$ by its average over an oscillation: $\langle\dot{\bar{\Theta}}^2\rangle = \rho_a$,

[25]We will be interested only in the zero-momentum mode of $\bar{\Theta}$; the higher momentum modes correspond to relativistic axions whose energy density rapidly red shifts away.

[26]We are assuming that $\dot{\bar{\Theta}}_1 \ll H$ initially.

where

$$\rho_a = (f_{PQ}/N)^2 \left[\dot{\bar{\Theta}}^2 + m_a(T)^2\bar{\Theta}^2\right]/2. \tag{10.42}$$

Multiplying (10.41) by $\dot{\bar{\Theta}}$, and replacing $\langle\dot{\bar{\Theta}}^2\rangle$ by ρ_a, (10.41) becomes

$$\dot{\rho}_a = \left[\frac{\dot{m}_a}{m_a} - 3H\right]\rho_a. \tag{10.43}$$

The solution to this equation is simple to write down:

$$\rho_a = \text{const}\,\frac{m_a(T)}{R^3}. \tag{10.44}$$

Once the axion mass settles down to its zero temperature value (at $T \ll \Lambda_{\rm QCD}$), the axion energy density varies as R^{-3}: The coherent axion oscillations behave just like NR matter. Further, when this occurs, $\bar{\Theta}$ decreases as $R^{-3/2}$. While the axion mass is still varying, ρ_a does not decrease as R^{-3}; however, the axion number density, $n_a = \rho_a/m_a(T)$, does. This simply means that the number of axions per comoving volume is conserved, even when the axion mass is varying. Using this fact it is simple to estimate the present axion number density. The initial energy density trapped in the misaligned axion field is $\rho_a = m_a(T_1)^2\bar{\Theta}_1^2(f_{PQ}/N)^2/2$. Taking the axion oscillations to begin at $T = T_1$, where T_1 is defined by $m_a(T_1) = 3H(T_1)$, the initial axion number density is

$$n_a(T_1) \simeq \rho_a(T_1)/m_a(T_1) \simeq m_a(T_1)\bar{\Theta}_1^2(f_{PQ}/N)^2/2. \tag{10.45}$$

The temperature T_1 is found by solving $m_a(T_1) = 3H(T_1)$. For an axion of mass 10^{-5}eV, T_1 is about 1 GeV, and in general T_1 scales as $m_a^{0.18}$.

Assuming that there has been no entropy production since the axion field began to oscillate, the ratio of axion number density to entropy density is conserved and is given by

$$\frac{n_a}{s} \simeq \frac{m_a(T_1)\bar{\Theta}_1^2(f_{PQ}/N)^2/2}{2\pi^2 g_* T_1^3/45} \simeq \frac{\bar{\Theta}_1^2 m_\pi^2 f_\pi^2}{g_*^{1/2} m_a^2 T_1 m_{Pl}}. \tag{10.46}$$

The energy density in axions today is then given by this constant ratio times the present entropy density, s_0, times the axion mass m_a. Since $T_1 \propto m_a^{0.18}$, we see that $\Omega_{\rm MIS}h^2 \propto m_a^{-1.18}$, where the unusual power of the mass traces to the way in which the axion mass turns on.

When this calculation is done very carefully (anharmonic effects taken into account, the motion of the axion field integrated precisely, and so on) the following expression results for the axion's contribution to the present energy density [36]:

$$\Omega_{\rm MIS}h^2 = 0.85 \times 10^{\pm 0.4}\Lambda_{200}^{-0.7}(m_a/10^{-5}{\rm eV})^{-1.18}, \tag{10.47}$$

where $\Lambda_{200} \equiv \Lambda_{\rm QCD}/200$ MeV, and the $10^{\pm 0.4}$ factor reflects the theoretical uncertainties. In deriving this formula, it has been assumed that there has been no significant entropy production since the epoch of axion production ($T \sim T_1$). If there has been significant entropy production, say the entropy in a comoving volume ($S \equiv R^3 s$) increased by a factor γ, then $\Omega_{\rm MIS}h^2$ is *reduced* by the same factor of γ.

Moreover, we have assumed that the initial misalignment angle of the axion field is just the *rms* average of a uniform distribution of initial values from $-\pi$ to π:

$$(\bar{\Theta}_1)_{rms} \equiv \left(\int_{-\pi}^{\pi} d\bar{\Theta}_1\, \bar{\Theta}_1^2/2\pi\right)^{1/2} = \pi/\sqrt{3}. \tag{10.48}$$

If the Universe never underwent inflation this is the reasonable thing to do: At the time the axion field began to oscillate the presently observable Universe was comprised of about 10^{30} or so causally distinct volumes, each of which should have an independently chosen value for $\bar{\Theta}_1$.

On the other hand, if the Universe underwent inflation, either after or during PQ symmetry breaking, then the entire observable Universe should be within a single inflationary patch, within which $\bar{\Theta}_1$ takes on the same value. That value is equally likely to be in any interval between $-\pi$ and π; i.e., $\bar{\Theta}_1$ is just as likely to be between 0.1 and 0.2, as it is to be between 1.5 and 1.6.[27] If one averages over all inflationary patches in the entire Universe, then the *rms* value of $\bar{\Theta}_1$ should be $\pi/\sqrt{3}$ (see Fig. 10.9). However, that tells us nothing about the initial value of $\bar{\Theta}$ in our "neck of the woods." In order to determine $\bar{\Theta}_1$, we would have to measure the mass density of axions and the axion mass! Restoring the $\bar{\Theta}_1$ dependence in the energy density of relic, coherent axions we find

$$\Omega_{\rm MIS}h^2 = 0.13 \times 10^{\pm 0.4}\Lambda_{200}^{-0.7} f(\bar{\Theta}_1)\bar{\Theta}_1^2(m_a/10^{-5}{\rm eV})^{-1.18}, \tag{10.49}$$

[27]Of course, it is fair to say that the *a priori* probability of $\bar{\Theta}_1$ being in the interval $[1, 1.5]$ is 100 times greater than that of it being in the interval $[0.01, 0.015]$.

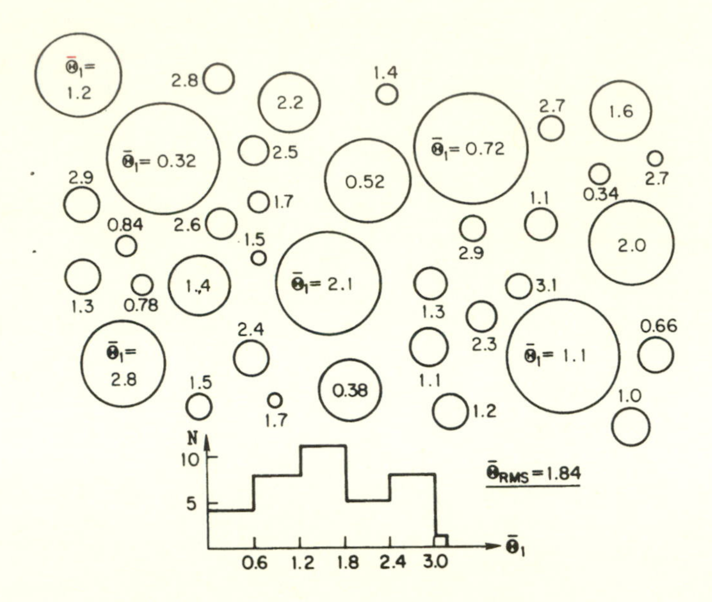

Fig. 10.9: Distribution of $|\bar{\Theta}_1|$ in an inflationary Universe.

where the function $f(\bar{\Theta}_1)$ accounts for anharmonic effects: $f(\bar{\Theta}_1)$ is monotonically increasing and $f(\bar{\Theta}_1 = 0) = 1.0$.

Note that the theoretical uncertainties inherent in $\Omega_{\rm MIS}$ are large: from particle physics a factor of $10^{\pm 0.4}$ and from cosmology a factor of h^2—all told, easily a factor of 10. For canonical values and $\bar{\Theta}_1 = \pi/\sqrt{3}$, an axion mass of about 10^{-5}eV or so corresponds to closure density in axions.

Axion production from the initial misalignment of $\bar{\Theta}$ is clearly highly non-thermal: $\Omega_{\rm MIS} \propto m_a^{-1.18}$, and $\Omega_{\rm MIS} \sim 1$ obtains for $m_a \sim 10^{-5}$eV (see Fig. 10.7).[28] Moreover, the axions produced this way are a condensate of zero-momentum axions. To see this, let us suppose that the axion field varied significantly on the scale of H_1^{-1} when the oscillations began.[29] The characteristic axion momentum at birth is $H_1 \sim m_a(T_1)$; that is, axions are born semi-relativistic (but at a temperature $\gg$ than their mass). From their initial momentum ($p_1 \sim H^{-1}$) and number density, cf. (10.45), we

[28]One might wonder if the axion interactions discusssed previously could possibly thermalize this coherent, non-thermal population. The answer is no, unless $m_a \gtrsim$ eV. Although axions are absorbed out of the condensate, their re-emission is strongly stimulated by the existence of the condensate itself [37].

[29]Of course, if the Universe inflated, then $\bar{\Theta}_1$ is smooth on all scales out to the present Hubble length and beyond.

can estimate their phase space occupancy:

$$f_a(\vec{\mathbf{p}}_a = 0) \sim \frac{m_a(T_1)(f_{\rm PQ}/N)^2}{p_1^3} \sim 10^{50}\,(m_a/{\rm eV})^{-2.7} \tag{10.50}$$

—truly a Bose condensate! Likewise, we can estimate the typical axion velocity today:

$$v_a/c \sim \left(\frac{p_a}{m_a}\right)_{\rm today} \sim \frac{p_1(3\,{\rm K}/T_1)}{m_a} \sim 10^{-22}\,(m_a/{\rm eV})^{-0.82}. \tag{10.51}$$

10.3.3 Axionic string decay

The final mechanism for axion production is even more intriguing: production through the decay of axionic strings [35]. In the case that the Universe never inflated, the initial value of $\bar{\Theta}$ within our present Hubble volume not only uniformly samples the interval $[-\pi, \pi]$, but can also have non-trivial topology. That is, the mapping of $\bar{\Theta}_1$ to three-dimensional space cannot, in general, be smoothly deformed to a uniform value of $\bar{\Theta}_1$ throughout space. The topological entities that arise are axionic strings. The reader who stayed awake while reading Chapter 7 must have anticipated this fact, since PQ SSB involves the spontaneous breakdown of a $U(1)$ symmetry.

In Chapter 7 we discussed in some detail the formation of the one-dimensional defects—strings—that arise when a $U(1)$ gauge symmetry is spontaneously broken. Strings also arise when a global $U(1)$ symmetry is broken—here $U(1)_{\rm PQ}$. The main difference between global and gauge strings is that the energy per length of a long straight string diverges logarithmically. This is because there is no gauge field contribution to cancel the $\partial_\theta\vec{\phi}$ piece of $D_\mu\vec{\phi}$ in the spatial gradient term; for a global symmetry $D_\mu = \partial_\mu$ (θ = angle around the string axis). The energy per length of two anti-parallel strings is finite and proportional to the logarithm of their separation. This is because at large distances $\vec{\phi}$ is constant and does not wind, so that $\partial_\mu\vec{\phi} = 0$. The energy per length of an axionic string is given by $\mu \sim f_{\rm PQ}^2 \ln(f_{\rm PQ}d)$, where d is the characteristic distance between strings.

Provided that the Universe did not inflate before, or during, PQ SSB, a network of axionic strings will fill the Universe after PQ symmetry breaking. Based upon our experience with a gauge string network, we expect that the axionic string network will rapidly approach a scaling solution

where

$$\rho_S \sim \frac{\mu}{t^2} \qquad \text{and} \qquad \frac{\rho_S}{\rho_R} \sim G\mu. \tag{10.52}$$

Recall that a string network is comprised of both long pieces of string and small loops, and even more importantly, that a string energy crisis is avoided by strings constantly cutting themselves into loops, with the loops politely dissipating their energy into gravitational waves. If the scaling solution for axionic strings is to be achieved, similar things must happen. However, there is one important difference: Axionic string dissipates its energy primarily by the radiation of axions, and not gravitational waves! The decay of the axionic string network leads to a third population of relic axions.

To maintain the scaling solution, in each Hubble time ($H^{-1} \sim t$), essentially the full energy density of axionic string must be dissipated by the radiation of axions, so that the change in the number density of axions per entropy density is given by

$$\Delta(n_a/s) \sim \frac{\mu/t^2}{\omega T^3}\Delta(Ht), \tag{10.53}$$

where ω is average energy per axion radiated. The total number of axions produced in a comoving volume is obtained by integrating the above equation:[30]

$$\frac{n_a}{s} \sim \mu \int_{T_1}^{f_{PQ}} \frac{dT}{\omega m_{Pl}^2}, \tag{10.54}$$

where T_1 is the temperature at which the axion mass becomes comparable to the expansion rate. For $T \lesssim T_1$, the axion mass becomes dynamically important, and the string network becomes a network of domain walls bounded by strings which then quickly decays.

In order to calculate axion production through string decay we must know $\omega(t)$, the average energy of an axion produced by string dissipation at time t. This is where the discussion becomes less than clear: Davis [35] argues that the axions produced have the longest wavelengths that they could be expected to have, of order the horizon, $\omega(t) \sim t^{-1}$; whereas Harari and Sikivie [38] argue that the radiated axions have a $1/k$ spectrum, which leads to $\omega \sim \ln(f_{PQ}t)t^{-1}$—a difference of a factor of $\ln(f_{PQ}t)$ or about 100!

---

[30] As before, it is the relic number density of axions that is important; any momenta that axions radiated by strings may have will have long since been red shifted away by the expansion.

Integrating (10.54), we find that the number of axions produced by string decay (per comoving volume) is

$$n_a/s \sim [1 \text{ or } \ln(f_{\rm PQ}t_1)]\frac{f_{\rm PQ}^2}{T_1 m_{Pl}}. \tag{10.55}$$

The form of this expression should be familiar: Up to a factor of $[1 \text{ or } \ln(f_{\rm PQ}t_1)]/\bar{\Theta}_1^2$, it is identical to the expression for the number of axions produced by the initial misalignment of the $\bar{\Theta}$ angle! Depending upon the spectrum of axions produced by the decay of axionic strings, these axions contribute either a comparable number of axions, or 100 times as many axions, as the misalignment mechanism does.

More precisely, if Davis is correct, the string-produced population of axions contributes a mass density

$$\Omega_S h^2 \simeq (m_a/10^{-3}\text{eV})^{-1.18}, \tag{10.56}$$

so that axions provide closure density for a mass of about 10^{-3}eV, rather than about 10^{-5}eV. If Harari and Sikivie [38] are correct, then string-produced axions only slightly increase the axion density over that due to the misalignment mechanism. Clearly the issue of the axion energy spectrum is a very important one. While much is known about the evolution of gauge string networks, owing to their non-localized energy density, very little is known about global string networks.

In any case, we see that the present energy density of axions produced by coherent processes increases with *decreasing* axion mass. Based upon our knowledge of the present age of the Universe, $\Omega_a h^2$ must be less than about 1. In the non-inflationary case, this restricts the axion mass to be greater than about 10^{-6}eV if Harari and Sikivie are correct, or greater than about 10^{-3}eV if Davis is correct. Because of the peculiar scaling of Ω_a with the axion mass, cosmological axion production leads to a *lower* limit to the axion mass.

In the inflationary case there are no string-produced axions (as the value of $\bar{\Theta}_1$ is uniform throughout the observable Universe). Moreover, the bound to m_a based upon the present mass density of axions depends upon $\bar{\Theta}_1$:

$$m_a \gtrsim 10^{-6}\,[\bar{\Theta}_1/(\pi/\sqrt{3})]^{1.7}\,\text{eV}. \tag{10.57}$$

These bounds are shown in Fig. 10.3.

To summarize our discussion of relic axions, thermal axions can probably never contribute closure density. However, the mass density of relic axions produced by misalignment, or by the decay of axionic strings, can be very significant. There are still substantial uncertainties in the calculation of the relic axion abundance: if the Universe inflated, the value of $\bar{\Theta}_1$; and if the Universe did not inflate, the spectrum of string-produced axions. In either case, there are additional uncertainties in the values of Λ_{QCD} and the Hubble constant. A precise determination of the value of the axion mass that leads to $\Omega_a \simeq 1$ is impossible at present. One thing is certain, if axions do indeed contribute $\Omega_a = 1$, they behave as cold dark matter because of their intrinsically small velocities.[31]

10.4 Isocurvature Axion Fluctuations

In an axion-dominated Universe ($\Omega_a \simeq 1$) that undergoes inflation after, or during, PQ symmetry breaking, isocurvature axion perturbations will arise in addition to the usual adiabatic perturbations associated with inflation [39]. These isocurvature fluctuations correspond to fluctuations in the local axion-to-entropy ratio (n_a/s). In contrast, the adiabatic (curvature) perturbations occur in all species (by the equivalence principle) and are characterized by $\delta(n_a/s) = 0$ (see Section 9.2). These isocurvature fluctuations arise due to de Sitter space induced quantum fluctuations in the axion field, which in turn correspond to fluctuations in the initial misalignment angle, $\bar{\Theta}_1$.

To understand the origin of the isocurvature fluctuations, express the fluctuations in $\bar{\Theta}_1$ in a Fourier expansion:

$$\frac{\delta\bar{\Theta}_1(\vec{\mathbf{x}})}{\bar{\Theta}_1} = (2\pi)^{-3}\int d^3k\,\delta_\Theta(k)\exp(-i\vec{\mathbf{k}}\cdot\vec{\mathbf{x}}). \qquad (10.58)$$

After the coherent axion oscillations begin ($T \lesssim \Lambda_{QCD}$), fluctuations in $\bar{\Theta}_1$ will result in fluctuations in n_a, since $n_a \propto \bar{\Theta}_1^2(x)$. In particular,

$$\delta_k = 2\delta_\Theta(k), \qquad (10.59)$$

where $\delta n_a/n_a = (2\pi)^{-3}\int d^3k\,\delta_k e^{-ikx}$, and n_a is the mean axion number density.

As with any massless, minimally coupled scalar field, the axion field will have de Sitter space zero-point fluctuations. The de Sitter space produced

[31] Using the results of Section 9.3.4 to compute the free-streaming length, it follows that $\lambda_{FS} \sim 10^{-6}$ Mpc for $m_a \sim 10^{-5}$ eV.

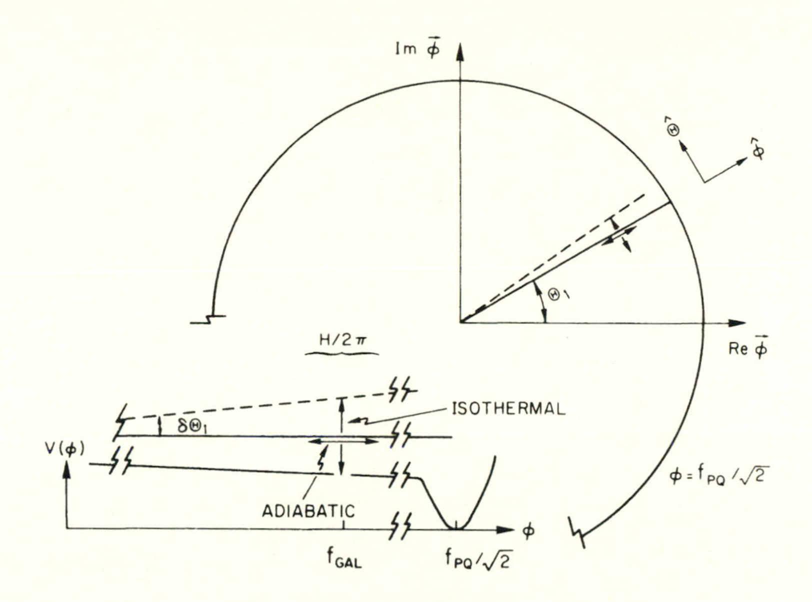

Fig. 10.10: Isocurvature and adiabatic perturbations in Pi's model.

quantum fluctuations in the axion field are characterized by

$$k^3|\delta_a(k)|^2/2\pi^2 = (H/2\pi)^2, \tag{10.60}$$

and so the fluctuations in $\bar{\Theta}_1$ are given by

$$k^3|\delta_\Theta(k)|^2/2\pi^2 = \frac{k^3|\delta_a(k)|^2/2\pi^2}{f_{\rm GAL}^2\bar{\Theta}_1^2} = \frac{(H/2\pi)^2}{f_{\rm GAL}^2\bar{\Theta}_1^2}. \tag{10.61}$$

Here H is the expansion rate during inflation, and $f_{\rm GAL}$ is the value of $|\vec{\phi}|$ when the scales of interest (galaxies, etc.) go outside the horizon during the inflationary epoch. If PQ SSB occurs during inflation, the value of $|\vec{\phi}|$ may still be evolving toward its SSB value, $f_{\rm PQ}/\sqrt{2}$, when the scales of cosmological interest cross outside the horizon.

Once a given mode crosses outside the horizon, the fluctuations on that scale "freeze in" as classical fluctuations in the $\bar{\Theta}$ field. After the coherent axion oscillations commence, these fluctuations correspond to fluctuations in the local axion number density, with the *rms* axion mass fluctuation on a given scale being

$$\left(\frac{\delta n_a}{n_a}\right)_\lambda \equiv k^{3/2}|\delta_k|/\sqrt{2}\pi \simeq \frac{H}{\pi}\frac{1}{\bar{\Theta}_1 f_{\rm GAL}}. \tag{10.62}$$

Note that $f_{\rm PQ}$ and $\bar{\Theta}_1$ are not independent: In an axion-dominated Universe ($\Omega_a h^2 \sim 1$) they are related by,

$$\bar{\Theta}_1 \simeq 1.3 \left[\frac{(f_{\rm PQ}/N)}{10^{12}\,{\rm GeV}}\right]^{-0.59}, \tag{10.63}$$

cf. (10.49).

Note that these fluctuations, like their adiabatic counterpart, are scale-invariant. In order to be relevant for galaxy formation, $\delta n_a/n_a$ must be of the order of 10^{-4} to 10^{-5}. These isocurvature perturbations also lead to anisotropies in the CMBR, and the observed isotropy of the CMBR constraints $\delta n_a/n_a$ to be less than a *few* $\times\ 10^{-4}$. If reheating after inflation is "good," $T_{\rm RH}^2 \simeq Hm_{Pl}$. In order that the PQ symmetry not be restored after inflation, $T_{\rm RH}$ must be less than $f_{\rm PQ}$. These two facts imply that $(H/f_{\rm PQ}) \lesssim (f_{\rm PQ}/m_{Pl})$. If reheating is good and $f_{\rm PQ} \simeq 10^{12}$ to 10^{13}GeV ($\bar{\Theta}_1$ order unity), then $\delta n_a/n_a \simeq 10^{-7}$—far too small to be of relevance for galaxy formation. However, if reheating is poor, $H/f_{\rm PQ}$ can be greater than $(f_{\rm PQ}/m_{Pl})$, or if $f_{\rm PQ} \gg 10^{13}$GeV (corresponding to $\bar{\Theta}_1 \ll 1$), $f_{\rm PQ}/H$ can be greater than 10^{-7}, and $\delta n_a/n_a \simeq 10^{-4}$ can easily be achieved.

An example of a model in which PQ symmetry breaking occurs during inflation is the model of Pi discussed in Chapter 8. In Pi's model PQ symmetry breaking is effected by the field responsible for inflation, i.e., the inflation field $\phi = |\vec{\phi}\,|$. During inflation $|\vec{\phi}\,|$ rolls in the $\bar{\Theta}_1$ direction; galactic-sized perturbations leave the horizon when $|\vec{\phi}\,| = f_{\rm GAL} \simeq f_{PQ}/300$. The de Sitter-space-produced quantum fluctuations in $\vec{\phi}$ can be decomposed onto the $\hat{\phi}$ and $\hat{\bar{\Theta}}$ directions; each have magnitude of $\mathcal{O}(H/2\pi)$. The fluctuations in the $\hat{\phi}$ direction lead to the usual adiabatic density perturbations, while the fluctuations in the $\hat{\bar{\Theta}}$ direction lead to the isocurvature axion perturbations. The orthogonality of the two modes (curvature and isocurvature) is manifest; see Fig. 10.10. In Pi's model $f_{PQ} \simeq 10^{18}$GeV and $\bar{\Theta}_1 \simeq 10^{-3}$, and $\delta n_a/n_a \simeq 10^{-4}$—about the size required to initiate structure formation.

The processed spectrum for isocurvature axion perturbations is flatter than the adiabatic one (see Fig. 9.4), and so with isocurvature axion perturbations galaxies form slightly later, and almost all scales go non-linear together. Whether or not the resulting scenario of structure formation is qualitatively different (than for adiabatic perturbations) is yet to be determined.

10.5 Detection of Relic Axions

While thermal axions probably cannot provide a significant fraction of the present mass density, the lifetimes of multi-eV axions are "well-matched" to the present age of the Universe, i.e., sufficiently long so that not all the relic axions have decayed, and sufficiently short so that a substantial fraction are decaying at present [40]. Since axion decay is a two-body process, the decay-produced photons are mono-energetic, with wavelength $\lambda_a = 2hc/m_a = 24800\,\text{Å}/(m_a/\text{eV})$. The line produced by axion decays will be Doppler broadened with a characteristic width $\Delta\lambda \simeq (v/c)\lambda_a$, where v is the typical velocity of axions in the system under study. Although multi-eV thermal axions will not contribute substantially to the present mass density of the Universe, they will find their way into the many gravitational potential wells that exist, e.g., in galaxies (including our own) and in clusters of galaxies. Taking into account the number of axions expected to find their way into our galaxy, the intensity of radiation from decaying axions within the halo of our own galaxy should be [41]:

$$\begin{aligned} I_{\text{halo}} &= 2\times 10^{-23}\,\text{erg}\,\text{cm}^{-2}\,\text{sec}^{-1}\,\text{arcsec}^{-2}\,\text{Å}^{-1} \\ &\quad \times (m_a/\text{eV})^{10}\left[(E_{PQ}/N - 1.95)/0.72\right]^2 J(\theta), \end{aligned} \tag{10.64}$$

where $J(\theta)$ is the angular dependence of the signal which traces to the fact that we do not reside at the center of our galaxy. The width of such a line is expected to be of order the virial velocity in our galaxy, or $\Delta\lambda \simeq 10^{-3}\lambda_a$.

When one turns a telescope to the blank night sky one sees many lines—not all from axion decays; rather from airglow! Shown in Fig. 10.11 is a high-resolution spectrum of the night sky [42] and the predicted axion line for two values of the axion mass. From the existing data it is clear that an axion mass of greater than 4 eV or so is definitely precluded, which closes the door on the possibility of thermal axions providing closure density.

Even more favorable in terms of detection is the line from relic thermal axions that reside in clusters [41]

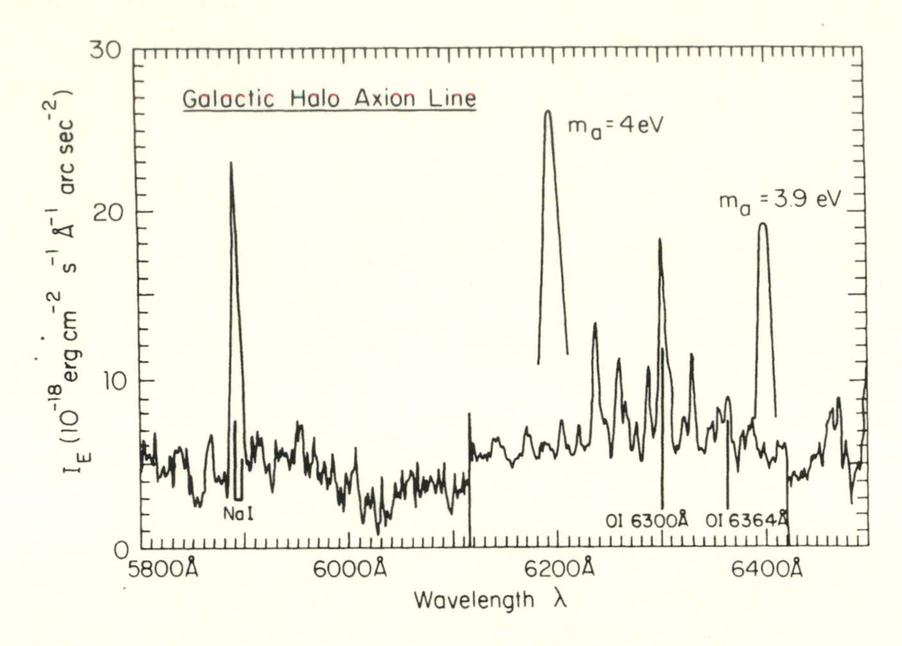

Fig. 10.11: A high-resolution spectrum of the night sky (from [42]) and the narrow line predicted from axion decays within our halo (for $m_a = 3.9$ eV and 4.0 eV).

$$I_{\rm cluster} = 2 \times 10^{-20}\,{\rm erg\,cm^{-2}\,sec^{-1}\,arcsec^{-2}\,\AA^{-1}} \times (m_a/{\rm eV})^7\,[(E_{PQ}/N - 1.95)/0.72]^2. \quad (10.65)$$

Here the line width is expected to be of order the virial velocity in a cluster, or $\Delta\lambda \simeq 10^{-2}\lambda_a$. Moreover, by observing a cluster one has two other advantages: First, one can remove many of the airglow lines by subtracting "off cluster" measurements from "on cluster" measurements; second, the wavelength of the cluster axion line depends upon the red shift of the cluster, $\lambda_a(\text{cluster}) = (1 + z_{\rm cluster})\lambda_a$, and by looking at two or more clusters with different red shifts one can further discriminate against night sky lines. Currently an observational effort is being mounted to search for line radiation from decaying cluster axions [43], and it is hoped that this effort will be sensitive to an axion as light as 2 eV. The rapidly decreasing strength of the axion line with decreasing axion mass, together with the increasing glow of the night sky at longer wavelengths, precludes searching for axion line radiation for axion masses smaller than about 2 eV.

10.5.1 Sikivie-type axion detectors

While the discovery of any relic from the early Universe would be an extremely important one, if that relic contributed the bulk of the present mass density, the discovery would have to rank with the discovery of the expansion of the Universe and of the CMBR. As discussed in Section 10.3, relic axions of mass 10^{-5}eV or so would likely provide the closure density. Since such axions are so very cold, they behave as cold dark matter and therefore should be present in the halo of our galaxy today. Based upon the measured rotation curve of our galaxy and the amount of luminous material inside of the orbit of the sun, the local halo density has been estimated to be

$$\rho_{\rm HALO} \simeq 0.5 \times 10^{-24}\,{\rm g\,cm^{-3}} \simeq 0.3\,{\rm GeV\,cm^{-3}}. \qquad (10.66)$$

If the halo material is composed of 10^{-5}eV axions, this corresponds to a local axion density of $n_a(\text{local}) \simeq 3 \times 10^{13}\,\text{cm}^{-3}$. Moreover, such axions should have gravitationally produced velocities of order 10^{-3}c, so that the local flux of axions should be $\mathcal{F}_a = n_a(\text{local})v_a \simeq 10^{21}\,\text{cm}^{-2}\,\text{sec}^{-1}$. These axions have energies $E_a = m_a(1 + v_a^2/2) \simeq 10^{-5}\text{eV}\,[1 + \mathcal{O}(10^{-6})]$.

The lifetime of these axions for free decay is excruciatingly long: $\tau_a \simeq 10^{42}$ yr. However, because of the axion's coupling to $\vec{\mathbf{E}} \cdot \vec{\mathbf{B}}$, they can be "coaxed" by a strong magnetic field to interconvert to photons of energy E_a, corresponding to frequencies of $f_a = 2\,\text{GHz}\,(m_a/10^{-5}\text{eV})$. Sikivie [44] has proposed a cosmic axion detector based upon this fact: a resonant microwave cavity immersed in a very strong magnetic field. Because of their enormous local number density and long de Broglie wavelengths, $\lambda_a = h/m_a v_a \simeq 10^4\,\text{cm}\,(10^{-5}\text{eV}/m_a)$, locally, axions can be described by a coherent, classical field—at least on length scales much less than 10^4cm. Provided that the cavity is smaller than this, the coupling of a given cavity mode to the galactic axion field is

$$g_{a\gamma\gamma}\, a\, \vec{\mathbf{B}} \cdot \int_{\rm Vol} \vec{\mathbf{E}}_{nl}\, d^3x, \qquad (10.67)$$

where $\vec{\mathbf{E}}_{nl}$ is the electric field amplitude for the mode labeled by nl, and $\vec{\mathbf{B}}$ is the constant, applied magnetic field. From this expression, it is clear that axions couple only to TM cavity modes. The power delivered to the cavity from axion–photon conversions is

$$P_{nl} \simeq 2 \times 10^{-22}\,\mathrm{W} \frac{\mathrm{Vol}}{10\,\mathrm{L}} \left(\frac{B}{6\,\mathrm{T}}\right)^2 C_{nl} \frac{m_a}{10^{-5}\mathrm{eV}} \mathrm{Min}\left[\frac{Q_{nl}}{Q_a}, 1\right] \tag{10.68}$$

where C_{nl} is a mode-dependent form factor of order unity, Q_{nl} is the Q of the TM_{nl0} mode, and $Q_a = E_a/\Delta E_a \simeq 10^6$ characterizes the intrinsic sharpness of the axion line. Needless to say, axion–photon conversion is not going to solve the world's energy crisis![32]

This signal must be detected above the intrinsic noise associated with the detector system, i.e, cavity, amplifier, coaxial feeds, graduate students, etc. The signal-to-noise ratio S/N is determined by the integration time t, the bandwidth B, and the system noise temperature T_n:[33]

$$\frac{S}{N} = \sqrt{t/B}\,\frac{P_{nl}}{T_n}. \tag{10.69}$$

Prototype resonant cavity detectors based upon Sikivie's idea have been built by three groups—at Brookhaven, Florida, and Japan [45]. Roughly speaking, all three detectors are characterized by the following parameters: volume of order 10 liters, magnetic field strength of order 6 Tesla, and cavity Q of order 10^5. Operating at bandwidth of $B \simeq f_a/Q_a \simeq 1000$ Hz, such detectors can achieve a signal-to-noise ratio of about 3 in a mere 10^6 sec. That would be sufficient sensitivity to carry out a definitive search, provided that one knew the axion mass to an accuracy of a part in 10^6. As we have discussed in Section 10.3, that is far from being the case. Operating with such a bandwidth, it would take about 2×10^{12} sec (60,000 yr) to scan an octave in axion mass. Therein lies the difficulty—not knowing where to look. In the one experiment that has already reported results, a collaborative effort of Rochester-Brookhaven-Fermilab [46], the actual integration times were of order seconds, corresponding to an expected signal-to-noise ratio of 3×10^{-3}—not sensitive enough to detect halo axions if they have the properties expected (but, theorists don't always get everything right the first time). However, the first cosmic axion search did scan several octaves in axion mass. Second generation detectors that are larger, have stronger magnetic fields, and have lower system noise temperatures are being planned. The search for cosmic axions goes on!

[32] However, it may come as close as cold fusion!

[33] Of course, in a real experiment it is not possible to deliver 100% of the power deposited in the cavity to the detector; typical efficiencies are of order 2/3.

Another experiment based on Sikivie's idea of magnetically induced axion–photon conversion has been proposed by van Bibber and his collaborators [47]. They would like to build a large, underground container filled with gas (H_2 or ^{4}He) and immersed in a strong magnetic field, to search for hadronic axions of mass 0.1 to 5 eV that are emitted by the sun. Solar axions that interconvert in their detector would produce keV photons.

The axion is a very compelling theoretical idea, perhaps the most minimal extension of the standard model. The search for the axion provides an immense challenge, with eighteen orders of magnitude in mass to be examined. While laboratory experiments have been able to explore only a few decades in mass, the quest for the axion has spurred astrophysicists and cosmologists to new heights of creativity: The Heavenly Laboratory has left but two windows for the axion: 10^{-6}eV to 10^{-3}eV, and for hadronic axions only, 2 to 5 eV, and provides the means of exploring both of these windows. Moreover, relic axions of mass 10^{-6} to 10^{-5}eV or 10^{-3}eV may well be the dark matter in the Universe!

10.6 References

1. G. 't Hooft, *Phys. Rev. Lett.* **37**, 8 (1976); *Phys. Rev. D* **14**, 3432 (1976); R. Jackiw and C. Rebbi, *Phys. Rev. Lett.* **37**, 172 (1976); C. G. Callan, R. Dashen, and D. Gross, *Phys. Lett.* **63B**, 334 (1976); A. Belavin, A. Polyakov, A. Schwartz, and Yu. Tyupkin, *Phys. Lett.* **59B**, 85 (1975). Recent reviews of the axion and the strong-CP problem include, R. D. Peccei, in *CP Violation*, ed. C. Jarlskog (WSPC, Singapore, 1989), p. 503; H.-Y. Cheng, *Phys. Rep.* **158**, 1 (1988); and J.-E. Kim, *Phys. Rep.* **150**, 1 (1987).

2. K. Choi, C. W. Kim, and W. K. Sze, *Phys. Rev. Lett.* **61**, 794 (1988); D. B. Kaplan and A. V. Manohar, *Phys. Rev. Lett.* **56**, 2004 (1986).

3. S. Weinberg, *Phys. Rev. D* **11**, 3583 (1975), and references therein.

4. R. D. Peccei and H. R. Quinn, *Phys. Rev. Lett.* **38**, 1440 (1977); *Phys. Rev. D* **16**, 1791 (1977).

5. S. Weinberg, *Phys. Rev. Lett.* **40**, 223 (1978); F. Wilczek, *Phys. Rev. Lett.* **40**, 279 (1978).

6. For a discussion of the properties of axions and their couplings to matter, see: D. Kaplan, *Nucl. Phys.* **B260**, 215 (1985); M. Srednicki,

Nucl. Phys. **B260**, 689 (1985); P. Sikivie, in *Cosmology and Particle Physics*, eds. E. Alvarez, et al. (WSPC, Singapore, 1986). Here we have followed the normalization conventions of Kaplan and Sikivie.

7. K. Choi, K. Kang, and J.-E. Kim, *Phys. Rev. Lett.* **62**, 849 (1989); M. Carena and R. D. Peccei, *Phys. Rev. D* **40**, 652 (1989).
8. J.-E. Kim, *Phys. Rev. Lett.* **43**, 103 (1979); M. A. Shifman, A. I. Vainshtein, and V. I. Zakharov, *Nucl. Phys.* **B166**, 493 (1980).
9. A. R. Zhitnitsky, *Sov. J. Nucl. Phys.* **31**, 260 (1980); M. Dine, W. Fischler, and M. Srednicki, *Phys. Lett.* **104B**, 199 (1981).
10. D. Kaplan, *Nucl. Phys.* **B260**, 215 (1985).
11. See, e.g., J.-E. Kim, *Phys. Rep.* **150**, 1 (1987).
12. D. A. Dicus, E. W. Kolb, V. L. Teplitz, and R. V. Wagoner, *Phys. Rev. D* **18**, 1829 (1978); K. Sato, *Prog. Theor. Phys.* **60**, 1942 (1978).
13. H. B. Nielsen and M. Ninomiya, *Phys. Rev. Lett.* **62**, 1429 (1989); K. Choi and R. Holman, *Phys. Rev. Lett.* **62**, 2575 (1989); J. Preskill, M. Wise, and S. P. Trivedi, *Phys. Lett.* **B**, in press (1989).
14. S. Chandrasekhar, *An Introduction to the Study of Stellar Structure* (Dover, New York, 1967); D. D. Clayton, *Principles of Stellar Evolution and Nucleosynthesis* (The University of Chicago Press, Chicago, 1983); M. Schwarzschild, *Structure and Evolution of the Stars* (Dover, New York, 1965).
15. R. Stothers, *Ap. J.* **155**, 935 (1969); *Phys. Rev. Lett.* **24**, 538 (1970).
16. K. Hirata, et al., *Phys. Rev. Lett.* **58**, 1490 (1987).
17. R. M. Bionta, et al., *Phys. Rev. Lett.* **58**, 1494 (1987).
18. S. Dimopoulos, J. Frieman, B. Lynn, and G. D. Starkman, *Phys. Lett.* **179B**, 223 (1986).
19. These rates are from: D. A. Dicus, E. W. Kolb, V. L. Teplitz, and R. V. Wagoner, *Phys. Rev. D* **22**, 839 (1980); M. Fukugita, S. Watamura, and M. Yoshimura, *Phys. Rev. D* **26**, 1840 (1982); and L. L. Krauss, J. Moody, and F. Wilczek, *Phys. Lett.* **144B**, 391 (1984). For screening corrections to these rates, see G. G. Raffelt, *Phys. Rev. D* **33**, 897 (1986).

20. J. Frieman, S. Dimopoulos, and M. S. Turner, *Phys. Rev. D* **36**, 2201 (1987).

21. J. N. Bahcall and R. K. Ulrich, *Rev. Mod. Phys.* **60**, 297 (1988); also see, J. Bahcall in [25].

22. F. Avignone, III, et al., *Phys. Rev. D* **35**, 2752 (1987).

23. G. G. Raffelt and D. S. P. Dearborn, *Phys. Rev. D* **36**, 2211 (1987).

24. D. S. P. Dearborn, D. N. Schramm, and G. Steigman, *Phys. Rev. Lett.* **56**, 26 (1986).

25. For a review of the standard picture of the cooling of a nascent neutron star and analysis of the KII and IMB data, see, e.g., J. N. Bahcall, *Neutrino Astrophysics* (Cambridge University Press, Cambridge, 1989); D. N. Schramm, *Comments on Nucl. and Part. Sci.* **17**, 239 (1987); D. Q. Lamb and T. Loredo, *Phys. Rev. D*, in press (1989).

26. A. Burrows, M. S. Turner, and R. P. Brinkmann, *Phys. Rev. D* **39**, 1020 (1989).

27. R. P. Brinkmann and M. S. Turner, *Phys. Rev. D* **38**, 2338 (1988).

28. M. S. Turner, *Phys. Rev. Lett.* **60**, 1797 (1988).

29. C. Alcock and A. Olinto, *Ann. Rev. Nucl. Part. Sci.* **38**, 161 (1988).

30. D. Gross, R. Pisarski, and L. Yaffe, *Rev. Mod. Phys.* **53**, 43 (1981).

31. P. Sikivie, *Phys. Rev. Lett.* **48**, 1156 (1982).

32. See P. Sikivie, in [6].

33. M. S. Turner, *Phys. Rev. Lett.* **59**, 2489 (1987); also see G. G. Raffelt and M. S. Turner, *Phys. Rev. D*, in press (1989).

34. J. Preskill, M. Wise, and F. Wilczek, *Phys. Lett.* **120B**, 127 (1983); L. Abbott and P. Sikivie, *ibid*, 133 (1983); M. Dine and W. Fischler, *ibid*, 137 (1983).

35. R. Davis, *Phys. Lett.* **180B**, 225 (1986); R. Davis and E. P. S. Shellard, *Nucl. Phys.* **B**, in press (1989). Also see A. Dabholkar and J. M. Quashnock, *Nucl. Phys.* **B**, in press (1989).

36. M. S. Turner, *Phys. Rev. D* **33**, 889 (1986).

37. J. M. Flynn and L. Randall, *Phys. Rev. D*, in press (1989).

38. D. Harari and P. Sikivie, *Phys. Lett.* **195B**, 361 (1987).

39. D. Seckel and M. S. Turner, *Phys. Rev. D* **32**, 3178 (1985).

40. T. Kephart and T. Weiler, *Phys. Rev. Lett.* **58**, 171 (1987).

41. M. S. Turner, *Phys. Rev. Lett.* **59**, 2489 (1987).

42. A. L. Broadfoot and K. R. Kendall, *J. Geophys. Res. Sp. Phys.* **73**, 426 (1968).

43. M. Bershady, M. T. Ressell, and M. S. Turner, work in progress (1989).

44. P. Sikivie, *Phys. Rev. Lett.* **51**, 1415 (1983).

45. *Proceedings of the 1989 BNL Workshop on Cosmic Axions*, ed. A. Melissinos (WSPC, Singapore, 1989).

46. S. De Panfilis, et al., *Phys. Rev. Lett.* **59**, 839 (1987); W. U. Wuensch, et al., *Phys. Rev. D*, in press (1989).

47. K. van Bibber, P. M. Mc Intyre, D. E. Morris, and G. G. Raffelt, *Phys. Rev. D* **39**, 2089 (1989).

11

TOWARD THE PLANCK EPOCH

11.1 Overview

In this Chapter we will discuss some of the most interesting, profound, and topical speculations about the earliest moments of the Universe: those concerning the Planck epoch, $t \lesssim 10^{-43}$ sec. Up to this point our discussion of the very early Universe ($t \ll 10^{-2}$ sec) has been based upon highly speculative, but nevertheless well formulated, ideas about the nature of the fundamental interactions at energies well beyond 1 TeV. For instance, the existence of a grand unified theory of the strong and electroweak interactions is speculation, but the construction of a spontaneously broken, Yang-Mills gauge theory based upon a specific gauge group is well understood. In contrast to our discussions of the post-Planck epoch, even the most promising ideas about the Planck epoch are based upon ideas that have yet to reach the stage of being well formulated theories. Because the underlying physics is still cloudy, it is not clear how (or even whether) one should approach the Planck time. In spite of the fact that the physics is not well formulated, we will attempt to give the reader a taste for some of the most exciting and potentially important research being pursued at present. The reader should understand that we are fishing in murky waters, and that the ideas we shall discuss here are both highly speculative and not yet unambiguously formulated. The faint-hearted reader may wish to skip to the Finale.

In order to understand the Planck epoch and answer one of the simplest cosmological questions one can ask—How did the Universe begin?—one must treat the gravitational degrees of freedom and interactions quantum mechanically. This is a tall order and a goal that is still far from being fulfilled. That is not to say that there is a lack of promising theoretical avenues. In fact, at this moment there may be more promising ideas than in the past several decades. Among the promising approaches to this most difficult problem are superstring theory, quantum cosmology, and quantum topology change.

Quantizing the gravitational degrees of freedom and interactions is a very formidable task. One mature and well formulated first step in this direction is the semi-classical approach. Here, the gravitational degrees of freedom are treated classically, and the quantum evolution of the other fields—in the simplest case a single scalar field—is studied in the classical, curved space-time background. We have discussed such an example in Chapter 8, when treating the origin of density perturbations in inflationary Universe models: The quantum fluctuations of the inflaton (scalar) field are computed in the classical de Sitter background. It is these fluctuations that eventually give rise to density perturbations, which in turn, may initiate structure formation in the Universe. A second, related, example is the quantum evolution of the inflaton field itself; while not discussed in Chapter 8, the quantum evolution of the inflaton field is the subject of the paper by Guth and Pi that is reprinted in *Early Universe: Reprints*. Another, even more well known, result of the semi-classical approach is the Hawking radiation associated with black holes [1]. Hawking showed that because of the event horizon associated with a black hole, a black hole should emit a thermal spectrum of particles, with a temperature $T_H = 1/8\pi GM$ (M is the mass of the hole). Owing to this radiation, black holes should evaporate in a time $t \simeq (2560\pi/g_*)G^2M^3$—for holes less massive than 10^{15}g, in less than the age of the Universe. If small ($M \lesssim 10^{15}$ g), primordial black holes exist, they have a myriad of extremely interesting and important astrophysical and cosmological consequences [2], consequences that could indeed provide the first experimental evidence for quantum gravitational effects. The general area of quantum fields in curved space-time is a mature subject in its own right, and we refer the interested reader to the monograph of Birrell and Davies [3].

One interesting, but more pedagogical, aspect of the semi-classical approach is that of the interpretation of quantum mechanics in the Universe—a setting in which one cannot resort to the usual crutch of differentiating between the observer and the system under study. This important issue has been thoroughly addressed in several reviews [4].

Several promising and not-unrelated approaches to quantizing the gravitational degrees of freedom fall under the general category of quantum cosmology. One direction that has been pursued is to derive, using the canonical quantization procedure, the analogy of a Schrödinger equation (more accurately, a Klein-Gordon equation) for the *wave function of the*

Universe, a wave function governing both the matter fields and the space-time geometry. The equation governing the wave function of the Universe is known as the Wheeler-DeWitt equation [5]. This approach, especially its path integral formulation, has been a lively avenue of investigation and appears to be one of the most promising approaches to quantum gravity. In the next Section we will go through the derivation of the Wheeler-DeWitt equation, and in Section 11.3 we will discuss some mini-superspace wave functions.

Another aspect of quantum cosmology is that of third quantization. Third-quantized operators create and destroy Universes complete with second-quantized fields. While today the single-Universe approximation seems to be quite adequate—When was the last time you saw a Universe created or destroyed?—the quantum effects of emitting or absorbing small Universes ("baby Universes") could play an important role in determining the fundamental parameters that we measure in our Universe, e.g., the value of the cosmological constant, particle masses, and coupling constants [6]. One can think of this process as being analogous to the running coupling constant in QCD—the coupling constant of a quark is renormalized by the sea of virtual particles that exist within Nature's vacuum. Baby Universes, wormholes, and space-time topology change is presently a very intriguing and potentially fruitful avenue of research. We will not discuss this subject here and refer the interested reader to [7].

An apparently very different approach to quantum gravity is through the unification of the interactions of Nature—strong, weak, electromagnetic, and gravitational. There are at least two paths that are being pursued; that there are two paths traces to the fact that our understanding of gravity is geometrical, while that of the other interactions involves gauge symmetries. Historically, the first path involves trying to formulate the non-gravitational interactions in a geometrical way, by introducing additional, compactified space-like dimensions. The isometries, or space-time symmetries, of these extra spatial dimensions are supposed to give rise to the gauge symmetries that are observed in Nature, e.g., seven additional compact dimensions are required to account for the $SU(3)\otimes SU(2)\otimes U(1)$ gauge symmetry of the standard model. The dynamics of these higher-dimensional theories is usually derived from a straightforward generalization of the Einstein-Hilbert action, and such theories are known as Kaluza-Klein theories. While any additional space-like dimensions must be very small today—much less than 10^{-16} cm—it could be that at early times they were on equal footing with the more familiar three space-like dimensions, and the fact that the Universe in its infancy was more than four dimensional might have had important consequences for the subsequent history of the Universe. While we will discuss extra-dimensional cosmology in this Chapter, this approach to unification does not look as promising as other approaches at the moment. The difficulties lie in constructing a consistent quantum field theory: Four-dimensional gravity theory is non-renomalizable, and higher-dimensional theories are even worse; further, such theories seem incapable of accomodating chiral fermions—and that of course is the way fermions occur in Nature.

.The other path to the unification of the interactions of Nature is almost opposite in philosophy: The geometric character of gravity is ignored, and the gravitational interaction is treated in the same way as the other interactions. This approach has encompassed the ideas of supersymmetry, supergravity, and superstrings. The most promising idea by far is that of the superstring. The superstring theory is a quantum theory of one-dimensional entities—open and closed strings—that unifies all the forces of Nature in a self-consistent, finite quantum field theory. Gravity arises because superstring theories *must* incorporate a massless spin-2 field if they are to be consistent. The superstring theory is the first self-consistent theory constructed that unifies all of the interactions of Nature. However, much work remains. At present, the theory is formulated only as an S-matrix; and it is still a mystery as to how the geometric character of gravity arises. Interestingly enough, the most elementary formulation of the superstring requires (for consistency) at least six (and perhaps twenty-two) extra-spatial dimensions.[1] Of course, it may be that the additional dimensions are a temporary construct that can be eventually discarded—or it may be that the dimensionality of space is greater than four, and that the extra dimensions are presently compactified. In this Chapter we will discuss a unique implication of superstring theory—the possibility that the Planck-epoch Universe was composed of string, which in turn implies a limiting temperature for the Universe. Furthermore, our discussion of cosmology with extra dimensions may be of some relevance for superstring theory, as some of the same issues arise—Why are three of the space-like dimensions large? and How do the compact dimensions get and stay that way? Our discussion of quantum cosmology may be relevant to this issue

[1]While both Kaluza-Klein and superstring theories involve the existence of extra spatial dimensions that today are compactified, the philosophies of the two approaches are very different. In the Kaluza-Klein approach it is the isometries of the compact dimensions that give rise to the gauge symmetries of Nature, while in superstring theory the gauge symmetries are fundamental and the additional dimensions are required only to formulate the theory.

too; it could be that the wave function of the Universe involves a path integral over spaces of different dimensionality and topology, and happens to be "sharply peaked" around four! Since we will only barely touch upon superstring theory, we refer the reader interested in superstrings to the monograph by Green, Schwartz, and Witten [8].

As we stated at the beginning of this lengthy overview, the Planck era, unlike the epochs that follow, involves theoretical ideas that are both very speculative and not yet well formulated. The window on the Planck era is a difficult one to open, as it necessarily involves treating gravity on a quantum-mechanical basis. There are a number of very promising and exciting developments at present—quantum cosmology, including the wave function of the Universe and topology change, and superstrings, and an understanding of this era may allow one to answer one of the simplest questions of cosmology: How did the Universe begin?

11.2 The Wheeler-DeWitt Equation

In the wave function of the Universe program the quantum nature of geometry enters through canonical quantization of the Hamiltonian formulation of general relativity. In this Section we will derive the Wheeler-DeWitt equation, and in the next Section we will discuss some simple solutions. We begin with a brief review of the Hamiltonian formulation. For a more thorough exposition of the Hamiltonian formulation, the reader is referred to Wald [9], or Misner, Thorne, and Wheeler [10].

In the usual Lagrangian formulation of general relativity the fundamental field is the metric tensor, $g_{\mu\nu}$, and the formulation is manifestly space-time covariant. The Einstein-Hilbert action is usually expressed as

$$S_{\text{E-H}} = -\frac{1}{16\pi G}\int d^4x\sqrt{-g}\,[\mathcal{R}(g) + 2\Lambda], \tag{11.1}$$

where $\mathcal{R}(g)$ is the Ricci scalar, which is a function of $g_{\mu\nu}$ and its first and second derivatives (see Chapters 2 and 3), and Λ is a possible cosmological constant (or vacuum energy). The Einstein equations follow from requiring the action to be stationary under variations of the metric. Because $\mathcal{R}$ involves second derivatives, if one considers a compact manifold with boundary ∂M, and allows variations of the metric that vanish on ∂M but whose normal derivatives do not, one must add a surface term. Including the necessary surface term, the action becomes

$$S = S_{\text{E-H}} + \frac{1}{8\pi G}\int_{\partial M} d^3x\sqrt{h}\,K, \tag{11.2}$$

where K is the trace of the extrinsic curvature K_{ij} of the boundary three surface, and h is the determinant of the metric induced on the three surface. These will be made explicit below.

In contrast, the Hamiltonian formulation is *not* space-time covariant. The Hamiltonian formulation of general relativity requires a $3+1$ split of the metric, and the dynamical degrees of freedom are the *spatial* components of the metric. The usual $3+1$ split is accomplished through the Arnowitt, Deser, Misner (ADM) decomposition of the space-time metric $g_{\mu\nu}$ in terms of a lapse function N, a shift vector N_i, and an induced spatial metric h_{ij}.[2] To this end, one begins by constructing hypersurfaces, Σ_t, parameterized by some global time-like variable t. Since eventually we will restrict our study to closed FRW models, we will take the three sphere (S^3) as a concrete example for Σ_t to illustrate the Hamiltonian formalism. Recall that the line element for S^3 can be written as $\vec{\mathbf{dl}}^2 = R^2(t)\left[d\chi^2 + \sin^2\chi\left(d\theta^2 + \sin^2\theta d\phi^2\right)\right]$. Successive hypersurfaces $\Sigma_t \to \Sigma_{t+\delta t}$ will be described by the same spatial metric with $R(t) \to R(t+\delta t)$. The choice of S^3 for Σ_t is an example of a synchronous system.

The spatial components of $g_{\mu\nu}$ on the hypersurfaces Σ_t induce a spatial metric (i.e., a three-dimensional Riemannian metric), given by $h_{ij} = -g_{ij} + n_i n_j$, where n_μ is the vector field normal to Σ_t. For our three-sphere example, the normal vector to S^3 is in the time direction, i.e., $n_i = 0$, $n_0 = 1$, and the induced metric on $\Sigma_t = S^3$ is simply the metric for a three sphere of radius $R(t)$.

The construction of the lapse function and the shift vector is illustrated in Fig. 11.1. Pick a point P_1 on Σ_t and construct the normal at that point. The normal intersects Σ_{t+dt} at the point P_2. The proper distance between points P_1 and P_2 defines the lapse function N: $d\tau = Ndt$. In general, the normal vector will take the spatial coordinates x^i on Σ_t to some other spatial coordinates on Σ_{t+dt}. Let P_3 denote the point on Σ_{t+dt} with the same spatial coordinates as the point P_1 on Σ_t. The vector from P_2 to P_3 defines the shift vector N^i.[3] The shift vector describes the distortion of

[2]One can question, and many have, whether the ADM split is the correct path to canonical quantization. The appropriate gravitational degrees of freedom to quantize is by no means a closed issue; see, e.g., [11].

[3]The traditional convention may be confusing to the neophyte; the lapse N is *not*

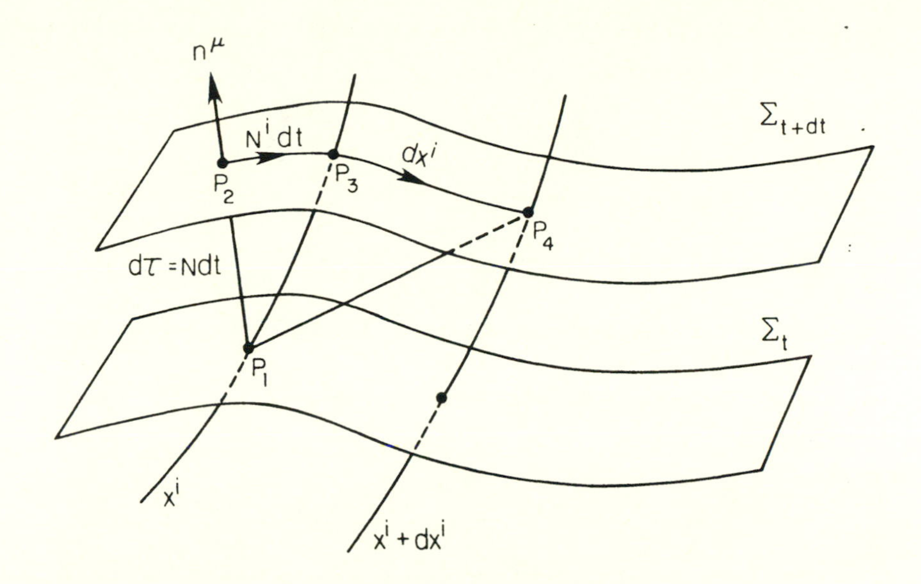

Fig. 11.1: The lapse function N and the shift vector N^i.

the surface Σ_t as it evolves in time. Later we will show that the shift and the lapse are not dynamical but are essentially Lagrange multipliers. For our S^3 example, n^μ has vanishing spatial components: The normal vector takes us from a point on one three sphere with radius $R(t)$ at time t to a point on another three sphere with radius $R(t+dt)$ at time $t+dt$ with the same spatial coordinates. Moreover, the lapse function is independent of t, with value $N = 1$.

Fig. 11.1 also demonstrates how the space-time metric can be constructed from N, N^i and h_{ij}. Take some other point, P_4, on Σ_{t+dt} with spatial coordinate $x^i + dx^i$. The proper length from P_1 to P_4 is defined in terms of the space-time metric $g_{\mu\nu}$ as

$$ds^2 = g_{\mu\nu} dx^\mu dx^\nu. \tag{11.3}$$

The distance from P_1 to P_4 can also be expressed in terms of the lapse and the shift as[4]

$$ds^2 = (Ndt)^2 - h_{ij}(N^i dt + dx^i)(N^j dt + dx^j). \tag{11.4}$$

the magnitude of the shift vector N_i; i.e. $N^2 \neq h_{ij}N^iN^j$.

[4]The metric on Σ_t is of course h_{ij}, which need not be equal to g_{ij}.

By equating (11.3) and (11.4) it is a simple exercise to express the spacetime metric and its inverse in terms of N, N_i, and h_{ij}:

$$g_{\mu\nu} = \begin{pmatrix} N^2 - N_i N_j h^{ij} & -N_j \\ -N_i & -h_{ij} \end{pmatrix},$$

$$g^{\mu\nu} = \begin{pmatrix} 1/N^2 & -N^j/N^2 \\ -N^i/N^2 & N^i N^j/N^2 - h^{ij} \end{pmatrix}. \tag{11.5}$$

Here h^{ij} is the inverse of h_{ij}, and $N_i = h_{ij}N^j$; moreover, it should be clear that $\sqrt{-g} = N\sqrt{h}$.

It is very useful to define the *extrinsic* curvature of Σ_t:

$$K_{ij} \equiv \frac{1}{2N}\left[N_{i|k} + N_{k|i} - \frac{\partial h_{ij}}{\partial t}\right], \tag{11.6}$$

where "|" denotes covariant differentiation with respect to the spatial metric h_{ij}. The extrinsic curvature measures the curvature of Σ_t relative to the enveloping 4-geometry. Continuing with our closed FRW example,

$$K_{ij}(\mathrm{FRW}) = -\frac{1}{2N}\frac{\partial h_{ij}}{\partial t} = -\frac{1}{N}\frac{\dot{R}}{R}h_{ij} = -\frac{\dot{R}}{R}h_{ij} \tag{11.7}$$

since $N = 1$ and $N_i = 0$. The trace of K_{ij} is denoted by $K \equiv K^i{}_i = h^{ij}K_{ij}$. For the closed FRW model, $K = -3\dot{R}/R = -3H$.

Using the metric in the form (11.5), it is straightforward to express the Ricci scalar $\mathcal{R}$ in terms of the lapse, the shift, and the induced metric:

$$\mathcal{R} = K^2 - K_{ij}K^{ij} - {}^3\mathcal{R}. \tag{11.8}$$

Up to a possible 4-divergence, the Ricci scalar calculated using (11.8) is identical to the Ricci scalar found by the "old-fashioned" method (from $g_{\mu\nu}$ calculate $\Gamma^\mu{}_{\nu\alpha}$, construct $R^\mu{}_{\nu\alpha\beta}$, and contract to find $\mathcal{R}$; see Chapters 2 and 3).[5] Equation (11.8) is the preferred form for construction of the Hamiltonian, since it contains products of first derivatives but no second derivative terms. To illustrate this, consider our closed-FRW example. The Ricci scalar from (11.8) is $\mathcal{R} = 6(\dot{R}/R)^2 - {}^3\mathcal{R} = 6(\dot{R}/R)^2 - 6/R^2(t)$.

[5] The addition of a 4-divergence to the Lagrangian does not change the equations of motion, as it contributes only a surface term to the action.

The Ricci scalar found in Chapters 2 and 3 by the "old-fashioned" method (denoted here by $\mathcal{R}'$) is $\mathcal{R}' = -6/R^2(t) - 6(\dot{R}/R)^2 - 6\ddot{R}/R$, and it follows that $\mathcal{R}' = \mathcal{R} - (-g)^{-1/2}d(6\dot{R}R^2)/dt$. When we construct the Hamiltonian for the closed-FRW model in the next Section, we will want an action that involves R and $\dot{R}$, but not $\ddot{R}$, so we will opt for (11.8).

Now express the gravitational Lagrangian density in terms of ${}^3\mathcal{R}$, K_{ij}, and K using (11.8):

$$\begin{aligned} \mathcal{L}[g_{\mu\nu}] &= -\sqrt{-g}\,\mathcal{R}/16\pi G, \\ \mathcal{L}[N, N_i, h_{ij}] &= -\sqrt{h}\,N\left[K^2 - K_{ij}K^{ij} - {}^3\mathcal{R}\right]/16\pi G. \end{aligned} \tag{11.9}$$

From the definition of the extrinsic curvature, cf. (11.6), we see that K_{ij} (and its trace) involves time derivatives of h_{ij} and spatial derivatives of N_i. The three-curvature ${}^3\mathcal{R}$ involves only spatial derivatives of h_{ij}. Since (11.9) does not contain time derivatives of N or N_i, the momenta conjugate to N and N_i vanish:

$$\begin{aligned} \pi &\equiv \frac{\delta\mathcal{L}}{\delta\dot{N}} = 0, \\ \pi^i &\equiv \frac{\delta\mathcal{L}}{\delta\dot{N}_i} = 0, \end{aligned} \tag{11.10}$$

where overdot denotes time derivative. These are referred to as "primary" constraints; moreover, it is clear that the lapse and the shift are not dynamical variables. The momenta conjugate to h_{ij} are

$$\pi^{ij} \equiv \frac{\delta\mathcal{L}}{\delta\dot{h}_{ij}} = \sqrt{h}\left(K^{ij} - h^{ij}K\right)/16\pi G. \tag{11.11}$$

The Hamiltonian is formed in the canonical way from the "coordinates" q^i and their conjugate momenta π^i: $H \equiv \int d^3x \sum_i \{\pi^i\dot{q}^i - \mathcal{L}[q^i]\}$. Therefore, the gravitational Hamiltonian is[6]

$$\begin{aligned} H &= \int d^3x\left(\pi^{ij}\dot{h}_{ij} + \pi^i\dot{N}_i + \pi\dot{N} - \mathcal{L}\right) \\ &= \int d^3x\left(N\mathcal{H}_G + N_i\mathcal{H}^i\right), \end{aligned} \tag{11.12}$$

[6]For open cosmological models there are additional surface terms that give rise to the total energy.

where

$$\begin{aligned}
\mathcal{H}_G &\equiv \sqrt{h}\left(K_{ij}K^{ij} - K^2 - {}^3\mathcal{R}\right)/16\pi G \\
&= \frac{16\pi G}{2\sqrt{h}}\left(h_{ik}h_{jl} + h_{il}h_{jk} - h_{ij}h_{kl}\right)\pi^{ij}\pi^{kl} - \sqrt{h}\,{}^3\mathcal{R}/16\pi G \\
&\equiv 16\pi G\, G_{ijkl}\pi^{ij}\pi^{kl} - \sqrt{h}\,{}^3\mathcal{R}/16\pi G, \qquad (11.13)
\end{aligned}$$

and

$$\mathcal{H}^i \equiv -2\pi^{ij}{}_{|j}/16\pi G. \qquad (11.14)$$

Since the primary constraints $\pi = \pi^i = 0$ hold at all times, we have $\dot{\pi} = \dot{\pi}^i = 0$. Writing the Poisson brackets for $\dot{\phi}$ and $\dot{\phi}^i$, we find

$$\begin{aligned}
\dot{\pi} &= -\{H, \pi\} = \frac{\delta H}{\delta N} = 0, \\
\dot{\pi}^i &= -\left\{H, \pi^i\right\} = \frac{\delta H}{\delta N_i} = 0. \qquad (11.15)
\end{aligned}$$

The vanishing of $\delta H/\delta N$ and $\delta H/\delta N_i$ results in the "secondary" constraints

$$\mathcal{H}_G = \mathcal{H}^i = 0. \qquad (11.16)$$

The secondary constraint $\mathcal{H}_G = 0$ can be expressed as

$$\mathcal{H}_G = 16\pi G\, G_{ijkl}\pi^{ij}\pi^{kl} - \sqrt{h}\,{}^3\mathcal{R}/16\pi G = 0, \qquad (11.17)$$

where G_{ijkl} is defined implicitly in (11.13).

Note that H is independent of N and N_i; as promised the lapse and the shift are not dynamical variables. The true dynamical variable is h_{ij}, and its evolution is given by (11.17). It can also be shown that the secondary constraint equations may be used in place of the Hamilton equations.

So far the development has been purely classical. The first step toward quantization is to identify the Hamiltonian constraint equation $\mathcal{H}_G = 0$ as the zero-energy Schrödinger equation $\mathcal{H}_G(\pi_{ij},\ h_{ij})\Psi[h_{ij}] = 0$, where the state vector Ψ is the wave function of the Universe. The second step is the canonical quantization procedure in which the momenta are replaced by derivatives of their corresponding coordinates: $\pi^i \to -i\partial/\partial q_i$, and

$$\pi^{ij} \to -i\left(\frac{1}{16\pi G}\right)^{3/2}\frac{\delta}{\delta h_{ij}}. \qquad (11.18)$$

The canonical quantization procedure leads to the Wheeler-DeWitt equation [5]

$$\left[\frac{G_{ijkl}}{(16\pi G)^2}\frac{\delta}{\delta h_{ij}}\frac{\delta}{\delta h_{kl}} + \frac{\sqrt{h}\,{}^3\mathcal{R}}{16\pi G}\right]\Psi[h_{ij}] = 0. \tag{11.19}$$

The generalization of the Wheeler-DeWitt equation with cosmological constant and matter fields (denoted generically by ϕ) is straightforward:

$$\left[\frac{G_{ijkl}}{(16\pi G)^2}\frac{\delta}{\delta h_{ij}}\frac{\delta}{\delta h_{kl}} + \frac{\sqrt{h}({}^3\mathcal{R} - 2\Lambda)}{16\pi G} - \mathcal{T}\right]\Psi[h_{ij},\, \phi] = 0, \tag{11.20}$$

where $\mathcal{T} = T^0_{\ 0}(\phi, -i\partial/\partial\phi)$, and $T_{\mu\nu}$ is the stress-energy tensor of the matter fields. We have ignored the subtleties in the factor ordering of quantum operators associated with the $\pi_i \dot{q}^i$ terms. For many choices of the factor ordering, the effect of factor ordering can be parameterized by a constant a, and the corresponding Hamiltonian is obtained by the substitution

$$\pi^2 \rightarrow -q^{-a}\left[\frac{\partial}{\partial q}q^a\frac{\partial}{\partial q}\right]. \tag{11.21}$$

The choice in (11.20) corresponds to $a = 0$. The factor ordering will not affect any of the semi-classical calculations done in the next Section, and we will let the choice of a be dictated by convenience.

A remarkable, and surprising, feature of the wave function of the Universe is the fact that it is *independent* of time: Ψ depends upon only the three-geometry h_{ij} and the matter field content ϕ. Because of this, the interpretation of Ψ is a matter of intense debate [12]. One well motivated interpretation of $\Psi[h_{ij}, \phi]$ is that it measures probabilistic correlations between h_{ij} and ϕ. In this way one can imagine using some function of ϕ, e.g., the energy density ρ_ϕ, as a surrogate time variable. The role of time, or even whether time has a role to play in quantum cosmology, is an issue that is far from settled. Likewise, the interpretation of Ψ has yet to be resolved. The only statement that most practitioners would agree with is that when $\Psi \sim \exp[iS_{\mathrm{E-H}}/\hbar]$, classical behavior ought to pertain.

It should also be emphasized that having an equation for the wave function of the Universe no more resolves the issue of the quantum evolution of the Universe, than the Schrödinger equation for an electron resolves the issue of the quantum evolution of said electron. The Schrödinger equation merely evolves the wave function, and there are many solutions. To be specific about the quantum evolution of an electron or of the Universe one needs information about the initial quantum state. As we shall see there are several well motivated proposals for the initial quantum state of the Universe, which lead to different predictions. Provided that the predictions are sufficiently sharp, present-day observations can, in principle, be used to test the various proposals for the initial state.

It may be that inflation provides a cosmological obstruction to such a program. As we have emphasized in Chapter 8, if inflation occurs, it occurs well after the Planck epoch, and presumably well into the epoch of classical cosmology. Furthermore, in dealing with regions comparable in size to our Hubble volume, the effect of inflation is to lessen severely the sensitivity of the present state of the Universe to the conditions that existed before inflation. In any case, the quantum cosmology era determines the pre-inflationary state of the Universe; whether or not, or by how much, inflation blurs the memory of that state remains to be seen.

11.3 The Wave Function of the Universe

The wave function $\Psi[h_{ij}, \phi]$ of the previous section is defined on an infinite-dimensional space of all possible 3-geometries and matter field configurations, known as *superspace*.[7] In practice, the infinite number of degrees of freedom of superspace make the problem intractable, and all but a finite number of degrees of freedom must be "frozen out." The resulting finite-dimensional superspace is known as *mini-superspace*. In this Section we consider a simple mini-superspace model in which the only degree of freedom is that of the scale factor R of a closed homogeneous and isotropic Universe, and a homogeneous, massive scalar field ϕ.

Moreover, to begin we will freeze out the degrees of freedom of the scalar field and assume that its only effect is to provide a vacuum energy density. For simplicity, we shall write the vacuum energy density associated with the scalar field as a cosmological term, $\Lambda = 8\pi G\rho_{\rm VAC}$. The classical evolution of the scale factor for this closed FRW model is governed by the Friedmann equation

$$\left(\frac{\dot{R}}{R}\right)^2 + \frac{1}{R^2} = \frac{\Lambda}{3}, \tag{11.22}$$

which has the familiar de Sitter solution

$$R(t) = R_0 \cosh(R_0^{-1}t), \tag{11.23}$$

[7]Superspace as used here bears no relation to the superspace of supersymmetry.

where $R_0 \equiv (\Lambda/3)^{-1/2}$. The de Sitter solution describes a Universe that in the infinite past was infinitely large, contracts to a minimum size R_0 at $t = 0$, and then expands to infinite size as $t \to +\infty$. Note too that the classical solution has a forbidden range for R: $0 < R < R_0$.

Although it classically is impossible to start from "nothing" (here defined as $R = 0$) and end up in de Sitter space, it may be possible when one considers the quantum-mechanical nature of the problem. To address this, we use the Wheeler-DeWitt equation. We could obtain the Wheeler-DeWitt equation from (11.19), but it is easier (and more instructive) to re-derive it for a mini-superspace model where the scale factor R is the only degree of freedom.

Since the shift vector N_i vanishes for the FRW metric, the Wheeler-DeWitt equation follows from the canonical quantization of $H = 0$, where H is the Hamiltonian. Writing the metric in the form $ds^2 = dt^2 - R^2(t)d\Omega_3^2$ where the line element for the unit 3-sphere is $d\Omega_3^2 = d\chi^2 + \sin^2\chi[d\theta^2 + \sin^2\theta d\phi^2]$, the action for our mini-superspace model becomes

$$\begin{aligned} S &= -\frac{1}{16\pi G}\int d^4x\sqrt{-g}\,[\mathcal{R} + 2\Lambda] \\ &= -\frac{6\cdot 2\pi^2}{16\pi G}\int dt\left[R\dot{R}^2 - R + R^3\Lambda/3\right], \end{aligned} \tag{11.24}$$

where the angular degrees of fredom have been integrated out. We see that the Lagrangian $L \equiv \delta S/\delta t$ is equal to $-(3\pi/4G)\left[R\dot{R}^2 - R + \Lambda R^3/3\right]$. The dynamical degree of freedom is R, with conjugate momentum $\pi_R \equiv \delta L/\delta\dot{R} = -3\pi R\dot{R}/2G$. Thus the resulting Hamiltonian is

$$H = \pi_R\dot{R} - L = -\frac{G}{3\pi}\frac{1}{R}\pi_R^2 + \frac{3\pi}{4G}R\left(1 - \Lambda R^2/3\right). \tag{11.25}$$

Following the canonical quantization prescription, $\pi_R \to i\partial/\partial R$, the Wheeler-DeWitt equation, $H\Psi(R) = 0$, becomes

$$\left[\frac{\partial^2}{\partial R^2} - \frac{9\pi^2}{4G^2}\left(R^2 - \frac{\Lambda}{3}R^4\right)\right]\Psi(R) = 0. \tag{11.26}$$

By choosing the factor ordering corresponding to $a = 0$, the equation has the familiar form of a one-dimensional Schrödinger equation for a particle

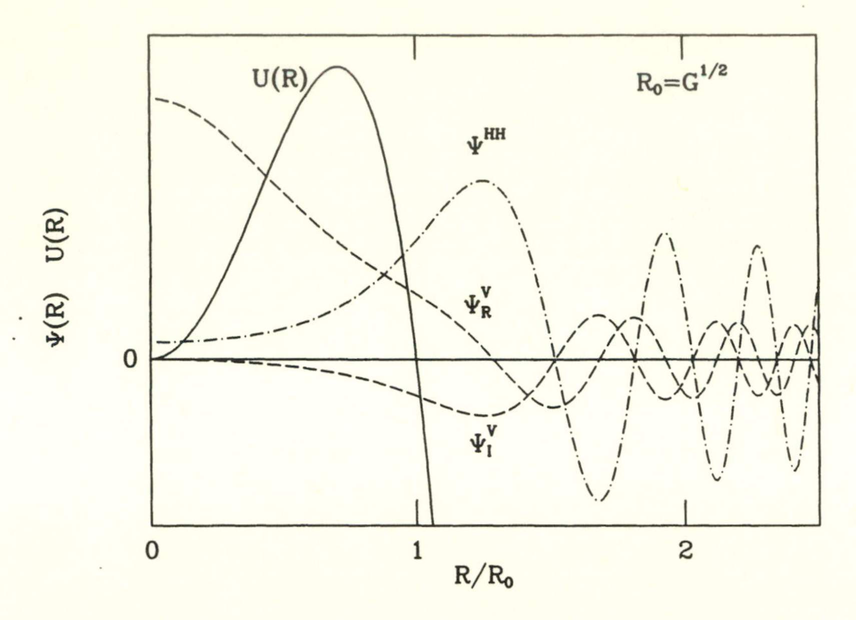

Fig. 11.2: The potential $U(R)$ (solid curve) and the mini-superspace wave functions for Hartle-Hawking (dot-dashed curve) and for Vilenkin boundary conditions (dashed curve). The real and imaginary parts of Vilenkin's wave function are so indicated. The Hartle-Hawking wave function is real.

with zero total energy moving in the potential

$$U(R) = \frac{9\pi^2 R_0^2}{4G^2}\left[\left(\frac{R}{R_0}\right)^2 - \left(\frac{R}{R_0}\right)^4\right]. \qquad (11.27)$$

The potential $U(R)$ is shown in Fig. 11.2. As expected, there is a classically forbidden region $0 < R < R_0$, a classically allowed region $R > R_0$, and of course, R_0 is the "turning point."

In the particle analogy there is a forbidden region $0 < R < R_0$; moreover, a zero-energy particle at $R = 0$ will be stuck there if one considers only its classical plight. However, there is a quantum-mechanical probability that the particle can tunnel through the barrier and emerge at the classical turning point ($R = R_0$), beyond which it can evolve classically. Recall from Chapter 7 that the WKB approximation for the tunnelling probability is proportional to $\exp(-S_E)$, where S_E is the Euclidean action. The Euclidean version of de Sitter space is a 4-sphere of radius $R_0 = (\Lambda/3)^{-1/2}$ and volume $V_4 = 8\pi^2 R_0^4/3 = 8\pi^2 3\Lambda^2$. From the Euclidean field equations it follows that $\mathcal{R} = -4\Lambda$, so $(\mathcal{R} + 2\Lambda) = -2\Lambda = \text{const}$, and the Euclidean action is simply

$$S_E = \frac{\mathcal{R} + 2\Lambda}{16\pi G} V_4 = -\frac{2\Lambda}{16\pi G} V_4 = -\frac{3\pi}{\Lambda G}. \tag{11.28}$$

Note that the Euclidean action calculated by the usual analytic continuation technique is *negative.* Hartle and Hawking [13] use (11.28) to calculate the tunnelling probability for the Universe to make the transition from $R = 0$ to $R = R_0$:

$$P_T \propto \exp(-S_E) = \exp(3\pi/G\Lambda). \tag{11.29}$$

Others [14] argue that the usual WKB result that leads to a tunneling probability proportional to $\exp(-S_E)$ does not apply since the gravitational energy is negative, and that the correct analytic continuation to imaginary time should yield a tunnelling probability proportional to $\exp(+S_E)$, or more generally $\exp(-|S_E|)$. With this prescription, the tunnelling probability is proportional to

$$P_T \propto \exp(-3\pi/G\Lambda). \tag{11.30}$$

We should emphasize that this disagreement is not a matter of someone making a mistake; rather, it is a matter of the initial state (or boundary conditions) as our discussions below will make clear. Put another way, within the context of the WKB approximation in quantum mechanics, in a classically forbidden region the wave function is proportional to $\exp[\pm \int^q (ip_{q'})dq']$, with the proper sign being determined by the boundary conditions.

The solutions of (11.26) may also be found by familiar WKB methods. With a different choice of factor ordering (which will have no effect on semi-classical calculations), the Wheeler-De Witt equation for our minisuperspace model can be cast in a form that can be solved exactly in terms of special functions. With the choice of factor ordering corresponding to $a = 1$, (11.26) becomes

$$\left[\frac{\partial^2}{\partial R^2} - \frac{1}{R}\frac{\partial}{\partial R} - \frac{9\pi^2}{4G^2}\left(R^2 - \frac{\Lambda}{3}R^4\right)\right]\Psi(R) = 0. \tag{11.31}$$

The solution to this equation is a linear combination of Airy functions $\mathrm{Ai}[z(R)]$ and $\mathrm{Bi}[z(R)]$, where $z(R) = (3\pi R_0^2/4G)^{2/3}(1 - R^2/R_0^2)$. Everyone agrees upon the solutions; the disagreement comes when one has to enforce boundary conditions to obtain the correct linear combination of the two solutions. Vilenkin [15] argues that the correct boundary condition is that of a purely expanding solution (i.e., $i\Psi^{-1}\partial\Psi/\partial R > 0$), corresponding to a purely outgoing wave. Vilenkin refers to this solution as "creation of the Universe from nothing," where "nothing" corresponds to the initial state $R = 0$. Hartle and Hawking [13] advocate the "no-boundary" boundary condition,[8] which in this context corresponds to a real wave function (equal mixtures of expanding and contracting solutions in the classically allowed region). The wave functions corresponding to the Vilenkin and Hartle-Hawking boundary conditions are

$$\begin{aligned} \Psi^V(R) &\propto \mathrm{Ai}[z(R)]\,\mathrm{Ai}[z(R_0)] + i\,\mathrm{Bi}[z(R)]\,\mathrm{Bi}[z(R_0)], \\ \Psi^{HH}(R) &\propto \mathrm{Ai}[z(R)]. \end{aligned} \tag{11.32}$$

These solutions are shown in Fig. 11.2.

Now let's extend the model by relaxing the assumption that the scalar field ϕ is frozen at a value corresponding to a vacuum energy $\rho_{\rm VAC} = \Lambda/8\pi G$. With the assumption that ϕ is spatially homogeneous, the action is

$$\begin{aligned} S &= \int d^4x\sqrt{-g}\left[-\frac{\mathcal{R}}{16\pi G} + \frac{1}{2}\partial^\mu\phi\partial_\mu\phi - V(\phi)\right] \\ &= 2\pi^2\int dt\left[-\frac{6R\dot{R}^2 - 6R}{16\pi G} + \frac{1}{2}R^3\dot{\phi}^2 - V(\phi)\right]. \end{aligned} \tag{11.33}$$

The new dynamical degree of freedom associated with the scalar field has conjugate momentum $\pi_\phi \equiv \delta L/\delta\dot{\phi} = 2\pi^2 R^3\dot{\phi}$. The Hamiltonian now becomes

$$\begin{aligned} H &= \pi_R\dot{R} + \pi_\phi\dot{\phi} - L \\ &= -\frac{G}{3\pi}\frac{1}{R}\pi_R^2 + \frac{1}{4\pi^2}\frac{1}{R^3}\pi_\phi^2 - \frac{3\pi}{4G}R + 2\pi^2R^3V(\phi). \end{aligned} \tag{11.34}$$

[8] More specifically, the Hartle-Hawking proposal for the initial state is $\Psi_0[h_{ij}] \propto \int\mathcal{D}g\exp[-S_E(g)]$, where the path integral over geometries extends over all compact manifolds (whose induced metric is h_{ij}), and $S_E(g)$ is the Euclidean action associated with the manifold. Because a compact manifold has no boundaries, this proposal is referred to as the "no-boundary" boundary condition.

After canonical quantization, $\pi_R \to i\partial/\partial R$ and $\pi_\phi \to i\partial/\partial\phi$, the Wheeler-Dewitt equation, $H\Psi(R,\phi)=0$, becomes

$$\left[R^{-a}\frac{\partial}{\partial R}R^{a}\frac{\partial}{\partial R}-\frac{1}{R^2}\frac{3}{4\pi G}\frac{\partial^2}{\partial\phi^2}-U(R,\phi)\right]\Psi(R,\phi)=0,$$

$$U(R,\phi)\equiv\frac{9\pi^2}{4G^2}\left[R^2-R^4\frac{8\pi G}{3}V(\phi)\right], \tag{11.35}$$

where here we have left the factor ordering (parameterized by a) undetermined.

If we make the loose-shoe approximation, $\phi=\phi_0=\text{const}$, the $V(\phi_0)$ term is equivalent to a cosmological constant $\Lambda=8\pi GV(\phi_0)$, and (11.26) is recovered. If the potential has several extrema,[9] say at $\phi=\phi_i$, then using the prescription $P_T\propto\exp(-|S_E|)$, cf. (11.30), quantum tunnelling will exponentially favor the maximum with the *largest* value of $V(\phi_i)$.[10] This result is welcome news for inflationary scenarios, as it is compatible with one of the most important requirements for inflation, namely that there be a scalar field that is displaced from the minimum of its potential. This raises the intriguing possibility that the necessary initial conditions for inflation may in fact be *predicted* by quantum cosmology. If instead one uses the prescription $P_T\propto\exp(-S_E)$, cf. (11.29), then quantum tunnelling to the minimum with the *smallest* value of $V(\phi_i)$ is most favored.[11]

The quantum cosmology program is an ambitious and promising one, and our brief discussion in this Chapter does not do it full justice. Several very fundamental issues need to be resolved before the program stands on firm enough ground to be tested. Among them is the question of the choice of the gravitational degrees of freedom to be quantized. The ADM split is but one of several possibilities; and one can question whether it isolates the correct degrees of freedom to be quantized. As we have emphasized earlier, just having the Schrödinger equation for the wave function of the Universe does not solve the problem. Finding the solutions is one thing—that's mathematics—selecting the appropriate solution is another thing—that's physics! Or put another way, what are the boundary conditions for the wave function of the Universe? Moreover, there is the question of interpretation—what does the wave function of the Universe mean? And

[9]The tunnelling calculation goes through even if ϕ_i is not an extremum, and only requires that $|dV/d\phi|\ll\max\{|V|/m_{Pl},\ m_{Pl}/R^{-2}\}$.

[10]It is not clear how to interpret this result if the potential is unbounded from above.

[11]The interpretation of this result is difficult for an extremum where $V(\phi_i)=0$.

of course, there is the issue of time. Does it have a fundamental role to play in quantum cosmology? Or can it simply be replaced by another variable—e.g., the energy density of some field, or the entropy?

11.4 Cosmology and Extra Dimensions

In the past few years the search for a consistent quantum theory of gravity and the quest for a unification of gravity with other forces have both led to a renewed interest in theories with extra spatial dimensions.[12] These extra spatial dimensions must be "hidden," and are assumed to be unseen because they are compact and small, presumably with typical dimensions of order the Planck length, $\mathcal{O}(10^{-33}\text{cm})$. If the "internal" dimensions are static and small compared to the large "external" dimensions, the only role they might play in the dynamics of the Universe today is in determining the structure of the physical laws. However, at the Planck time, the characteristic size of *both* internal and external dimensions are likely to have been the same, and the internal dimensions may have had a more direct role in the dynamics of the evolution of the Universe. This Section presents some speculations about the role of extra dimensions in cosmology.

Theories that have been formulated with extra dimensions include Kaluza-Klein theories [18], supergravity theories [18], and superstring theories [19]. The motivation and goals of these different approaches are quite disparate (see Section 11.1); but for their cosmological applications, they all have several common features. With this observation as a motivation, we will keep our discussion of higher-dimensional cosmology as generic as possible. Among the common features of theories with extra dimensions are:

• *There are large and small spatial dimensions.* If some of the dimensions are compact and smaller than the three large dimensions, it is possible to dimensionally reduce the theory by integrating over the extra dimensions and obtain an "effective" $3+1$-dimensional theory. Accelerator experiments have probed matter at distances as small as 10^{-16}cm without finding evidence of extra dimensions. This is not surprising, as the extra dimensions in most theories are expected to have a size characteristic of the Planck length. The large dimensions may also be compact. If so, their characteristic size is certainly comparable to, or greater than, the present

[12]Theories with extra spatial dimensions date back to the seminal work of Kaluza [16] and Klein [17]. Theories with additional time-like dimensions appear to be plagued with ghosts, see, e.g., [18].

Hubble distance, 10^{28}cm. This disparity of about 61 orders of magnitude is somewhat striking, and seems to beg for an answer to the question: "What makes the extra dimensions so small?" However, if gravity has anything to do with the size of spatial dimensions, the only reasonable size *is* the Planck length, and perhaps a more appropriate question to ask is: "What makes the observed dimensions so large?" As discussed in Chapter 8, one possible answer to the last question is inflation.

• *The effective low-energy theory depends upon the internal space.* In Kaluza-Klein theories the low-energy gauge group is determined by the continuous isometries of the internal manifold. In superstring theories, the structure of the internal space determines the number of generations of chiral fermions, whether there a is low-energy supersymmetry, etc. Any change in the character of the internal space *could* lead to drastically different low-energy physics.

• *The fundamental constants we observe are not truly fundamental.* In theories with extra dimensions the truly fundamental constants are those of the higher-dimensional theory. The constants that appear in the dimensionally reduced theory are the result of integration over the extra dimensions. If the volume of the extra dimensions should change, the value of the constants we observe in the dimensionally reduced theory would change. Exactly how "our" constants of Nature vary with the size of the compactified dimensions depends upon the theory.

In Kaluza-Klein theories, the gauge symmetries arise from continuous isometries of the internal manifold; while in superstring theories the gauge symmetries are part of the fundamental theory. In all theories, the gravitational constant is inversely proportional to the volume of the internal manifold. In the most general case, the size of the internal manifold is not specified by a single length. However, for the sake of simplicity we will assume that there is a single radius, b, that characterizes the size of the internal manifold.

• *The internal dimensions are static.* If the internal dimensions change with time, the fundamental constants also do so. Limits to the time variability of the fundamental constants can be translated to limits to the time variability of the extra dimensions.

Primordial nucleosynthesis is a sensitive probe of changes in α, because the neutron–proton mass difference $Q = m_n - m_p = 1.293$ MeV has an electromagnetic component, and the primordial yield of ^{4}He is sensitive to the value of Q. Although there is no detailed understanding of the neutron-proton mass difference, it is reasonable to assume that the electromagnetic contribution is comparable to the mass difference (but probably of the opposite sign). With this assumption $\alpha/\alpha_0 = Q/Q_0$, where subscript 0 denotes the values today.

The neutron–proton ratio at freeze out is given in terms of the freeze-out temperature T_f by $(n/p)_f \simeq \exp(-Q/T_f)$, so n/p is very sensitive to small changes in Q. Therefore, if the internal dimensions had a different size at the time of primordial nucleosynthesis, Q would be different, and the primordial mass fraction of ^{4}He produced would be different. As discussed in Chapter 4, there are limits to the primordial abundance of ^{4}He. These limits can be used to restrict the range over which fundamental constants could have varied between nucleosynthesis and now, which in turn limits the amount by which b could differ from its present value (denoted as b_0). For the superstring theory, the primordial helium abundance lies within an acceptable range only if $1.005 \geq b_N/b_0 \geq 0.995$, where b_N is the value of the internal radius at the time of primordial nucleosynthesis. For Kaluza-Klein models, the constraint is slightly less stringent: $1.01 \geq b_N/b_0 \geq 0.99$. In either case, by the time of primordial nucleosynthesis the internal dimensions were already very close to the size that they are today [20].

All theories with extra dimensions must have some mechanism to keep the internal dimensions static. In the absence of such a mechanism, the extra dimensions would either contract or expand. The origin of the vacuum stress responsible for this is unknown.

• *The ground state geometry does not have all the symmetries of the theory.* It is generally assumed that the ground state geometry of the space-time is of the form $M^4 \times B^D$, where M^4 is four-dimensional Minkowski space,[13] and B^D is some compact D-dimensional space. In general we might expect that the symmetries of the ground state are not as large as the symmetries of the full theory due to spontaneous symmetry breaking. One of the implications of SSB is the existence of a massless (at least at the classical level) Nambu-Goldstone boson, which is sometimes called the *dilaton.*

• *The particle spectrum contains an infinite number of massive states.* If the radius of the internal space is b, then b^{-1} sets the scale for excitations of the internal dimensions. In our $(3+1)$-dimensional world, these excitations behave as massive particles. The spectrum of the massive states depends upon the type of theory and the structure of the internal manifold. Since b is expected to be close to the Planck length, the massive states should have masses close to m_{Pl}.

[13] The assumption of M^4 is not quite correct in a cosmological context, and should be replaced by $R^1 \times S^3$ for the closed model, $R^1 \times H^3$ for the open model.

11.4.1 A ten-dimensional example

In theories with extra dimensions new types of interactions may arise. As a concrete example, consider the Chapline-Manton action [8, 21], which is an $N = 1$ supergravity and an $N = 1$ super-Yang-Mills theory in ten space-time dimensions. This theory is thought to be the point-like field-theory limit of a ten-dimensional superstring theory. Whether the ten-dimensional field theory limit of the superstring ever makes sense is an open question. This is because the ten-dimensional field theory description makes sense only in a realm between two similar length scales. The lower limit to the range of validity is determined by the string tension. For distances smaller than this scale, it is necessary to include the massive excitations of the string. Fundamental physics in this regime is "stringy" and any point-like field theory description is inadequate. The largest distance for which the ten-dimensional field theory is valid is set by the compactification scale, which is determined by the radius of the internal space. At distances much larger than the compactification scale—energies much smaller than the inverse of the radius of the internal space—the theory should be well described by a $(3+1)$-dimensional field theory. Thus, the ten-dimensional field theory should be valid for distances larger than that set by the string tension but smaller than that set by the compactification scale. Since these two scales are expected to be of the same order of magnitude, it is not clear if the ten-dimensional field theory description ever applies. Nevertheless, it offers a convenient and interesting starting point for an exploration of cosmology in extra dimensions.

The Chapline-Manton Lagrangian contains an $N = 1$ supergravity multiplet $\{e_M^A;\ \psi_M;\ B_{MN};\ \lambda;\ \phi\}$, where e_M^A is the vielbein, ψ_M is the Rarita-Schwinger field, B_{MN} is the Kalb-Ramond field, λ is known as the sub-gravitino, and ϕ is the dilaton; and a super Yang-Mills multiplet $\{G_{MN};\ \chi\}$, where G_{MN} is the Yang-Mills field strength tensor and χ is the gaugino field. With the definition $\sigma = (3/4)\ln(8\pi\bar{G}\phi^2)$, and for the moment setting $8\pi\bar{G} = 1$ where $\bar{G}$ is the ten-dimensional gravitational constant, the Lagrangian is[14]

$$e^{-1}\mathcal{L} = -\frac{1}{2}\mathcal{R} - \frac{1}{2}\bar{\psi}_M\Gamma^{MPS}D_P\psi_S - \frac{3}{4}\exp(-\sigma)H_{MNP}H^{MNP}$$

[14]The following notation will be used: D is the number of extra dimensions; $M,\ N,\ P,\ Q,\ \cdots$ run from 0 to $D+3$; $\mu,\ \nu,\ \rho,\ \cdots$ are indices in the extra dimensions; and $m,\ n,\ p,\ q,\ \cdots$ are indices in the ordinary three spatial dimensions.

$$-\frac{1}{4}\partial_M\sigma\partial^M\sigma - \frac{1}{\sqrt{2}}\bar{\psi}_M \not{\partial}\sigma\Gamma^M\lambda - \frac{1}{2}\bar{\lambda}\not{D}\lambda$$

$$+\frac{\sqrt{2}}{16}\exp(-\sigma/2)H_{MNP}\left[\bar{\psi}_Q\Gamma^{QMNPR}\psi_R + 6\bar{\psi}^M\Gamma^N\psi^P\right.$$

$$\left.-\sqrt{2}\bar{\psi}_R\Gamma^{MNP}\Gamma^R\lambda + \mathrm{Tr}\,\bar{\chi}\Gamma^{MNP}\chi\right] - \frac{1}{2}\mathrm{Tr}\,\bar{\chi}\not{D}\chi$$

$$-\frac{1}{4}\exp(-\sigma/4)\mathrm{Tr}\left[\bar{\chi}\Gamma^M\Gamma^{NP}\left(\psi_M + \frac{2}{\sqrt{12}}\Gamma_M\lambda\right)G_{NP}\right]$$

$$-\frac{1}{4}\exp(-\sigma/2)\mathrm{Tr}\,G_{MN}G^{MN} + \cdots, \tag{11.36}$$

where $\Gamma^{MNP} = \Gamma^{[M}\Gamma^N\Gamma^{P]}$, $\not{D} = \gamma_P D^P$, $\not{\partial} = \gamma_P\partial^P$, and $H_{MNP} = \partial_{[M}B_{NP]}$ (square brackets indicate that the indicies are to be antisymmetrized). Four-fermion terms have been omitted for brevity.

The generalized Einstein equations follow from $R_{MN} = 8\pi\bar{G}(T_{MN} - g_{MN}\mathcal{T}/8)$ and are straightforward to obtain:

$$\begin{aligned} R_{MN} \;=\; & \frac{9}{2}\exp(-\sigma)\left(H_{MPQ}H_N^{\;PQ} - \frac{1}{12}g_{MN}H_{PQR}H^{PQR}\right) \\ & -\exp(-\sigma/2)\left(\mathrm{Tr}G_{MP}G_N^{\;P} - \frac{1}{16}g_{MN}\mathrm{Tr}G_{PQ}G^{PQ}\right) \\ & -\frac{1}{2}\partial_M\sigma\partial^M\sigma - \frac{1}{8}\left(\mathrm{Tr}\bar{\chi}\Gamma_{PQR}\chi\right)\left(\bar{\lambda}\Gamma^{PQR}\lambda\right)g_{MN} \\ & -\frac{3}{16}\left(\mathrm{Tr}\bar{\chi}\Gamma_{PQR}\chi\right)^2 g_{MN} + \frac{9}{2}\exp(-\sigma/2)H_M^{\;PQ}\mathrm{Tr}\bar{\chi}\Gamma_{NPQ}\chi \\ & -\frac{3}{16}\exp(-\sigma/2)g_{MN}H_{PQR}\mathrm{Tr}\bar{\chi}\Gamma^{PQR}\chi + \cdots. \end{aligned} \tag{11.37}$$

The task at hand is to solve (11.37) to find the evolution of the scale factor(s) of the Universe as they evolve toward the expected ground state: D static dimensions and three dimensions expanding as in the standard FRW cosmology. This is a tall order—even for the (3+1)-dimensional Einstein theory we are far from knowing the most general cosmological solution.

Before we can proceed to find solutions, it is necessary to choose background field configurations for the gauge and matter fields. But, what are the symmetries of the metric? What are the vacuum (background) values of H_{MNP}? of G_{MN}? of $\bar{\chi}\Gamma\chi$? of $\bar{\lambda}\Gamma\lambda$? of σ? In general, many (possibly infinitely many) solutions of the field equations are expected, even if there is but one ground state that describes the microphysics of our Universe. How is the ground state to be found? Is it unique? How does the Universe evolve to the ground state? Perhaps when the true string nature of the equations is taken into account there will be but one solution to the string equations even if there are many solutions to the field theory. Perhaps something in the evolution of the Universe prefers a unique or small number of possibilities.

Such questions are reminiscent of the questions we considered when discussing inflation. If the conditions in some region of the Universe are favorable for inflation, that region of the Universe will grow (in physical size) relative to a region where the conditions do not favor inflation. One might imagine that the Universe starts in a state with no particular background field configuration, but in a quantum state described by a wave function Ψ that describes the probability of a given configuration, Ψ(field configurations). If in some region of the Universe the wave function is peaked about a particular configuration that leads to inflation of some of the spatial dimensions, that region will grow relative to the non-inflating regions. All that is required to produce the Universe we observe is some region where the initial conditions are such that three spatial dimensions inflate to a very large size and the other D spatial dimensions remain static.

It could very well be then that the theory (Chapline-Manton or otherwise) does not have a unique ground state, and that on scales much larger that the present horizon, different regions of the Universe are today very different—different numbers of "large" and "small" dimensions, and different topologies for the internal space, thereby leading to very different microphysics in different parts of the Universe. All of this is a result of the fact that initial conditions may not have been uniform throughout the Universe and that some mechanism, like inflation, allowed some regions to grow exponentially.

Before we become completely unglued and conclude that physics and cosmology are environmental sciences, or even worse, become anthropocentric, we should turn back to physics and discuss some possible mechanisms for the stabilization of the internal space. To wit, we will consider several toy models where the rhs of (11.37) is simplified by considering only one or two contributions out of the many possible.

To conclude, the purpose of this Subsection was to illustrate a possible, well motivated starting point and to convince the reader (by brute force!)

of the utility of considering simple toy models. The toy models discussed in the next Subsection will allow us to isolate the effects of individual contributions of the terms on the rhs of (11.37).

11.4.2 Toy models with extra dimensions

For simplicity, the metric will be taken to have the symmetry $R^1 \times S^3 \times S^D$,

$$g_{MN} = \begin{pmatrix} 1 & & \\ & -a^2(t)\tilde{g}_{mn} & \\ & & -b^2(t)\tilde{g}_{\mu\nu} \end{pmatrix}, \tag{11.38}$$

where $\tilde{g}_{mn}$ is the metric for an S^3 of unit radius so that the scale factor $a(t)$ is the actual radius of our ordinary three space, and $\tilde{g}_{\mu\nu}$ is the metric for an S^D of unit radius so that the scale factor $b(t)$ is the actual radius of the internal space. The components of the $(D+4)$-dimensional Ricci tensor are

$$\begin{aligned} R_{00} &= -3\frac{\ddot{a}}{a} - D\frac{\ddot{b}}{b}, \\ R_{mn} &= -\left[\frac{\ddot{a}}{a} + 2\frac{\dot{a}^2}{a^2} + D\frac{\dot{a}}{a}\frac{\dot{b}}{b} + \frac{2}{a^2}\right] g_{mn}, \\ R_{\mu\nu} &= -\left[\frac{\ddot{b}}{b} + (D-1)\frac{\dot{b}^2}{b^2} + 3\frac{\dot{a}}{a}\frac{\dot{b}}{b} + \frac{D-1}{b^2}\right] g_{\mu\nu}. \end{aligned} \tag{11.39}$$

Writing the generalized Einstein equations, we have

$$R_{MN} = 8\pi\bar{G}\left[T_{MN} - \frac{1}{D+2}g_{MN}T^P{}_P - \frac{1}{D+2}\frac{\Lambda}{8\pi\bar{G}}g_{MN}\right], \tag{11.40}$$

where $\bar{G}$ is the gravitational constant in $D+4$ dimensions,[15] and Λ is a cosmological constant in $D+4$ dimensions. The form of the rhs of (11.40) allows us to consider all of the different contributions to the rhs of (11.37).

As with the usual RW metric, the symmetries of the metric dictate the form of the stress-energy tensor:

$$T^0{}_0 \equiv \rho,$$

[15] $\bar{G}$ is related to Newton's constant G by $\bar{G} = GV_D^0$, where V_D^0 is the volume of the internal space today. The dimensions of $\bar{G}$ are $(\text{length})^{D+2}$.

$$\begin{aligned} T^m{}_n &\equiv -p_3\tilde{g}^m{}_n, \\ T^\mu{}_\nu &\equiv -p_D\tilde{g}^\mu{}_\nu, \end{aligned} \tag{11.41}$$

and it follows that $T^M_M = \rho - 3p_3 - Dp_D$. In terms of ρ, p_3, p_D, and $\rho_\Lambda = \Lambda/8\pi\bar{G}$, the Einstein equations become

$$\begin{aligned} 3\frac{\ddot{a}}{a} + D\frac{\ddot{b}}{b} &= -\frac{8\pi\bar{G}}{D+2}\left[(D+1)\rho + 3p_3 + Dp_D - \rho_\Lambda\right], \\ \frac{\ddot{a}}{a} + 2\frac{\dot{a}^2}{a^2} + D\frac{\dot{a}}{a}\frac{\dot{b}}{b} + \frac{2}{a^2} &= \frac{8\pi\bar{G}}{D+2}\left[\rho + (D-1)p_3 - Dp_D + \rho_\Lambda\right], \\ \frac{\ddot{b}}{b} + (D-1)\frac{\dot{b}^2}{b^2} + 3\frac{\dot{a}}{a}\frac{\dot{b}}{b} + \frac{D-1}{b^2} &= \frac{8\pi\bar{G}}{D+2}\left[\rho - 3p_3 + 2p_D + \rho_\Lambda\right]. \end{aligned} \tag{11.42}$$

Let us now consider different possibilities for the rhs of (11.40).

•R_{MN} = *Nothing.* The simplest possible form for the right hand side is zero. Provided that the curvature terms, i.e., the lhs of (11.37), are more singular than the matter terms, i.e., the rhs of (11.37), the vacuum solution may be a good approximation at very early times. To explore the consequences of this possibility, let us abandon the choice of $R^1 \times S^3 \times S^D$, and consider a $D+3$ torus for the ground-state geometry. The $D+3$ spatial coordinates can be chosen to take the values $0 \leq x^i \leq L$, where L is a parameter with dimensions of length. The general cosmological solutions of the vacuum Einstein equations are the Kasner solutions, with metric

$$ds^2 = dt^2 - \sum_{i=1}^{D+3}\left(\frac{t}{t_0}\right)^{2r_i}(dx^i)^2. \tag{11.43}$$

The Kasner metric is a solution to the vacuum Einstein equations provided the Kasner conditions are satisfied:

$$\sum_{i=1}^{D+3} r_i = \sum_{i=1}^{D+3} r_i^2 = 1. \tag{11.44}$$

The physical volume of a comoving volume element varies as $(t/t_0)^n$, where $n = \sum_i r_i = 1$. Thus the mean scale factor, which is proportional to the $(D+3)$-root of the physical volume, grows as $t^{1/(D+3)}$.

In order to satisfy the Kasner conditions at least one of the r_i must be negative. It is possible to have three spatial dimensions expanding isotropically and D dimensions contracting isotropically, with the choice [22]

$$r_1 = r_2 = r_3 \equiv r = \frac{3 + (3D^2 + 6D)^{1/2}}{3(D+3)},$$

$$r_4 = \cdots = r_{3+D} \equiv q = \frac{D - (3D^2 + 6D)^{1/2}}{D(D+3)}. \qquad (11.45)$$

Provided that $D > 0$, $r > 0$, and $q < 0$ as desired. With this choice the metric may be written as in (11.38),

$$ds^2 = dt^2 - a^2(t)d\vec{x}^2 - b^2(t)d\vec{y}^2, \qquad (11.46)$$

where x^i are the spatial coordinates of the three expanding dimensions, and y^i are the spatial coordinates of the D contracting dimensions. The two scale factors evolve as $a(t) = (t/t_0)^r$, $b(t) = (t/t_0)^q$.

In three dimensions the Kasner models are well known and well studied. Both the Kasner model and the flat FRW model are specific examples of Bianchi I cosmologies. The closed FRW model is a special case of the Bianchi IX solutions. In general, the Bianchi IX vacuum solutions have the interesting feature that the approach to the singularity[16] is "chaotic" [23]. As the initial singularity is approached, the scale factors in different spatial directions undergo a series of oscillations, i.e., alternating phases of contraction and expansion. The oscillation of the scale factors is well described by a sequence of Kasner models in which expanding and contracting dimensions are interchanged during "bounces." The question of whether such a chaotic approach to the initial singularity is present in more than three spatial dimensions has been considered. It has been shown that chaotic behavior occurs only for models with between three and nine spatial dimensions [24]. If the vacuum Einstein equations provide a good description of the dynamics at very early times, then this fact might be of some importance.

The Kasner solutions do not have a static internal space; and if they are ever relevant, it must be at early times—recall nucleosynthesis provides a stringent limit to the change in the size of the internal space since that epoch. Thus at the very least, at late times the right-hand side must be more complicated than nothing. Next to nothing, the simplest thing to

[16]When we refer to the "approach to the singularity" we mean the time reversal of the usual expansion from the singularity.

consider on the right-hand side is a free scalar field.[17] Before discussing the effect of a free scalar field on the evolution of the Universe it is necessary to discuss the regularization of a scalar field in a non-trivial background geometry.

The free energy of a non-interacting, spinless boson of mass μ is given by [25]

$$F = T\frac{1}{2}\ln \mathrm{Det}\left(-\Box_{4+D} + \mu^2\right). \tag{11.47}$$

Since we are interested in finite-temperature effects, we take the time coordinate to be periodic in imaginary time, with period of $1/T$. The relevant geometry is $S^1 \times S^3 \times S^D$, where the radii of the spheres are $1/2\pi T$, a, and b. The eigenvalues of "$\Box$" on a compact space are discrete and are given by the triple sum (hereafter μ will be set to zero)

$$\begin{aligned} 2F &= T\sum_{r=-\infty}^{\infty}\sum_{m,n=0}^{\infty} D_{mn}\ln\Big[r^2(2\pi T)^2 + m(m+2)a^{-2} \\ &\quad + n(n+D-1)b^{-2}\Big], \end{aligned} \tag{11.48}$$

where D_{mn} is the factor that counts the degeneracy of states: $D_{mn} = (m+1)^2(2n+D-1)(n+D-2)!/(D-1)!n!$.

The free energy given by (11.48) is, of course, infinite due to the usual infinite zero-point energy. To deal with the infinities and extract the finite physical energy density, a regularization scheme must be implemented. For the purpose of regularization, each term in the sum can be expressed as an integral using the formula [25][18]

$$\ln X = \left(\frac{d}{ds}X^s\right)_{s\to 0} = \frac{d}{ds}\left(\frac{1}{\Gamma(-s)}\int_0^\infty dt\, t^{s-1}\exp(-tX)\right)_{s\to 0}. \tag{11.49}$$

Using this prescription, we find that the finite part of the free energy is

[17] This does not imply that a free scalar field is next to nothing.

[18] The basic idea of regularization is to extract the finite piece by taking the difference of the result for a compact space at finite temperature and that for flat space at zero temperature. The regularization used here is valid only for D odd. The D even case may have an additional logarithmic correction.

given by

$$2T^{-1}F = \frac{d}{ds}\left[\frac{1}{\Gamma(-s)}\int_0^\infty dt\, t^{-s-1}\sigma_1(4\pi^2T^2t)\sigma_3(a^{-2}t)\sigma_D(b^{-2}t)\right]_{s\to 0}, \tag{11.50}$$

where the functions σ_i are given by

$$\sigma_i(x) = \sum_{n=0}^{\infty}\frac{(2n+i-1)(n+i-1)!}{(i-1)!n!}\exp[-n(n+i-1)x]. \tag{11.51}$$

The full expression for the free energy is quite difficult to evaluate, but it has several interesting limits. In the "flat-space" limit, that is, where the radius of S^3 is much larger than the radius of S^D ($a \gg b$), $\sigma_3 \to (\sqrt{\pi}/4)a^3t^{-3/2}$. In this limit the free energy can be approximated by

$$F = \frac{\Omega_3 a^3}{b^4}\left[c_1 - c_2(bT)^4 - c_3(bT)^{D+4}\right], \tag{11.52}$$

where $\Omega_i = 2(\pi)^{(i+1)/2}/\Gamma[(i+1)/2]$ is the volume of the unit i-sphere. (As discussed in Chapter 2, $\Omega_3 = 2\pi^2$.) The temperature-independent term proportional to c_1 is the Casimir term.[19]

For temperatures much less than the energy scale set by the radius of the internal space, $T \ll b^{-1}$, the leading-order, temperature-dependent term is that proportional to c_2. Further, $c_2 = \pi^2/90$ and $F/\Omega_3 a^3 = -\pi^2T^4/90$—just the free energy density of a massless, spin-zero boson in $3+1$ dimensions. In this regime, the excitations of the scalar field in the internal space are "frozen out."[20] For temperatures much larger than the energy scale set by the radius of the internal space, $T \gg b^{-1}$, the leading-order, temperature-dependent term is that proportional to c_3. In this case $c_3 = (2\zeta(D+4)/\pi^{3/2})\Gamma[(D+4)/2]/\Gamma[(D+1)/2]$, and $F/\Omega_3 a^3\Omega_D b^D = -c_3T^{D+4}/\Omega_D$—just the free energy density of a massless, spin-zero boson in $D+4$ dimensions. In this regime the excitations of the scalar field in the internal space are relevant.

[19] c_1 is c_N of Candelas and Weinberg [26].

[20] The components of the momentum in the compact dimensions are quantized in units of b^{-1}. For temperatures $T \ll b^{-1}$, these components must be in the lowest quantum state ($n = 0$), and so the degrees of freedom in the internal space are said to be frozen out.

	Casimir $T=0$	Low Temperature $0 \ll T \ll b^{-1}$	High Temperature $T \gg b^{-1}$
ρ	$c_1/\Omega_D b^{4+D}$	$(\pi^2/30)T^4/\Omega_D b^{4+D}$	$(D+3)c_3 T^{D+4}/\Omega_D$
p_3	$-c_1/\Omega_D b^{4+D}$	$(\pi^2/90)T^4/\Omega_D b^{4+D}$	$c_3 T^{D+4}/\Omega_D$
p_D	$4c_1/D\Omega_D b^{4+D}$	0	$c_3 T^{D+4}/\Omega_D$
T^M_M	0	0	0

Table 11.1: Contributions to thermodynamic quantities.

The internal energy is given in terms of the free energy, the temperature, and the entropy,

$$S = -\left[\frac{\partial F}{\partial T}\right]_{a,b}, \tag{11.53}$$

by $U = F + TS$. The thermodynamic quantities ρ, p_3, and p_D are defined in terms of the internal energy:

$$\begin{aligned}
\rho &= \frac{U}{\Omega_3 \Omega_D a^3 b^D}, \\
p_3 &= -\frac{a}{3\Omega_3\Omega_D a^3 b^D}\left[\frac{\partial U}{\partial a}\right]_{b,S}, \\
p_D &= -\frac{b}{D\Omega_3\Omega_D a^3 b^D}\left[\frac{\partial U}{\partial b}\right]_{a,S}.
\end{aligned} \tag{11.54}$$

The important thermodynamic quantities are given in Table 11.1 in the zero-temperature, low-temperature, and high-temperature limits. Much can be learned from Table 11.1. In the zero-temperature and in the low-temperature limits, dimensional reduction is possible. Upon integration over the internal dimensions the effective energy density and pressure in (3+1)-dimensions is obtained by multiplying by $V_D = \Omega_D b^D$. After dimensional reduction the Casimir terms are proportional to $c_1 b^{-4}$ (temperature independent). Thus, at low temperatures, $0 \ll T \ll b^{-1}$, after dimensional reduction, $\rho = 3p_3 \rightarrow (\pi^2/30)T^4$ and $p_D = 0$, which is precisely what one expects for a spinless boson in $3+1$ dimensions. In the high-temperature limit dimensional reduction does not make sense.

It is straightforward to perform a similar analysis for particles of higher spin. The technical details are more difficult, but the physics is quite similar.

•$\hat{R}_{MN}$ = *Radiation* [27]. This case corresponds to the "flat-space" ($a \gg b$), "high-temperature" ($T \gg b^{-1}$) limit with vanishing cosmological constant. In this limit T_{MN} is isotropic, with $p_3 = p_D \equiv p$ (see Table 11.1). The Einstein equations are

$$\begin{aligned} 3\frac{\ddot{a}}{a} + D\frac{\ddot{b}}{b} &= -8\pi\bar{G}\rho, \\ \frac{\ddot{a}}{a} + 2\frac{\dot{a}^2}{a^2} + D\frac{\dot{a}}{a}\frac{\dot{b}}{b} &= 8\pi\bar{G}p, \\ \frac{\ddot{b}}{b} + (D-1)\frac{\dot{b}^2}{b^2} + 3\frac{\dot{a}}{a}\frac{\dot{b}}{b} + \frac{D-1}{b^2} &= 8\pi\bar{G}p. \end{aligned} \qquad (11.55)$$

In keeping with the flat-space assumption we have neglected the curvature term $(2/a^2)$ in (11.42). The equation of state is that for a generalized relativistic fluid: $p = \rho/N$, where $N = D + 3$. Conservation of stress energy, $T^{MP}{}_{;P} = 0$, implies that $\rho\bar{\sigma}^{N+1} = \text{const}$, where $\bar{\sigma} \propto (a^3 b^D)^{1/N}$ is the mean scale factor. Moreover, since $\rho \propto T^{N+1}$, there is a conserved quantity $S_N = (\bar{\sigma}T)^N = a^3 b^D T^N$. We recognize S_N as the entropy per comoving volume in $N = D + 3$ spatial dimensions.

Equations (11.55) supplemented by the evolution of ρ and p can be integrated to give $a(t)$ and $b(t)$. A typical solution is shown in Fig. 11.3. Both scale factors emerge from an initial singularity. The scale factor for the internal space reaches a maximum value and recollapses to a second singularity. As b approaches the second singularity, a is driven to infinity. The evolution of the temperature is also shown in Fig. 11.3. There is a rather striking feature in Fig. 11.3: As the second singularity for the internal space is approached, *both* $a(t)$ and $T(t)$ increase. Such an occurence seems paradoxical; recall that in the standard cosmology $T \propto R(t)^{-1}$. The explanation is simple: T must increase so that $S_N = \sigma^N T^N$ remains constant. In the time interval where both $a(t)$ and $T(t)$ increase, the mean scale factor is actually decreasing—b^D decreases more rapidly than a^3 increases, and so T must increase to keep S_N constant.

During the period of increasing scale factor a and temperature T, the entropy per comoving volume in the three expanding dimensions, $S_3 \equiv a^3T^3$, increases. Of course the *total* entropy in a comoving volume is conserved, but in the approach to the second singularity entropy is "squeezed out" of the contracting dimensions and into the expanding dimensions.

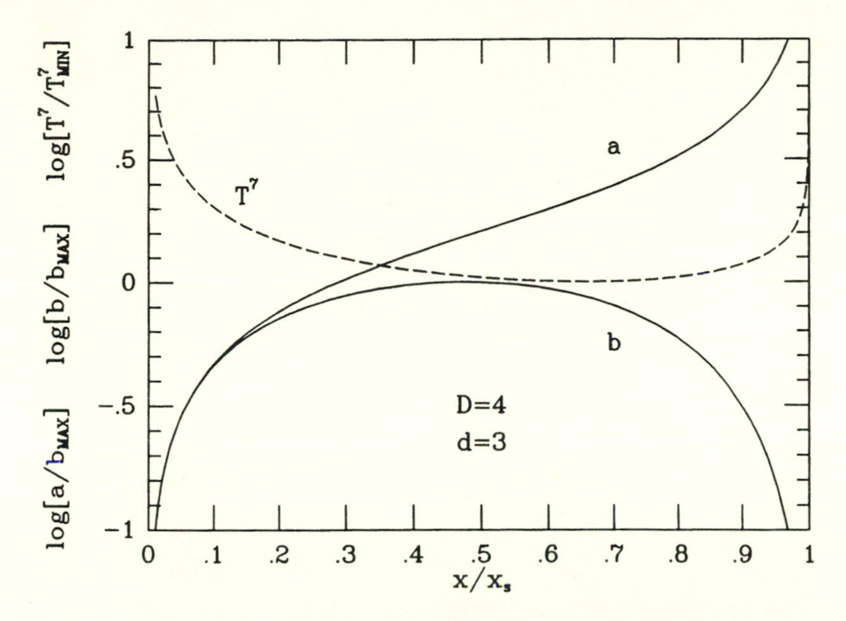

Fig. 11.3: Evolution of the scale factors a and b and the temperature for the $R_{MN} =$ *radiation* extra-dimensional cosmological solution. The parameter x/x_s is a measure of the time in units of the time necessary to reach the second singularity.

Increasing entropy per comoving volume in the three expanding dimensions sounds interesting—recall, entropy production is the mechanism by which inflation "solves" the shortcomings of the standard cosmology. Consider the entropy contained within the horizon in the three expanding dimensions, $S_{\rm HOR} \equiv (4\pi/3)d_{H3}^3T^3$, where d_{H3} is the physical distance to the horizon in the expanding 3-space:

$$d_{H3} = a(t)\int_0^t dt'/a(t'). \tag{11.56}$$

In the approach to the second singularity, d_{H3}, T, and $S_{\rm HOR}$ all approach infinity.[21] That is, as $b(t) \to 0$, the entropy within a horizon volume diverges. This is certainly a good start toward resolving the horizon problem.

To solve the horizon problem, the entropy within the horizon must be greater than 10^{88}; i.e., $S_{\rm HOR}$ must exceed 10^{88}. However $S_{\rm HOR}$ only diverges as $b(t) \to 0$. Before the second singularity is reached, two unfortunate things necessarily happen. First, there must be a mechanism

[21]Recall in the standard cosmology, at early times $S_{\rm HOR} \sim (m_{Pl}/T)^3 \to 0$ as $T \to \infty$.

that stabilizes the internal dimensions preventing b from becoming arbitrarily small. Second, the high-temperature assumption, $T \gg b^{-1}$, breaks down. This is because the decrease of b outpaces the increase in T, and so eventually T falls below b^{-1}. When this occurs, it is necessary to use the "low-temperature" limit of the free energy where the only dynamical effect of the extra dimensions is to control the values of fundamental constants. Both effects—the ultimate stability of $b(t)$ and the temperature falling below b^{-1}—serve to quench the increase in $S_{\rm HOR}$. The conditions required to allow $S_{\rm HOR}$ to grow to a value in excess of 10^{88} have been carefully studied [27]: Without resort to very special initial conditions, or extrapolation beyond the point where the high-temperature assumption breaks down, it is not possible to squeeze sufficient entropy into our three spatial dimensions to solve the horizon problem.

The case of radiation domination had three features in common with the vacuum Kasner solutions: (i) some expanding and some contracting dimensions; (ii) no mechanism to stabilize the internal dimensions; and (iii) a limited realm of applicability. The case of radiation domination also had a new, very interesting feature—the possibility of squeezing entropy out of the internal dimensions and into our three spatial dimensions.

•R_{MN} = *Casimir* + Λ [26]. The combination of Casimir forces plus a cosmological constant has one important new feature, the possibility of a classically stable ground state. With ρ, p_3, and p_D from column 1 of Table 11.1, the Einstein equations, cf. (11.42), become

$$\begin{aligned} 3\frac{\ddot{a}}{a} + D\frac{\ddot{b}}{b} &= -\frac{8\pi\bar{G}}{D+2}\left[\frac{(D+2)c_1}{\Omega_D}b^{-4-D} - \rho_\Lambda\right], \\ \frac{\ddot{a}}{a} + 2\frac{\dot{a}^2}{a^2} + D\frac{\dot{a}}{a}\frac{\dot{b}}{b} &= -\frac{8\pi\bar{G}}{D+2}\left[\frac{(D+2)c_1}{\Omega_D}b^{-4-D} - \rho_\Lambda\right], \\ \frac{\ddot{b}}{b} + (D-1)\frac{\dot{b}^2}{b^2} + 3\frac{\dot{a}}{a}\frac{\dot{b}}{b} &= \frac{8\pi\bar{G}}{D+2}\left[\frac{4(D+2)c_1}{D\Omega_D}b^{-4-D} + \rho_\Lambda\right] \\ &\quad -\frac{D-1}{b^2}. \end{aligned} \tag{11.57}$$

Again we have neglected the curvature of S^3 ($1/a^2 \to 0$), and the curvature term for S^D, $(D-1)/b^2$, has been moved to the right hand side of the $\mu\nu$ equation.

To find a static solution we set $b = b_0 =$ const (i.e., $\dot{b} = \ddot{b} = 0$) and

$\dot{a} = \ddot{a} = 0$.[22] The first two equations determine b_0 in terms of ρ_Λ:

$$b_0^{-4-D} = \frac{\Omega_D}{(D+2)c_1}\rho_\Lambda. \qquad (11.58)$$

Remembering that $\bar{G} = GV_D$, we can then use the third equation to determine b_0 in terms of the Planck length $l_{Pl} = m_{Pl}^{-1}$:

$$b_0^2 = \frac{8\pi c_1(4+D)}{D(D-1)} l_{Pl}^2. \qquad (11.59)$$

In general, there may be other interesting solutions to this system of equations. For instance in the limit where *both* a and b go to infinity, the right-hand sides of all three equations approach a constant, given by

$$H^2 = \frac{D(D-1)}{4+D} b_0^{-2}. \qquad (11.60)$$

In this limit, the cosmological solution is $a(t) = b(t) = \exp(\pm Ht/\sqrt{3})$. This solution describes exponentially growing scale factors for both S^3 and S^D—driven of course by the cosmological constant.

The static minimum $b = b_0$ is stable against small perturbations, since a linear perturbation analysis reveals that $\delta b(t) = b(t) - b_0$ has no exponentially growing modes. However, the existence of the exponentially growing solutions for a and b suggests that the static minimum is not stable against arbitrarily large excursions from $b = b_0$.

•R_{MN} = *Monopole* $+\Lambda$ [28]. In the previous model *quantum* Casimir effects "working" against the cosmological constant stabilized the extra dimensions. It is also possible to balance the cosmological constant by the effects of a *classical* field. To illustrate this, consider the Einstein-Maxwell theory in six space-time dimensions. The action for the model is given by

$$S = -\frac{1}{16\pi\bar{G}} \int d^6x \sqrt{-g_6} \left[\mathcal{R} + \frac{1}{4} F_{MN} F^{MN} + 2\Lambda \right], \qquad (11.61)$$

where F_{MN} is the field strength tensor in six dimensions. The effect of the Maxwell field in the Einstein equations is through its contribution to the

[22]The "large dimensions" are taken to be static also, because in the absence of any matter or radiation, or a curvature term for the large dimensions, one expects $\dot{a} = \ddot{a} = 0$. A realistic model would also include ordinary matter and radiation, which would drive the expansion of the ordinary three spatial dimensions.

stress tensor:

$$T_{MN} = F_{MQ}F_N^{\ Q} - \frac{1}{4}g_{MN}F_{PQ}F^{PQ}. \qquad (11.62)$$

We will take the ground-state geometry to be $R^1 \times S^3 \times S^2$, where as usual, we will assume that $a \gg b$. Further, we will take a magnetic monopole form for F^{MN}. The magnetic monopole ansatz has vanishing components of F_{MN} except for indices in the internal space, $F_{\mu\nu} = \sqrt{-g_2}\,\varepsilon_{\mu\nu} f(t)$, where $f(t)$ is a function of time and g_2 is the determinant of the S^2 metric. This ansatz, of course, satisfies the field equations for F_{MN}. The Bianchi identities can be used to express $f(t)$ in terms of the radius of the S^2, $f(t) = f_0/b(t)$, where f_0 is a numerical constant.

With the monopole ansatz for F_{MN} the non-vanishing components of the electromagnetic stress tensor are

$$T^0{}_0 = \frac{1}{2}\frac{f_0^2}{b^4}; \qquad T^m{}_n = -\frac{1}{2}\frac{f_0^2}{b^4}\tilde{g}^m{}_n; \qquad T^\mu{}_\nu = \frac{1}{2}\frac{f_0^2}{b^4}\tilde{g}^\mu{}_\nu, \qquad (11.63)$$

which can be expressed in terms of ρ, p_3, and p_2, where $\rho = p_2 = -p_3 = f_0^2/2b^4$. The Einstein equations for cosmological constant plus monopole are

$$\begin{aligned}
3\frac{\ddot{a}}{a} + 2\frac{\ddot{b}}{b} &= -2\pi\bar{G}\left[\frac{f_0^2}{b^4} - \rho_\Lambda\right], \\
\frac{\ddot{a}}{a} + 2\frac{\dot{a}^2}{a^2} + 2\frac{\dot{a}}{a}\frac{\dot{b}}{b} &= -2\pi\bar{G}\left[\frac{f_0^2}{b^4} - \rho_\Lambda\right], \\
\frac{\ddot{b}}{b} + \frac{\dot{b}^2}{b^2} + 3\frac{\dot{a}}{a}\frac{\dot{b}}{b} &= 2\pi\bar{G}\left[3\frac{f_0^2}{b^4} + \rho_\Lambda\right] - \frac{1}{b^2}. \qquad (11.64)
\end{aligned}$$

As in the previous example, by setting $\dot{b} = \ddot{b} = 0$ and $\dot{a} = \ddot{a} = 0$ we can solve for the radius b_0 of the static solution,

$$b_0 = \frac{f_0^{1/2}}{\rho_\Lambda^{1/4}}; \qquad b_0 = \sqrt{8\pi}\, f_0 l_{Pl}. \qquad (11.65)$$

In addition to the static solution with $b = b_0$, there is a quasi-static solution with $b = \sqrt{3}\, b_0$ and a increasing exponentially: $a = a_0 \exp(Ht)$, where $H = \sqrt{2}/3b_0$. Finally, there is the solution where *both* a and $b \to \infty$, with both scale factors increasing exponentially with rate $H = 1/2\sqrt{5}\, b_0$.

•$R_{MN} = \mathcal{R}^2 + \Lambda$ [29]. The Casimir, monopole, and cosmological constant terms can arise in the Chapline-Manton action. Although terms such as $\mathcal{R}^2$, $R_{MN}R^{MN}$, and $R_{MNPQ}R^{MNPQ}$ do not explicitly appear in the Chapline-Manton action, they are expected to be present in the low-energy expansion of the superstring theory,[23] and probably in all other higher-dimensional theories as well. Consider the theory whose gravitational action is given by[24]

$$\begin{aligned} S &= -\frac{1}{16\pi\bar{G}}\int d^{4+D}x\sqrt{-g_{4+D}}\left[\mathcal{R} + 2\Lambda + a_1\mathcal{R}^2 + a_2 R_{MN}R^{MN}\right. \\ &\quad \left. + a_3 R_{MNPQ}R^{MNPQ}\right]. \end{aligned} \tag{11.66}$$

There is a static $M^4 \times S^D$ solution to the generalized Einstein equations derived from this action, provided that the following conditions are met:

$$\begin{aligned} 0 &< D(D-1)a_1 + (D-1)a_2 + 2a_3, \\ 0 &< (D-1)a_2 + 2a_3, \\ 0 &< a_3, \\ \Lambda &= \frac{1}{4}\frac{D(D-1)}{D(D-1)a_1 + (D-1)a_2 + 2a_3}. \end{aligned} \tag{11.67}$$

For this solution, the value of b is given by

$$b_0^2 = 2D(D-1)a_1 + 2(D-1)a_2 + 4a_3. \tag{11.68}$$

This theory is more difficult to analyze because of the higher-derivative terms in the equations of motion. Nevertheless, it has been shown that in addition to the static solution, there is a solution with b constant and a increasing exponentially [29]. Unlike the corresponding solutions for Casimir or monopole plus cosmological term where b = constant and a grows exponentially, this solution is not stable. This suggests that the $M^4 \times S^D$ static solution is stable against large dilations of the internal space and is the true ground state.

[23]The Ricci tensor has dimensions of (energy)^2, and so "$\mathcal{R}^2$" terms have dimensions of (energy)^4. Thus the coefficient of any $\mathcal{R}^2$ term relative to the usual $\mathcal{R}$ term must have dimensions of (energy)$^{-2}$. One naturally expects this energy scale to be of order m_{Pl}, so after the Planck epoch the effect of $\mathcal{R}^2$ terms is suppressed by m_{Pl}^{-2}.

[24]For $D = 6$, the action is ghost free provided that $a_3 = -a_2/4$.

By simplifying the rhs of the Einstein equations and isolating the effects of one or two individual contributions, we have been able to explore the behavior of cosmological solutions to higher-dimensional theories. With the help of our simple toy models we have seen an interesting parade of solutions—from chaotic behavior to static solutions to de Sitter-like solutions. None of our toy models has really resolved the fundamental issues—Why do three dimensions get large while the others remain small? and What physical mechanism stabilizes the size of the internal space? These two very important questions remain unanswered in all theories with extra dimensions.

11.4.3 Remnants from extra dimensions

Another aspect of cosmology with extra dimensions is the possibility that some remnant associated with the extra dimensions, for instance a massive monopole or a stable massive particle species, survives until the present. Before discussing specific particles it is useful to review our discussion of Chapter 5 about the survival of thermal relics. Once a particle species with thermal abundance becomes non-relativistic, its abundance in a comoving volume must decrease if it is to remain in equilibrium. For a stable (or long-lived) species, only annihilations can serve to keep the species in equilibrium. However, the expansion of the Universe quenches the annihilation of a species of mass m at a temperature T_f given by $x_f \equiv m/T_f \sim \ln(m_{Pl}m\sigma_0)$, where σ_0 is related to the annihilation cross section σ_A by $\langle|v|\sigma_A\rangle = \sigma_0(m/T)^{-n}$. As the reader no doubt appreciates by now, it is useful to compare the density of any particle species under consideration to the entropy density. Provided the expansion is adiabatic, after annihilations cease, the ratio of the number density of the particle species ψ to the entropy density remains constant and is given by

$$Y_\psi \sim \frac{x_f^{n+1}}{m_{Pl}m\sigma_0}. \tag{11.69}$$

In general $\sigma_0 \propto m^{-a}$. Since the effective annihilation cross section decreases with mass, the more massive the particle is, the more likely it is to survive annihilation. For masses close to the Planck mass and $\sigma_0 \simeq m^{-2}$, annihilations are never effective and a particle survives with $Y_\psi \sim 1$, i.e., about as abundant as photons. This would be a great embarrassment, since it would result in a contribution to Ω_0 from that species of about 10^{26} or so. Creation of entropy, as in inflation, could circumvent this difficulty. If inflation occurs and the Universe is reheated to a temperature of $T_{RH} \ll m$, the relic abundance of ψ's would not be determined by freeze out, but instead by their production during and just after reheating. It is likely that this number is too small to be interesting today, but it is possible, if m is small enough, that interesting numbers of ψ's would be produced.

Here "interesting" means a value large enough to one day be detectable, but small enough not to be already ruled out. The most general limit to the abundance of massive stable particles follows from the overall mass density of the Universe. For a particle of mass m, the limit $\Omega_0 h^2 \lesssim 1$ implies $Y_\psi \leq 3 \times 10^{-28}(m_{Pl}/m)$, or $n_\psi \leq 9 \times 10^{-25}(m_{Pl}/m)$ cm^{-3}. A more useful way to express these limits is as a limit to the average particle flux: $F_\psi \lesssim 2 \times 10^{-18}(m_{Pl}/m)$ cm^{-2}sr^{-2}sec^{-1}. One would expect a massive relic to behave as cold dark matter, in which case such particles would be concentrated in the halo of our galaxy, and could have a larger local flux. Using the local halo density, about 0.5×10^{-24}g cm^{-3}, and the virial velocity, we obtain a limit to the local flux of $\psi's$: $F_\psi \lesssim 6 \times 10^{-14}(m_{Pl}/m)$cm^{-2}sr^{-1}sec^{-1}.

Now for the candidates:

- *Pyrgons* [31]. In Kaluza-Klein theories there is an infinite tower of four-dimensional particle states. These states, known as pyrgons, correspond to excitations of the particle species with non-zero momenta in the internal space.

In the five-dimensional theory the mass spectrum of pyrgons is a tower of spin-2 particles with mass $m_k = kb^{-1}$, where k is an integer and b is the radius of the internal space[25] (in the five-dimensional theory the internal space is a circle). In this theory the $k = 1$ pyrgons are stable. This is because the charge operator is proportional to the mass operator. The zero modes are neutral, and the kth mode has charge $e_k = ke_1$. The kth pyrgon can decay to k charge-1 pyrgons, but the $k = 1$ pyrgons cannot decay to zero modes because the $k = 0$ states are neutral.

In more realistic Kaluza-Klein theories the mass spectrum is more complicated, but the general features remain, namely that in four dimensions there are zero modes and massive modes with masses proportional to the inverse of the radii of the internal space. The question of stability of the

[25]This fact is easy to see. The energy of a five-dimensional massless particle is $E^2 = \vec{\mathbf{p}}^2 + p_5^2$, where $\vec{\mathbf{p}}$ is the component of the momentum in the three large dimensions and p_5 is the component of the momentum in the internal space. The momentum in the compact internal space must be quantized in units of b^{-1}: $p_5 = k/b$ (k integer), and so $E^2 = \vec{\mathbf{p}}^2 + (k/b)^2$. Viewed from four dimensions, the pyrgon behaves as a particle of mass k/b.

pyrgons is, however, more subtle. In general, there will be selection rules that prevent some of the massive modes from decaying. Such a selection rule is present in $N = 8$ supergravity models with S^7 as an internal space.[26]

The only reason for massive pyrgons to be stable is their being endowed with a quantum number that cannot be represented by combinations of zero-mode states (which here are assumed to include only the observed particles). As an example, suppose a massive pyrgon state breaks the usual $SU(3)$ relationship of electric charge and triality. This could occur if the pyrgon is color neutral with a fractional electric charge, or fractionally charged but a color singlet. In either case, it could not decay to the known particles (so long as $SU(3)$ of color is unbroken). Another possibility is that the pyrgon carries a quantum number that none of the observed particles in Nature carries.

In superstring theories, gauge symmetries are fundamental, but there still could be excitations of the extra dimensions that are stable. There could also be excited-string states that are stable. In the heterotic superstring [32] for instance, there are $8,064$ zero modes, $18,883,584$ $k = 1$ modes, $6,209,272,160$ $k = 2$ modes, $\cdots$ (remember, the increase is exponential!). Some of these massive modes might be stable. For instance, in the $SO(32)$ heterotic superstring there is a stable, massive fermion.

• *Monopoles.* Just as GUT monopoles correspond to topological defects in the orientation of the vacuum expectation value of a Higgs field, there are magnetic monopoles in Kaluza-Klein theories that correspond to topological defects in compactification [33]. The Kaluza-Klein monopoles satisfy the Dirac quantization condition $ge = 1/2$ and have masses $m_M \sim m_{Pl}/e \sim 10^{20}$GeV. The cosmological production of Kaluza-Klein monopoles is uncertain because there is direct analogue to the Kibble mechanism [34]. Because the physical mechanism for compactification is unknown, it remains to be seen whether the compactification transition is analogous to a SSB transition; that is, all spatial dimensions on equal footing above some critical temperature, and D compactified dimensions below the critical temperature.[27] Since the symmetry breaking that gives rise to the Kaluza-Klein monopoles is topological in nature, it does not seem likely that its restoration can be studied by classical methods.

[26] $N = 8$ supergravity in 3+1 dimensions is equivalent to $N = 1$ supergravity in eleven dimensions. The case of $D = 7$ is interesting as at least seven additional dimensions are required to account for an $SU(3) \otimes SU(2) \otimes U(1)$ gauge symmetry.

[27] In this case, SSB corresponds to the process of compactification; i.e., the symmetry breaking $\mathrm{Diff}^{D+4} \rightarrow \mathrm{Diff}^4 \times \mathcal{I}$ where Diff^n is the diffeomorphism group in n dimensions and $\mathcal{I}$ is the isometry group of the internal space.

In theories where a gauge symmetry is broken by a topological mechanism (rather than by the usual Higgs mechanism) [35], there are additional topologically stable excitations, including magnetic monopoles and particles with fractional electric charge [36].

Relics provide an important, if not singular, probe of the earliest history of the Universe. Some of the relics predicted by higher dimensional theories are so unusual that, if detected, they could be uniquely identified as such. This fact alone argues convincingly for searching for very massive particles in the cosmic rays.

Whether or not extra dimensions are relevant to the Planck epoch remains to be determined. We hope that at the very least we have given the reader a glimpse of the very rich cosmological consequences of theories with extra dimensions.

11.5 Limiting Temperature in Superstring Models

String theory was originally introduced as a theory of the strong interactions [37]. As a result, the thermodynamic properties of string theories have been studied for many years; see, e.g., [38]. A characteristic feature of string theories is a density of states, $\rho(m)dm$ is the number of states with mass between m and $m+dm$, that increases exponentially with mass for large mass.[28] In the large-mass limit the density of states can be written as

$$\rho(m) = Cm^{-A}\exp(Bm). \tag{11.70}$$

The constant C will not be of interest here. The constants A and B depend upon the theory. Some examples are

$$\begin{array}{lll} \text{Open String}: & A=9/2, & B=\pi\sqrt{8}(\alpha')^{1/2}; \\ \text{Closed String}: & A=10, & B=\pi\sqrt{8}(\alpha')^{1/2}; \\ \text{The Heterotic String}: & A=10, & B=\pi(2+\sqrt{2})(\alpha')^{1/2}; \end{array} \tag{11.71}$$

where α' is the "Regge slope," the inverse of the string tension. For the superstring theory α' is expected to be of order m_{Pl}^{-2}. For the original hadronic string, $\alpha' \sim 1/\text{GeV}^2$.

If one imagines that the Universe began in a "stringy" state, then the exponential spectrum of string states has profound implications, which

[28]These states simply correspond to classical excitations of a relativistic string (endowed with the appropriate internal symmetries).

follow simply from the thermodynamics of such a theory. Some of the implications are reminescent of the pre-quark model days, where based upon the rapidly increasing number of hadron resonances, Hagedorn postulated a density of states of the form (11.70) with $A = 5/2$, and showed that the Universe should have a limiting temperature of about 200 MeV [39]. The present situation is similar; however, the scale has been increased by about nineteen orders of magnitude.

The usual way to discuss the thermodynamics of strings is to start with the canonical ensemble. The partition function for the canonical ensemble is

$$\begin{aligned} \ln Z &= \frac{V}{(2\pi)^9}\int dm\,\rho(m)\int d^9k\,\ln\left[\frac{1+\exp\left[-(k^2+m^2)^{1/2}/T\right]}{1-\exp\left[-(k^2+m^2)^{1/2}/T\right]}\right] \\ &\simeq V\sum_{n=0}^{\infty}\left[\frac{1}{2n+1}\right]^5\int_{\eta}^{\infty} dm\, m^{-A}\exp(Bm)m^5 \\ &\quad \times K_5[(2n+1)m/T], \end{aligned} \tag{11.72}$$

where V is the (nine-dimensional) spatial volume, η is the mass below which the exponential approximation for ρ breaks down, and K_n is a modified Bessel function of the second kind. Using the limiting form $K_n(x) \to x^{-1/2}\exp(-x)$, we can express the partition function in terms of the incomplete gamma function

$$\ln Z \simeq \left(\frac{TT_0}{T_0 - T}\right)^{-A+11/2}\Gamma\left[-A+\frac{11}{2}, \left(\frac{T_0 - T}{TT_0}\right)\eta\right], \tag{11.73}$$

where $T_0 = B^{-1}$.

From this expression it is clear that the partition function diverges for $T \geq T_0$, independent of A. The pressure p, average energy $\langle E\rangle$, and specific heat C_V are given in terms of $\ln Z$ by

$$p = T\frac{\partial \ln Z}{\partial V}; \quad \langle E\rangle = T^2\frac{\partial \ln Z}{\partial T}; \quad C_V = \frac{d\langle E\rangle}{dT}. \tag{11.74}$$

For $A \leq 13/2$, they too diverge as $T \geq T_0$, hence the notion of a limiting temperature. For $A > 13/2$, p and $\langle E\rangle$ approach a constant as $T \to T_0$, and for $A > 15/2$, C_V also approaches a constant. If all thermodynamic quantities remain finite as $T \to T_0$, T_0 is *not* a limiting temperature [40].

Therefore, the open string has a limiting temperature $T_0 \simeq m_{Pl}$, but the closed or heterotic string does not.

What actually occurs in the case $A > 15/2$ is that the partition function diverges as $T \to T_0$, and the canonical-ensemble description breaks down. Physically, fluctuations become so large that a thermodynamic description based upon the canonical ensemble becomes nonsensical. One can get around this difficulty by using the microcanonical ensemble. When the microcanonical ensemble is used, it is found that the most likely configuration is one string carrying most of the energy of the system, with the remaining strings accounting for very little energy. Moreover, the specific heat for such a configuration is *negative.*

The fact that the specific heat is negative has interesting consequences. For example, it implies that a string system cannot come into thermal equilibrium with an ordinary heat bath. One might recall that black holes—and self-gravitating systems in general—are also characterized by negative specific heat.

It is somewhat ironic that in 1989, we end our treatment of cosmology in almost the same way as Weinberg [41] ended his treatment of cosmology in 1972—with a discussion of a limiting temperature for the Universe. In Weinberg's case it was a limiting temperature of only 0.2 GeV, in our case it is a limiting temperature of 10^{19} GeV. Perhaps it is some indication of progress in the fields of particle physics and cosmology that in the intervening seventeen years we have been able to add seven chapters, covering nearly twenty orders of magnitude in temperature.

11.6 References

1. S. W. Hawking, *Nature* **248**, 30 (1974); *Commun. Math. Phys.* **43**, 199 (1975).

2. B. J. Carr, in *Observational and Theoretical Aspects of Relativistic Astrophysics and Cosmology*, eds., J. L. Sanz and L. J. Giocoechea (World Scientific, Singapore, 1985); *Ap. J.* **201**, 1 (1975); *ibid* **206**, 8 (1976).

3. N. D. Birrell and P. C. W. Davies, *Quantum Fields in Curved Space* (Cambridge Univ. Press, Cambridge, 1982).

4. H. Everett, *Rev. Mod. Phys.*, **29**, 454 (1957); B. DeWitt and N. Graham, *The Many Worlds Interpretation of Quantum Mechanics*, (Princeton Univ. Press, Princeton, 1973); M. Gell-Mann and J. B.

Hartle, in *Complexity, Entropy, and Information (Vol. VIII in the SFI Studies in the Science of Complexity)*, ed., W. Zurek (Addison-Wesley, Redwood City, Calif., 1990).

5. B. S. DeWitt, *Phys. Rev.* **160**, 1113 (1967); J. A. Wheeler, in *Batelle Rencontres*, eds., C. DeWitt and J. A. Wheeler (Benjamin, New York, 1968).

6. S. Coleman, *Nucl. Phys.* **310**, 643 (1988).

7. See e.g., A. Strominger, in *Proceedings of the 1988 Theoretical Advanced Studies Institute* (Brown Univ.), (World Scientific, Singapore, 1989) in press.

8. M. B. Green, J. H. Schwarz, and E. Witten, *Superstring Theory* (Cambridge Univ. Press, Cambridge, 1987).

9. R. M. Wald, *General Relativity* (Univ. of Chicago Press, Chicago, 1984), Chs. 9 and 10 and Appendix E.2.

10. C. W. Misner, K. Thorne, and J. A. Wheeler, *Gravitation* (Freeman, San Francisco, 1973), Ch. 21.

11. A. Ashtekar, *Phys. Rev. Lett.* **57**, 2244 (1986).

12. For a listing of over 200 papers (containing well over 200 opinions) on the subject, see J. J. Halliwell, *A Bibliography of Papers on Quantum Cosmology*, ITP–UCSB preprint NSF-ITP-88-132. For an excellent discussion of the intricacies of quantum cosmology, see J. B. Hartle, in *Highlights in Gravitation and Cosmology*, eds, B. Iyer, A. Kanbajvi, J. V. Narlikar, and C. B. Vishveshuara (Cambridge Univ. Press, Cambridge, 1988).

13. J. B. Hartle and S. W. Hawking, *Phys. Rev. D* **28**, 2960 (1983).

14. A. D. Linde, *JETP* **60**, 211 (1984); A. Vilenkin, *Phys. Rev. D* **30**, 549 (1984); **37**, 888 (1988); V. A. Rubakov, *Phys. Lett.* **148B**, 280 (1984); Ya. B. Zel'dovich and A. A. Starobinskii, *Sov. Astron. Lett.* **10**, 135 (1984).

15. A. Vilenkin, *Phys Rev. D* **37**, 888 (1988).

16. T. Kaluza, *Preus. Acad. Wiss.* **K1**, 966 (1921).

17. O. Klein, *Zeit. Phys.* **37**, 895 (1926); *Nature* **118**, 516 (1926).

18. M. J. Duff, B. E. W. Nilsson, and C. N. Pope, *Phys. Rep.* **130**, 1 (1986).

19. J. H. Schwarz, *Superstrings* (World Scientific, Singapore, 1985).

20. E. W. Kolb, M. J. Perry, and T. P. Walker, *Phys. Rev. D* **33**, 869 (1986).

21. G. Chapline and N. Manton, *Phys. Lett.* **120B**, 105 (1983).

22. A. Chodos and S. Detweiler, *Phys. Rev. D* **21**, 2176 (1980).

23. J. Barrow, *Phys. Rep.* **85**, 1 (1982).

24. A. Hosoya, L. G. Jensen, and J. A. Stein-Schabes, *Nucl. Phys.* **B283**, 657 (1987).

25. S. Randjbar-Daemi, A. Salam, and J. Strathdee, *Phys. Lett.* **135B**, 388 (1984).

26. P. Candelas and S. Weinberg, *Nucl. Phys.* **B237**, 397 (1984).

27. E. Alvarez and M. Beleu-Gavela, *Phys. Rev. Lett.* **51**, 931 (1983); D. Sahdev, *Phys. Lett.* **137B**, 155 (1984); E. W. Kolb, D. Lindley, and D. Seckel, *Phys. Rev. D* **30**, 1205 (1984); R. B. Abbott, S. Barr, and S. Ellis *Phys. Rev. D* **30**, 720 (1984).

28. Y. Okada, *Phys. Lett.* **150B**, 103 (1985).

29. Q. Shafi and C. Wetterich, *Phys. Lett.* **129B**, 387 (1983).

30. F. S. Accetta, M. Gleiser, R. Holman, and E. W. Kolb, *Nucl. Phys.* **B276**, 501 (1986).

31. E. W. Kolb and R. Slansky, *Phys. Lett.* **135B**, 378 (1984).

32. D. A. Gross, J. A. Harvey, E. Martinec, and R. Rohm, *Phys. Rev. Lett.* **54**, 502 (1985).

33. R. Sorkin, *Phys. Rev. Lett.* **51**, 87 (1983); D. Gross and M. J. Perry, *Nucl. Phys.* **B226**, 29 (1983).

34. J. A. Harvey, E. W. Kolb, and M. J. Perry, *Phys. Lett.* **149B**, 465 (1984).

35. Y. Hosotani, *Phys. Lett.* **129B**, 193 (1983).

36. X.-G. Wen and E. Witten, *Nucl. Phys.* **B261**, 651 (1985).

37. See, for example, P. M. Frampton, *Dual Resonance Models* (Benjamin-Cummings, Reading, Mass. 1974).

38. K. Huang and S. Weinberg, *Phys. Rev. Lett.* **25**, 895 (1970); S. Frautschi, *Phys. Rev. D* **3**, 2821 (1971).

39. R. Hagedorn, *Astron. Astrophys.* **5**, 184 (1970).

40. M. Bowick and S. Wijewardhana, *Phys. Rev. Lett.* **54**, 2485 (1985).

41. S. Weinberg, *Gravitation and Cosmology* (Wiley, New York, 1972), Ch. 15.

FINALE

We conclude this monograph with our personal views on the future of the particle physics–cosmology interface. Speculation about the future is often preceded by a review of the path to the present; by taking a few steps back it is possible to get a running start for the leap into the future. In contrasting the past with the present, the most striking difference is one of attitude. Prior to the mid-1960's, twentieth-century cosmologists were hampered time and time again by a lack of confidence in their own sound theoretical predictions, which led to numerous missed opportunities. In contrast, since the mid-1960's, cosmologists have shown a growing confidence in theoretical predictions.

Shortly after Einstein introduced the field equations in 1915, he constructed the first modern cosmological models. Although Einstein had revolutionized the concepts of space, time, and geometry, he could not extricate himself from his 19th-century view of a static Universe. While his elegant cosmological models demanded an expanding (or contracting) Universe, he chose to introduce the cosmological term to ensure static (but unstable) solutions. In so doing, Einstein (and others) missed one of the greatest opportunities ever in theoretical physics: the prediction of the expansion of the Universe.[1] Some ten years later the expansion of the Universe was discovered by Hubble, and it was left to Lemaître to relate the Hubble expansion explicitly to the cosmological solutions found by Friedmann.

This lack of confidence in sound cosmological predictions continued after Gamow's prediction of a relic background radiation in the 1940's. As Weinberg recounts in *The First Three Minutes* (Basic Books, New

[1] One could argue that this apparent lack of confidence was not limited to cosmology alone: Dirac first tried to identify his negative energy solution with the proton, rather than predicting the existence of the unseen positron.

York, 1977), this radiation *could* have been discovered—and was all but discovered—much earlier had Gamow's prediction been taken seriously. Instead, it was not until 1964 that Penzias and Wilson accidentally discovered the background radiation. It is somewhat ironic that their discovery came shortly before a Princeton experiment designed to actually look for the CMBR came on line—and would have found it. A very sensible cosmological prediction, the existence of a relic background of photons, was ignored because cosmologists did not have faith in their own predictions.

Soon after the discovery of the microwave background, several groups finally took heed of Gamow and collaborators' suggestion that conditions in the Universe seconds after the bang were just right for the synthesis of the light elements D, He, and Li. Even so, it wasn't until the mid-1970's—and early 1980's for Li—that the cosmological origin for these elements became generally accepted. The successful predictions of primordial nucleosynthesis remain the most solid evidence that the standard cosmology provides an accurate description of the Universe at times as early as 10^{-2}sec after the bang, as well as a sensible framework for discussing the evolution of the Universe at times as early as 10^{-43}sec.

Fortunately for present-day cosmologists, the standard hot big bang model does not explain everything—yet! As we emphasized in Chapter 8, there are a small number of cosmological facts that the standard model accommodates but does not explain. These facts include the observed predominance of matter over antimatter, the origin of the seed inhomogeneities required to initiate structure formation, the nature of the ubiquitous dark matter, the smallness of any cosmological term, and the large-scale isotropy, homogeneity, and flatness of the Universe. In addition, there are several very big questions—What caused the bang? and Did the Universe begin from a singularity?

In the twenty-five years since the discovery of the microwave background a new confidence has emerged. Most—though not all—cosmologists now subscribe to the view that the standard cosmology is basically correct, and that the above-mentioned shortcomings will eventually be solved within the context of some extended version of the model, e.g., inflation. Moreover, it is almost universally believed that the modification of the standard cosmology will involve the input of fundamental physics in a crucial way. The appreciation that fundamental physics plays an important role in cosmology is another important theme in the most recent history of cosmology—which is also evidenced by the increasing numbers of physicists who are turning their research efforts to cosmology.

The importance of physics input to cosmology is well illustrated by a simple, although somewhat dated, example. In the late 1960's and early 1970's the success of primordial nucleosynthesis was somewhat dashed by the specter of a Universe with a limiting temperature of only a few hundred MeV, owing to the apparently exponentially rising number of hadronic states. The introduction of the quark/parton model in the 1970's, and the emergence of QCD with its asymptotic freedom and small number of fundamental constituents—quarks and gluons—as *the* theory of the strong interactions shattered that barrier and opened the door to the study of the very early history of the Universe ($t \ll 10^{-2}$sec).

The focus has now turned to the application of the most current—and often still very speculative—ideas in modern particle theory to the earliest moments of the Universe. These ideas include the well established $SU(3)_C \otimes SU(2)_L \otimes U(1)_Y$ gauge theory of the strong, weak, and electromagnetic interactions, and numerous well motivated ideas about fundamental physics at energies exceeding 1 TeV, e.g., supersymmetry, grand unification, supergravity, quantum cosmology, and superstrings. Moreover, the application of modern particle theory to cosmology has already begun to shed light on many of the aforementioned loose ends of the standard cosmology. It appears likely that the baryon number of the Universe evolved dynamically through interactions that are predicted to exist in almost all schemes of the unification of the forces—interactions that do not conserve B, C, and CP and operated with strength equal to that of the other known forces at very early times ($t \sim 10^{-34}$sec). Once we understand the physics behind the unification of quarks and leptons—so-called grand unification—we should be able to calculate the baryon number of the Universe with the same ease that we now calculate the primordial abundances of the light elements.

Inflation holds the promise of resolving the isotropy, homogeneity, flatness, and inhomogeneity puzzles. If inflation does not explain the origin of the primeval inhomogeneity required to trigger structure formation, then perhaps an early Universe phase transition will—for example, through the production of cosmic string. The most attractive hypothesis for the nature of the dark matter is that it consists of relic, elementary particles left over from the earliest moments—and particle physicists have been very generous in providing candidate relics.

The confidence of present-day cosmologists in their predictions and their fervant application of modern particle theory to the early Universe has revitalized the study of structure formation. Structure formation through the gravitational amplification of primeval inhomogeneities is certainly a part of the standard cosmology. However, until recently progress had been hindered by ignorance of the "initial data" for this problem: the quantity and composition of matter in the Universe and the nature of the primeval inhomogeneity. Early-Universe scenarios have provided well founded ideas as to the appropriate initial data for the structure formation problem. One such scenario—cold dark matter with inflation-produced density perturbations—is the most detailed picture of structure formation ever formulated. While it remains to be seen whether or not this current paradigm proves correct, one would be hard pressed to find any cosmologist who disagreed with the proposition that the key to understanding structure formation ulitimately traces to the earliest moments of the Universe and fundamental physics. If cold dark matter should falter, there are other attractive, but less well developed, ideas based upon cosmic strings and other early-Universe scenarios waiting in the wings.

An optimist (e.g., *either* one of the authors) has no other choice than to conclude that we are well on our way to answering many (if not all) of the most pressing questions in cosmology. Even a less than optimistic person would have to conclude that the answers to many of the pressing questions must lie in understanding the earliest history of the Universe, which in turn necessarily involves the application of physical theory at the most fundamental level to the cosmological setting.

Now to the future. Inflation has literally revolutionized the way cosmologists think about the earliest history of the Universe, and it seems all but certain that inflation, or at least one of its offshoots, will continue to be a very promising avenue to pursue. The past decade has seen a renaissance in the study of structure formation—with among other things the development of the first paradigm for structure formation: cold dark matter. Not only will there be increased opportunities for testing this paradigm, but other competing scenarios as well as they become more fully developed.

While the past two decades have seen the frontiers of cosmology pushed back almost ten orders of magnitude in temperature, the Planck scale continues to stand as a barrier. It is not clear how to surmount the Planck barrier, but we know that it won't be done without an understanding of the quantum nature of gravity. As we discussed in Chapter 11, until very recently, there has been a lack of promising ideas about the quantization of gravity. While there are now a number of promising ideas to pursue, none has yet reached the state of being "well formulated." That situation is likely to change in the decade ahead. Superstring theory is probably the most promising approach to unification of gravity with all the forces. Superstring cosmology—which is still in its infancy—opens a Pandora's box of new challenges—Whence came geometry? Was there an initial singularity? Are there extra spatial dimensions, and if so, why did some remain small and others grow? If superstrings are correct, cosmologists may again face the specter of a limiting temperature—albeit 10^{19}GeV.

Unlike in the first fifty years of modern cosmology, where theory played a more passive role, during last twenty-five years theory has led the way. However, this pattern cannot continue indefinitely. If the first fifty years of modern cosmology suffered from a lack of confidence in well motivated theoretical predictions, it can be stated with equal fervor that cosmology in the last decade has begun to suffer from a lack of experimental and observational tests of exciting theoretical predictions. Even in cosmology, the ultimate test of theory is grounded in experimental and/or observational verification. The problem confronting contemporary cosmologists is a simple mismatch in red shift. Much of the most promising cosmological speculation involves very high red shifts: The red shift of the Planck era is $z_{Pl} = 10^{32}$, while observations are most easily done at relatively low red shifts $z \lesssim 4 \simeq \ln[\ln(z_{Pl})]$. Optimists that we are, we are certain that we see the light at the end of the tunnel. We firmly believe that experiments and observations that *will* be done in the next decade will begin to restore the balance between theoretical speculation and experimental reality. We see numerous "low-z" observations and "$z = 0$" (laboratory) experiments on the immediate horizon that will provide windows to the Universe at "high-z"—and in the process windows to fundamental physics at very high energies and short distances.

By the end of this century NASA should have four space-based observatories in place that will view the Universe with unprecedented sensitivity and resolution in the optical and ultraviolet (Hubble Space Telescope),[2] in the infrared (SIRTF), in the x ray (AXAF), and in the gamma ray (GRO). These observatories in space, together with the new generation of large-aperture, ground-based telescopes (from 8 m to 16 m in diameter) and special-purpose cosmology telescopes, should provide cosmologists with the data necessary to resolve important issues such as the value of the Hubble constant (and perhaps even the deceleration parameter), the epoch of galaxy formation, the clustering properties of both galaxies and rich clusters, the topology of the galaxy distribution, etc. With some good fortune and hard work, the LIGO project (or similar efforts in Europe or Australia) may open a new window to the Universe—gravitational wave

[2] For those handful of readers who wagered that this monograph would be finished before the Hubble Space Telescope was launched, write NASA Headquarters, Code EZ to collect your winnings.

astronomy. As we have discussed, there are many cosmological sources of gravitational radiation—inflation and cosmic strings to mention two.

At present, our "highest-z" windows on the early Universe are the CMBR—the Universe at red shift $z_{dec} \simeq 1100$ and age few hundred-thousand years—and the abundances of the light elements—the Universe at a red shift of $z \sim 10^{12}$ and age 1 sec. Here too, there should be dramatic improvements. The Cosmic Background Explorer (COBE) is scheduled to be launched soon, and a host of new experiments to probe the spectrum and isotropy of the CMBR with significantly greater precision are either underway or in the works. If the cold dark matter (or hot dark matter) scenario is correct, a detectable anisotropy is just around the corner.

With regard to primordial nucleosynthesis, the deuterium abundance is known only to within a factor of 2, and only in our immediate neighborhood. There is still debate as to whether the primordial abundance of Li has been determined. And even the abundance of ^{4}He is known only to a precision of about 5%. The last parameter of the standard model that affects primordial nucleosynthesis—the number of neutrino families—should be determined in precise measurements of the width of the Z^0. All of these issues should be settled in the coming decade, further strengthing the status of the standard cosmology and sharpening its precision as a unique laboratory for studying particle physics—or perhaps finding a crack in the cosmic egg.

Because of the linkage between the frontiers of cosmology and of high-energy physics—the inner space/outer space connection—experiments in particle physics will also undoubtedly play an important role in cosmology. At present, there are two powerful proton–antiproton colliders operating at center-of-mass energies of 2 and 0.5 TeV (the Tevatron at Fermilab and the S$p\bar{p}$S collider at CERN), and two new $e^\pm$ colliders operating at a center-of-mass energy of about 100 GeV (SLC at SLAC and LEP at CERN). The electron–proton collider (26 GeV on 820 GeV) at DESY (HERA) should begin operations soon, and within the next decade both the Super Conducting SuperCollider (SSC), a 20 TeV on 20 TeV proton–proton collider, and the UNK (at Serpukhov), a 3 TeV on 3 TeV proton–proton collider, should come on line. In addition, there are other smaller projects in the works—the heavy-ion collider at Brookhaven (RHIC) and proposals for B and τ factories.

Experiments done at these facilities have the potential to discover the top quark, evidence for supersymmetry or neutrino masses, the Higgs particle, or even a dark matter candidate (or at the least, to narrow the list). Discovery of the top quark and the electroweak Higgs particle would provide the last parameters necessary to specify the scalar potential for electroweak symmetry breaking, and in turn, this would allow one to understand the details of the electroweak phase transition in the early Universe. The electroweak interaction provides the only Higgs system accessible to laboratory experiment in the near future, and although the electroweak Higgs cannot be the scalar field that drives inflation, the discovery of the Higgs would reassure us that the Higgs mechanism has something to do with Nature—or if the Higgs is not found, rudely awaken a generation of particle theorists.

QCD is universally accepted as the theory of the strong interactions, and it is generally believed that there should be a cosmological phase transition associated with deconfinement and/or chiral symmetry restoration, at a temperature of a few hundred MeV. However, our knowledge of the details of the phase transition are far from complete—present-day computational power has thus far proved no match for the non-perturbative aspects of QCD. With a little optimism one can hope that the details of the QCD deconfinement and chiral symmetry breaking transitions will be sorted out by numerical experiments soon—among other things, settling the issue of whether baryon number inhomogeneity can circumvent the important nucleosynthesis bound to Ω_B.

These are some of the important experiments and observations that should be done in the separate disciplines of cosmology and particle physics. In the past decade, the blossoming of the field of particle cosmology has brought with it an experimental component that is neither purely astronomy nor particle physics. Talented experimentalists have built and are building experiments to detect relic dark matter particles—axions, photinos, cosmions, and heavy neutrinos. The discovery of particle dark matter would easily rank in importance with that of the CMBR and the expansion of the Universe. Others have built detectors to search for magnetic monopoles and other exotic relics in the cosmic rays. Included in the category of experiments, often referred to as non-accelerator experiments, are neutrino mass experiments and large, well instrumented underground detectors such as the Kamiokande II and the Irvine-Brookhaven-Michigan water Cherenkov detectors, which have been used to detect neutrinos from SN 1987A and the sun, to search for magnetic monopoles and neutrinos from dark matter annihilations in the sun, and, oh yes, to search for proton decay.

Most certainly we have not included in our list the most important or exciting experimental and observational discoveries that will be made in the coming years. As always, those discoveries will be the surprises that await us! Even so, we can try to mention some startling examples of the kind of surprises that might lie ahead. For example, a definitive determination that Ω_0 is not unity—undermining the theoretical prejudice of a generation of cosmologists; or that $H_0 t_0$ is greater than unity—necessitating a cosmological constant. Lack of a detectable cosmological anisotropy in the CMBR at level of 10^{-6} would shoot down essentially every existing theory of structure formation.

Whatever future cosmologists write about cosmology in the decades following the discovery of the CMBR, we can be certain that they will not criticize contemporary cosmologists for failure to take their theoretical ideas—and sometimes wild speculations—seriously enough. Perhaps future cosmologists will laugh at our naïveté. But, if they do, we can hope they will admire our courage and boldness in attacking problems once thought to be beyond the reach of human comprehension. The extent to which we shall be rewarded for our courage and boldness remains to be seen. These authors remain ever optimistic!

APPENDIX A

A.1 Units

The system of units one selects for a problem often reveals much about the underlying physics of the problem and one's approach to it. Cosmology today involves scales from the most microscopic, distances $\ll 1$ fermi $= 10^{-13}$ cm, to the most macroscopic, distances $\gg$ Mpc $\sim 3 \times 10^{24}$ cm. The focus of this monograph is the early Universe, and that focus is reflected in our choice of units, so-called natural, or high energy physics units. In this system the fundamental constants $\hbar = c = k_B = 1$, and there is one fundamental dimension, energy, so that

$$[\text{Energy}] = [\text{Mass}] = [\text{Temperature}] = [\text{Length}]^{-1} = [\text{Time}]^{-1}.$$

We will usually take the unit of energy to be a GeV $= 10^3$ MeV $= 10^6$ keV $= 10^9$ eV. Of course, at times we will find it more convenient to use other units—e.g., Mpc, $M_\odot$, erg s^{-1}, etc.

A.1.1 Conversion factors

Energy:	1 GeV $= 1.6022 \times 10^{-3}$erg
Temperature:	1 GeV $= 1.1605 \times 10^{13}$K
Mass:	1 GeV $= 1.7827 \times 10^{-24}$g
Length:	1 GeV^{-1} $= 1.9733 \times 10^{-14}$cm
Time:	1 GeV^{-1} $= 6.5822 \times 10^{-25}$sec
Power:	1 GeV^2 $= 2.4341 \times 10^{21}$erg sec^{-1}

Number density:	$1\ \text{GeV}^3 = 1.3014 \times 10^{41}\text{cm}^{-3}$
Mass density:	$1\ \text{GeV}^4 = 2.3201 \times 10^{17}\text{g cm}^{-3}$
Volume emissivity:	$1\ \text{GeV}^5 = 3.1678 \times 10^{62}\text{erg cm}^{-3}\text{sec}^{-1}$
Cross section:	$1\ \text{barn} = 10^3\text{mb} = 10^{-24}\text{cm}^2$
	$1\ \text{mb} = 2.5681\ \text{GeV}^{-2}$
	$1\ \text{acre} = 4.0469 \times 10^{31}\text{barns}$
Wavelength/energy:	$\lambda = 12398.4\ \text{Å}/E(\text{eV})$
Astronomical unit:	$1\ \text{AU} = 1.4960 \times 10^{13}\text{cm} = 7.5812 \times 10^{26}\text{GeV}^{-1}$
Parsec:	$1\ \text{pc} = 3.2615\ \text{light} - \text{yr} = 3.0856 \times 10^{18}\text{cm}$
Megaparsec:	$1\ \text{Mpc} = 10^6\text{pc} = 3.0856 \times 10^{24}\text{cm}$
	$= 1.5637 \times 10^{38}\text{GeV}^{-1}$
Sidereal day:	$1\ \text{da (sidereal)} = 86,164.091\ \text{sec}$
	$= 1.3090 \times 10^{29}\text{GeV}^{-1}$
Sidereal year:	$1\ \text{yr (sidereal)} = 3.1558 \times 10^7\text{sec}$
	$= 4.7944 \times 10^{31}\text{GeV}^{-1} = 7.0000\ \text{dog yrs}$
Magnetic field:	$1\ \text{Tesla} = 10^4\ \text{Gauss}$
Field energy:	$1\ (\text{Gauss})^2/8\pi = 1.9084 \times 10^{-40}\text{GeV}^4$
	$1\ (\text{Tesla})^2/2 = 1.9084 \times 10^{-32}\text{GeV}^4$
Lorentz force:	$1\ e'\ \text{Gauss} = 5.9157 \times 10^{-21}\text{GeV}^2$
	$1\ e\ \text{Tesla} = 5.9157 \times 10^{-17}\text{GeV}^2$
Current:	$1\ \text{Amp} = 4.1083 \times 10^{-6}\text{e GeV}$

A.1.2 Fundamental constants

Planck's constant:	$h = 6.6261 \times 10^{-27}\text{cm}^2\,\text{g sec}^{-1}$
	$\hbar = 1.0546 \times 10^{-27}\text{cm}^2\,\text{g sec}^{-1}$
Speed of light:	$c = 2.9979 \times 10^{10}\text{cm sec}^{-1}$
Boltzmann's constant:	$k_B = 1.3807 \times 10^{-16}\text{erg K}^{-1}$
Electron charge (CGS):	$e = 4.8032 \times 10^{-10}\text{esu} = 0.30282$
Electron charge (MKS):	$e' = 1.6022 \times 10^{-19}\text{Coulomb} = 0.085425$
Weak mixing angle:	$\sin^2\theta_W = 0.23 \pm 0.005$

Fermi constant:	$G_F = 1.1664 \times 10^{-5}\mathrm{GeV}^{-2}$
	$= (292.80\ \mathrm{GeV})^{-2}$
Newton's constant:	$G = 6.6720 \times 10^{-8}\mathrm{cm}^3\,\mathrm{g}^{-1}\mathrm{sec}^{-2} \equiv m_{Pl}^{-2}$
Planck energy:	$m_{Pl} \equiv (\hbar c^5/G)^{1/2} = 1.2211 \times 10^{19}\mathrm{GeV}$
Planck mass:	$m_{Pl} \equiv (\hbar c/G)^{1/2} = 2.1768 \times 10^{-5}\mathrm{g}$
Planck time:	$t_{Pl} \equiv (\hbar G/c^5)^{1/2} = 5.3904 \times 10^{-44}\mathrm{sec}$
Planck length:	$l_{Pl} \equiv (\hbar G/c^3)^{1/2} = 1.6160 \times 10^{-33}\mathrm{cm}$
Planck density:	$\rho_{Pl} \equiv c^5/\hbar G^2 = 5.1584 \times 10^{93}\mathrm{g\ cm}^{-3}$
Fine-structure const:	$\alpha_{\mathrm{EM}} = 1/137.036 \equiv e^2/4\pi \equiv e'^2$
Rydberg:	$1\ \mathrm{Ry} = \alpha_{EM}^2 m_e c^2/2 = 13.606\ \mathrm{eV}$
Bohr radius:	$\mathrm{a}_0 = (\hbar/m_e c)/\alpha_{EM} = 5.2918 \times 10^{-9}\mathrm{cm}$
	$= 2.6817 \times 10^5\mathrm{GeV}^{-1}$
Bohr magnetron:	$\mu_B = e\hbar/2m_e$
	$= 5.7884 \times 10^{-14}\mathrm{GeV\ Tesla}^{-1}$
Nuclear magnetron:	$\mu_N = e\hbar/2m_p$
	$= 3.1525 \times 10^{-17}\mathrm{GeV\ Tesla}^{-1}$
Thomson cross section:	$\sigma_T = 8\pi\alpha_{\mathrm{EM}}^2/3m_e^2 = 6.6524 \times 10^{-25}\mathrm{cm}^2$
Wien law const:	$\lambda_{MAX}\,T = 0.28978\ \mathrm{cm\ K} = 1.2655$
Electron mass:	$m_e = 0.5110\ \mathrm{MeV}$
Neutron mass:	$m_n = 939.566\ \mathrm{MeV}$
Proton mass:	$m_p = 938.272\ \mathrm{MeV}$
Mass $^{12}\mathrm{C}/12$:	$1\ \mathrm{amu} = 931.494\ \mathrm{MeV}$
Avogadro's number:	$N_A = 6.0220 \times 10^{23}$
Radiation density const:	$a = \pi^2 k_B^4/15\hbar^3 c^3 = 0.65797$
	$= 7.5646 \times 10^{-15}\mathrm{erg\ cm}^{-3}\mathrm{K}^{-4}$
Stefan-Boltzmann const:	$\sigma = ac/4 = \pi^2 k_B^4/60\hbar^3 c^2 = 0.16449$
	$= 5.6705 \times 10^{-5}\mathrm{erg\ cm}^{-2}\mathrm{sec}^{-1}\mathrm{K}^{-4}$
Degree:	$1\ \mathrm{deg} = 60\ \mathrm{arc\ min} = 3600\ \mathrm{arc\ sec}$
Radian:	$1\ \mathrm{rad} = 57.296\ \mathrm{deg}$
Arc sec:	$1\ \mathrm{arc\ sec} = 4.8481 \times 10^{-6}\mathrm{rad}$
Steradian:	$1\ \mathrm{sr} = 3.2828 \times 10^3\mathrm{deg}^2$
	$= 4.2545 \times 10^{10}(\mathrm{arc\ sec})^2$

Furthermore, we use Heaviside-Lorentz electromagnetic units, so that $\alpha_{\rm EM} = e^2/4\pi$, the magnitude of the charge of the electron is $(4\pi\alpha_{\rm EM})^{1/2} = 0.30282$, and $\mathcal{L} = -(1/4)F^{\mu\nu}F_{\mu\nu}$ is the Lagrangian density for the free Maxwell field, where $F^{\mu\nu}$ is the electromagnetic field strength tensor. In this system, magnetic field strength is measured in Tesla, the EM energy density is $(T^0_{\ 0})_{EM} = (\vec{\mathbf{E}}^2 + \vec{\mathbf{B}}^2)/2$, and the potential due to a point charge q is $\phi = q/4\pi r$. For reference, in CGS EM units, $\alpha_{EM} = e'^2$, the magnitude of the charge of the electron is $e' = \alpha_{EM}^{1/2} = 0.085425$, and the Lagrangian density for the free Maxwell field is $\mathcal{L} = -(1/16\pi)F^{\mu\nu}F_{\mu\nu}$. In this system, magnetic field strength is measured in Gauss, the EM energy density is $(T^0_{\ 0})_{EM} = (\vec{\mathbf{E}}^2 + \vec{\mathbf{B}}^2)/8\pi$, and the potential due to a point charge q is $\phi = q/r$. Note that while a magnetic field of strength 1 Tesla is equivalent to one of strength 10^4 Gauss, 1 Tesla = 10^4 Gauss is not a conversion factor in the usual sense, e.g., $(1 \text{ Tesla})^2/2 = (10^4 \text{ Gauss})^2/8\pi$.

A.2 Physical Parameters

The following is a list of useful physical parameters for astrophysics and cosmology.

A.2.1 Astrophysical parameters

Solar mass: $M_\odot = 1.989 \times 10^{33}\text{g} = 1.116 \times 10^{57}\text{GeV}$
$= 1.189 \times 10^{57}\text{protons}$

Solar radius: $R_\odot = 6.9598 \times 10^{10}\text{cm} = 3.5270 \times 10^{24}\text{GeV}^{-1}$

Solar luminosity: $\mathcal{L}_\odot = 3.90 \times 10^{33}\text{erg sec}^{-1} = 1.60 \times 10^{12}\text{GeV}^2$

Earth radius: $R_\oplus = 6.3782 \times 10^{8}\text{cm} = 3.2223 \times 10^{22}\text{GeV}^{-1}$
($R_\oplus$ = earth equatorial radius)

Earth mass: $M_\oplus = 5.977 \times 10^{27}\text{g} = 3.357 \times 10^{51}\text{GeV}$

Luminosity: $\mathcal{L} = 3.02 \times 10^{35}\text{erg sec}^{-1}10^{-2M_b/5}$
(M_b = absolute bolometric magnitude)

Energy flux: $\mathcal{F} = 2.52 \times 10^{-5-0.4m_b}\text{erg cm}^{-2}\text{sec}^{-1}$
(m_b = apparent bolometric magnitude)
$d\mathcal{F}/d\lambda\,(\lambda \simeq 4400\text{Å})$
$= 6.76 \times 10^{-9-0.4m_B}\text{erg cm}^{-2}\text{Å}^{-1}\text{sec}^{-1}$
(m_B = apparent blue magnitude)

Solar magnitudes: $m_{b\odot} = -26.85 \qquad M_{b\odot} = 4.72$

Distance modulus: $m - M = 5\log(D/10\text{ pc})$

Specific intensity: $I_\lambda \equiv d\mathcal{F}/d\lambda d\Omega$

$I_E \equiv d\mathcal{F}/dEd\Omega = (\lambda/E)I_\lambda$

Brightness: $m_b \text{ mag (arc sec)}^{-2} = m_b\,\mu$

$= 2.52 \times 10^{-5-0.4m_b}\text{erg cm}^{-2}(\text{arc sec})^{-2}\text{sec}^{-1}$

$m_B \text{ mag (arc sec)}^{-2} = m_B\,\mu_B$

$= 6.76 \times 10^{-9-0.4m_B}\text{erg cm}^{-2}(\text{arc sec})^{-2}\text{Å}^{-1}\text{sec}^{-1}$

Night sky: $22\mu_B \simeq 1.1 \times 10^{-17}\text{erg cm}^{-2}(\text{arc sec})^{-2}\text{Å}^{-1}\text{sec}^{-1}$

Jansky: $1\text{ Jy} = 10^{-23}\text{ erg cm}^{-2}\text{sec}^{-1}\text{Hz}^{-1}$

$= 2.4730 \times 10^{-48}\text{ GeV}^3$

A.2.2 Cosmological parameters

In the following, the subscript "0" will denote the present value of a quantity, and $T_{2.75} = T_{\gamma 0}/2.75$ K.

Hubble const: $H_0 = 100h\text{ km sec}^{-1}\text{Mpc}^{-1}$

$= 2.1332h \times 10^{-42}\text{GeV} \qquad (0.4 \lesssim h \lesssim 1)$

Hubble time: $H_0^{-1} = 3.0856 \times 10^{17}h^{-1}\text{sec} = 9.7776 \times 10^9 h^{-1}\text{yr}$

Hubble distance: $cH_0^{-1} = 2997.9h^{-1}\text{Mpc} = 9.2503 \times 10^{27}h^{-1}\text{cm}$

Critical density: $\rho_C \equiv 3H_0^2/8\pi G = 1.8791h^2 \times 10^{-29}\text{g cm}^{-3}$

$= 8.0992h^2 \times 10^{-47}\text{GeV}^4 = 1.0540h^2 \times 10^4\text{eV cm}^{-3}$

Photons: $T_{\gamma 0} = 2.75T_{2.75}\text{K} = 2.3697 \times 10^{-13}T_{2.75}\text{ GeV}$

$n_{\gamma 0} = 421.84\,T_{2.75}^3\text{cm}^{-3}$

$\rho_{\gamma 0} = 2.0747 \times 10^{-51}T_{2.75}^4\text{GeV}^4$

$= 4.8135 \times 10^{-34}T_{2.75}^4\text{g cm}^{-3}$

Neutrinos: $T_{\nu 0} = (4/11)^{1/3}T_{\gamma 0} = 1.9629T_{2.75}\text{K}$

$= 1.6914 \times 10^{-13}T_{2.75}\text{GeV}$

$n_{\nu 0} = \frac{3}{11}n_{\gamma 0} = 115.05\,T_{2.75}^3\text{cm}^{-3}$ (per species)

$\rho_{\nu 0} = \frac{21}{8}(4/11)^{4/3}\rho_{\gamma 0} = 0.68132\rho_{\gamma 0}$ (3 species)

Entropy: $s_0 = [43\pi^4/11 \cdot 45 \cdot \zeta(3)]n_{\gamma 0} = 7.0394\,n_{\gamma 0}$

$= 2969.5\,T_{2.75}^3\text{ cm}^{-3}$

Decoupled species:	$T_{X0} = [3.9091/g_{*S}(T_D)]^{1/3} T_{\gamma 0}$
Scale factor:	$R/R_0 = 3.7329 \times 10^{-13} T_{2.75} g_{*S}^{-1/3} (\mathrm{GeV}/T)$
Radiation-matter equality:	$R_{EQ} = 4.3069 \times 10^{-5} (\Omega_0 h^2)^{-1} T_{2.75}^4 R_0$
	$T_{EQ} = 5.5021\, (\Omega_0 h^2)\, T_{2.75}^{-3}\, \mathrm{eV}$
	$t_{EQ} = 4.3608 \times 10^{10}\, (\Omega_0 h^2)^{-2}\, T_{2.75}^6\, \mathrm{sec}$
Decoupling:	$R_{dec} \equiv R_0/1100$
	$T_{dec} = 0.26\, T_{2.75\mathrm{K}}\, \mathrm{eV}$
	$t_{dec} = 5.6384 \times 10^{12} (\Omega_0 h^2)^{-1/2}\, \mathrm{sec}$
	(valid if $\Omega_0 h^2 \gtrsim 0.04\, T_{2.75}^4$)
Age of Universe:	$t = 0.30118 g_*^{-1/2} m_{Pl}/T^2 \quad (T \geq T_{EQ})$
	$= 2.4207 \times 10^{-6} g_*^{-1/2} (\mathrm{GeV}/T)^2 \mathrm{sec}$
	$= 1.7372 \times 10^{19} g_{*S}^{2/3} g_*^{-1/2} (1+z)^{-2} \mathrm{sec}$
	$= 0.40157 g_*^{-1/2} m_{Pl}/T^{3/2} T_{EQ}^{1/2} \quad (T \leq T_{EQ})$
	$= 2.0571 \times 10^{17}\, (\Omega_0 h^2)^{-1/2} (1+z)^{-3/2} \mathrm{sec}$
	$= 7.5039 \times 10^{11} (\Omega_0 h^2)^{-1/2} T_{2.75}^{3/2} (T/\mathrm{eV})^{-3/2} \mathrm{sec}$
Rel. degrees of freedom (ρ_R):	$g_* = \sum_{i=boson} (T_i/T)^4 g_i + \frac{7}{8} \sum_{i=fermion} (T_i/T)^4 g_i$
	$g_{*0} = 2 + \frac{42}{8} (4/11)^{4/3} = 3.3626$ (3 ν species)
Rel. degrees of freedom (s):	$g_{*S} = \sum_{i=boson} (T_i/T)^3 g_i + \frac{7}{8} \sum_{i=fermion} (T_i/T)^3 g_i$
	$g_{*S0} = 2 + \frac{42}{8} (4/11) = 3.9091$ (3 ν species)
Ω(relativistic):	$\Omega_{\gamma\nu\bar{\nu}} h^2 = 4.3069 \times 10^{-5} T_{2.75}^4$
Ω(baryons):	$\Omega_B h^2 = 3.7278 \times 10^7 T_{2.75}^3 \eta$
	$= 2.6241 \times 10^8 T_{2.75}^3 n_B/s$
Entropy in the horizon:	$S_{\mathrm{HOR}} \equiv (4\pi/3) t^3 s$
	$= 5.01 \times 10^{-2} g_*^{-1/2} (m_{Pl}/T)^3 \quad (t \leq t_{EQ})$
	$= 2.94 \times 10^{87} (\Omega_0 h^2)^{-3/2} (1+z)^{-3/2} \quad (t \geq t_{EQ})$
Baryons in the horizon:	$N_{\mathrm{B-HOR}} = (n_B/s) S_{\mathrm{HOR}}$
	$= 1.91 \times 10^{-10} (\Omega_B h^2) g_*^{-1/2} (m_{Pl}/T)^3 \quad (t \leq t_{EQ})$
	$= 1.12 \times 10^{79} (\Omega_B/\Omega_0^{3/2} h)(1+z)^{-3/2} \quad (t \geq t_{EQ})$

Baryon mass in the horizon:	$M_{\rm B-HOR} = m_N N_{\rm B-HOR}$
	$= 2.93 \times 10^{-10} (\Omega_B h^2) g_*^{-1/2} ({\rm GeV}/T)^3 {\rm M}_\odot \quad (t \leq t_{EQ})$
	$= 9.42 \times 10^{21} (\Omega_B / \Omega_0^{3/2} h)(1+z)^{-3/2} M_\odot \quad (t \geq t_{EQ})$
Mass within scale λ:	$M(\lambda) \equiv \pi \lambda^3 \rho_{NR}/6$
	$= 1.45 \times 10^{11} (\Omega_0 h^2) \lambda_{\rm Mpc}^3 {\rm M}_\odot$
Physical size of scale λ:	$R\,\lambda/R_0 = 1.15 \times 10^{12} \lambda_{\rm Mpc} ({\rm GeV}/T) g_{*S}^{-1/3}$ cm
	$= 5.83 \times 10^{25} \lambda_{\rm Mpc} ({\rm GeV}/T) g_{*S}^{-1/3} {\rm GeV}^{-1}$
Horizon crossing:	$T_{\rm HOR}(\lambda) = 63 g_*^{-1/2} g_{*S}^{1/3} \lambda_{\rm Mpc}^{-1} {\rm eV}$
	$(\lambda \lesssim \lambda_{EQ} \equiv 13 (\Omega_0 h^2)^{-1} {\rm Mpc})$
	$t_{\rm HOR}(\lambda) = 6.1 \times 10^8 g_*^{1/2} g_{*S}^{-2/3} \lambda_{\rm Mpc}^2 {\rm sec}$
	$T_{\rm HOR}(\lambda) = 948 (\Omega_0 h^2)^{-1} \lambda_{\rm Mpc}^{-2} {\rm eV} \quad (\lambda \gtrsim \lambda_{EQ})$
	$t_{\rm HOR}(\lambda) = 2.6 \times 10^7 (\Omega_0 h^2) \lambda_{\rm Mpc}^3 {\rm sec}$
Angle subtended by Hubble radius:	$\Delta\theta(H^{-1}, z \gg 1) = \Omega_0^{1/2} z^{-1/2}/2$
	$= 0.91^\circ\ \Omega_0^{1/2} (z/1000)^{-1/2}$
Angle subtended by scale λ:	$\Delta\theta(\lambda, z \gg 1) = (R_0 \lambda / H_0^{-1}) \Omega_0 / 2$
	$= 34.4''\ (\Omega_0 h) \lambda_{\rm Mpc}$
	$= 65.4'' (\Omega_0^{2/3} h^{1/3}) (M/10^{12} M_\odot)^{1/3}$

$T_{\rm HOR}(\lambda)$ and $t_{\rm HOR}(\lambda)$ refer to the temperature and age of the Universe when the comoving scale λ had a physical size equal to that of the age of the Universe ($R\lambda = ct$). The scale that crosses the horizon at matter-radiation equality is $\lambda_{EQ} = 13(\Omega_0 h^2)^{-1}$ Mpc, as computed by using the matter-dominated expression for t_{EQ}. (Using the radiation-dominated expression for t_{EQ} leads to the change $13 \rightarrow 9.8$.) Finally, the quantities $\Delta\theta(H^{-1}, z \gg 1)$ and $\Delta\theta(\lambda, z \gg 1)$ are the angles subtended on the sky today by the Hubble distance H^{-1} and the comoving scale λ at a red shift $z \gg 1$, respectively.

APPENDIX B

At early times the Universe was very hot and very dense, and thus fundamental physics must play a key role in understanding its earliest history. Particle physicists have their own standard model, the $SU(3)_C \otimes SU(2)_L \otimes U(1)_Y$ gauge theory of the strong, weak, and electromagnetic interactions. It is every bit as successful as the standard cosmology, and accounts for all known particle physics at energies below about 1 TeV in a fundamental way. Moreover, their standard model, in the splendor of its success, allows one to ask a new set of questions and to make speculations about physics at energies much greater than 1 TeV, corresponding to distances much less than 10^{-17} cm. In this short "primer" on modern particle theory we will describe the standard model and some of the interesting speculations—grand unification, supersymmetry, supergravity, superstrings, and composite models—that attempt to go beyond it. In the process we will emphasize three of the most important concepts in modern particle theory: symmetry, both as an organizing principle and as a principle for describing dynamics; gauge theory, the mathematical framework that incorporates symmetry into particle theory; and spontaneous symmetry breaking, the mechanism that allows Nature to have a much richer underlying symmetry than we are able to perceive. Our primer is by no means complete and is intended to introduce the neophyte to current ideas in particle theory. It is not meant to be a review of modern particle theory; for the reader interested in such we suggest the more complete expositions given in [1–15].

B.1 Quarks, Leptons, and Gauge Bosons

At the most fundamental level yet probed, the constituents of matter are quarks and leptons. Leptons are spin-1/2 particles that interact through the electromagnetic and weak interactions (electroweak interaction), but not the strong (color) interaction. Leptons come in pairs: the electron and its neutrino (e^-, ν_e), the muon and its neutrino (μ^-, ν_μ), and the tau lepton and its neutrino (τ^-, ν_τ). Quarks are spin-1/2 particles that interact through the strong and electroweak interactions, and they too come in pairs: up and down (u, d), charm and strange (c, s), and top and bottom (t, b). The "up" version of each pair has charge $+2/3$, and the "down" version of each pair has charge $-1/3$. Each quark type (or "flavor") comes in three colors. Color is a "three-dimensional" analogue of charge, and the force between colored particles is so strong that the only states that exist in Nature are color singlet combinations: quark triplets (one quark of each color) known as baryons, which include the proton (uud), the neutron (udd), the Λ (uds), and their myriad of friends; and quark-antiquark states known as mesons, the pi mesons ($\pi^+ = u\bar{d}$, etc.), the K mesons ($K^+ = u\bar{s}$, etc.), and their myriad of friends.

The energy of an isolated, colored particle is believed to be infinite; put another way, at large separations, $r \gg 10^{-13}$ cm, the potential energy of a quark-antiquark pair increases with r, roughly as $V(r) \sim \text{GeV}^2\, r$. The fact that free colored states of finite energy do not exist is known as *confinement*. At high temperature (or high baryon-number density) a phase transition from the confined phase to a deconfined phase (where the energy of an isolated colored particle is finite) is expected to take place.[1] The phase transition temperature is thought to be about 200 MeV. The confinement/deconfinement transition is currently being studied in lattice simulations of quantum chromodynamics (the theory of the color force).

Each quark-lepton pair is referred to as a generation or family. At present there is evidence for three complete families, with the important exception of the top quark. To date searches for the top quark have been unsuccessful and have set a lower limit to its mass of about 70 GeV. There is no fundamental understanding as to the number of families; and it could be that there are more than three. Since the Z^0 boson can decay into quark-antiquark and lepton-antilepton final states, its decay width depends upon the number of quark-lepton generations. Recent preliminary measurements of the Z^0 width at SLC and LEP are coming very close to ruling out the possibility of a fourth generation. Moreover, primordial nucleosynthesis constrains the number of light (mass $\lesssim$ MeV) neutrino species to be less than or equal to 4 (see Section 4.6). Thus it seems increasingly likely

[1] Actually, two phase transitions are expected (which are perhaps related): confinement/deconfinement and chiral symmetry breaking. It is still possible that one or both transitions are not really phase transitions at all, but rather a smooth transition like recombination.

that there are but three families of quarks and leptons.

The interactions of quarks and of leptons are mediated (or transmitted) by the exchange of spin-1 particles called *gauge bosons.* The photon (γ) mediates the electromagnetic interaction, the $W^{\pm}$ and Z^0 bosons mediate the weak interactions, and eight gluons (G) mediate the color interaction. Within the context of the standard model, all of these forces operate at the quark-lepton level. The so-called strong nuclear force that binds nucleons together is now interpreted as being the residual force between color neutral objects that are made of quarks, in the same way that the van der Waals force is the residual force between charge neutral objects that are made of charged particles.

All of the elementary particles yet discovered are either leptons, quark triplet states, quark-antiquark states, or gauge bosons.[2] The fact that the fundamental particles occur in multiplets—8 gluons, 3 pairs of quark color triplets, and 3 lepton pairs reflects the underlying symmetries of Nature that serve to organize the particle states. Moreover, these same symmetries also describe the dynamics of the fundamental interactions.

B.2 The Standard Model

By the early 1960's it was realized that symmetry plays an important role in elementary particle physics as an organizing principle. The flavor $SU(3)$ quark model,[3] in which the u, d, and s quarks are the triplet of fundamental constituents, organized all the then known hadrons (strongly interacting particles) into family multiplets—an octet of spin-0 mesons (which includes the π and K mesons), an octet of spin-1/2 baryons (which includes the neutron and proton), a decuplet of spin-3/2 baryons (which includes the $\Delta(3,3)$ resonance), and so on.

By the 1970's it was realized that the role of symmetry was much richer, describing both the dynamics as well as organizing the fundamental fields into family multiplets. Theories that implement this grand role for symmetry are known as *Yang-Mills gauge theories.* The gauge theory that describes the strong, weak, and electromagnetic interactions of the quarks

[2]It is also expected that color neutral objects comprised of gluons, called glueballs, should exist, and there is some evidence for such bound states.

[3]Flavor $SU(3)$ is now appreciated to be an *approximate,* global symmetry which arises accidentally because the three lightest quark species, u, d, and s, are nearly massless, and therefore approximately interchangeable entities for purposes of the strong interaction. The symmetries of most relevance today are *exact,* albeit, spontaneously broken.

and leptons at energies below about 1 TeV is the $SU(3)_C \otimes SU(2)_L \otimes U(1)_Y$ gauge theory. The $SU(3)_C$ part describes the strong (color) interaction and is known as quantum chromodynamics (QCD); the $SU(2)_L \otimes U(1)_Y$ part describes the electroweak interaction. Here C refers to color, L to left, and Y to (weak) hypercharge, and $SU(N)$ is the group of special unitary transformations on N objects.

To illustrate the basics of gauge symmetry and gauge theory, consider the familiar $U(1)_{EM}$ gauge theory of electromagnetism (quantum electrodynamics, or QED). The Lagrangian density for QED is

$$\mathcal{L}_{QED} = i\bar{\psi}\gamma_\mu D^\mu \psi - m_e \bar{\psi}\psi - \frac{1}{4}F^{\mu\nu}F_{\mu\nu}, \tag{B.1}$$

where the gauge-covariant derivative is $D_\mu = \partial_\mu + iqA_\mu$, the electromagnetic field-strength tensor is $F_{\mu\nu} = \partial_\mu A_\nu - \partial_\nu A_\mu$, A_μ is the electromagnetic potential (or gauge field), ψ is the spinor that represents the electron, q and m_e are the charge and mass of the electron, and γ_μ are the Dirac gamma matrices. This theory is called a gauge theory because the Lagrangian is invariant under the *position-dependent* gauge transformations

$$\begin{aligned} \psi &\rightarrow \exp[iq\chi(x)]\psi, \\ A_\mu &\rightarrow A_\mu - \partial_\mu \chi(x), \end{aligned} \tag{B.2}$$

where $\chi(x)$ is a scalar function of the space and time coordinates. Because $\chi(x)$ is permitted to depend upon on the space-time coordinate, this is referred to as a *local* $U(1)$ (or phase roatation) symmetry. Local gauge invariance necessitates the existence of the gauge field A^μ. Had χ been restricted to be *position independent*, the gauge field A^μ would not have been required; that is, the Lagrangian $\mathcal{L} = i\bar{\psi}\gamma_\mu\partial^\mu\psi - m_e\bar{\psi}\psi$ is invariant under *position-independent* phase rotations. Position-independent symmetry transformations are called *global* symmetries. The "promotion" of a global symmetry to a local symmetry requires the introduction of the gauge field; moreover, the existence of the gauge field leads to a force mediated by it.

The invariance of the Lagrangian under local transformations of an arbitrary gauge group is a straightforward generalization:

$$\begin{aligned} D_\mu &= \partial_\mu + igG^a W_\mu^a, \\ \psi &\rightarrow \exp[igG^a\chi_a(x)]\psi, \\ W_a^\mu &\rightarrow W_a^\mu - \partial^\mu\chi_a(x) - gc_{abc}\chi_b(x)W_c^\mu; \end{aligned} \tag{B.3}$$

where G^a are the generators of the group, $[G_a, G_b] = ic_{abc}G^c$, and c_{abc} are called the structure constants of the group. The field-strength tensor is

$F^a_{\mu\nu} = \partial_\mu A^a_\nu - \partial_\nu A^a_\mu - gc_{abc}A^b_\mu A^c_\nu$ and g is the gauge coupling constant. To summarize, the key features of local gauge invariance are: (i) The necessity of gauge fields to implement local gauge transformation invariance; (ii) the existence of gauge fields results in gauge interactions, i.e., interactions between particles that carry the charge of the gauge group; and (iii) the grouping of particles into multiplet representations of the gauge group.

The concept of multiplets, or families, of particles can be illustrated with a familiar example: the scalars, vectors, and tensors in ordinary three-dimensional space. Under the group of ordinary spatial rotations, known as $O(3)$, different quantities transform in different ways. Scalar quantities are unaffected by rotations, and are said to be singlets under $O(3)$; the three components of a vector transform amongst themselves, and are said to be the fundamental representation of $O(3)$; the components of a two-index tensor also transform amongst themselves and form a tensor representation (or realization) of the $O(3)$ group; and so on. In the same way the particles of Nature are grouped into multiplets or representations of the gauge group. Under gauge transformations different members of the multiplet transform into one another; stated in the language of "interactions," the gauge bosons (the quanta of the gauge fields) mediate interactions that transform multiplet members amongst themselves.

All the gauge bosons lie in the *adjoint* representations of the gauge group: The adjoint representation of $SU(3)_C$ has eight components, corresponding to the eight gluons; the adjoint representation of $SU(2)_L$ has three members, corresponding to the $W^\pm W^0$ bosons; and that of $U(1)_Y$ has one component, corresponding to the B^0 boson. (As we shall see soon the photon and Z^0 boson are linear combinations of the W^0 and B^0 bosons.) Under $SU(3)_C$, the quarks occur in triplets—the fundamental representation of $SU(3)$—and the leptons occur in color singlets—the scalars of $SU(3)$. So far as $SU(3)_C$ is concerned, there are six color triplet quarks (plus antiquarks), and six color singlet leptons. The gluons carry color, but not in the same way as the quarks do: Gluons carry a kind of color–anticolor, so that a gluon interacting with a quark results in the quark changing its color (red quark + blue–anti-red gluon $\rightarrow$ blue quark).

Before discussing the multiplet structure of the electroweak part of the theory we have to mention a very interesting feature of Nature: handedness or chirality. The fields that participate in the electroweak interaction are the left- and right-handed components of the quark and lepton fields:

$$\psi_L = \frac{1-\gamma_5}{2}\,\psi, \qquad \psi_R = \frac{1+\gamma_5}{2}\,\psi, \qquad \psi = \psi_L + \psi_R. \tag{B.4}$$

In the ultrarelativistic limit the fields $\psi_{R,L}$ have a simple physical interpretation: A left-handed fermion is one whose spin is anti-parallel to its momentum vector (negative helicity), and a right-handed fermion is one whose spin is parallel to its momentum vector (positive helicity).

Remarkably, under the electroweak gauge group the left- and right-handed projections of a fermion transform differently! Under $SU(2)_L$, the right-handed projections are singlets, while the left-handed projections form doublets; hence the notation $SU(2)_L$. For example, $(\nu_e)_L$, e_L^- form a doublet; u_L, d_L form a doublet; and so on. That the left- and right-handed components sit in different representations is responsible for parity violation (the parity operator transforms $\psi_R \leftrightarrow \psi_L$). Moreover, the fact that right-handed fields do not participate in $SU(2)_L$ interactions means that parity is violated maximally. The charge associated with $U(1)_Y$ is referred to as (weak) hypercharge and is just an analogue of ordinary charge. The left- and right-handed components of quarks and leptons have different hypercharge assignments so that parity is violated here too (though not maximally). Hypercharge is related to the third component of weak isospin t_3 and ordinary charge Q by

$$Q = t_3 + Y/2 \tag{B.5}$$

where $t_3 = 1/2$ for $u_L, c_L, t_L, (\nu_e)_L, (\nu_\mu)_L, (\nu_\tau)_L$—the upper components of the doublets, $t_3 = -1/2$ for $d_L, s_L, b_L, e_L^-, \mu_L^-, \tau_L^-$—the lower components of the doublets, and $t_3 = 0$ for the right-handed singlet fields. Note that the right-handed neutrino is a singlet under all the interactions of $SU(3)_C \otimes SU(2)_L \otimes U(1)_Y$ as it is a color singlet, $t_3 = 0$, $q = 0$, and $Y = 0$. So far as the standard model goes, ν_R does not exist!

The gauge bosons of $SU(2)_L$, $W^\pm W^0$, also carry weak isospin, and transform as a triplet ($I = 1$), with $t_3(W^+) = 1$, $t_3(W^0) = 0$, and $t_3(W^-) = -1$. The gauge boson of $U(1)_Y$, B^0, does not carry hypercharge or weak isospin. (This is a generic feature of Abelian[4] gauge symmetries: the gauge boson is uncharged.) The $W^\pm$ bosons transform members of weak doublets into one another: $u_L + W^- \to d_L$, $e_L^- + W^+ \to \nu_L$ (note how weak isospin is conserved in each interaction). The full structure of the standard model is shown in Table B.1.

An interesting and important feature of the standard model that we will not discuss in any detail is that of Cabibbo-Kobayashi-Maskawa (CKM) mixing. Namely, the fact that the quark flavor eigenstates and quark mass

[4]An Abelian group is one whose group elements commute. $U(1)$ transformations commute; $SU(N)$ transformations do not.

Particle	Q	t_3	Y	Color
u_L, c_L, t_L	2/3	1/2	1/3	triplet
d_L, s_L, b_L	$-1/3$	$-1/2$	1/3	triplet
$(\nu_e)_L$, $(\nu_\mu)_L$, $(\nu_\tau)_L$	0	1/2	-1	singlet
e_L^-, μ_L^-, τ_L^-	-1	$-1/2$	-1	singlet
u_R, c_R, t_R	2/3	0	4/3	triplet
d_R, s_R, b_R	$-1/3$	0	$-2/3$	triplet
$(\nu_e)_R$, $(\nu_\mu)_R$, $(\nu_\tau)_R$	0	0	0	singlet
e_R^-, μ_R^-, τ_R^-	-1	0	-2	singlet
8 gluons	0	0	0	octet
$W^\pm$	± 1	± 1	0	singlet
W^0	0	0	0	singlet
B^0	0	0	0	singlet
Φ	1, 0	$\pm 1/2$	1	singlet

$$\Leftarrow SU(3)_C \Rightarrow \qquad \Leftarrow SU(3)_C \Rightarrow \qquad \Leftarrow SU(3)_C \Rightarrow$$

$$SU(2)_L \Updownarrow \quad \begin{pmatrix} u_L^r & u_L^g & u_L^b \\ d_L^r & d_L^g & d_L^b \end{pmatrix} \begin{pmatrix} c_L^r & c_L^g & c_L^b \\ s_L^r & s_L^g & s_L^b \end{pmatrix} \begin{pmatrix} t_L^r & t_L^g & t_L^b \\ b_L^r & b_L^g & b_L^b \end{pmatrix}$$

$$SU(2)_L \Updownarrow \quad \begin{pmatrix} e_L \\ (\nu_e)_L \end{pmatrix} \qquad \begin{pmatrix} \mu_L \\ (\nu_\mu)_L \end{pmatrix} \qquad \begin{pmatrix} \tau_L \\ (\nu_\tau)_L \end{pmatrix}$$

Table B.1: The family structure and charge assignments of the standard model. The superscripts r, g, and b refer to the three different colors. Each quark and lepton has a corresponding antiparticle, related to it by the charge conjugation operator C. Since C is violated in the electroweak theory, it is most useful to discuss the CP conjugates of the quarks and leptons: Under CP, $u_L \to \bar{u}_R$, $e_L^- \to e_R^+$, and so on. The CP conjugates sit in precisely the same multiplets: i.e., e_R^+, $\overline{\nu}_R$ form an $SU(2)_L$ doublet, and so on.

eigenstates do not coincide. Because of this the weak interaction leads to transistions between quarks of the different generations: e.g., $c \to d$; $t \to s$; and $t \to d$.

As the mathematician Emmy (Amalie) Noether pointed out more than fifty years ago, there is an intimate relationship between symmetries and conservation laws: For every symmetry of the Lagrangian there is a corresponding conservation law. For example, the time- and space-translation invariance of the Lagrangian lead to energy-momentum conservation; and rotational invariance leads to angular momentum conservation. The standard model possesses enormous "internal" symmetry,[5] that of $SU(3)_C \otimes SU(2)_L \otimes U(1)_Y$. Correspondingly, the charges associated with the generators of the gauge group—color, weak isospin, and hypercharge—are conserved. (As we shall discuss in the next Section, $SU(3)_C \otimes SU(2)_L \otimes U(1)_Y$ is spontaneously broken to $SU(3)_C \otimes U(1)_{EM}$, so that only color and ordinary charge are exactly conserved.)

In addition to the quantities that are conserved because of gauge symmetry, the theory has several global symmetries that lead to additional conserved quantities. Baryon number (B) and the three lepton numbers (L_e, L_μ, L_τ) are exactly conserved, as the theory is invariant under global phase rotations associated with these quantities.[6] (Were it not for the fact the quark flavor and mass eigenstates are not identical the baryon number associated with each generation would be separately conserved.) Further, because the color interaction cares not about quark flavor, the strong interaction conserves flavor—that is, u-ness, d-ness, strangeness, charm, bottomness, and topness—and flavor change must occur through the weak interaction. For example, $n \to p + e^- + \bar{\nu}_e$, $\Lambda \to \pi^- + p$, and $\Lambda_C^+ \to \bar{K}^0 + p$ are all weak processes ($\pi^- = \bar{u}d$, $\Lambda_C^+ = udc$, and $\bar{K}^0 = \bar{d}s$).

One notices several interesting features of the quark and lepton family structure. (1) "Repetitiveness:" The pattern of quark pair, lepton pair is repeated three times. That is, there are three identical families of quarks and leptons. (2) There are two types of fundamental constituents—colored quarks and colorless leptons—which both participate in the electroweak interactions in the same way. (3) The "non-existence" of right-handed

[5]Gauge symmetries are often referred to as internal symmetries because gauge transformations correspond to rotations in an internal space, rather than in ordinary space.

[6]Within the standard model B conservation is violated by quantum mechanical effects associated with instantons (which arise from "triangle diagrams"). Because B is conserved classically but not quantum mechanically, baryon number is said to be an anomalous symmetry. Except possibly at temperatures of order 100 GeV, these effects are unimportant; see Section 6.8.

neutrinos: In the standard model right-handed neutrinos are without interactions; i.e., they are singlets under the complete gauge group. As far as the standard model goes, they are dispensible! (4) The interactions are "sewn together;" i.e., the gauge group is a product of three groups rather than a single, simple group. As we shall discuss in Section B.4, all of these facts are shortcomings associated with the $SU(3)_C \otimes SU(2)_L \otimes U(1)_Y$ theory, which like their counterparts in cosmology, point to something beyond the standard model.

B.3 The Higgs Sector

The standard model as just described is mathematically beautiful, but bears no resemblance to reality in several key respects: It predicts that the $W^{\pm}$ and Z^0 bosons are massless and that the weak force is long range—whereas $M_W \simeq 80$ GeV and $M_Z \simeq 91$ GeV and the weak force is very short range (less than 10^{-16}cm); it cannot accommodate the masses of the known fermions: $m_u \simeq 5$ MeV, $m_d \simeq 9$ MeV, $m_c \simeq 1.5$ GeV, $m_s \simeq 200$ MeV, $m_t \geq 70$ GeV, $m_b \simeq 5$ GeV, $m_e = 0.5$ MeV, $m_\mu = 106$ MeV, $m_\tau = 1.8$ GeV, $m_{\nu_e} \leq 13$ eV, $m_{\nu_\mu} \leq 0.25$ MeV, and $m_{\nu_\tau} \leq 35$ MeV. These difficulties delayed the emergence of the eletroweak model by more than a decade.[7]

The crucial feature of the standard model that allows the rich underlying symmetry of the model to be reconciled with the apparent lower degree of symmetry of the world we live in is spontaneous symmetry breaking (SSB). Not only does spontaneous symmetry breaking play a crucial role in the standard model, but it also plays a central role in virtually all theories that go beyond the standard model. Before we discuss the Higgs mechanism—the standard means of implementing SSB—we should warn the reader that this aspect of the standard model is the one for which there is essentially no direct experimental confirmation. All particle theorists would agree that there must be some means to degrade the underlying gauge symmetry. Whether it is precisely the Higgs mechanism described below is still an open question.

In discussing the Higgs mechanism and spontaneous symmetry breaking, we will focus on the $SU(2)_L \otimes U(1)_Y$ part of the standard model, as that is where SSB occurs; the $SU(3)_C$ part of the gauge symmetry is unbroken.[8] In our world, only the $U(1)_{EM}$ symmetry of electromagnetism

[7]The earliest version of the standard model, proposed by Glashow in 1961, resembled the model described in the previous Section.

[8]While the color force seems qualitatively different than the electromagnetic force, it

remains manifest: $SU(2)_L \otimes U(1)_Y$ is said to be broken to $U(1)_{EM}$, in the conventional notation, $SU(2)_L \otimes U(1)_Y \rightarrow U(1)_{EM}$. This feat is accomplished by a scalar field known as the Higgs field, for which there is as of yet no experimental evidence.[9]

The Higgs field required to break the electroweak symmetry to just that of electromagnetism is a complex $SU(2)_L$ doublet with hypercharge $Y = \pm 1$. From the relationship between hypercharge, charge, and isospin, we see that two of the four Higgs fields must have electric charge ± 1, and that the other two must be neutral (see Table B.1). The Higgs part of the Lagrangian is given by

$$\mathcal{L}_{\rm Higgs} = (D^\mu \Phi)(D_\mu \Phi) - V(|\Phi|), \tag{B.6}$$

where the non-derivative terms are collectively put in $V(|\Phi|)$ and referred to as the Higgs (or scalar) potential. Such non-derivative terms are not permitted for the fermion fields (they are forbidden by gauge and Lorentz symmetry), but are permissible for a scalar field. Moreover, the gauge symmetry of the electroweak theory does not specify the exact form of $V(|\Phi|)$. Supposing that it contains no terms higher than $|\Phi|^4$, as is required for renormalizability, the Higgs potential must be of the general form

$$V(|\Phi|) = -m^2|\Phi|^2 + \lambda|\Phi|^4 \tag{B.7}$$

where

$$\Phi = \frac{1}{\sqrt{2}}\begin{pmatrix} \phi_1 + i\phi_2 \\ \phi + i\phi_3 \end{pmatrix}, \tag{B.8}$$

$$|\Phi|^2 = \Phi^\dagger \Phi = (\phi_1^2 + \phi_2^2 + \phi^2 + \phi_3^2)/2, \tag{B.9}$$

ϕ_1, ϕ_2, ϕ_3, and ϕ are four real scalar fields, and the mass scale m^2 and the coupling λ are a priori arbitrary (also see Section 7.1.3). In fact, with foresight, we have chosen the sign of the quadratic term in V to be negative; that sign is essential to SSB.

In constructing a quantum theory one starts by finding the classical ground state (vacuum), and then expands all the fields about it. Here that entails minimizing the potential $V(|\Phi|)$. *Provided* that $-m^2 < 0$, the potential is minimized for a value

---

is not because of SSB; $SU(3)_C$ is unbroken and the 8 gluons are massless. The difference arises due to the non-Abelian nature of $SU(3)_C$, and in particular, the fact that the color force increases in strength at large distances.

[9]Nor has it been ruled out. The search for the Higgs boson is at the top of the priority list for the present and the next generation of accelerators.

$$|\Phi|^2 = m^2/2\lambda. \tag{B.10}$$

That is, in the classical vacuum state $|\Phi|$ takes on a non-zero value. As we shall see the consequences of this are manifold. (We should emphasize that had we chosen a positive coefficient for the $|\Phi|^2$ term, the classical vacuum state would be $|\Phi| = 0$.) Since Φ is a field that is a second-quantized operator, the fact that $|\Phi| \neq 0$ implies that the vacuum is filled with an indeterminate number of Higgs particles. That the vacuum is filled with a condensate of Higgs particles leads to spontaneous symmetry breaking.

Because $\langle|\Phi|\rangle \neq 0$, some of the gauge bosons of $SU(2)_L \otimes U(1)_Y$ acquire a mass. That this should occur is simple to understand, although we will not go through all the details. The key is the structure of the gauge-covariant derivative,

$$D_\mu \Phi = \partial_\mu \Phi + ig\frac{\tau_a}{2} W_\mu^a \Phi + ig'\frac{Y}{2} B_\mu \Phi, \tag{B.11}$$

where g is the $SU(2)_L$ coupling constant, g' is the $U(1)_Y$ coupling constant, W_μ^a are the three gauge fields associated with $SU(2)_L$, B_μ is the gauge field associated with $U(1)_Y$, and τ_a are the Pauli matrices. In squaring $D_\mu \Phi$, terms of the general form $|\Phi|^2 W_\mu W^\mu$ and $|\Phi|^2 B_\mu B^\mu$ arise; when $|\Phi|$ is replaced with its vacuum-expectation value, these terms correspond to a mass term for some of the gauge fields. This is a rather miraculous feat: We started with a doublet scalar field coupled in a gauge-invariant manner to *massless* gauge bosons, and because $\langle|\Phi|\rangle \neq 0$, some of the gauge bosons acquire masses.[10]

When all the dust settles the $W^\pm$ bosons acquire a mass of

$$M_W^2 = \frac{1}{4} g^2 \sigma^2, \tag{B.12}$$

the Z^0 boson a mass of

$$M_Z^2 = \frac{1}{4}(g^2 + g'^2)\sigma^2 = \frac{M_W^2}{\cos^2\theta_W}, \tag{B.13}$$

where $\sigma^2 = m^2/\lambda = (246 \text{ GeV})^2$ and $g'/g = \tan\theta_W$. The photon (A^μ) remains massless. In terms of the original neutral gauge fields W_μ^3 and B_μ, the photon (A^μ) and Z^0 boson (Z^μ) are given by

---

[10] One might ask why mass terms like $m^2 W_\mu W^\mu$ and $m^2 B_\mu B^\mu$ were not just included in the first place. They weren't because such terms spoil the renormalizability of the theory and thereby make it inconsistent. The miracle of the Higgs mechanism is that one can have the cake and eat it too: The "vacuum-induced" mass terms that arise *do not* spoil renormalizability.

$$Z^\mu = \cos\theta_W W^{3\mu} - \sin\theta_W B^\mu, \tag{B.14}$$

$$A^\mu = \sin\theta_W W^{3\mu} + \cos\theta_W B^\mu. \tag{B.15}$$

The electric charge of the positron is $e = g\sin\theta_W$, and the Fermi constant (over $\sqrt{2}$) $G_F/\sqrt{2} = g^2/8M_W^2 = m^2\lambda/2$. Moreover, there is one neutral Higgs boson of (tree-level) mass $M_H = \sqrt{2}\,m$. Remember that m is a priori arbitrary; as we discuss in Section 7.1.3 the Higgs mass could plausibly be anywhere between about 10 GeV and a few TeV. Through the Higgs mechanism, the four original Higgs fields have become the longitudinal components of the $W^\pm$ and Z^0 bosons and the neutral Higgs particle.

The Higgs mechanism allows the underlying gauge symmetry to be "hidden," and the $SU(2)_L \otimes U(1)_Y$ theory is said to be broken to $U(1)_{EM}$. The symmetry is broken because the vacuum—i.e., the lowest energy state—does not respect the full gauge symmetry: The state characterized by $|\Phi|^2 = m^2/\lambda$ is not invariant under $SU(2)_L$ or $U(1)_Y$, only $U(1)_{EM}$. Analogous phenomena occur in condensed matter physics; in fact the Higgs mechanism is borrowed from that discipline. Consider a ferromagnet; its free energy[11] can be written in terms of the bulk magnetization $\langle|\mathcal{M}|\rangle$. At low temperatures, the free energy is minimized by $\langle|\mathcal{M}|\rangle \neq 0$, that is, by having the individual spins aligned. The fact that $\langle|\mathcal{M}|\rangle \neq 0$ spontaneously breaks the $O(3)$ invariance of Maxwell's equations. At high temperatures (above the Curie temperature) entropy wins out over energetics, and the free energy of a ferromagnet is minimized by $\langle|\mathcal{M}|\rangle = 0$, that is, spins unaligned and $O(3)$ symmetry restored.

An even closer analogue to SSB in modern particle theory is that of superconductivity. Here the effective theory (Ginzburg-Landau theory) is given by a potential of the same form as (B.7), where Φ is a single complex scalar field and corresponds to the Cooper pair wave function. At very low temperatures $V(|\Phi|)$ is minimized by $\langle|\Phi|\rangle \neq 0$ and $U(1)_{EM}$ is spontaneously broken. The implication of $\langle|\Phi|\rangle \neq 0$ is that the vacuum is filled with a condensate (indefinite number) of Cooper pairs. At high temperatures $V(|\Phi|)$ is minimized by $\langle|\Phi|\rangle = 0$, and the gauge symmetry of electromagnetism is restored.

SSB in particle physics is analogous to that in condensed matter physics in essentially all respects. Just as the $O(3)$ invariance in a ferromagnet or the $U(1)_{EM}$ invariance in a superconductor is restored by finite-temperature effects, the full $SU(2)_L \otimes U(1)_Y$ invariance of the electroweak

[11] At zero temperature, the free energy of a gauge theory is given by $V(\Phi)$; at high temperatures, by the thermally-corrected, effective potential $V_T(\Phi)$; see Chapter 7.

theory should be restored at high temperatures. For the electroweak theory the symmetry restoration temperature is of the order of $\sigma \simeq 246$ GeV. (Symmetry restoration and its implications for cosmology are addressed in some detail in Chapter 7.)

Through SSB then, the electroweak theory is broken to $U(1)_{EM}$, and the underlying symmetry of the electroweak force is hidden: Only the gauge boson of electromagnetism remains massless, with the weak force mediated by the very massive $W^{\pm}$ and Z^0 bosons, which accounts for its short range. At distances much larger than about 10^{-16} cm (energies much less than 100 GeV), the weak interaction is effectively point-like and the theory can be written as an effective current–current or four–point interaction:

$$\mathcal{L}_{WEAK} = -\frac{G_F}{\sqrt{2}} J^{\mu\dagger}_{\rm neutral} J^{\rm neutral}_{\mu} - \frac{4G_F}{\sqrt{2}} J^{\mu\dagger}_{\rm charged} J^{\rm charged}_{\mu}, \tag{B.16}$$

where

$$\begin{aligned} J^{\mu}_{\rm charged} &= \overline{\psi}_L(\nu_e)\gamma^\mu\psi_L(e^-) + \overline{\psi}_L(u)\gamma^\mu\psi_L(d) + \cdots, \\ J^{\mu}_{\rm neutral} &= \overline{\psi}_L(\nu_e)\gamma^\mu\psi_L(\nu_e) - \overline{\psi}_L(e^-)\gamma^\mu\psi_L(e^-) + \overline{\psi}_L(u)\gamma^\mu\psi_L(u) \\ &\quad -\overline{\psi}_L(d)\gamma^\mu\psi_L(d) + \cdots - 2\sin^2\theta_W J^{\mu}_{EM}, \qquad \text{(B.17)} \\ J^{\mu}_{EM} &= \frac{2}{3}\overline{\psi}(u)\gamma^\mu\psi(u) - \frac{1}{3}\overline{\psi}(d)\gamma^\mu\psi(d) - \overline{\psi}(e)\gamma^\mu\psi(e) + \cdots. \end{aligned}$$

The charged-current interaction piece of the weak Lagrangian is the so-called Fermi theory, or $V - A$ theory; it is responsible for the charge-changing weak processes like $\nu_e + n \to e^- + p$. The neutral current piece is a prediction of the electroweak theory, one that was experimentally verified by experiments at CERN and Fermilab in 1973. It is responsible for charge-preserving interactions like $\nu_e + e^- \to \nu_e + e^-$ or $\nu_i + D \to n + p + \nu_i$, and neutral current processes can interfere with ordinary electromagnetic processes. The interference between the weak neutral current and the ordinary electromagnetic current has been seen in $e^- D$ scattering experiments at SLAC, in $e^{\pm}$ scattering experiments at SLAC, DESY, and KEK, and in parity-violating atomic transitions (in cesium).

From the form of $\mathcal{L}_{\rm WEAK}$, it is clear that all weak cross sections should be proportional to G_F^2 (at energies well below M_W, M_Z). Since G_F^2 has dimensions of $(\text{energy})^{-4}$ and a cross section has dimensions of $(\text{energy})^{-2}$, on the basis of dimensional considerations it follows that $\sigma \propto G_F^2 E^2$, where E is some characteristic energy in the problem. For example, the thermally-averaged cross section for $e^+e^- \to \nu\overline{\nu}$ varies as $G_F^2 T^2$ (for $T \gg m_e$).

Finally, let us consider the problem of fermion masses. A fermion mass term is of the form $m\overline{\psi}\psi$ and can also be written as $m\overline{\psi}\psi = m\overline{\psi}_L\psi_R + m\overline{\psi}_R\psi_L$. That is, at the operator level the effect of a mass term is to create a left-handed particle and destroy a right-handed particle (or vice versa).[12] Because the electroweak theory is chiral—right-handed and left-handed components sit in different representations—such a term is forbidden by gauge symmetry: Under $SU(2)_L$ ψ_L is a doublet, while ψ_R is a singlet.

The Higgs comes to the rescue! A gauge singlet term can be formed from ψ_L, ψ_R, and Φ,

$$\mathcal{L}_{\text{fermion}} = -h\overline{\psi}_L\Phi\psi_R. \tag{B.18}$$

Such a term is an $SU(2)_L$ singlet because both Φ and ψ_L are $SU(2)_L$ doublets. The coupling constant h is called the fermion's Higgs or Yukawa coupling and is a priori arbitrary. When the Higgs doublet takes on its vacuum expectation value, $\mathcal{L}_{\text{fermion}}$ corresponds to a fermion mass term, and the fermion acquires a mass

$$m_f = \frac{h\sigma}{\sqrt{2}} = \frac{h}{g}\sqrt{2}\,M_W. \tag{B.19}$$

We see that the Higgs mechanism can also give fermions mass, albeit of size depending upon the arbitrary Yukawa couplings h.

One can also see why, in the context of the electroweak theory, neutrinos are massless: There is no ν_R. It is possible in an extended version of the electroweak model—the so-called Majoron model—to give neutrinos mass, through a term

$$-h\,\left[\psi_L(\nu)\,\Sigma\,\psi_L(\nu) + \overline{\psi}_L(\nu)\,\Sigma^\dagger\,\overline{\psi}_L(\nu)\right], \tag{B.20}$$

where Σ is a Higgs triplet. Such a term turns a neutrino (ν_L) into an antineutrino ($\overline{\nu}_R$), or equivalently creates two neutrinos or antineutrinos, and thus violates lepton number. This type of mass is referred to as a Majorana mass; in the "economical" Majorana representation for the neutrino $\nu = \overline{\nu}$, and neutrino and antineutrino are distinguished *only* by their handedness. Charged fermions cannot have a Majorana mass, as charge conservation forbids the possibility of a mass term like $m\overline{\psi}(e^-)_L\overline{\psi}(e^-)_L$.

[12] That a mass term should do this is straightforward to understand: For a massive particle the spin projection along the momentum can be changed by a Lorentz boost antiparallel to the particle's momentum. For a massless particle this is not true as no Lorentz boost can "turn" the direction of the particle around.

Spontaneous symmetry breaking is a key element of the standard model. However, the ultimate experimental test of the Higgs mechanism, the discovery of the Higgs particle, still remains. Further, SSB is also a key element of essentially every model proposed that goes beyond the standard model—grand unification, supersymmetry, superstrings, family symmetry, etc. Much rides on this clever but untested idea.

In concluding our discussion of the standard model we should warn the reader that there are a number of important features of the standard model that we have not discussed; they include CP violation, CKM mixing, flavor-changing neutral currents (FCNC), the GIM mechanism for suppressing FCNC, nucleon structure functions, the computation of the mass spectrum of the bound states of QCD, and the experimental tests of the model. We refer the interested reader to the other more complete expositions in [1–15].

B.4 Beyond the Standard Model

The standard model of particle physics is a remarkable achievement. It provides a fundamental theory of the non-gravitational interactions of quarks and leptons valid up to energies of order 1 TeV. Moreover, it is consistent with the great wealth of experimental data that exists at present. In the glory of its success it allows one to ask a host of new and even more profound questions and, upon closer examination it has a number of shortcomings. This is very reminescent of the standard cosmology.

Included in its shortcomings and in the list of questions that it allows one to ask are the following.

- The model is not really unified as there are three coupling constants, g, g', and g_S. This is because the gauge group is a product of three factors.
- Quarks and leptons are not unified as they appear as separate entities, in different multiplets. Because of this fact, there is no a priori reason to expect the magnitude of the electron and proton charges to be equal, yet they are to many significant figures.
- The model contains at least 21 parameters—quark masses, the $\bar{\Theta}$ angle of QCD, CKM mixing angles, θ_W, etc.—which are not predicted by the theory itself and must be put in by hand.
- There is no explanation for the family structure of three (or more) generations; as I. I. Rabi put it, "who ordered the muon?" Moreover, so far as everyday life goes, the first generation is all that we require. Related, is the question of fermion masses; in the standard model masses are "dialed in" as Yukawa couplings.
- Why is the theory left-handed?

- Where does the weak scale, $\sigma \simeq 246$, come from, why is it so small compared to the other energy scale in physics, the Planck scale, and how is it to be stabilized against radiative corrections?
- Where does gravity fit in?
- Why is the vacuum energy so small today: $\rho_{\rm VAC} \lesssim 10^{-46}\ {\rm GeV}^4$?

Just like the standard cosmology these shortcomings/questions point to a grander theory that goes beyond the standard model. Moreover, the mathematical tools at hand—gauge theories, spontaneous symmetry breaking, and more recently supersymmetry and superstrings—are so powerful that particle physicists have proposed a host of very attractive theories that go beyond the standard model and address most, if not all, of these issues.

B.4.1 Minimal extensions

To begin, there are a handful of simple extensions of the standard model that have been put forth to address very specific issues. They include the following.

Peccei-Quinn (PQ) symmetry. As we discuss in Chapter 10 in great detail, PQ symmetry was proposed to solve the strong-CP problem of QCD, namely the fact that $\bar{\Theta} \lesssim 10^{-10}$. In order to implement PQ symmetry one additional Higgs doublet is required; in many unification schemes a second Higgs doublet arises automatically.

Majoron models. As we discussed in the previous Section, adding an additional Higgs triplet to the electroweak theory allows one to give the neutrinos mass. This Higgs too acquires a vacuum expectation value, but typically it must be very small, less than about 10 keV! Another scale to explain.

Right-handed interactions. It seems unfair that the lefties have all the fun. Often as an attempt to restore left-right symmetry, an analogous right-handed structure is added to the theory, with additional W and Z bosons for the right-handed interactions. In some schemes of grand unification these additional right-handed interactions arise automatically. In any case, the masses of the "right-handed" W and Z bosons must be in excess of about 1 TeV to be consistent with both laboratory and cosmological data.[13]

---

[13]In such theories the effective number of neutrino types doubles as there are both left- and right-handed neutrinos. Only the $(T_X/T)^4$ factor can keep such models consistent with the nucleosynthesis bound $N_\nu \leq 4$, cf. (4.40).

B.4.2 Grand unification

This is the modest goal for the Yang-Mill gauge theories that unify the strong, weak, and electromagnetic interactions as well as the quarks and leptons. Collectively, such theories, which are based upon some simple group G that is broken to $SU(3)_C \otimes SU(2)_L \otimes U(1)_Y$ by the Higgs mechanism, are referred to as GUTs, for grand unified theories.

At first sight such a unification might seem very implausible as the intrinsic strengths of the three interactions are very different; in terms of their coupling strengths, at laboratory energies: $\alpha_S \gg \alpha_2 \gg \alpha_{EM}$, where $\alpha_S = g_S^2/4\pi$ is the $SU(3)_C$ color "fine structure constant," $\alpha_2 = g^2/4\pi$, and $\alpha_{EM} = e^2/4\pi \simeq 1/137$. However, owing to vacuum polarization effects coupling constants are not constant, but "run" (vary) with energy. The variation of a coupling constant with energy (or distance) is given by a renormalization group equation

$$\beta \equiv \frac{\partial g}{\partial \ln \sqrt{-q^2/\mu^2}},$$

$$\beta = -b_0 \frac{g^3}{16\pi^2} - b_1 \frac{g^5}{(16\pi^2)^2} - \cdots, \tag{B.21}$$

where $q^2 < 0$ is the momentum transfer (which sets the energy/distance scale) and μ is some convenient energy scale. Using this, it follows that

$$\alpha_{EM}(q^2) = \frac{\alpha_{EM}(m_e^2)}{1 - [\alpha_{EM}(m_e^2)/3\pi]\ln(-q^2/m_e^2)}; \tag{B.22}$$

from which we see that α_{EM} *increases* logarithmically with energy. That is, at short distances the electromagnetic interaction strengthens.

The situation for non-Abelian gauge theories can be very different; for QCD the evolution of the coupling strength is given by

$$\alpha_S(q^2) = \frac{\alpha_S(\mu^2)}{1 + [(33 - 2N_f)\alpha_S(\mu^2)/12\pi]\ln(-q^2/\mu^2)}, \tag{B.23}$$

where N_f is the number of quark flavors. The sense of the evolution is opposite that of α_{EM} (provided that $N_f < 17$): $\alpha_S(q^2)$ tends to zero as $-q^2 \to \infty$—known as asymptotic freedom—and diverges for some small value of $-q^2$—known as infrared slavery. (The scale at which the QCD coupling constant becomes very large is denoted by Λ_{QCD}; the value of Λ_{QCD} is between 100 and 400 MeV.) This qualitatively different behavior owes to the fact that in non-Abelian gauge theories the gauge bosons themselves carry the charge of the theory and thus contribute to vacuum polarization (with the opposite sign).

Because the coupling constants in gauge theories "run" with energy, and at different rates, it is possible for the three coupling constants—α_S, α_2, and α_{EM}—to evolve to a common value; and they do, at an energy of about 3×10^{14} GeV and value $\alpha_{GUT} \simeq 1/45$.[14] Moreover, since the weak mixing angle θ_W is given by $\tan^2\theta_W = \alpha_1/\alpha_2$, it too evolves with energy, increasing to a value of $\sin^2\theta_W = 3/8$ at the "unification scale."

The hint that unification of the strong, weak, and electromagnetic with a Yang-Mills gauge theory is on the right track is the convergence of the three coupling constants at an energy scale of about 3×10^{14} GeV. The convergence is independent of the unification group: The only assumption made is that only the known quarks, leptons, gauge and Higgs bosons contribute to vacuum polarization (and thereby the running of the couplings). The convergence of the couplings is a necessary condition for unification; in addition, one needs a model that can accommodate the known particles.

As it turns out there are many simple groups G that will do the trick, e.g., $SU(5)$, $SO(10)$, $E6$, $E8$, $\cdots$. Here we will focus on the simplest and most economical unification scheme: $SU(5)$. In spite of the fact that its prediction for the proton lifetime has been falsified, we use $SU(5)$ as an example because of its simplicity and the lack of a better candidate.

In $SU(5)$ each generation of quarks and leptons is assigned to the reducible 15-dimensional representation $\overline{\mathbf{5}} + \mathbf{10}$ where $\mathbf{5}$ is the fundamental representation. The quarks and leptons of the first family are accommodated as follows

$$\begin{aligned} \overline{\mathbf{5}} &\supset \overline{d}^1_L,\ \overline{d}^2_L,\ \overline{d}^3_L,\ e^-_L,\ (\nu_e)_L, \\ \mathbf{10} &\supset u^1_L,\ u^2_L,\ u^3_L,\ d^1_L,\ d^2_L,\ d^3_L,\ \overline{u}^1_L,\ \overline{u}^2_L,\ \overline{u}^3_L,\ \overline{e}_L, \end{aligned} \tag{B.24}$$

where $i = 1, 2, 3$ is the color index and $\overline{e} = e^+$. The fifteen right-handed CP conjugate fermion fields are accommodated in a $\mathbf{5} + \overline{\mathbf{10}}$ representation. Note that each generation fits snugly into a $\overline{\mathbf{5}} + \mathbf{10}$ representation with no room for any new particles.[15]

---

[14]It is not actually the coupling constants α_S, α_2, and α_{EM} that converge; rather, it is α_S, α_2, and $\alpha = (5/3)\alpha_1$ where $\alpha_1 = g'^2/4\pi$. ($5\alpha_1/3$ is used instead of α_1 for technical reasons.) At the unification scale, where the coupling constants converge, $\sin^2\theta_W$ must be precisely 3/8. A convenient means of testing the convergence is to predict $\sin^2\theta_W$ at low energies. The prediction, $\sin^2\theta_W 0.216 \pm 0.003$, is very close to the present experimental value, $\sin^2\theta_W = 0.226 \pm 0.005$.

[15]In some GUTs new fermions are predicted, e.g., in $SO(10)$ the right-handed neutrino N_R.

The gauge bosons are assigned to the 24-dimensional adjoint representation. Since the standard model has but twelve gauge bosons (eight gluons, $W^{\pm}Z^0$, and γ) there are twelve new gauge bosons in $SU(5)$, dubbed the X, Y gauge bosons. The X, Y gauge bosons are rather odd creatures, sometimes referred to as diquarks or leptoquarks. Classified according to their $SU(3)_C \otimes SU(2)_L \otimes U(1)_Y$ quantum numbers they are

$$\begin{pmatrix} Y^{-1/3} \\ X^{-4/3} \end{pmatrix} \quad \text{isodoublet, color triplet, } Y = -5/3,$$

$$\begin{pmatrix} \bar{X}^{4/3} \\ \bar{Y}^{1/3} \end{pmatrix} \quad \text{isodoublet, color anti-triplet, } Y = +5/3, \tag{B.25}$$

These new gauge bosons have fractional electric charge and carry color. They are sometimes called leptoquarks or diquarks because they mediate transitions between quarks and leptons or quarks and antiquarks within a given multiplet; e.g.,

$$\begin{aligned} e_L^- &\rightarrow \bar{d}_L + X^{-4/3}, \\ \bar{u}_L &\rightarrow u_L + X^{-4/3}, \\ (\nu_e)_L &\rightarrow \bar{d}_L + Y^{-1/3}, \\ \bar{d}_L &\rightarrow u_L + Y^{-1/3}, \\ u_L &\rightarrow \bar{e}_L + Y^{-1/3}. \end{aligned} \tag{B.26}$$

It is impossible to assign the X, Y gauge bosons lepton and baryon numbers so that these new interactions conserve B and L; thus, the new interactions in $SU(5)$ violate B and L. In fact this prediction is generic to grand unification: Because quarks and leptons sit in the same representations, there must be new gauge bosons that mediate processes that violate B and L conservation.

In our $SU(5)$ example, processes mediated by X, Y exchange like

$$\begin{aligned} \bar{u} + \bar{u} &\rightarrow d + e^- \quad (\Delta B = \Delta L = 1), \\ \bar{u} + \bar{d} &\rightarrow d + \nu_e \quad (\Delta B = \Delta L = 1), \end{aligned} \tag{B.27}$$

violate B and L by one unit.[16] (Due to an accidental, global symmetry $SU(5)$ gauge interactions do conserve $B - L$.) Because of these new interactions the proton is unstable! It can decay through X exchange by

[16]In some GUTs, processes that violate $B - L$ and B by two units are predicted and give rise to neutron-antineutron oscillations. Such an interaction is equivalent to

$$p \rightarrow \pi^0 + e^+, \tag{B.28}$$

which at the quark level is $uu(d) \rightarrow e^+\overline{d}(d)$ where the quark in parentheses goes along as a "spectator."

Of course the full symmetry of the GUT cannot be manifest; if it were the proton would decay in 10^{-24} sec. The gauge group G must be spontaneously broken to $SU(3)_C \otimes SU(2)_L \times U(1)_Y$.[17] For $SU(5)$, this is accomplished by a 24-dimensional Higgs representation, which leads to masses of order the unification scale for the twelve X, Y gauge bosons. Thus, like the weak interaction, at energies below 10^{14} GeV or so the processes mediated by X, Y boson exchange can be treated as a four-fermion interaction with strength $G_{\Delta B} \simeq \alpha/M^2 \simeq 10^{-31}\ \mathrm{GeV}^{-2}$, where $M \simeq 3 \times 10^{14}$ GeV is the unification scale. Since $G_{\Delta B}$ is extremely small compared to $G_F \simeq 1.17\times10^{-5}\,\mathrm{GeV}^{-2}$, these new B, L violating interactions are extremely weak at energies below 10^{14} GeV. The decay rate of the proton must be proportional to $G^2_{\Delta B}$, so that on dimensional grounds the proton lifetimes must be

$$\tau_p = \Gamma_p^{-1} \sim (G^2_{\Delta B} m_N^5)^{-1} = \alpha^{-2} M^4 m_N^{-5} \sim 10^{31}\ \mathrm{yr}. \tag{B.29}$$

When the various theoretical uncertainties are taken into account, the $SU(5)$ prediction for the proton lifetime is $10^{30\pm1.5}$ yr. The current limits to proton longevity are in excess of 10^{32} yr, which seems to rule out the $SU(5)$ GUT.

Additional Higgs bosons are required in order to give masses to the quarks and leptons, at the very least one complex 5-dimensional Higgs. The 5-dimensional Higgs contains the usual doublet Higgs required for electroweak SSB and a color triplet Higgs $H^{\pm1/3}$, which can also mediate B, L violation. The triplet component must acquire a mass comparable to M to guarantee the proton's longevity, while the doublet component must acquire a mass of order a few 100 GeV to trigger electroweak SSB at the appropriate scale. We shall return to this dilemma.

In GUTs the quarks and leptons are assigned to common multiplets and so relations among the fermion masses are predicted; as before the overall Yukawa couplings are still arbitrary. In $SU(5)$ there is an apparently

---

giving the neutron a Majorana mass (δm); the $n - \bar{n}$ oscillation time $\tau_n = \delta m^{-1}$. On dimensional grounds the nuclear annihilation rate due to $n - \bar{n}$ oscillations is $\Gamma \sim \delta m^2/m_N$. The current limit to the nucleon lifetime implies that $\tau_n \gtrsim 10^8$ sec or $\delta m \lesssim 10^{-23}$ eV.

[17]In general, there can be various intermediate stages of symmetry breaking, e.g., $SO(10) \rightarrow SU(4) \otimes SU(2)_L \otimes SU(2)_R \rightarrow SU(3)_C \otimes SU(2)_L \otimes U(1)_Y$.

successful prediction for the b-quark mass: At the unification scale $m_b = m_\tau$; however, Yukawa couplings too run with energy and the low energy prediction for m_b is 5 to 5.5 GeV.

In the simplest version of $SU(5)$ neutrinos remain massless. However, in more complicated versions of $SU(5)$ and in GUTs in general the neutrino can develop a small mass. The scheme for such, known as the "see-saw mechanism," is very clever. In $SU(5)$ there is no place for the right-handed neutrino, here denoted by N_R; however, it can be added as an $SU(5)$ singlet. In other GUTs N_R fits into the theory as an $SU(3)_C \otimes SU(2)_L \otimes U(1)_Y$ singlet. Once the theory has a right-handed neutrino a Dirac mass term $(m\,\overline{\nu}_L N_R)$ for the neutrino is possible. Moreover, because N_R is an $SU(3)_C \otimes SU(2)_L \otimes U(1)_Y$ singlet it can have a large Majorana mass term: $M\,N_R N_R$.[18] The neutrino mass terms can be written in matrix form as

$$(\nu_L\ \overline{N}_R)\begin{pmatrix} 0 & m \\ m & M \end{pmatrix}\begin{pmatrix} \nu_L \\ \overline{N}_R \end{pmatrix}. \tag{B.30}$$

When diagonalized, the result is two mass eigenstates: $m_1 = m^2/M$ and $m_2 \simeq M$, where ν_1 is ν_L with a small, $\mathcal{O}(m^2/M^2)$, admixture of N_R, and ν_2 is N_R with the same small admixture of ν_L.

One would expect the Dirac mass m to be of the same size as the other fermion masses—anywhere from m_e^2 to m_t^2—and the Majorana mass M can be as large as the unification scale—or in some schemes equal to some "intermediate" scale between the weak and GUT scales. Thus, one expects left-handed neutrinos to have masses that are very small compared to other fermion masses, smaller by a factor of m/M,

$$m_\nu \simeq \frac{m^2}{M} \sim \frac{(m/\mathrm{GeV})^2}{(M/10^9\mathrm{GeV})}\ \mathrm{eV}. \tag{B.31}$$

The see-saw mechanism explains why, if neutrinos do have mass, their masses are so small (relative to other fermion masses) and suggests that they are tantalizingly close to being of cosmological interest!

In sum, the successes of grand unification are:

- The unification of the strong and electroweak interactions.
- The unification of the quarks and leptons, thereby explaining why $|q(e^-)| = |q(p)|$.

---

[18]In the absence of an $SU(2)_L$ isotriplet of Higgs, ν_L cannot have a Majorana mass term.

- The successful "prediction" of $\sin^2\theta_W$.[19]
- The successful prediction of the b-quark mass.
- Explanation of why neutrino masses are so small (if they are not zero).

Moreover, GUTs have important cosmological implications. As we discuss in Chapter 6, the prediction of B-violating interactions leads to a framework for understanding the origin of the baryon asymmetry. In addition, in GUTs cosmologically interesting topological beasts such as monopoles, strings, and domain walls can arise.

On the other hand, GUTs provide no particular insight as to the resolution of the generation problem, the origin of CP violation, the unification of gravity, nor are the number of required "input parameters" substantially reduced. In addition, there is no particular prediction for the unification gauge group, and $SU(5)$, the simplest GUT, seems to be ruled out.

B.4.3 Supersymmetry/Supergravity

Supersymmetry is the symmetry that relates fermions and bosons. Moreover, when supersymmetry is "promoted" from a global to a local symmetry, known as supergravity, it leads to a theory with general coordinate invariance. General coordinate invariance is of course the hallmark of general relativity. Thus, supersymmetry in its form as a local symmetry offers the promise of unifying gravity with the other forces!

The particle multiplet structure of a supersymmetry theory consists of boson-fermion pairs: (0, 1/2), (1/2, 1), (3/2, 2). Unfortunately, the known particles cannot be paired in such a way. Thus, if supersymmetry has anything to do with our world it must be spontaneously broken and supersymmetric partners—or superpartners—must exist for all the known particles. Since "spartners" have not yet been discovered, they must be sufficiently massive so that they would not have been found (typically more massive than 50 to 100 GeV).

The simplest supersymmetric theory is a supersymmetrized version of the standard model. Supersymmetric GUTs can also be constructed. To supersymmetrize the standard model one must add: six spin-0 partners for the known quarks—called squarks for scalar quarks; six spin-0 partners for the known leptons—called sleptons for scalar leptons; twelve spin-1/2 partners for the gauge bosons—called gauginos; spin-1/2 partners for the Higgs bosons—called higgsinos; and a spin-3/2 partner for the graviton—

[19] Although the prediction and the measured value differ by about 2σ.

called the gravitino. The masses of these new particles are expected to be of order the weak scale for reasons described below. Moreover, because of a global symmetry often imposed upon supersymmetric theories (R parity),[20] the lightest neutral sparticle—called the neutralino—must be stable.[21] Because the scale of supersymmetry breaking is of order the weak scale, the sparticles interact with ordinary matter with roughly weak strength. This makes the neutralino an ideal dark matter candidate (see Sections 5.2, 5.6).

Supersymmetric theories have a technical feature that makes them attractive for another reason: stabilizing the hierarchy problem. The hierarchy problem involves the existence of at least three apparently fundamental energy scales in particle physics: the weak scale, 300 GeV; the GUT scale, 10^{14} GeV; and the Planck scale, 10^{19} GeV. Radiative corrections to the masses of scalar particles are quadratically divergent, and one would naïvely expect the masses of all scalar particles to be driven to the highest possible energy scale—the Planck scale. This of course would be disastrous. (It is possible, by very "delicate" adjustment of the parameters of the theory, to ensure that even in the face of radiative corrections the electroweak Higgs doublet remains light. Such a "fine tuning" of parameters is seen as abhorrent by many particle theorists.[22])

In supersymmetric theories, the radiative corrections due to fermions and bosons cancel miraculously, thereby stabilizing any mass hierarchies that exist. Unfortunately, supersymmetry does little—or nothing—to account for the apparent hierarchy of energy scales in the first place. In a supersymmetric theory you can set the mass hierarchy and forget it; radiative corrections will not upset it. In order to stabilize the weak scale, supersymmetry must remain unbroken down to an energy scale of order the weak scale, which is why the scale of supersymmetry breaking is expected to be of order the weak scale. The wish that supersymmetry be broken at the weak scale explains why the sparticle masses are expected to be comparable to the weak scale. Moreover, the fact that the known world of elementary particles does not exhibit supersymmetry, at least at

---

[20] R parity is imposed to forbid the existence of a class of dimension-4 operators that would lead to very rapid proton decay.

[21] In general the neutralino is a linear combination of the photino, zino, and higgsino fields; in some of the simplest supersymmetric theories the neutralino is almost purely a photino.

[22] Fine tuning refers to the fact that to achieve the necessary cancellations parameters have to be chosen with extreme precision. Put another way, the final result for the Higgs mass is unstable to very small changes in any of the parameters.

mass scales up to 100 GeV or so, implies that the scale of supersymmetry breaking must be comparable to the weak scale or larger.

In addition, in supersymmetric theories the fermionic and bosonic contribution to the vacuum energy also cancel. So long as supersymmetry is "in force" one has a reason why $\rho_{VAC} = 0$. Since supersymmetry must be broken at a scale greater than about 100 GeV, supersymmetry can at best guarantee that $\rho_{VAC} \lesssim (100\,\mathrm{GeV})^4$, which misses solving the cosmological constant problem by some 54 orders of magnitude—although it could alleviate the problem by 68 orders of magnitude.

B.4.4 Superstring theory

As in we discuss in Section 11.1, the superstring is a quantum field theory of one-dimensional entities (strings) that unifies *all* the forces of Nature (including gravity) in a self-consistent and finite[23] quantum theory. Quite an achievement. Superstring theories incorporate both supersymmetry and unified gauge theories and must be formulated in 10 or 26 dimensions (6 or 22 additional spatial dimensions). The most attractive versions of the superstring are based upon the gauge groups $E8 \otimes E8$ and $SO(32)$. At "large" distances (much larger than 10^{-33} cm) and "low" energies (much less than 10^{19} GeV) it is expected that the theory reduces to a unified quantum field theory of point-like quarks and leptons; it is hoped that superstring theory will predict *the* GUT. The mathematics underlying string theory is very rich and beautiful (translation: complicated and not fully understood), and so thus far this hope has yet to be realized, although some strides have been made. For example, in some versions of the superstring the number of families and the values of fermion Yukawa couplings are determined by topological considerations. Whether or not superstring theory has anything to do with reality remains to be seen. In any case the superstring is a landmark achievement as it is the first example of a self-consistent theory that unifies all the interactions of Nature.

B.4.5 Kaluza-Klein theories

As we discuss in Section 11.1, Kaluza-Klein theories realize Einstein's dream of unifying the interactions of Nature geometrically. These theories are based upon a higher-dimensional version of general relativity. The extra-spatial dimensions form a compact manifold whose characteristic size must be less than $\mathcal{O}(10^{-16}\,\mathrm{cm})$ to have escaped detection, and they are

[23]Gauge theories are not finite; infinities arise and must be renormalized away.

typically expected to be $\mathcal{O}(10^{-33}$ cm). The isometries (space-time symmetries) of the compactified dimensions give rise to the gauge symmetries of four dimensions. In order to explain the standard model, seven additional spatial dimensions are required. These theories have fallen out of favor for several reasons. First, a quantum formulation of the four-dimensional Einstein theory is plagued by infinities that cannot be renormalized away, and the problem is worse in higher dimensions. Second, such theories do not seem to be able to accomodate chiral fermions, and the known quarks and leptons exhibit handedness.

B.4.6 Composite models

The hundreds of hadronic resonances that exist motivated the idea that hadrons were made of something more fundamental—quarks. Three generations (perhaps more) and the existence of two kinds of fundamental entities—quarks and leptons—suggests to some that quarks and leptons may be made of more fundamental building blocks, called by some, *preons.* Numerous such models have been proposed that are able to explain the "spectroscopy" of the known quark and lepton states; however, dynamics has posed serious problems for preon models. A fundamental hurdle is the fact that an object that has structure on the scale d, should naturally have a mass of order d^{-1}. Current experiments indicate that any substructure of quarks and leptons must be on scales smaller than about 10^{-17} cm, corresponding to a mass scale of more than a TeV. While there are ways of protecting composite quarks and leptons from acquiring such large masses (supersymmetry, chiral symmetry, or as Nambu-Goldstone states), composite models have not met with any great success. (These models also face the same problems that family symmetry does with regard to FCNC; see below.)

Another variation on this theme is the idea that only the Higgs particles are composite entities, bound states of techni-quarks held together by a stronger version of the color force called technicolor. The idea is attractive in that it eliminates the Higgs sector in favor of another gauge sector (recall that the gauge sector of a theory has little or no arbitrariness once the gauge group is specified). Technicolor models can account for essentially all the features of the standard model *except* fermion masses. In order to account for fermion masses the models have to become very complicated. For this reason, technicolor models have not met with any great success. Technicolor models do have the great virtue that they can be tested with the current and next generation of particle accelerators: A technicolor version of the Higgs should have very different properties than the standard Higgs.

B.4.7 Family symmetry

One of the fundamental questions in particle theory is that of the family structure of quarks and leptons. Composite models address this problem through quark and lepton substructure. The approach of family symmetry is very different. Quarks and leptons are supposed to be fundamental with *family or horizontal* symmetries relating the different generations. Both global and gauge symmetries have been proposed, e.g., $SU(2)$, $O(3)$, $SU(3)$ and $SU(3) \otimes SU(3)$.[24] In either case, new interactions arise that mediate transitions between the families, e.g., $e^- \leftrightarrow \mu^- \leftrightarrow \tau^-$ or $d \leftrightarrow s \leftrightarrow b$, so-called flavor-changing neutral currents (FCNC).[25] Such processes, if they exist, are very rare. For example, consider the limits to the branching ratios for the following FCNC processes: $\mu^- \rightarrow e^-\gamma$ ($< 5 \times 10^{-11}$); $\mu^- \rightarrow e^-e^+e^-$ ($< 10^{-12}$); $K_L \rightarrow \mu^+\mu^-$ ($< 10^{-8}$); $K^+ \rightarrow \pi^+\nu\bar{\nu}$ ($< 1.4 \times 10^{-7}$). Because FCNC processes are very rare and because the masses of the different generations are very different, any family symmetry that might exist must be spontaneously broken at a very large energy scale (typically greater than 10^9 GeV). In the case that the family symmetry is a global symmetry, new massless scalar (or pseudo-scalar) particles are predicted to exist. (That such bosons should arise traces to Goldstone's theorem: For each broken generator of a global symmetry there must be a massless boson, called a Nambu-Goldstone boson.) These particles, sometimes referred to as *familons*, are very weakly coupled (coupling proportional to the inverse of the family symmetry breaking scale) and can lead to interesting processes, e.g., $\nu \rightarrow \nu' + f$. Like composite models, family symmetry has yet to meet with any great success.

B.5 References

1. I. J. R. Aitchison and A. J. G. Hey, *Gauge Theories in Particle Physics* (Adam Higler Ltd, Bristol, 1982).

[24]The original flavor $SU(3)$ symmetry is a primitive example of a family symmetry.

[25]The weak interactions also change flavor, e.g., $u \leftrightarrow d$ or $\nu_e \leftrightarrow e^-$; however, these processes involve charged currents, i.e., both the flavor and charge state of the quark or lepton change.

2. D . H. Perkins, *Introduction to High Energy Physics* (Addison-Wesley, Reading, MA, 1982).

3. C. Quigg, *Gauge Theories of the Strong, Weak, and Electromagnetic Interactions* (Benjamin/Cummings, Reading, MA, 1983).

4. G. G. Ross, *Grand Unified Theories* (Benjamin/Cummings, Reading, MA, 1984).

5. P. Ramond, *Field Theory: A Modern Primer* (Benjamin/Cummings, Reading, MA, 1981).

6. J. C. Taylor, *Gauge Theories of Weak Interactions* (Cambridge Univ. Press, Cambridge, 1976).

7. H. Georgi, *Weak Interactions and Modern Particle Theory* (Benjamin/Cummings, Menlo Park, CA, 1984).

8. H. Georgi, *Lie Algebras in Particle Physics* (Benjamin/Cummings, Reading, MA, 1982).

9. E. D. Commins and P. H. Bucksbaum, *Weak Interactions of Leptons and Quarks* (Cambridge Univ. Press, Cambridge, 1983).

10. L. B. Okun, *Leptons and Quarks* (North-Holland, Amsterdam, 1982).

11. J. Wess and J. Bagger, *Supersymmetry and Supergravity* (Princeton Univ. Press, Princeton, 1983).

12. P. G. O. Freund, *Introduction to Supersymmetry* (Cambridge Univ. Press, Cambridge, 1986).

13. P. Langacker, *Phys. Rep.* **72**, 185-385 (1981).

14. H. E. Haber and G. L. Kane, *Phys. Rep.* **117**, 75-263 (1985).

15. M. B. Green, J. H. Schwarz, and E. Witten, *Superstring Theory* (Cambridge Univ. Press, Cambridge, 1987).

INDEX

Abell catalogue, 23–24
Abundances of light elements, 15–16, 92–95
  observations, 100–107
  predictions, 96–99, 466.
  See also Deuterium; Helium; Lithium
Abundance of relic particle species, 119–130, 151–152
  of decaying particles, 130–136
Action, 38–39, 47–48, 217–218, 277, 459, 479
  Chapline-Manton, 467, 481
  Einstein-Hilbert, 47–48, 449, 451–452
  Euclidean, 202–207, 460–461
  Nambu, 228
Adiabatic density perturbations, *See* Curvature perturbations
Affine connection $\Gamma^{\alpha}_{\beta\gamma}$, 29–31, 37, 116, 358
Age of the Universe t, 52–60, 64
  at decoupling, 80, 353
  at matter-radiation equality, 51, 77, 322, 366
  at recombination, 78
  by globular-cluster dating, 12–13
  by radioactive dating, 12–13
  from the expansion, 52
  from white dwarfs, 12–13
  matter dominated, 53–55
  present t_0, 12–14, 53–58, 310
  radiation dominated, 55, 64, 118
  vacuum dominated, 55–56
Angular diameter distance d_A, 44
Angular diameter vs redshift relation, 8, 44
Anisotropic cosmologies, 303–309, 471–472
Annihilation catatrophe, 127, 159
Anthropic principle, 269, 315
Average energy per particle, 62–63
Axions, 401–443
  axionic strings, 222, 424, 433–436
  bremsstrahlung, 406, 418–419
  couplings to matter, 404–408, 410–411, 418–419, 441
  decay to two photons, 407, 439–440
  detection, 439–443
    Sikivie-type detectors, 441–443
  domain walls, 422–424
  DSFZ, 406–407, 411–417
  hadronic, 406–407, 415
  inflation, 296, 424, 431–432
  mass, 403, 405, 422, 432
  mass limits, 407–408, 408–421, 435
  potential, 422–423
  production, 424–436
    misalignment, 427–433
    string decay, 433–436
    thermal, 424–427
  relic density, 424–436
  stars, 408–421
    axion emission rates, 410–412, 418–420
    red giants, 415–417
    Sun, 411–415

SN 1987A, 417–421
symmetry breaking scale f_{PQ}, 403, 405, 422

B^0, 512, 513, 518
$B+L$, 178–179, 188–189
$B-L$ 178–179, 181, 188–189, 526
Baryogenesis, 157–193
inflation, 273, 281
simple model, 168–176
sphalerons, 184–189
spontaneous, 190–191
Baryon asymmetry, 127–128, 157–193, 168, 281
damping, 171–173, 176–179
evidence for, 158–159
non-thermalizing modes, 177–179, 181, 188–189
sphalerons, 184–189
Baryon number of the Universe B, 67–68, 81–82, 84, 159, 169, 172–176, 183, 191, 348
Baryon number violation, 157, 160–165, 525–527
monopole catalysis, 244, 249–254
sphalerons, 184–189
Baryon-to-photon ratio η, 16, 68, 78, 81, 88–89, 96, 97–99, 101, 103–104, 105–107, 146
See also Baryon number of the Universe.
β decay, 90–91
Bianchi identity, 50, 480
Bianchi models, 303–306, 472
Biasing, 23, 311, 392–395
Big bang nucleosynthesis (BBN), *See* Primordial nucleosynthesis
Big crunch, 83
Binding energy, 88, 92
Black holes, 365, 448
Boltzmann equation, 71, 115–119, 120–121, 124, 137, 168–176, 182, 187, 351–355, 386–387
axions, 425–427
baryogenesis, 168–176
origin of species, 119–130
Bose-Einstein statistics, 61–62, 117–118, 144
Bubble nucleation, 201–207, 205, 240

C & CP violation, 160–161, 162–165, 167–168, 171, 184, 190, 192, 314, 512, 514
strong CP problem, 401–404
Catalogues of galaxies, 10–11, 22, 24–25
APM, 22
CfA, 22–25, 331, 371
IRAS, 10–11, 380
UGC, 24
Zwicky, 22–23
Chemical equilibrium, 61, 77–78, 87, 89, 159, 169
Chemical potential μ, 61, 66, 70, 87, 89, 169, 180, 190–191
Clusters of galaxies, 18, 335–336, 376–377, 439–440
Virgo, 19, 158
Coherent oscillations, 272, 278–281
axions, 427–433, 436
Cold dark matter, *See* Dark matter
Collisional damping, 353–355
Silk scale, 355, 363, 367–369
Collisionless damping, 351–353
Collision operator, 116–118
Color, *See* Quantum chromodynamics
Comoving coordinates, 30, 33–34, 38, 40, 43, 65, 83–84, 325
Composite models, 531–532
Confinement, 508

Conformal time η, 35, 37, 38, 276, 346
Conformal flatness, 35, 46
Conservation of stress-energy, 41, 66, 277, 279, 359, 476
Correlation function ξ, 333–336, 376–377, 385–388, 393–394
  cluster-cluster, 23, 334, 373, 377
  galaxy-galaxy, 23, 333–334, 335, 369, 371–372, 377, 393–394
  J_3, 334–336
Correlation length ξ, 237–238, 240
  scalar field, 216
Cosmic microwave background radiation (CMBR), 8–10, 14–15, 60
  isotropy (anisotropy), 8, 14–15, 262, 287–288, 294, 311–312, 321–322, 383–390
    autocorrelation function, 385–388
    cosmic strings, 231
    density perturbations, 321–322, 374, 383–390
    dipole, 380
    domain walls, 220
    gravitons, 290, 310, 312–313
    Sachs-Wolfe effect, 383–384
  spectral distortion, 14–15, 144–146
Cosmic no-hair theorem, 303–309
Cosmic rays, 158, 232, 485
Cosmic scale factor $R(t)$, 4, 8, 30, 33–34, 47, 59–60, 267, 322, 325, 327–328, 353
  cosmic strings, 227–228
  domain walls, 219–220
  extra dimensions, 470–482
  inflation, 272
  matter-radiation equality, 51, 322, 348, 370
  Taylor expansion of, 41–42
Cosmic strings, 15, 220–233, 391
  axionic strings, 222, 424, 433–436
  evolution, 226–232
    intercommutation, 228–230
    scaling solution, 229–231, 433–434
  gravitational effects, 223–226
    lensing, 225, 231–232
    radiation, 228–231
  mass per unit length μ, 222–225, 433
  metric, 224
    deficit angle $\Delta\theta$, 224–225
  structure formation, 226, 231–231, 369–370, 391
  superconducting, 232, 396
  wakes, 226, 231
  winding number, 221–222
Cosmic virial theorem, 19
Cosmological constant Λ, 20, 47, 56, 267–268, 305, 314, 390–391, 451, 458–461, 478–482
  See also Vacuum energy density
Cosmological principle, 33
Cowsik-McClelland bound, 123
Critical density ρ_C, 6, 50
  See also Ω
Critical temperature T_C, 187, 198–199, 212, 239–240, 270, 292
Cross section σ, 71, 74, 95, 120–121, 124, 126–127, 137, 151–152, 166, 171, 408–409, 426, 482
  neutrino, 74, 128–129, 409
  strong, 249–250
  weak, 519–520
Curvature, 30
  extrinsic, 454–455

of the Universe, 50
three, 30, 51, 455
See also Riemann curvature tensor; Ricci scalar; Ricci tensor
Curvature perturbations, 337–341, 358, 363, 366
inflation produced, 283–288, 301, 438
Curvature radius, *See* Radius of curvature

Dark matter, 16–21, 57, 110, 139–141, 151, 310–311
baryonic, 18–19
candidates, 323
CMBR fluctuations, 389
cold, 15, 124–130, 312, 323–324, 374–378, 390, 529
perturbation spectrum, 366
detection, 152, 239–254, 396, 439–443, 483
hot, 15, 122–124, 311–312, 323–324, 370–374
perturbation spectrum, 366, 367–368, 370–371
See also Neutrinos; Relic particle species
Decaying particles, 130–136, 139–150, 279–281, 407, 439–440
Decay rate, 91, 96–97, 165–168, 170
Deceleration parameter q_0, 4–6, 41, 43–44, 51–52
Decoupling of particle species, 69–70, 76, 115, 119–130
neutrinos, 74–76
photons, 74, 77–81, 136–139, 321, 353–355, 383, 386
relativistic particle, 122–124
Degenerate particles, 62, 107, 180, 416, 419
Degrees of freedom, 68
effectively massless g_* & g_{*S}, 64–65, 67, 75, 76, 93, 97–98, 108–109, 123, 426
internal g, 61, 117, 121
Density contrast δ, 324, 329, 338, 343
biasing, 392–395
Fourier expansion, 324–325, 329–336, 345–348, 359–362, 378–380
growing mode, 347, 362, 364
horizon crossing, 364
Newtonian treatment, 341–348
particle horizon, 325–326
relativistic treatment, 355–362
See also Density perturbations; Structure formation
Density perturbations, 8, 14, 263
amplitude, 288, 312
axions, 436–439
curvature-dominated Universe, 348
evolution, 341–364
Newtonian treatment, 341–348
relativistic treatment, 355–362
Fourier expansion, 263
gauge modes, 286, 360–362
gaussian fluctuations, 264, 329
inflation, 283–291, 293–295, 341, 362, 436–439
linear regime, 326, 335, 341–364
mass within a sphere, 326–327
non-linear regime, 327, 335
peculiar velocities, 378–380, 394
dark matter, 382
power spectrum, 264, 329–336, 380
correlation function, 333–336
dark matter, 368–369
rotational modes, 345–346

spectrum, 288, 323, 364–369, 438–439
spherical collapse, 327–329, 367
temperature fluctuations, 338–340, 383–390
wavenumber k (& wavelength λ), 325–326
window function, 264, 330–331, 335, 380–381
See also Curvature perturbations; Density contrast; Isocurvature perturbations; Structure formation

de Sitter phase, 267, 270–272, 283, 287, 305–306, 458–459
See also Inflation

Detailed balance, 121

Deuterium (D), 15–16, 92, 98–99, 100–101, 102–103
mythical bottleneck, 92

Development angle, 59

Diameter of the Universe D_0, 85–86

Dilaton, 466, 467

Domain walls, 213–220, 422–424
gravitational effects, 215–216
thickness, 214, 422–424

Dwarf spheroidal galaxies, 373

Dynamical timescale, 344, 375

Effective potential,
finite-temperature, 198–200, 211–213, 270, 297–298
one-loop, 199, 208–210, 292, 295
Coleman-Weinberg, 292

Einstein equations, 47, 49, 267, 451, 479
D-dimensional, 468, 470–482
perturbed, 357–362

Energy density ρ, 48, 61–64, 76, 132–133, 198–199, 304–308
axions, 429–430
cosmic strings, 230
decaying particles, 131–132, 139–140
domain wall, 215, 218–220
gravitons, 289–291, 310, 312–313
See also Density perturbations; Vacuum-energy density

Energy-momentum tensor, *See* Stress-energy tensor

Entropy, 65–70, 81–82, 84, 85, 92, 262, 266, 272–273, 475
per comoving volume S, 65–70, 476–477
production, 130–136, 146, 181–184, 192, 477–478
inflation, 272–273, 281–282

Entropy density s, 66–67, 76, 118, 122–123, 348

Equation of state, 48–49, 267, 338–341, 476
cosmic strings, 227–228
domain walls, 218–220

Euclidean equation of motion, 201–207
bounce solution, 201–207
escape point, 202–203
See also Bubble nucleation

Eulerian equation of motion, 342
perturbed, 342–343, 344–348

Expansion of the Universe, 2–7, 39, 65, 118, 133
anisotropic, 303–309

Expansion rate, *See* Hubble parameter

Extra dimensions, 449–450, 464–485, 531
compactification, 316, 449–450, 464–465, 467, 484
fundamental constants, 465–466, 478
inflation, 316, 469
internal space, 465

radius of, 471
magnetic monopoles, 484–485
metric, 470
Kasner, 471–472
Extragalactic background radiations, 142–144
from neutrino decays, 142–144, 148
x ray, 9–10, 321
See also Cosmic microwave background radiation

Faber-Jackson relation, 5, 381
False vacuum, 185–187, 204, 212, 214, 221, 315–316
decay of, 185–187, 200–207
Familon, 532–533
Fermi-Dirac statistics, 61–62, 117–118
Fine tuning, 529
Finite-temperature field theory, 185–187, 195–200, 207, 211–213, 237, 422–424, 473–475
See also Effective potential
Flatness problem, 265–267, 273, 281–283
Flux, photon, 40–41
Free energy, 185–186, 198–200, 473–475
See also Effective potential
Free-streaming length, 351–353, 363, 367–368, 371
See also Collisionless damping
Freeze out, 72, 119–130, 173–174, 191
temperature, 125, 129, 138, 426
See also Decoupling of particle species
Friedmann equation, 47–52, 82, 133, 458

Galactic evolution, 4–6
Galaxy formation, 21, 373
See also Structure formation
Galaxy number count vs. redshift test, 6, 9, 43–44, 390
Gauge bosons, 71–72, 109, 161–162, 168, 175, 208, 233, 507–509, 513, 525–526
Gauge-invariant quantity ζ, 286–287, 355–357, 362
Gauge modes, 286, 360–362
Gauge theories, *See* Yang-Mills gauge theories
Geodesic motion, 36, 37, 39
Gibbons-Hawking temperature, 284–286
Global symmetry, 510-511, 528, 532
Grand Unified Theories (GUTs), 157, 160–161, 168, 174–176, 189, 213, 236, 268, 270, 316, 523–528
$SU(5)$, 178–179, 181, 239–242, 291–296, 524–526
See also Phase transitions; Spontaneous symmetry breaking
Gravitational radiation, 288–291, 310, 312–313, 359
cosmic strings, 228–231
Gravitinos, 135–136, 298, 302, 529
Gravitons, 76, 135, 288–291, 310, 312–313
Great attractor, 382

H_0, *See* Hubble constant
Halos, 17–18, 25, 245, 375–377, 439–443
Hamiltonian formulation of general relativity, 451–458
lapse function, 452–456
shift vector, 452–456, 458
Hamilton-Jacobi equation, 38, 39
Hamlet, 396
Harrison-Zel'dovich spectrum,

264–265, 287, 310–312, 364–367, 381
See also Curvature perturbations
Hawking radiation, 365, 448
Helium four (^{4}He), 15–16, 92–93, 95, 97, 98–99, 104–107, 147, 466
stellar, 415–416
Helium three (^{3}He), 15, 92–93, 98–99, 101–103
Hidden symmetry, *See* SSB
Higgs boson, 162, 168, 175, 208–213, 221, 233, 518
Coleman-Weinberg mass, 210–213
Higgs mechanism, *See* Spontaneous symmetry breaking
Homogeneity, 8–12, 29, 261–263, 276, 308–309
Horizon, *See* Particle horizons
Horizon problem, 261–263, 477–478
Hot dark matter, *See* Dark matter
Hubble constant H_0, 3–4, 14, 41–42, 50, 265, 310
Hubble diagram, 5–7
Hubble distance, 4, 262
Hubble parameter H, 50, 64, 70, 91, 119, 133, 267
radiation-dominated Universe, 64, 91
inflation, 272, 275, 278–279, 304
Hubble radius, 50, 266, 282, 285–286
Hubble's law, 4–5, 40, 42–43
Hubble time, 4, 12, 50, 52
Hypercharge (weak), 512–513

Inflation, 49, 241, 261–317
and baryogenesis, 181, 273, 281, 297, 301
chaotic, 299–300, 315–316
density perturbations, 283–291, 293–295, 301–302, 311, 341, 362, 369–370, 436–439
double, 313
flatness problem, 265–267, 273, 281–283
gravitinos, 136, 298, 302
gravitons, 288–291, 310, 312–313
horizon problem, 261–263
initial conditions, 303–309, 317, 458, 463, 469
magnetic monopoles, 238, 241
models, 291–303
new vs. old, 275
no-hair theorem, 303–309
number of e-folds, 278, 282, 284, 287, 293, 304, 307–308
observational tests, 309–313
reheating, 135, 182, 184, 272–274, 278–281, 294, 295, 297, 301–302, 438
$SU(5)$, 291–296
supersymmetry, 296–300
slow rollover 272–275, 277–278, 300–301
smoothness problem, 261–263, 281–282, 284, 293
successful, 300–303
See also Scalar field
Inhomogeneous cosmology, 308–309
Initial conditions, 447–487
Instanton, 184, 200–207, 422, 514
Interaction rate Γ, 65, 70, 74, 77, 95, 115–116, 121–122, 128, 165–167
axions, 425–426
weak, 90–91, 96
Isocurvature perturbations, 265, 313, 337–341, 362, 363, 366, 369
axion, 436–439
baryon, 192

Isothermal perturbations, *See* Isocurvature perturbations
Isothermal sphere, 375
Isotropy, 8–12, 29, 261–263, 303–309

Jeans instability, 263, 322, 342–351, 352, 363
  in an expanding Universe, 344–348
  wavenumber k_J, 343–344, 346, 349–350
Jeans mass M_J, 344, 348–351, 355, 363

Kaluza-Klein theories, *See* Extra dimensions
Kepler's third law, 17
Kibble mechanism, 237–239, 239–242
Kinematic tests, 6, 20, 310, 390
Kinematics, *See* Particle kinematics
Kinetic equilibrium, 61, 87, 169, 242–243

Lagrangian density $\mathcal{L}$, 162, 184–185, 196, 214, 221, 233, 276, 510, 516, 517, 519, 520
  axion, 404, 405, 428
  Chapline-Manton, 467–470
  gravitational, 455, 459
  QCD, 402–404
Large scale structure, 21–26
  clusters of galaxies, 18, 335–336, 376–377, 439–440
  superclusters, 24
  voids, 24, 374, 377
  See also Structure formation
Lee-Weinberg bound, 129
Lepton number of the Universe L, 178–179, 180–181, 189
Leptons, 507–509, 512, 513
Light elements, abundances of, *See* Abundances of light elements
Line element, *See* Robertson-Walker metric
Liouville operator, 116
Lithium (Li), 15–16, 95–96, 97, 98–99, 103–104, 110–111
Local symmetry, 510–511, 528, 532
Luminosity, 40–41, 412–413, 416
Luminosity distance d_L, 3–4, 40–43

Magnetic fields, 243–244, 245–249, 290–291
Magnetic monopoles, 157–158, 233–237, 239–254, 266, 302, 479–480, 484–485
  annihilations, 239–242
  catalysis of nucleon decay, 244, 249–254
  Dirac charge, 234
  flux limits, 239–240, 244–254
    mass density, 239–240, 245
    neutron stars, 250–252, 254
    Parker limit, 245–245, 254
  hedgehog configuration, 234
  mass, 235–236
  number density, 238, 239–240, 249
  Prasad-Sommerfield limit, 235–236
  thermal production, 242
  velocity dispersion, 242–244, 246, 252–253
Majoron model, 520–521, 522
Mass density, 16–17, 57
  of relic particles, 123, 126, 151–152
Mass fluctuation, 330–336, 385
  J_3, 334–336
  See also Density contrast; Density perturbations; Structure formation

Mass fraction, 88–89, 92–93
Mass-to-light ratio, 10–11
Matrix elements, 90–91, 117, 418–419
  for baryogenesis, 168, 171
Matter-dominated Universe, 49, 51, 53–55, 140, 346–347, 350
Matter–radiation equality, 51, 59–60, 74, 76–77, 141–142, 322
  and structure formation, 337, 348, 386
  See also Red shift; Age of the Universe; Cosmic scale factor
Maxwell-Boltzmann statistics, 117–118
Mean free path, 79, 353–354, 388–389, 408
Metric, *See* Robertson-Walker metric
Million red shift project, 22, 396–397
Minkowski line element, 35

Nambu-Goldstone boson, 403, 466, 532–533
Neutrinos, 16, 20, 60, 74–76, 128–130, 139–150, 311, 409–410
  and nucleosynthesis, 98–99, 105–106, 107, 108–109, 146–147, 180–181
  chemical potential, 89, 180–181
  cosmological limits, 123, 129, 139–150
  decoupling, 94, 122–124
  Dirac, 140, 520, 527
  free streaming, 353, 371
  Majorana, 140, 520–521, 527
  masses, 520–521, 527
  mass-lifetime limits, 140–150
  relic density, 122–124, 128–130, 151–152
  stellar, 414–421
  structure formation, 367–368, 371–374
  unstable, 139–150
Neutron–antineutron oscillations, 526
Neutron half life $\tau_{1/2}(n)$, 91, 96–97, 108
Neutron to proton ratio, 89–91, 94, 96, 107, 108, 147, 466
No-hair theorems, 303–309
Non-thermalizing modes, 177–179, 181, 188–189
Nontopological solitons, 195
Nuclear statistical equilibrium (NSE), 87–89, 91–92, 408–409
Nucleon decay, 157, 244, 249–254, 526
Number density, 61–64, 67, 70, 116, 120–121, 122–123, 169–171, 182–183, 190–191
  of axions, 425, 430, 434–435, 441
  of magnetic monopoles, 238
  of nuclei, 87–88
  of photons, 14, 67, 76
  cold dark matter, 126

Ω, 17, 50, 123, 126, 239–240, 265–266, 290, 312–313
Ω_B, 16, 18, 21, 101, 107, 310, 349, 369, 389
Ω_0, 6, 14, 16–21, 107–108, 265, 367, 379–380
  age of the Universe, 52–60
  axions, 427, 431, 435
  inflation, 273, 283, 310–311
  luminous matter, 17–18
  rotation curves, 17–18, 244–245
  smooth component, 20, 390–395

Virgo infall, 19
See also Kinematic tests
Ω problem, 20, 377, 390–395
Out-of-equilibrium decay, 130–136

Pancakes, 371–374
Particle–antiparticle asymmetry, 63, 127–128, 180–181
baryon, 157–193
Particle horizons, 36, 82–86, 262, 283–285, 477
density contrast, 337
horizon problem, 261–263
Kibble mechanism, 237–238
mass within, 84, 265, 350
Particle kinematics, 36–39
Peccei-Quinn symmetry, 296, 401–408, 422–423, 522
Peculiar acceleration, 379–380
Peculiar velocity, 11–12, 37, 378–382, 394
dark matter, 382
Phase space distribution, 61, 69, 90, 115–119, 169, 373, 375, 433
Bose-Einstein, 61, 144
Fermi-Dirac, 61
Phase transitions, 72, 195–200, 268, 314, 518–519
electroweak, 72, 187–188, 208–213, 519
first order, 199–200, 212, 240
GUT, 72, 239–242, 270–275, 292
quark/hadron, 72–74, 110, 508
second order, 199
See also Spontaneous Symmetry Breaking
Planck epoch, 72, 85, 447–487
Pop I, 12, 102
Pop II, 12, 102
Pop III, 12, 102
Power spectrum, 264, 329–336, 380, 384, 388
correlation function, 333–336
dark matter, 368–369
Preons, 531
Pressure, 61–64, 198–199, 223, 267, 276, 340, 344, 475–476, 486
Primordial abundances, *See* Abundances of light elements
Primordial nucleosynthesis, 15–16, 18, 74, 87–111, 232, 262, 298, 465–466
as a probe, 107–109
initial conditions, 89–93
neutrinos, 89, 94, 98–99, 105–106, 107, 108–109, 146–147
non-standard, 74, 110–111
See also Abundances of light elements
Proper distance, 36
Proton decay, *See* Nucleon decay
Pyrgons, 323, 483–484

q_0, *See* Deceleration parameter
Quantum chromodynamics (QCD), 401–404, 508, 509, 510
Quantum cosmology, 269, 448–449
inflation, 317, 458, 463, 469
scalar field, 462–464, 473–475
Quantum gravity, 448–451
baby Universes, 449
semi-classical approximation, 448, 460–462
Quarks, 507–509, 512, 513
Quasi-Stellar Objects (QSOs), 2, 10, 372, 394–395
absorption line systems, 25, 101, 394–395

Radiation-dominated Universe, 48–49, 51, 55, 140, 347–348, 349–350
Radius of curvature $R_{\rm curv}$, 51, 265–266, 273, 282, 310–311

Reaction rate, *See* Interaction rate
Recombination, 77–81, 136–139, 262, 349
  residual ionization, 78–79, 136–139, 354–355, 388–389
Red shift z, 3–4, 39–40, 41–43, 83–84
  at decoupling, 81, 322, 353
  at matter-radiation equality, 51, 77, 322, 348
  at recombination, 81, 349
  of particle momentum, 37–38, 69, 277, 352
  surveys, 22, 371
    CfA, 22–25, 331, 371
    IRAS, 10–11, 380
    Million red shift project, 22, 396–397
Reheating, 131, 134, 183–184, 272–273, 280–281, 297, 300, 301–302, 438, 483
Relic particle species, 57–58, 119–130, 239–254, 266–267, 273, 288, 482–485
  abundances, 151–152, 239–254, 424–436, 482–485
    gravitons, 76, 288–291, 310, 312–313
  decaying particles, 130–136, 139–150, 391, 407, 439–440
  inflation, 302
  structure formation, 338–341
  See also Axions; Dark matter; Magnetic Monopoles; Neutrinos
ρ_C, *See* Critical density
Riemann curvature tensor $R^{\mu\nu\alpha\beta}$, 29–30
Riccati equation, 122
Ricci scalar $\mathcal{R}$, 30–31, 49, 451, 454–455
Ricci tensor $R^{\mu\nu}$, 30–31, 49, 358
  D-dimensional, 468, 470–482
Robertson-Walker (RW) metric, 29–46, 49, 116, 261, 275–276, 459
  for a two sphere, 31–34
  for a three sphere, 34–36, 51, 452–453, 470
  perturbed, 357–362
Rotation curves, 17–18, 244–245
r-process elements, 13

Saha equation, 77–78
Scalar field, 196–207, 208–213, 214, 458, 462, 473–475
  coherent oscillations, 272, 278–281
  decay width, 277
  equation of motion, 271, 277–279, 304, 429
  inflation, 275–283, 289
    inflaton, 287, 295–296, 302–303
    smoothness, 287, 295–296
  lagrangian, 196, 214, 276
    potential, 196–207, 291–292, 291, 422–423
    See also Effective potential
  quantum fluctations, 276, 283–291, 293–294, 300–301, 309, 315–316, 436–438, 448
  vacuum expectation value, 216
Scale factor, *See* Cosmic scale factor
Silk damping, *See* Collisional damping
Singularity, initial, 50
Smoothness problem, 261–263, 281–282, 284, 293
Sound speed v_s, 343, 346, 349–350
Spatial curvature ${}^3\mathcal{R}$, 30, 51, 455
Sphalerons, 184–189
Spherical collapse, 327–329, 367
Spiral galaxies, 17–18
Spontaneous symmetry breaking

(SSB), 72, 195–254, 268, 292, 316, 515-521
condensed-matter phenomena, 518–519
axions, 403–404, 422–423, 436–439
domain walls, 213–220
electroweak, 72, 208–213
GUT, 72, 270–275
high temperature symmetry restoration, 195–200
See also Inflation; Phase transitions
Standard candle, 4, 43
Standard model of particle physics, 1, 71–72, 86, 89, 109, 128, 184–189, 192, 195, 315, 507–521
electroweak, 208–213, 515–521
degrees of freedom for, 64–65
QCD, 401–404, 508–509
Stellar evolution, 408–410, 412–413
and axions, 410–421
Stress-energy tensor $T_{\mu\nu}$, 47–48, 267, 358, 457, 470–471, 480
cosmic strings, 223, 227–228
domain wall, 214–215, 218–219
perfect fluid, 48, 276–277, 351
scalar field, 197
See also Conservation of $T_{\mu\nu}$
Structure formation, 21, 263–265, 321–397
biased, 392–395
cold dark matter, 374–378, 390
cosmic strings, 226, 231–232
gauge problems, 325–326, 355–357, 360–362, 364
hot dark matter, 370–374
initial data, 322–323, 364–369, 378–390
magnetic monopoles, 243
perturbation spectrum, 323
power law, 329–332, 364–369
spherical collapse, 327–329, 367
See also Density perturbations
Sunyaev-Zel'dovich effect, 383, 389–390
Supernovae, 147–150, 417–421
neutrinos, 147–150, 417–421
SN 1987A, 103, 148–150, 409–410, 417–421
Supergravity, 528–530
Superspace, 458
mini-, 458–464
Superstrings, 35–36, 450–451, 464, 467, 484, 530–531
limiting temperature, 485–487
Supersymmetry (SUSY), 295, 484, 528–530
inflation, 296–300, 316
Polyoni field, 299, 302
superpotential, 296–300
Surface brightness, 22, 44–45
Synchronous gauge, 357, 362, 364–365

Technicolor, 531–532
Thermal equilibrium, 60–65, 67, 70, 190, 424–427
departures from, 70, 115, 130–136, 151, 160–161, 279–281
baryogenesis, 165–166, 172, 181–184
local (LTE), 65–67
Thermal fluctuations, 185–187
Thermal history of the Universe, 70–81
Thermodynamics, 60–65
first law, 48, 66, 132
partition function, 486–487
second law, 65, 132
superstrings, 485–487
Θ-vacuum,
electroweak, 184–185
See also Sphalerons.

QCD, 402–404, 422, 427–433
Thin-wall aproximation, 204, 217–218
Topological defects, 195, 213–254
cosmic strings, 220–233, 433–436
domain walls, 213–220
formation of, 237–239
magnetic monopoles, 233–237, 484–485
Topology, 35–36, 45–46, 220, 233, 236, 402–403, 433
See also Topological defects.
Top quark, 208–210, 508
Tractrix, 34
Tully-Fisher relation, 5, 381
Tunnelling, 184, 200–207, 211, 270–271, 460–461
quantum, 184, 200–207, 463
thermal, 200, 207

Vacuum, 197, 205
QCD, 402–404
Vacuum-dominated Universe, 49, 55–56, 305–307
See also Inflation
Vacuum-energy density, 56, 197–198, 267–268, 270–272, 279, 305, 390–391, 458–461, 530
See also Cosmological constant
Vacuum-expectation value, 234, 238
Variance Δ, 330–332, 336
Vector bosons, *See* Gauge bosons
Violent relaxation, 328, 374–375
Virgo infall, 19
Virial theorem, 18–19, 328
cosmic, 19
Voids, 24, 374, 377
Volume of a sphere, 33, 35

$W^{\pm}W^{0}$, 509, 512, 513, 518
Wave function of the Universe Ψ, 269, 448–449, 451, 456–464
boundary conditions, 461–462, 463
creation from nothing, 462
no-boundary proposal, 462
Weak interactions, 89–91, 94, 96, 519–520
See also Cross section; Interaction rates; Neutrinos
Weakly Interacting Massive Particles, *See* WIMPs
Weyl curvature tensor $C^{\alpha\beta}_{\ \ \gamma\delta}$, 35
Wheeler-DeWitt equation, 449, 451–464
White dwarfs, 12–13, 150, 250
WIMPs, 21, 310–312, 322–324, 367–369, 386
and CMBR, 386
See also Axions; Neutrinos; Dark Matter
Winding number, 221–222, 402
Window function, 264, 330–331, 335, 380–381, 387–388

Yang-Mills gauge theories, 509, 510-511
Abelian, 512
non-Abelian, 512

Z^{0}, 109, 128, 508, 512, 513, 518
width, 109, 508